MEDICAL APPLICATIONS OF ARTIFICIAL INTELLIGENCE

MEDICAL APPLICATIONS OF ARTIFICIAL INTELLIGENCE

Edited by
Arvin Agah

CRC Press is an imprint of the
Taylor & Francis Group, an **informa** business

MATLAB® is a trademark of The MathWorks, Inc. and is used with permission. The MathWorks does not warrant the accuracy of the text or exercises in this book. This book's use or discussion of MATLAB® software or related products does not constitute endorsement or sponsorship by The MathWorks of a particular pedagogical approach or particular use of the MATLAB® software.

CRC Press
Taylor & Francis Group
6000 Broken Sound Parkway NW, Suite 300
Boca Raton, FL 33487-2742

First issued in paperback 2017

ISBN-13: 978-1-4398-8433-1 (hbk)
ISBN-13: 978-1-138-07227-5 (pbk)

Library of Congress Cataloging-in-Publication Data

Medical applications of artificial intelligence / editor, Arvin Agah.
p. ; cm.
Includes bibliographical references and index.
ISBN 978-1-4398-8433-1 (hardcover : alk. paper)
I. Agah, Arvin, editor of compilation.
[DNLM: 1. Artificial Intelligence. 2. Medical Informatics. W 26.55.A7]

R859.7.A78
610.285'63--dc23 2013038097

Visit the Taylor & Francis Web site at
http://www.taylorandfrancis.com

and the CRC Press Web site at
http://www.crcpress.com

Contents

Preface

As tools and techniques in artificial intelligence are being further enhanced, their applications to medicine are expanding. This book is an attempt to capture the breadth and depth of medical applications of artificial intelligence.

This book is composed of 28 chapters, written by 82 authors, with 50 unique affiliations, from 17 countries. The first five chapters provide a general overview of artificial intelligence, followed by 22 chapters that focus on projects that apply artificial intelligence to the medical domain. The final chapter provides a list of pertinent resources on artificial intelligence. Over 1200 entries are provided in the index.

The editor thanks all the authors for contributing to this book.

The editor acknowledges the efforts of the staff at CRC Press/Taylor & Francis in helping with this project from the initial concept to the printed book. Special thanks go to T. Michael Slaughter (Executive Editor), Laurie Schlags (Project Coordinator), Rachael Panthier (Project Editor), Michele Smith (Editorial Assistant—Engineering), and Amor Nanas (Manila Typesetting Company).

Editor

Dr. Arvin Agah is associate dean for research and graduate programs in the School of Engineering and professor of electrical engineering and computer science at the University of Kansas. His research interests include applied artificial intelligence and autonomous robots. He has published more than 150 refereed articles in these areas. Dr. Agah has been a co-investigator on projects that represent more than $32 million in research funding. He has supervised 13 PhD students and 40 master's degree students and has received multiple honors for his teaching excellence. Prior to joining the University of Kansas, he spent two years at the Bio-Robotics Division of Mechanical Engineering Laboratory in Tsukuba, Japan. He has been a researcher at the IBM Los Angeles Scientific Center, and Xerox Research Center in Rochester, New York. In addition to teaching courses at the University of Kansas, Dr. Agah has taught at the Department of Engineering Systems at University of Tsukuba in Japan, the Center of Excellence in Remote Sensing Education and Research at Elizabeth City State University in North Carolina, and the Department of Mechanical Engineering at the University of Canterbury in New Zealand. He is a senior member of IEEE and ACM.

Dr. Agah received his bachelor's degree in computer science with highest honors from the University of Texas at Austin, master's degree in computer science from Purdue University, and one in biomedical engineering and PhD in computer science from the University of Southern California.

Contributors

Mariëlle P. J. Aarts
Department of the Built Environment
Eindhoven University of Technology
Eindhoven, Netherlands

Arvin Agah
Department of Electrical Engineering and Computer Science
University of Kansas
Lawrence, Kansas

Alexander A. Aksenov
Mechanical and Aerospace Engineering
University of California, Davis
Davis, California

Myriam B. C. Aries
Department of the Built Environment
Eindhoven University of Technology
Eindhoven, Netherlands

Balazs Benyo
Department of Control Engineering and Information Technology
Budapest University of Technology and Economics
Budapest, Hungary

Nathalie Bricon-Souf
IRIT (UMR 5505)-UPS
University of Toulouse
Castres, France

Kriti Chakdar
Argus Health Systems, Inc.
Kansas City, Missouri

Pablo Charlón Cardeñoso
Institute of Ophthalmology Gómez-Ulla
Santiago de Compostela, Spain

J. Geoffrey Chase
Department of Mechanical Engineering
Centre for Bio-Engineering
University of Canterbury
Christchurch, New Zealand

Gerhard W. Cibis
Cibis Eye Care
Kansas City, Missouri

Patrick G. Clark
Department of Electrical Engineering and Computer Science
University of Kansas
Lawrence, Kansas

Emmanuel Conchon
IRIT (UMR 5505)-ISIS
University of Toulouse
Castres, France

Cristina E. Davis
Mechanical and Aerospace Engineering
University of California, Davis
Davis, California

Thomas Desaive
Hemodynamics Research Centre, Institute of Physics
University of Liege
Liege, Belgium

Ashok W. Deshpande
Berkeley Initiative in Soft Computing Environment Management Systems Special Interest Group
University of California
Berkeley, California

and

Viswamitra Research Institute (VRI)
Chicago, Illinois

and

College of Engineering Pune (COEP)
Pune, India

Jennifer L. Dickson
Department of Mechanical Engineering
Centre for Bio-Engineering
University of Canterbury
Christchurch, New Zealand

Lian Duan
Department of Information Systems
New Jersey Institute of Technology
Newark, New Jersey

Marcio Eisencraft
Centro de Engenharia, Modelagem e Ciências Sociais Aplicadas
Universidade Federal do ABC (UFABC)
Santo André, São Paulo, Brazil

Manuel Fernández-Carmona
Integrated Systems Engineering (ISIS) Group
Higher Technical School of Telecommunication Engineering
Campus de Teatinos
Málaga, Spain

Liam Fisk
Department of Mechanical Engineering, Centre for Bio-Engineering
University of Canterbury
Christchurch, New Zealand

Christopher M. Gifford
Asymmetric Operations Department
Applied Physics Laboratory
Johns Hopkins University
Laurel, Maryland

Jerzy W. Grzymala-Busse
Department of Electrical Engineering and Computer Science
University of Kansas
Lawrence, Kansas

and

Department of Expert Systems and Artificial Intelligence
University of Information Technology and Management
Rzeszow, Poland

Chengfan Gu
School of Materials Science and Engineering
The University of New South Wales
Sydney, Australia

Thomas Harlan
Iatric Systems
Victoria, British Columbia, Canada

Zdzislaw S. Hippe
Department of Expert Systems and Artificial Intelligence
University of Information Technology and Management
Rzeszow, Poland

Marcos Ortega Hortas
VARPA Group, Department of Computer Science
University of A Coruña
A Coruña, Spain

Yong Hu
Institute of Business Intelligence
Guangdong University of Foreign Studies
and
Sun Yat-sen University
Guangzhou, People's Republic of China

Jun Huan
Department of Electrical Engineering and Computer Science
University of Kansas
Lawrence, Kansas

Wen-Liang Hung
Department of Applied Mathematics
National Hsinchu University of Education
Hsin-Chu, Taiwan

Attila Illyes
Department of Anesthesia and Intensive Care
Gyula County Hospital
Gyula, Hungary

David O. Johnson
Faculty of Industrial Engineering, Human Technology Interaction (HTI)
Eindhoven University of Technology
Eindhoven, Netherlands

Wanda Larson
College of Nursing
University of Arizona
and
University of Arizona Medical Center
Tucson, Arizona

Jiuyong Li
School of Information Technology and Mathematical Sciences
University of South Australia
Mawson Lakes, Australia

Clodoaldo A. M. Lima
School of Arts, Sciences and Humanities
University of São Paulo
São Paulo, Brazil

Lin Liu
School of Information Technology and Mathematical Sciences
University of South Australia
Mawson Lakes, Australia

Mei Liu
Department of Computer Science
New Jersey Institute of Technology
Newark, New Jersey

Ana González López
VARPA Group
Department of Computer Science
University of A Coruña
A Coruña, Spain

Gang Luo
Department of Biomedical Informatics
University of Utah
Salt Lake City, Utah

Renata C. B. Madeo
School of Arts, Sciences and Humanities
University of São Paulo
São Paulo, Brazil

Yashar Madjidi
Department of Mechanical and Aerospace Engineering
Monash University
Clayton, Victoria, Australia

Michael E. Matheny
Geriatric Research Education and Clinical Care
Veterans Health Administration
and
Division of General Internal Medicine
Departments of Biomedical Informatics and Biostatistics
Vanderbilt University
Nashville, Tennessee

Genevieve B. Melton
Institute for Health Informatics
and
Department of Surgery
University of Minnesota
Minneapolis, Minnesota

Meenakshi Mishra
Department of Electrical Engineering and Computer Science
University of Kansas
Lawrence, Kansas

Siamak Najarian
Biomedical Engineering
Amirkabir University of Technology
Tehran, Iran

Ali Niknejad
Control Theory and Applications Centre (CTAC)
Faculty of Engineering and Computing
Coventry University
Coventry, United Kingdom

Takashi Omori
Brain Science Institute
Tamagawa University
Tokyo, Japan

Eimei Oyama
Intelligent Systems Research Institute
The National Institute of Advanced Industrial Science and Technology
Tsukuba, Japan

Pedram Pahlavan
Department of Biomedical Engineering
Amirkabir University of Technology
Tehran, Iran

Alberto Pasamontes
Department of Mechanical and Aerospace Engineering
University of California, Davis
Davis, California

Daniel J. Peirano
Department of Mechanical and Aerospace Engineering
University of California, Davis
Davis, California

Manuel Gonzalez Penedo
VARPA Group
Department of Computer Science
University of A Coruña
A Coruña, Spain

Sophie Penning
Hemodynamics Research Centre
Institute of Physics
University of Liege
Liege, Belgium

Sarajane Marques Peres
School of Arts, Sciences and Humanities
University of São Paulo
São Paulo, Brazil

Dobrila Petrovic
Control Theory and Applications Centre (CTAC)
Faculty of Engineering and Computing
Coventry University
Coventry, United Kingdom

Lukasz Piatek
Department of Expert Systems and Artificial Intelligence
University of Information Technology and Management
Rzeszow, Poland

Brian Potetz
Department of Electrical Engineering and Computer Science
University of Kansas
Lawrence, Kansas

and

Google, Inc.
Los Angeles, California

Manaswini Pradhan
Post Graduate (PG) Department of Information and Communication Technology
Fakir Mohan University
Orissa, India

Ranjit Kumar Sahu
Kalinga Institute of Medical Science (KIMS)
Kalinga Institute of Industrial Technology (KIIT) University
Bhubaneswar, Orissa, India

Jerome Scheuring
Arroki Inc.
Lawrence, Kansas

Sylvia Tidwell Scheuring
Department of Psychology and Research
University of Kansas
and
Arroki Inc.
Lawrence, Kansas

Björn Schräder
Technik and Architektur
Lucerne University of Applied Sciences and Arts
Horw, Switzerland

Geoffrey M. Shaw
Department of Intensive Care
Christchurch Hospital
Christchurch, New Zealand

Bijan Shirinzadeh
Department of Mechanical and Aerospace Engineering
Monash University
Melbourne, Australia

Naoji Shiroma
Department of Intelligent Systems
Ibaraki University
Hitachi, Japan

Matthew K. Signal
Department of Mechanical Engineering
Centre for Bio-Engineering
University of Canterbury
Christchurch, New Zealand

Julian Smith
Department of Surgery
Monash Medical Centre
Monash University
Clayton, Victoria, Australia

Noeimi Szabo-Nemedi
Department of Anesthesia and Intensive Care
Gyula County Hospital
Gyula, Hungary

Chunqiang Tang
IBM T.J. Watson Research Center
Yorktown Heights, New York

Selena B. Thomas
Advanced Care Staffing
Brooklyn, New York

Paul Thompson
Dartmouth College
Hanover, New Hampshire

Cristina Urdiales
Integrated Systems Engineering (ISIS) Group
Higher Technical School of Telecommunication Engineering
Campus de Theatines
Málaga, Spain

Joost van Hoof
Fontys University of Applied Sciences
Eindhoven, Netherlands

and

ISSO, Dutch Building Services Research Institute
Rotterdam, Netherlands

Kavishwar B. Wagholikar
Division of Biomedical Statistics and Informatics
Department of Health Science Research
Mayo Clinic
Rochester, Minnesota

Yan Wang
Institute for Health Informatics
University of Minnesota
Minneapolis, Minnesota

Norifumi Watanabe
School of Computer Science
Tokyo University of Technology
Tokyo, Japan

Harold T. G. Weffers
LaQuSo, Laboratory for Quality Software
Eindhoven University of Technology
Eindhoven, Netherlands

Adriana C. Westerlaken
TNO, Netherlands Organisation for Applied Scientific Research
Delft, Netherlands

Eveline J. M. Wouters
Fontys University of Applied Sciences
Eindhoven, Netherlands

Hua Xu
School of Biomedical Informatics,
University of Texas Health Science Center at Houston
Houston, Texas

Miin-Shen Yang
Department of Applied Mathematics
Chung Yuan Christian University
Chung-Li, Taiwan

Rui Zhang
Institute for Health Informatics
University of Minnesota
Minneapolis, Minnesota

Yongmin Zhong
School of Aerospace, Mechanical and Manufacturing Engineering
RMIT University
Bundoora, Victoria, Australia

1

Introduction to Medical Applications of Artificial Intelligence

Arvin Agah

CONTENTS

1.1 Introduction

Artificial intelligence (AI) concepts, techniques, and tools have been utilized in medical applications for over four decades. The overall goal has been to benefit health care by assisting health care professionals in improving their effectiveness, productivity, and consistency. Improvements in accuracy and efficiency of AI techniques have steadily increased AI's viability as a choice for tackling problems in medicine. The availability of AI software has played a significant role in the further adoption of AI for medical applications.

AI techniques have been successfully applied to the medical domain, and AI systems are being integrated into health care. Applications of AI span a wide and diverse range of fields in medicine. In addition to more traditional medical applications of AI (diagnosis, therapy, automatic classification, rehabilitation), more recent applications include disease genes, wearable computing, hospital scheduling, visualization, medical robotics, surgery simulation, artificial consciousness, and much more.

Medical applications of AI are so prevalent that literature surveys have been conducted in order to study the use of AI techniques in medical applications. One such study reported that neural networks were the most commonly used analytical tool (Ramesh et al. 2004). Another study designated fuzzy logic–neural networks as the most often used AI technique, stating that there have been intensive interests in AI techniques in genetics, cardiology, radiology, and so forth (Yardimci 2007).

There are numerous published works (books, journals, conference proceedings) on medical applications of AI, covering a very wide range, from nutrition (Buisson 2008), to data mining (Ciosa and Moore 2002), to traditional Chinese medicine (Feng et al. 2006). In addition to publications that provide a general coverage of different AI techniques for a variety of medical applications, the more specific publications on medical applications of AI can be categorized into two classes. The first set focuses on a specific field in AI, as it is applied to different areas in medicine. The second set concentrates on an explicit area in medicine and covers the utilization of a number of AI techniques in that area. Examples of books in the first set include those with foci on fuzzy logic (Barro and Marin 2010), expert systems (Fieschi 1990), software agents (Moreno and Nealon 2004), and genetic computation (Smith and Cagnoni 2011). Examples of books in the second set include medical informatics (Kelemen et al. 2008; Yoshida et al. 2010), medical imaging (Schaefer et al. 2009), medical diagnosis (Schmitt et al. 2010), and book series on computational intelligence in health care (Springer 2007–2011).

Many survey papers provide a strong starting point for those interested in AI in medicine. Examples of survey papers in the first set include AI applications in the intensive care unit (Hanson and Marshall 2001), medical applications of case-based reasoning (Holt et al. 2006), fuzzy logic in medicine (Torres and Nieto 2006), wearable computing (Lukowicz 2008), evolutionary computation (Pena-Reyes and Sipper 2000), and the use of smart and adaptive systems in different areas (Abbod et al. 2002), comprising emergency and intensive care, general and surgical medicine, pathology, and medical imaging. Examples of survey papers in the second set include a survey of medical documents' summarization (Afantenos 2005), hospital scheduling (Spyropoulos 2000), bioinformatics (Valentini 2009), and brain pathology (Hemanth et al. 2009).

1.2 A Bit of History

A number of early research efforts foresaw the immense potential of using computers, and eventually AI, to assist health care professionals. Medical data collection of physiological variables and use of digital data were discussed in the work of Ax (1960). Questioning the type of activities that could be supplanted by computers in assisting clinical psychologists was described in the work of Holtzman (1960). It was suggested that computers could assist in the evaluation of the actions during stages of the diagnostic testing process, calculating the alternative diagnostic probabilities (Ledley and Lusted 1960).

Medical applications of AI began in the 1960s. One of the first published works in medical applications of AI that included "artificial intelligence" in the title is by Hunt (1968), addressing deductive and inductive problem solving and decision-making capabilities of AI in the field of psychology. Reports on positive reactions of patients, when interfaced in conversation with a computer that modeled a physician interviewer, were presented in the work of Slack and Van Cura (1968). Discussions on the impact of the "intellectual" use of the computer in health care were included in the work of Schwartz (1970). The issues raised were social, psychological, organizational, legal, economic, and technical, and it was argued that addressing the challenges required new interactions among medicine and information sciences and new attitudes on the part of policy makers (Schwartz 1970). It is interesting to note that all such matters still hold true today, over 40 years since the issues were raised. Other examples of the earliest work on medical applications of AI include automated diagnosis of thyroid dysfunction (Nordyke et al. 1971) and therapy with interactive advice giving with physicians, including the reasoning for decision making, to serve as a tutorial and a consultant (Shortliffe et al. 1973).

A selection of symposia papers on artificial intelligence in medicine was published in 1982, referring to the area as "artificial intelligence in medicine" (AIM) (Szolovits 1982). It stated that the field emerged in the early 1970s in response to increases in demand in quality of medical services and growth of medical knowledge. It was proposed that AI systems could assist health care professionals in diagnosis, therapy, and prognosis.

Research highlights for the first decade of AIM were reported in the work of Clancey and Shortliffe (1984), where AIM is stated to be focused on AI programs for diagnosis and therapy recommendations. The AIM design features that physicians would consider important were discussed, including explaining the diagnostic and treatment decisions, being portable and flexible, improving cost efficiency, and autonomously learning from medical experts. In addition, a number of AIM challenges were identified by Clancey and Shortliffe (1984), which still hold true: methods to acquire and process data; knowledge acquisition and representation; capability to provide explanation; and integrations of AIM systems into the working environment of health care professionals, including the human–computer interface.

In the first decade of AI in medicine, the proper evaluation of the techniques was given consideration; this was needed in order to establish the quality of work in AI in medicine. One such evaluation (Chandrasekaran 1983) identified a number of challenges in evaluating AI systems: focusing only on the final results, instead of intermediate stages; comparing results against absolute standards of correctness, instead of work of other clinicians; not knowing the correct answer as clinicians may disagree; and difficulty in determining the ability to scale. Such issues are still impacting the evaluation of medical AI systems.

The adolescence of AI in medicine is discussed in the work of Shortliffe (1993), during the second decade of AI in medicine. The coming-of-age of AI in medicine is proposed to have started in 2009, that is, the third decade, offering perspectives from a number of experts in AIM (Patel et al. 2009). It was argued that the AIM field is robust, although less visible than AI's heyday. Progress has been made by researchers, and additional research grant funding would be beneficial. Goals are identified for a better understanding of the errors and risk taking, that is, the resiliency of AIM systems (Patel et al. 2009). Challenges identified included improved data capture and processing, reliable mechanisms for patients' confidentiality, and better modeling techniques. Some argue that AIM systems for discovery should not attempt to be autonomous and instead should be semiautonomous support systems for discovery. The following are identified to be important topics in AIM: knowledge representation, systems modeling, effective use of information in decision making, data analysis, and interdisciplinary education programs (Patel et al. 2009).

AIM was strong in its third decade, with numerous research efforts. It even garnered enough attention to result in online courses, namely, the medical artificial intelligence course, which was offered in 2005 (MIT OpenCourseWare 2005).

We are now in the fourth decade of AI in medicine. The research is strong and is continuing. Many books have been written to showcase the results, for instance, the work of Miller (2011). The advantages offered by AI in medicine are that such systems can offer accuracy (computers are less likely to make mistakes), cost and efficiency (no fatigue), and replication (ability to make numerous copies) (Butler 2011). Overall, AI technologies have not been integrated into medicine, as it was once predicted. Major problems have been unrealistic expectations and usability (Bond 2010). AI researchers have focused on building effective and usable software tools that can be used for medical applications. However, most AI systems in medicine are still semiautonomous, as some form of human supervision is still required to ensure proper diagnosis and treatment. There is interest from health care professionals to have systems that can assist them, not replace them. AI systems can monitor patients, make suggestions, and help mine the data. The challenge is in developing hybrid systems that can effectively and efficiently combine the experience of medical care professionals with attributes that AI software can provide. The interaction challenge requires more attention.

Research on AI in medicine is very strong, albeit focused. Currently, a search on PubMed (U.S. National Library of Medicine 2013) for the phrase "artificial intelligence" produces 18,328 results, and this will continue to grow.

The Conference on Artificial Intelligence in Medicine (AIME)—sponsored by the European Society for Artificial Intelligence in Medicine—is held biennially in odd years (Artificial Intelligence in Medicine Society 2013). Papers on theory, techniques, and applications of AI in medicine are published in the proceedings. The journal *Artificial Intelligence in Medicine* publishes nine issues a year on the theory and practice of AI in medicine (Elsevier 2013).

There are online sources to get information on AI in medicine. An introduction is provided in OpenClinical (2013). Pros and cons of AI in medicine are discussed in Healthinformatics (2013). AITopics (AAAI 2013) is an information portal for the science and applications of AI, developed and maintained by the Association for the Advancement of Artificial Intelligence (AAAI), providing history, overview, and detailed information on numerous AI subjects, including machine learning, natural languages, speech, robotics, and so forth.

For those interested in conducting AI experiments, there are now a number of online software resources. The Waikato Environment for Knowledge Analysis (Weka) (University of Waikato 2013) is a popular machine learning software package developed by the Machine Learning Group at the University of Waikato in New Zealand. Weka can be utilized using the online interface or can be integrated into Java programs. Weka includes tools for processing, classifying, clustering, and visualization of data.

Currently, in 2013, there are two global challenges related to medical applications of AI, with multi-million-dollar awards (XPrize 2013). The first competition focuses on hardware and software sensors to capture and interpret body metrics for individuals to improve their health (Nokia 2013). The second competition is based on diagnostic technologies for precise and reliable diagnoses for use by consumers in homes, integrating fields of AI, wireless sensing, and imaging (Qualcomm 2013). These types of global challenges illustrate the advances in technology, paving the way for the next phase of medical applications of AI, incorporating the technologies in everyday devices used by consumers in their homes, offices, and so forth.

In 2013, Telemedicine has entered a new stage, as U.S. Food and Drug Administration (FDA) clearance is given to Remote Presence Virtual Independent Telemedicine Assistant (RP-VITA)—a remote presence robot—for providing remote patient critical care for assessments and examinations (InTouch Technologies 2013; iRobot Corporation 2013). The design has been based on user-friendliness to encourage adoption by health care professionals. This is an important aspect to be considered by all AI systems for medical applications.

With all that has been done in the past four decades, and all these ongoing activities, we can look forward to further accomplishments in medical applications of AI in the future decades.

1.3 Chapters in This Book

This book is organized into 28 chapters. Chapters 2–5 provide an overview of the AI concept, methods, theories, tools, and technologies. Chapter 2 introduces major concepts in

AI, including support vector machines, neural networks, naïve Bayesian classifiers, hidden Markov models, *k*-means clustering, and principal component analysis.

Prominent machine learning and data mining methods are explained in Chapter 3, along with their medical applications. The techniques include logistic regression, *k*-nearest neighbor, *K**, decision trees, random forests, rule learners, neural networks, and evolutionary algorithms. The concept of hybrid systems is also introduced where an ensemble of classifiers or a multi-classifier is built using a number of classifiers. Chapter 4 provides another review of AI techniques with medical applications, covering fuzzy logic, genetic algorithm, and neural networks. Chapter 5 examines the information retrieval aspects of AI, drawing comparisons between medicine and law.

Chapters 6–27 provide in-depth coverage of medical applications of AI. Chapters 6–10 focus on classification problems in medicine. Image processing is the common theme in Chapters 11–14. Chapter 15–17 emphasize medical decision making. Use of AI in diagnosis is covered in Chapters 18–19, and assisted living is included in Chapters 20–21. Chapters 22–27 present a variety of medical applications of AI.

In Chapter 6, genetic algorithms are utilized to characterize soft tissue in terms of predicting the viscoelastic response. The experimental results are shown to be accurate and time efficient compared with other existing methods. The application of support vector machines and wavelet transform to classification of electroencephalogram (EEG) signals is the subject of Chapter 7. The work considers developing compact feature vectors in order to make the process more efficient.

The applications of naïve Bayesian classifiers are described in Chapter 8, with emphasis on two problematic characteristics of data sets, namely, high dimensionality and small size. The application domain is microarray gene expression for different diseases. Chapter 9 focuses on prediction of chemical toxicity profiles, using random forest and naïve Bayes methods. The performance of the computational techniques are analyzed and compared. The theme of classification continues in Chapter 10, where neural networks are used for cancer prediction. The classifier and dominant gene prediction methodologies are used for predicting tumors.

Chapter 11 presents the application of data mining for melanoma diagnosis. Algorithms are evaluated based on their dermatoscopic score, using basic melanoma data sets. AI applications to optical coherence tomography images are discussed in Chapter 12. The main layers of the retina are extracted, and 2-D and 3-D surfaces are reconstructed. Support vector machines and deep belief networks are applied to the analysis of pap smears in Chapter 13. The machine learning techniques are used to identify discriminative visual features that can be predictive of cancer and precancer grading. Chapter 14 discusses a fuzzy clustering approach to MRI segmentation, forming clusters of similar-intensity sets in images. MRI segmentation is used for detection in clinical diagnosis.

Chapter 15 presents a therapeutic decision-making approach, quantifying risk and uncertainty for optimal treatment. The methodology is applied to hyperglycemic management of critically ill patients. Clinical decision making is surveyed in Chapter 16, describing the practices and challenges. A narrative of a mock patient is used to contextualize and categorize the process. Clinical decision making is further explored in Chapter 17, based on the fuzzy naïve Bayesian approach. Other fuzzy set theoretic approaches are presented.

Chapter 18 emphasizes clinical diagnostics, namely, methodologies in metabolomic studies. AI methodologies of neural networks, genetic algorithms, and self-organizing maps are applied to spectrometry-like instruments to aid patient care and disease diagnosis and management. The amblyopia (lazy eye) vision disorder is the subject of Chapter 19.

The AI methodology of learning from examples is used with a variety of discretization techniques for rule induction.

Ambient assisted living and ambient intelligence are introduced in Chapter 20. Hardware components such as sensors, control devices, and actuators are discussed, along with standards and protocols. Use of intelligent light therapy for older adults is covered in Chapter 21, with an emphasis on treatment of people with dementia. The light therapy is based on the use of sensors and ambient intelligence in nursing to improve quality of life. Chapter 22 presents the concept of context awareness and its medical applications. A variety of research projects and surveys on the topic of context awareness are discussed, identifying research trends in health care. A survey of natural language processing in electronic health record systems is the focus of Chapter 23. Natural language processing tools, resources, and clinical systems are described.

Chapter 24 focuses on the intelligent personal health record, discussing its benefits and challenges. The AI techniques of expert systems, search technology, natural languages, and signal processing are integrated into intelligent personal health records. Application of AI in minimally invasive surgery and artificial palpation is covered in Chapter 25, illustrating the benefits of AI. The utilization of a neural network to enhance a tactile sensory system in a palpation-based medical assessment system is described.

Chapter 26 introduces the notion of a wearable behavior navigation system, using video conferencing and augmented reality technologies. First-aid treatment, specifically, making an arm sling using a bandage, is conducted by participants, following guidance from an expert. Chapter 27 presents AI applications to drug safety, focusing on surveillance and analysis of adverse drugs. The field of pharmacovigilance is presented, along with relevant AI applications, including statistical methods and data mining algorithms.

Chapter 28 concludes this book, presenting a list of AI resources. Select journals are listed that are related to medical applications of AI. A number of related conferences are also listed. Select tools for conducing AI experiments are introduced, including open-source software, products sold by companies, and other related resources.

References

AAAI (2013). AITopics. aitopics.org.

Abbod, M.F., Linkens, D.A., Mahfouf, M., and Dounias, G. (2002). Survey on the use of smart and adaptive engineering systems in medicine. *Artificial Intelligence in Medicine*, Vol. 26, 179–209.

Afantenos, S., Karkaletsis, V., and Stamatopoulos, P. (2005). Summarization from medical documents: a survey. *Artificial Intelligence in Medicine*, Vol. 33, 157–177.

Artificial Intelligence in Medicine Society. (2013). *Conference on Artificial Intelligence in Medicine*. aimedicine.info/aime.

Ax, A.F. (1960). Computers and psychophysiology in medical diagnosis. *IRE Transactions on Medical Electronics*, Vol. ME–7, Iss. 4, 263–264.

Barro, S. and Marin, R. (2010). *Fuzzy Logic in Medicine (Studies in Fuzziness and Soft Computing)*. Physica-Verlag, Heidelberg, Germany.

Bond, A. (2010). Reality checkup: Medical artificial intelligence still a hard sell in the clinic. *Scientific American*, http://www.scientificamerican.com/article.cfm?id=artificial-intelligence-medical-tests-software-diagnosis, January 12.

Buisson, J.-C. (2008). Nutri-Educ, a nutrition software application for balancing meals, using fuzzy arithmetic and heuristic search algorithms. *Artificial Intelligence in Medicine*, Vol. 42, 213–227.

Butler, L.M. (2011). *How Artificial Intelligence Doctors Will Change Medicine*. EzineArticles.com.

Chandrasekaran, B. (1983). On evaluating AI systems for medical diagnosis. *AI Magazine*, Vol. 4, No. 2, 34–48.

Ciosa, K.J. and Moore, G.W. (2002). Uniqueness of medical data mining. *Artificial Intelligence in Medicine*, Vol. 26, 1–24.

Clancey, W.J. and Shortliffe, E.H. (1984). *Readings in Medical Artificial Intelligence: The First Decade*. Addison Wesley, Reading, MA.

Elsevier. (2013). *Artificial Intelligence in Medicine Journal*. www.journals.elsevier.com/artificial-intelligence-in-medicine.

Feng, Y., Wu, Z., Zhou, X., Hou, Z., and Fan, W. (2006). Knowledge discovery in traditional Chinese medicine: state of the art and perspectives. *Artificial Intelligence in Medicine*, Vol. 38, 219–236.

Fieschi, M. (1990). *Artificial Intelligence in Medicine: Expert Systems*. Chapman & Hall, London.

Hanson, C.W. and Marshall, B.E. (2001). Artificial intelligence applications in the intensive care unit. *Critical Care Medicine*, Vol. 29, No. 2, 427–435.

Healthinformatics. (2013). *Artificial Intelligence in Medicine*. http://healthinformatics.wikispaces.com/Artificial+Intelligence+in+Medicine.

Hemanth, D.J., Vijila, C.K.S., and Anitha, J. (2009). A survey on artificial intelligence based brain pathology identification techniques in magnetic resonance images. *International Journal of Reviews in Computing*, Vol. 4, 30–45.

Holt, A., Bichindaritz, I., Schmidt, R., and Perner P. (2006). Medical applications in case-based reasoning. *The Knowledge Engineering Review*, Vol. 20, No. 3, 289–292.

Holtzman, W.H. (1960). Can the computer supplant the clinician? *Journal of Clinical Psychology*, Vol. 16, Iss. 2, 119–122.

Hunt, E. (1968). Computer simulation: artificial intelligence studies and their relevance to psychology. *Annual Review of Psychology*. Vol. 19, 135–168.

InTouch Technologies. (2013). *RP-VITA Remote Presence Robot*. www.intouchhealth.com.

iRobot Corporation. (2013). *RP-VITA Remote Presence Robot*. www.irobot.com.

Kelemen, A.A., Abraham, A., and Liang, Y. (2008). *Computational Intelligence in Medical Informatics (Studies in Computational Intelligence)*. Springer-Verlag, Berlin, Germany.

Ledley, R.S. and Lusted, L.B. (1960). The use of electronic computers in medical data processing: aids in diagnosis, current information retrieval, and medical record keeping. *IRE Transactions on Medical Electronics*, Vol. ME–7, 31–47.

Lukowicz, P. (2008). Wearable computing and artificial intelligence for healthcare applications. *Artificial Intelligence in Medicine*, Vol. 42, 95–98.

Miller, P.L. (2011). *Selected Topics in Medical Artificial Intelligence (Computers and Medicine)*. Springer-Verlag, New York.

MIT OpenCourseWare. (2005). MIT Course Number: HST.947 Medical Artificial Intelligence. ocw.mit.edu/courses/health-sciences-and-technology/hst-947-medical-artificial-intelligence-spring-2005

Moreno, A. and Nealon, J.L. (2004). *Applications of Software Agent Technology in the Health Care Domain*. Birkhäuser.

Nokia. (2013). *Nokia Sensing X challenge*. www.nokiasensingxchallenge.org.

Nordyke, R.A., Kulikowski, C.A., and Kulikowski C.W. (1971). A comparison of methods for the automated diagnosis of thyroid dysfunction. *Computers and Biomedical Research*, Vol. 4, 374–389.

OpenClinical. (2013). Artificial intelligence in medicine: an introduction. www.openclinical.org/aiinmedicine.html.

Patel, V.L., Shortliffe, E.H., Stefanelli, M., Szolovits, P., Berthold, M.R., Bellazzi, R., and Abu-Hanna, A. (2009). The coming of age of artificial intelligence in medicine. *Artificial Intelligence in Medicine*, Vol. 46, 5–17.

Pena-Reyes, C.A. and Sipper, M. (2000). Evolutionary computation in medicine: an overview. *Artificial Intelligence in Medicine*, Vol. 19, 1–23.

Qualcomm. (2013). Qualcomm Tricorder X Prize. www.qualcommtricorderxprize.org.

Ramesh, A.N., Kambhampati, C., Monson, J.R., and Drew, P.J. (2004). Artificial intelligence in medicine. *Annals of the Royal College of Surgeons of England*. Vol. 86, No. 5, 334–338.

Schaefer, G., Hassanien, A.E., and Jiang, J. (2009). *Computational Intelligence in Medical Imaging: Techniques and Applications*. Chapman and Hall/CRC, Boca Raton, Florida.

Schmitt, M., Teodorescu, H.-N., Jain, A., Jain, A., and Jain, S. (2010). *Computational Intelligence Processing in Medical Diagnosis (Studies in Fuzziness and Soft Computing)*. Physica-Verlag, Heidelberg, Germany.

Schwartz, W.B. (1970). Medicine and the computer: the promise and problems of change. *New England Journal of Medicine*, Vol. 283, No. 23, 1257–1264.

Shortliffe, E.H. (1993). The adolescence of AI in medicine: will the field come of age in the 90's? *Artificial Intelligence in Medicine*, Vol. 5, 93–106.

Shortliffe, E.H., Axline, S.G., Buchanan, B.G., Merigan, T.C., and Cohen, S.N. (1973). *Computers and Biomedical Research*, Vol. 6, Iss. 6, 544–560.

Slack, W.V. and Van Cura, L.J. (1968). Patient reaction to computer-based medical interviewing. *Computers and Biomedical Research*, Vol. 1, Iss. 5, 527–531.

Smith, S.L. and Cagnoni, S. (2011). *Genetic and Evolutionary Computation: Medical Applications*. John Wiley & Sons, Ltd., West Sussex, UK.

Springer. (2007–2011). *Advanced Computational Intelligence Paradigms in Healthcare – Series (1 to 6) (Studies in Computational Intelligence)*. Springer-Verlag, Berlin, Germany.

Spyropoulos, C.D. (2000). AI planning and scheduling in the medical hospital environment. *Artificial Intelligence in Medicine*, Vol. 20, 101–111.

Szolovits, P. (1982). *Artificial Intelligence in Medicine*. Westview Press, Inc., Boulder, Colorado.

Torres, A. and Nieto, J.J. (2006). Fuzzy logic in medicine and bioinformatics. *Journal of Biomedicine and Biotechnology*, Vol. 2006, 1–7.

University of Waikato. (2013). Weka (Waikato Environment for Knowledge Analysis). www.cs.waikato.ac.nz/ml/weka.

U.S. National Library of Medicine. (2013). PubMed. www.ncbi.nlm.nih.gov/pubmed.

Valentini, G. (2009). Computational intelligence and machine learning in bioinformatics. *Artificial Intelligence in Medicine*, Vol. 45, 91–96.

XPrize. (2013). X PRIZE Foundation. www.xprize.org.

Yardimci, A. (2007). A survey on use of soft computing methods in medicine. In *Proceedings of the 17th International Conference on Artificial Neural Networks*, Porto, Portugal, 69–79.

2

Overview of Artificial Intelligence

David O. Johnson

CONTENTS

2.1 Introduction to Artificial Intelligence

Artificial intelligence (AI) is a branch of computer science that aims to create computer software that emulates human intelligence. John McCarthy (2007), who coined the term in 1955, defines it as "the science and engineering of making intelligent machines." AI software emulates many aspects of human intelligence, such as reasoning, knowledge, planning, learning, communication, perception, and the ability to move and manipulate objects. There are a number of tools that AI uses to emulate these areas of human intelligence. In this chapter, we will look at a set of these tools loosely called "machine learning."

2.2 Machine Learning

Machine learning classifiers take as input empirical data and predict the features of the data. As an example, consider anticancer drug design. One method of anticancer drug design is to create compounds and then test how well they kill cancer cells in laboratory petri dishes. Thousands of compounds may be created and tested before one is found that kills cancer cells. Clearly, the process of creating the compound and then testing it is a lengthy one. Machine learning has been applied to speed this process up by identifying the compounds that are more likely to kill cancer cells. Those compounds are then created and tested.

How is this done? Before a compound is created, the chemical attributes of it can be determined (e.g., is it an acid or a base?). What the attributes are does not really matter. The idea is to find those compounds that have attributes similar to those of compounds known to kill cancer. Thus, the empirical data that is the input to the machine learning classifier are the attributes of the compounds, and the feature that the machine learning classifier predicts is whether the compound will kill cancer cells or not. This type of machine learning is called "supervised learning" because there is a set of data (e.g., the known cancer-killing compounds) that we can use to "train" the machine learning classifier to recognize which attributes are important in predicting whether a compound is cancer killing or not. There is another type of machine learning called "unsupervised learning," which we will talk about later in the chapter.

There are a number of metrics for measuring how well a particular machine learning classifier works. Everyone who proposes a new machine learning classifier can use these metrics to compare their classifier with other classifiers. All of these metrics are based on a "confusion matrix." The best way to explain a confusion matrix is with an example. Assume we have 100 compounds, and we are predicting whether they kill cancer or not. Suppose 30 of the compounds actually kill cancer, and 70 do not. Now suppose that the machine learning classifier correctly predicts that 20 of the cancer-killing compounds are cancer killing but incorrectly predicts that 10 of them are not cancer-killing compounds. Also suppose that the machine learning classifier correctly predicts that 60 of the non-cancer-killing compounds are non-cancer killing but incorrectly predicts that 10 of the non-cancer-killing compounds are cancer killing. We can represent this in a confusion matrix as shown in Table 2.1.

Confusion matrices are always shown in this way, and each cell has a generic name, as shown in Table 2.2.

The values in these cells are then used to calculate the metrics that are used to compare machine learning classifiers. There are many metrics that are used, but the most common

TABLE 2.1

Example of a Confusion Matrix for a Machine Learning Classifier That Predicts Whether a Compound Is Cancer Killing or Not

	Actually Kill Cancer	Actually Do Not Kill Cancer
Predicted to Kill Cancer	20	10
Predicted Not to Kill Cancer	10	60

TABLE 2.2

Standard Confusion Matrix Nomenclature

	Actually Kill Cancer	Actually Do Not Kill Cancer
Predicted to Kill Cancer	True positive (TP)	False positive (FP)
Predicted Not to Kill Cancer	False negative (FN)	True negative (TN)

are precision (P), recall (R), true positive rate (TPR), and false positive rate (FPR). These metrics are calculated as follows:

$$P = \mathrm{TP}/(\mathrm{TP} + \mathrm{FP})$$

$$R = \mathrm{TP}/(\mathrm{TP} + \mathrm{FN})$$

$$\mathrm{TPR} = \mathrm{TP}/(\mathrm{TP} + \mathrm{FN}) = R$$

$$\mathrm{FPR} = \mathrm{FP}/(\mathrm{TN} + \mathrm{FP})$$

In the example above, the TPR (and the R) of the machine learning classifier would be calculated as follows:

$$\mathrm{TPR} = R = \mathrm{TP}/(\mathrm{TP} + \mathrm{FN}) = 20/(20 + 10) = 0.667$$

2.3 Support Vector Machines

One of the easiest machine learning classifiers to understand is the support vector machine (SVM) (Cortes and Vapnik 1995). The SVM can be explained best with an example. Continuing with our earlier example, assume that each of the compounds we are going to classify has only two features: x and y. We know the values of x and y for our training set, which contains both cancer-killing compounds and ones that do not kill cancer. First, we will plot the compounds using the two features x and y, as shown in Figure 2.1.

The black dots represent compounds that are known not to kill cancer, and the white dots are known to kill cancer. Clearly, the cancer-killing compounds are clustered away from those that do not kill cancer. An SVM works by drawing a line between the two clusters and then using that line to predict if an unknown compound will kill cancer or not. To predict an unknown compound's cancer-killing capability, we plot the features, x and y, of the unknown compound. If the point falls on the cancer-killing side of the line, then

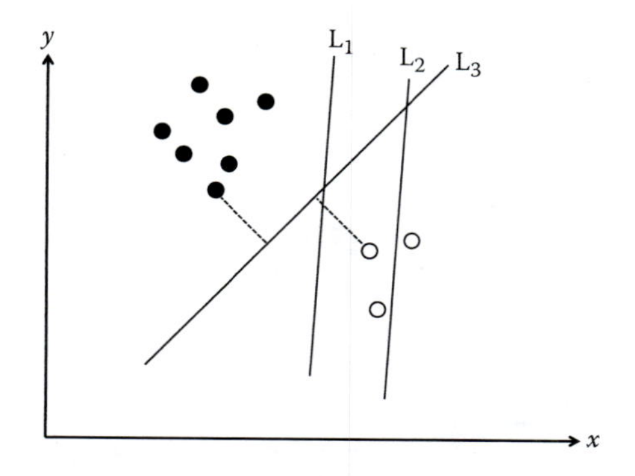

FIGURE 2.1
Support vector machine (SVM) example.

we predict that the unknown compound will kill cancer. Otherwise, we predict that it will not. The key to the SVM is to draw a line that separates the two clusters the best. In Figure 2.1, L_1 and L_3 separate the two clusters, but L_2 does not. An infinite number of lines could be drawn to separate the two clusters. L_3 is the line drawn by an SVM classifier. As shown by the dashed lines, it is equidistant from the two extreme points in each cluster.

Real compounds would have more than two features, the feature space would have a dimension for each feature, and the line would be a multidimensional plane. You can imagine this for three dimensions, but it is difficult for more than three dimensions. And, the mathematics for determining the dividing "line" is much more complicated. But, in principle, it works the same as our simple two-dimensional example above.

2.4 Neural Networks

Neural networks (Hopfield 1982) are another machine learning technique that works best when the dividing line between two classes is not straight, as shown in Figure 2.2. (Note: For this example, we are labeling the features x_1 and x_2, instead of x and y.)

A typical neural network is illustrated in Figure 2.3.

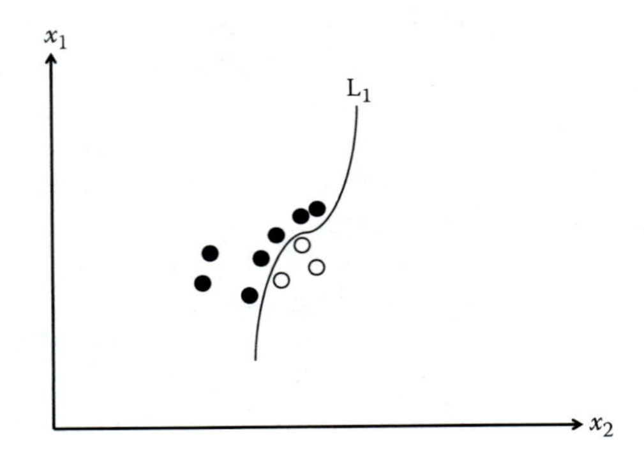

FIGURE 2.2
Neural network classification example.

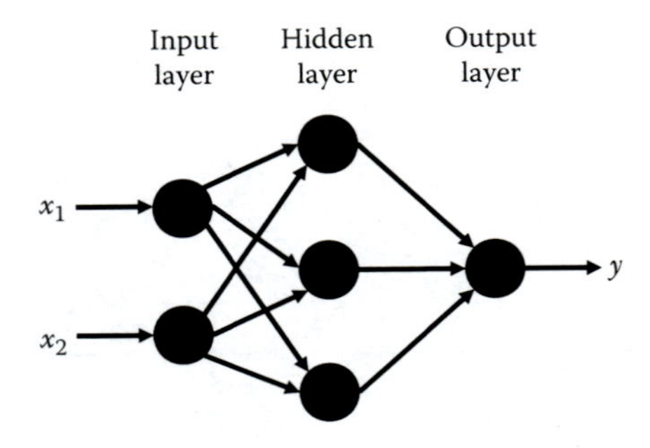

FIGURE 2.3
Typical neural network configuration.

A neural network is composed of multiple layers of nodes: an input layer, one or more hidden layers, and an output layer. Figure 2.3 shows one output, y, but a neural network may have more than one output. The inputs to each node generate an output that is fed to the next node in the network. The values of the features we are trying to classify, x_1 and x_2, are fed into the network as the initial inputs. The output y determines how the features are classified. The output y is normalized so that it varies between 0 and 1. Usually, if y is greater than 0.5, the features are classified one way (e.g., cancer killing), and if it is less than 0.5, it is classified the other way (e.g., non-cancer killing). If the output is exactly 0.5, then the classification is undetermined. There is also a strict classification method where the features are classified one way if y is greater than 0.9, the other way if it is less than 0.1, and are indeterminate otherwise.

The output of a node that is fed to the next node is usually calculated using a simple equation. There are a variety of equations that are used, but the most common is

$$f(x) = wx + b$$

Each node (in any layer) sums its inputs (x), multiplies it by a weight (w), adds a bias (b), and feeds the results ($f(x)$) to the next node. Each node has its own weight and bias. The weights and biases are determined by running the training data through the neural network multiple times and adjusting them, until the output matches the training data as close as possible. Then, theoretically, any unclassified input should be classified the same way the training data were classified.

2.5 Naïve Bayesian Classifier

The Naïve Bayesian classifier works like the SVM classifier shown in Figure 2.1, except the dividing line is determined using probability theory. As an example, consider the case of classifying a compound's cancer-killing capability based on a single feature. The extension to multiple features will be clear later. Assume we have a training set of cancer-killing compounds and know the values for a single feature, x, which can vary from 0 to 1. We can plot these as shown in Figure 2.4, where the x-axis is the value of the single feature and the y-axis is the percentage of compounds with that value, which are cancer killing.

The y-axis is also the probability that the value of feature x for a cancer-killing compound will be equal to the value on the x-axis, or $p(x)$. In other words, if the training sample contains 10 compounds whose value of feature x is 0.1, 0.2, 0.2, 0.2, 0.3, 0.5, 0.1, 0.2, 0.4, and 0.3, then the probability that a compound with a feature x will have the value of 0.1 is 20%. The probabilities that it will have a value of 0.2, 0.3, 0.4, and 0.5 are 40%, 20%, 10%, and 10%, respectively.

Now assume we have a training set of compounds that do not kill cancer and know the values for a single feature, x, which can vary from 0 to 1. We can plot these as shown in Figure 2.5, where the x-axis is the value of the single feature and the y-axis is the percentage of compounds with that value, which are cancer killing.

Now we have two probability density functions: one for cancer-killing compounds and one for those that do not kill cancer. How do we use this to classify whether a new compound is likely to be cancer killing or not? As an example, assume we have a compound whose feature x has a value of 0.6. From the plot in Figure 2.6, we see that there is a 3.6% chance it is a cancer-killing compound and 1.0% chance it is not a cancer-killing compound.

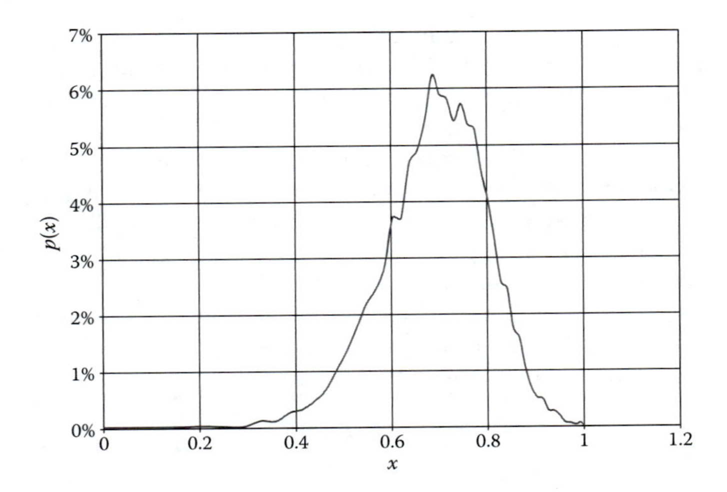

FIGURE 2.4
Plot of the value of feature x versus the percentage of the values of feature x for known cancer-killing compounds.

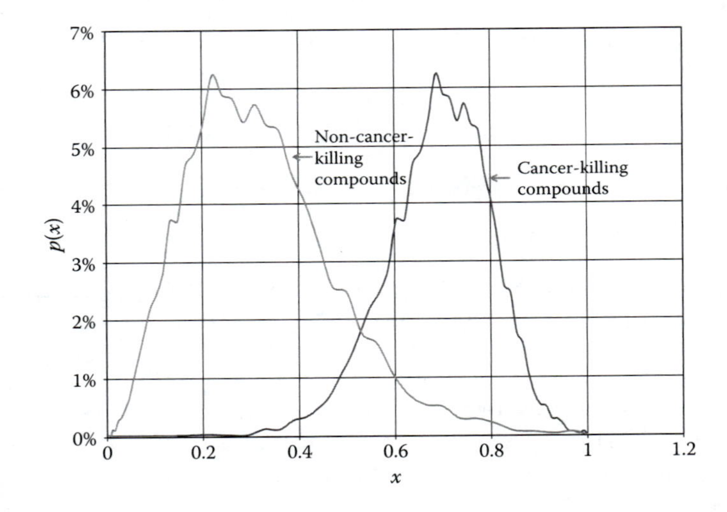

FIGURE 2.5
Plot of the value of feature x versus the percentage of the values of feature x for known non-cancer-killing compounds (left). Plot of the value of feature x versus the percentage of the values of feature x for known cancer-killing compounds (right).

Thus, the Bayesian classifier would classify the compound as cancer killing because it is more likely to kill cancer (3.6% probability) than not to (1.0% probability). Likewise, as shown in Figure 2.7, a compound with a feature x value of 0.4 has a 0.3% probability of being a cancer-killing compound and 4.2% probability of not being a cancer-killing compound.

Thus, the Bayesian classifier would classify this compound as a non-cancer-killing one because it is less likely to kill cancer (0.3% probability) than not to (4.2% probability).

How do we classify a compound without looking at the training data plots every time? First, we assume that the probability distribution functions are Gaussian, or normal, (also

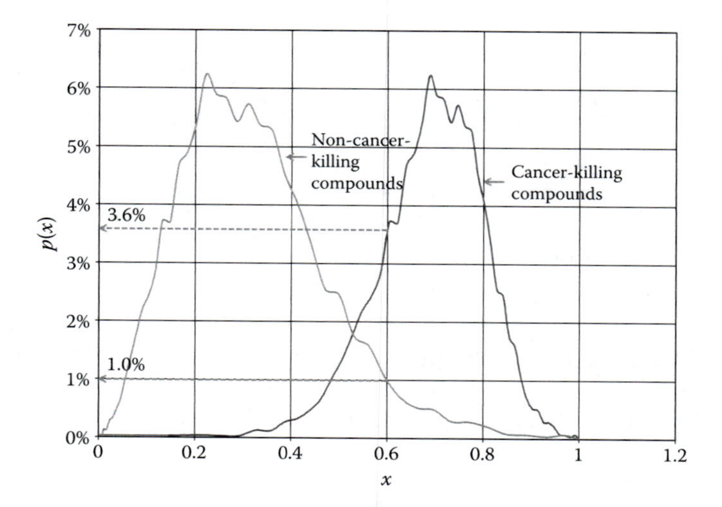

FIGURE 2.6
A new compound with the feature x value of 0.6 has a 3.6% probability of being a cancer-killing compound (top dashed line) and a 1.0% probability of being a compound that does not kill cancer (bottom dashed line).

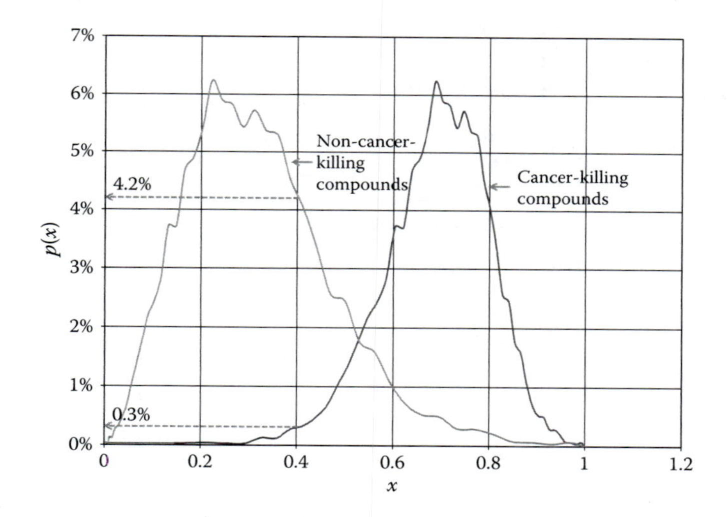

FIGURE 2.7
A new compound with the feature x value of 0.4 has a 0.3% probability of being a cancer-killing compound (bottom dashed line) and a 4.2% probability of being a compound that does not kill cancer (top dashed line).

known as bell curve). We can do that for two reasons: (1) most things in the universe are Gaussian, and (2) the central limit theorem says, "The cumulative effect of a large number of small random effects will be approximately normal, so the Gaussian assumption is valid in a large number of situations." Once we assume that it is a Gaussian distribution, we can use a "simple" calculation to determine which class (cancer killing or not) a compound belongs to:

$$d_j = \mathrm{sqrt}[(x - \mu_j)^2]$$

where
μ_j = jth mean training value
j = 1 for cancer-killing compounds
j = 2 for non-cancer-killing compounds

The classification rule is as follows:

If $d_1 < d_2$, then x is a cancer-killing compound.
Else x is a non-cancer-killing compound.

Figure 2.8 shows how we could use the classification rule to classify our two example compounds.

For the case with multiple features, we can expand the rule:

$$d_j = \text{sqrt}\left[\sum_i (x_i - \mu_{ij})^2\right]$$

where
μ_{ij} = jth mean training value of the ith feature
j = 1 for cancer-killing compounds
j = 2 for non-cancer-killing compounds

The classification rule is still the same:

If $d_1 < d_2$, then x is a cancer-killing compound.
Else x is a non-cancer-killing compound.

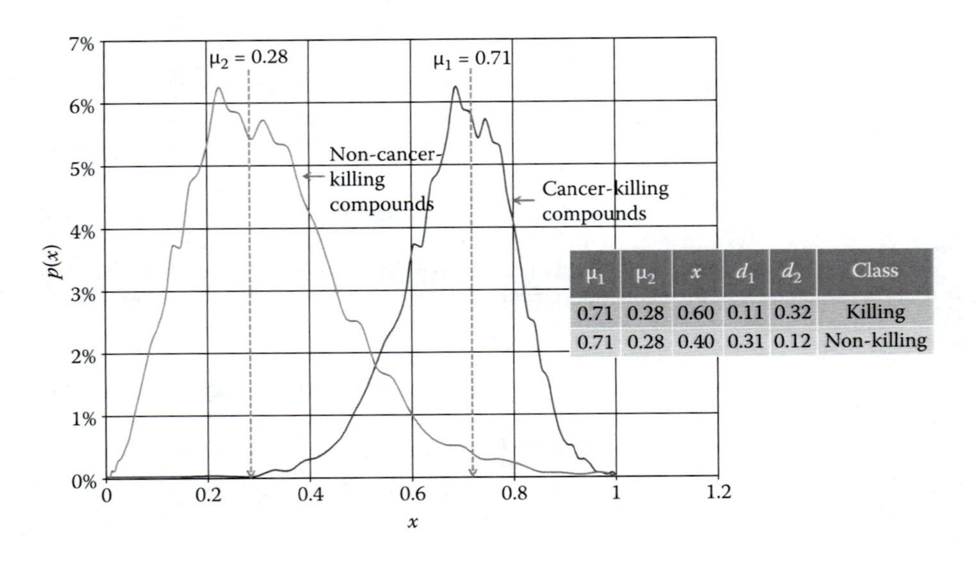

μ_1	μ_2	x	d_1	d_2	Class
0.71	0.28	0.60	0.11	0.32	Killing
0.71	0.28	0.40	0.31	0.12	Non-killing

FIGURE 2.8
Classification of two compounds with x feature values of 0.60 and 0.40 using the Bayesian classification rule.

Although Bayesian classifiers are not always the best as measured by the metrics we discussed earlier, they are the "gold standard" by which all other classifiers are measured.

2.6 Hidden Markov Models

Hidden Markov models (HMMs) are machine learning classifiers that are known for their application in temporal pattern recognition such as speech, handwriting, and bioinformatics (Baum & Petrie 1966; Baum & Eagon 1967; Baum & Sell 1968; Baum et al. 1970; Baum 1972). For example, an HMM is often used to recognize the sequence of sounds that make a specific spoken word.

A generalized HMM is depicted in Figure 2.9.

An HMM is composed of multiple states (x). Each state may transition to one or more states ($z_1, z_2, \ldots, z_n$) or back to itself. Each state may be transitioned to by one or more states ($y_1, y_2, \ldots, y_n$). The probabilities that a state will transition to another state or to itself must sum to one:

$$p(z_1) + p(z_2) + \ldots + p(z_n) + p(x) = 1$$

Each time the HMM transitions from one state to another or back to itself, an emission occurs. There is a one-to-one correspondence between state transitions and emissions, but two state transitions may have the same emission. In other words, emission a always occurs when the HMM transitions from x to z_1, but it may also occur when it transitions from x to z_2. The emissions, which are measured, are the input to the HMM, and the output is the sequence of state changes in temporal order. What kind of problems this might solve might be unclear at first, so we will look at an example. In this example, we have software that processes video frames and can track various objects and human body parts in each frame. By watching the body parts and objects move, we want to determine if the human in the video is grasping an object. We can model a human grasping an object with an HMM, as shown in Figure 2.10 (Johnson and Agah 2011).

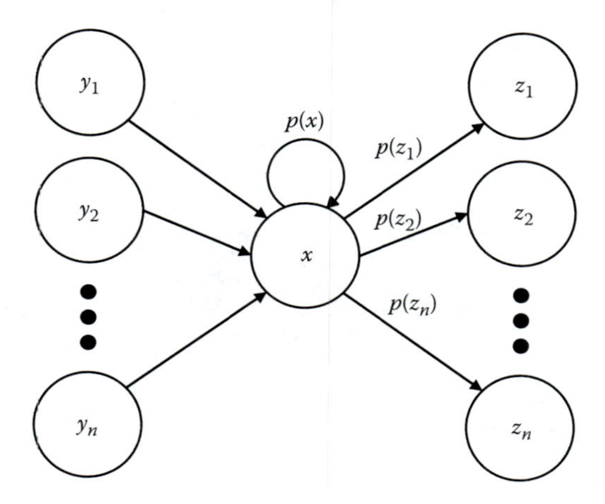

FIGURE 2.9
Generalized Hidden Markov model (HMM).

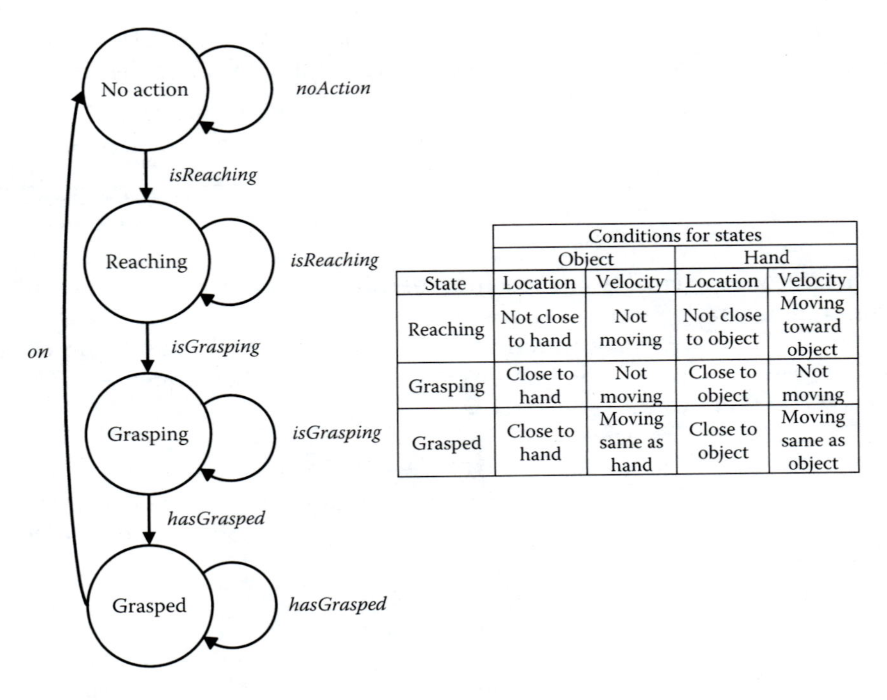

	Conditions for states			
	Object		Hand	
State	Location	Velocity	Location	Velocity
Reaching	Not close to hand	Not moving	Not close to object	Moving toward object
Grasping	Close to hand	Not moving	Close to object	Not moving
Grasped	Close to hand	Moving same as hand	Close to object	Moving same as object

FIGURE 2.10
HMM of a human grasping an object.

The transitions between the states are labeled with the emissions instead of the probabilities. The HMM has four states: no action, reaching, grasping, and grasped. The state of the HMM depends on the position and velocity of the hand and of the object that the hand is grasping. Figure 2.11 shows how the HMM works as a classifier.

The inputs are the emissions detected by the software from the video frames, and the output is the sequence of states. To determine whether the human is grasping an object, we just need to find the sequence {1, 2, 3, 4} in the output. The state transition matrix gives the probabilities of the HMM transitioning from one state to another (i.e., $p(x)$, $p(y_1)$, $p(y_2)$, ..., $p(y_n)$, $p(z_1)$, $p(z_2)$, ..., $p(z_n)$), as illustrated in Figure 2.12.

The probabilities in the state transition matrix are determined with a set of training data. In our example, each frame of a training video of people grasping objects was analyzed manually to determine the state transition matrix.

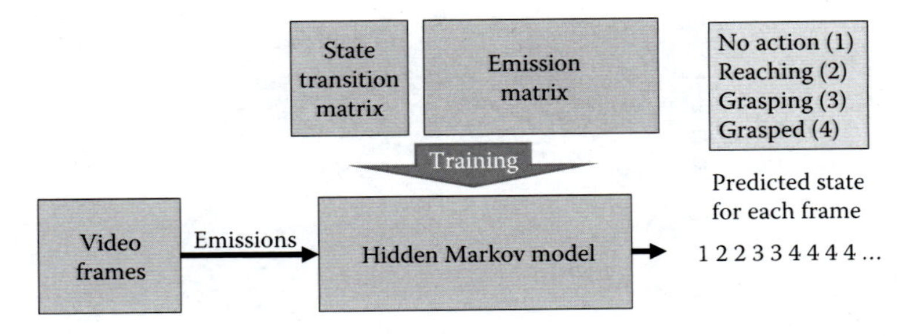

FIGURE 2.11
Example of an HMM classifier of a human grasping an object.

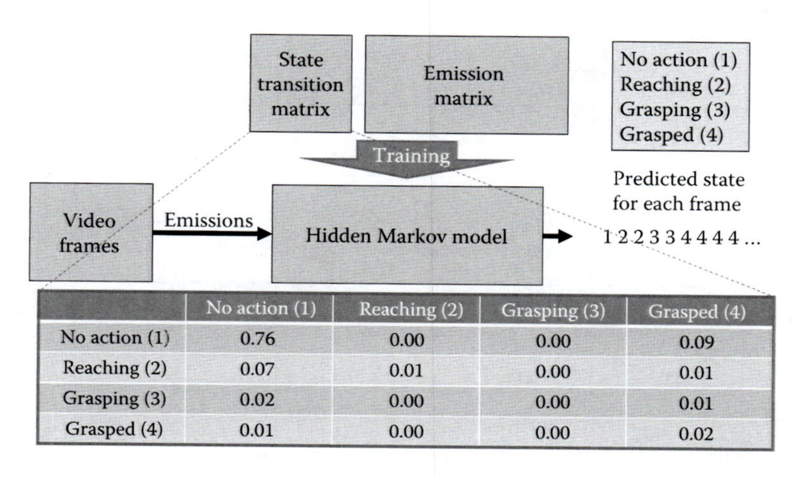

	No action (1)	Reaching (2)	Grasping (3)	Grasped (4)
No action (1)	0.76	0.00	0.00	0.09
Reaching (2)	0.07	0.01	0.00	0.01
Grasping (3)	0.02	0.00	0.00	0.01
Grasped (4)	0.01	0.00	0.00	0.02

FIGURE 2.12
Illustration of state transition matrix.

The emission matrix is the probability of an emission occurring during a state. we can define the emissions by starting with the conditions shown in Figure 2.10 and formalizing them as follows:

isReaching = $C_1 \wedge C_2 \wedge C_3$
C_1 = hand moving toward object
C_2 = object not moving
C_3 = hand not close to object
isGrasping = $C_4 \wedge C_5 \wedge C_6$
C_4 = hand close to object
C_5 = hand not moving
C_6 = object not moving
hasGrasped = $C_7 \wedge C_8$
C_7 = hand close to object
C_8 = hand moving same as object

Noise may cause more than one condition to be detected in each frame. To allow for noise, we can define eight emissions, one for each of the seven combinations of the three conditions and one for the emission where no condition is detected, as illustrated in Figure 2.13.

The probabilities in the emission matrix, as illustrated in Figure 2.14, are determined with a set of training data, too.

With the state transition and emission matrices determined from training data, the classifier can now predict the state of the HMM by measuring the emissions from a test video.

Another example of how an HMM might be used can be found in genetics. Genes are sequences of amino acids in strands of DNA and RNA (Watson et al. 2008). Amino acids are composed of nucleotides. There are four nucleotides. The nucleotides are abbreviated with the letters A, U/T, G, and C. RNA uses U, and DNA uses T instead. Each amino acid is composed of a codon of three nucleotides. That means there could be a maximum of 4^3

	Condition detected		
Emission	isReaching	isGrasping	hasGrasped
000	FALSE	FALSE	FALSE
001	FALSE	FALSE	TRUE
010	FALSE	TRUE	FALSE
011	FALSE	TRUE	TRUE
100	TRUE	FALSE	FALSE
101	TRUE	FALSE	TRUE
110	TRUE	TRUE	FALSE
111	TRUE	TRUE	TRUE

FIGURE 2.13
Definition of eight emissions based on the combination of three conditions.

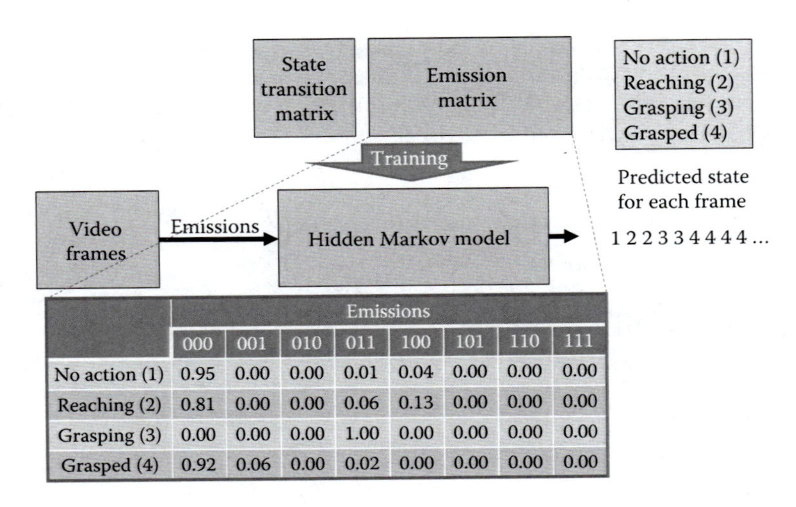

	Emissions							
	000	001	010	011	100	101	110	111
No action (1)	0.95	0.00	0.00	0.01	0.04	0.00	0.00	0.00
Reaching (2)	0.81	0.00	0.00	0.06	0.13	0.00	0.00	0.00
Grasping (3)	0.00	0.00	0.00	1.00	0.00	0.00	0.00	0.00
Grasped (4)	0.92	0.06	0.00	0.02	0.00	0.00	0.00	0.00

FIGURE 2.14
Illustration of emission matrix.

or 64 amino acids. However, some amino acids have more than one codon. There are also stop codons that indicate the end of a gene. The beginning of a gene starts with a start codon. Unlike stop codons, the start codon alone is not sufficient to indicate the beginning of a new gene. Other codons must be nearby the start codon.

Determining the codons in a strand of DNA or RNA would be simple, if one knew where the strand started. But, that is not the case. So the sequence of {A, U, G} could be a codon or the last nucleotide of one codon and the first two of the next codon. An HMM could be used to determine the most likely sequence of codons given a string of nucleotides. Once the sequence of codons is determined, another HMM could be used to determine the start codon of a new gene based on the sequence of nearby codons.

2.7 *k*-Means Clustering

In the previous sections, we have looked at examples of supervised machine learning. In supervised machine learning, we always have training data where we know how it is classified. There are cases where we do not know how the data should be classified. In these cases, we use unsupervised machine learning techniques to determine how the data should be classified. In this section, we will look at *k*-means clustering, which is a popular unsupervised machine learning technique (Steinhaus 1957; Forgy 1965; MacQueen 1967; Lloyd 1982).

2.7.1 *k*-Means Clustering When *k* is Known

As an example of *k*-means clustering, we will use our earlier example shown in Figure 2.1, only this time, we will assume we do not know to which class each point belongs, as shown in Figure 2.15.

The only input that the *k*-means clustering algorithm requires is the number of classes, which is defined as *k*. For this example, we will assume $k = 2$. Later, we will examine methods for determining the appropriate *k*. The algorithm works essentially by trying different groupings, or clusters, of the points and finding the one where the points in each group are the closest to each other. For example, Figure 2.16 shows a possible grouping of the points in Figure 2.15.

The algorithm determines how close the points in each cluster are by first calculating the centroid of the cluster and then taking the average of the distances from each point to the centroid. The centroid of a cluster is calculated by taking the average of the x values and the average of the y values, as shown by the small dots in Figure 2.17.

The distance from a point to the centroid is calculated using the standard distance formula as illustrated in Figure 2.17.

$$\text{distance} = \text{sqrt}[(x_{\text{centroid}} - x_{\text{point}})^2 + (y_{\text{centroid}} - y_{\text{point}})^2]$$

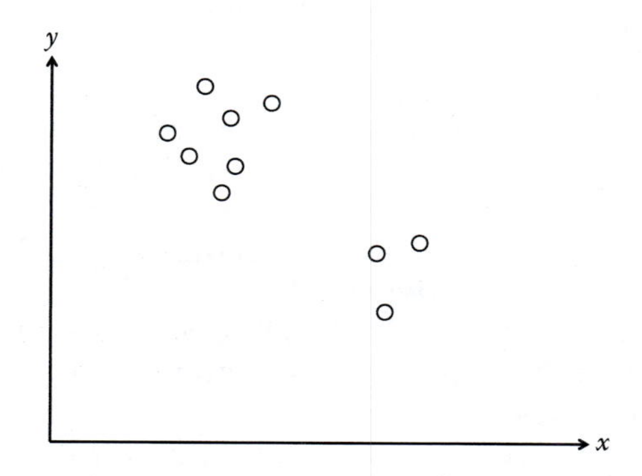

FIGURE 2.15
Example of *k*-means clustering.

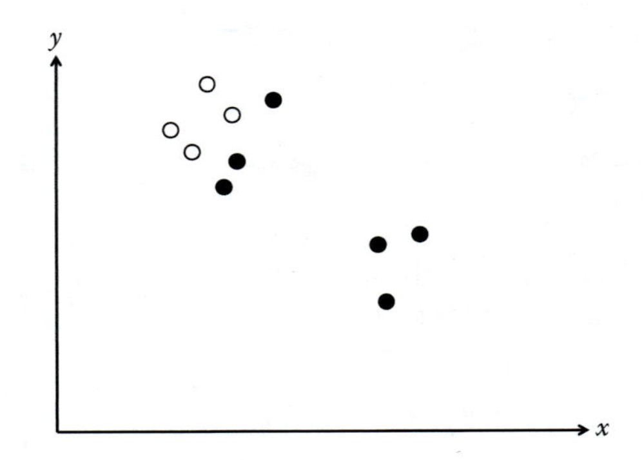

FIGURE 2.16
Possible cluster of points in Figure 2.15.

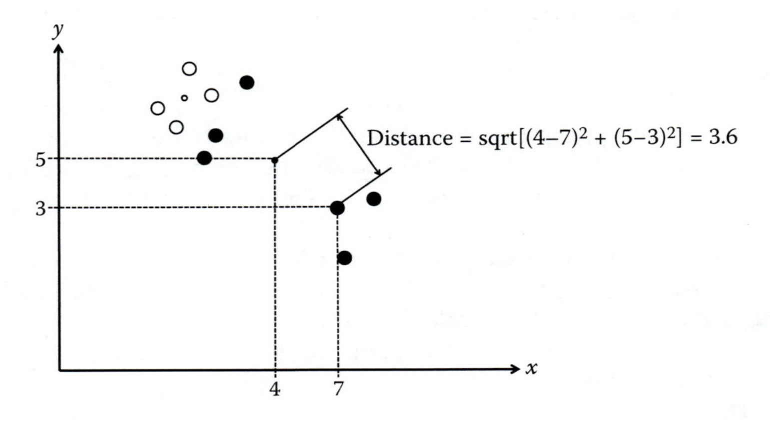

FIGURE 2.17
Centroids and distance for possible clusters in Figure 2.16.

The algorithm measures "closeness" by summing the average distances for the *k* clusters. It then outputs the clustering that produces the smallest "closeness." The algorithm could try all possible combinations of clusters, but even for our simple example above, that might take too long. Most *k*-means clustering problems have hundreds of points, more than two features, and more than two clusters, making the problem impractical to solve by trying all possible combinations. To get around this problem, implementations start by randomly distributing the centroids of the *k* clusters. Then, they calculate the "closeness" metric. Next they use a sophisticated algorithm to move one centroid. Then, they recalculate the "closeness" metric. If it is smaller than the original "closeness," then they leave the centroid there. If it is larger, then they put the centroid back to its original position. Then, they repeat the process of moving a centroid. The process is repeated until the "closeness" metric is below a threshold, usually set by the user of the *k*-means clustering software, or a maximum number of iterations occur. The end result is *k* clusters of the data that can then be used to train one of the supervised machine learning tools described earlier in the chapter.

The algorithm can be expanded to multiple features by just using a multidimensional distance formula. For example, if we had three features, the distance formula would be

$$\text{distance} = \text{sqrt}[(x_{\text{centroid}} - x_{\text{point}})^2 + (y_{\text{centroid}} - y_{\text{point}})^2 + (z_{\text{centroid}} - z_{\text{point}})^2]$$

2.7.2 *k*-Means Clustering When *k* is Unknown

The only question we have not answered is the following: What is the number of clusters or value of *k* to use? In this example, the number of clusters is obvious because we are looking for two classes of compounds: those that kill cancer cells and those that do not. For an example where the number of classes is not known, we will first look at a famous data set used as a benchmark for machine learning tools. In 1936, Sir Ronald Fisher created the Iris flower data set (Fisher 1936), which is a data set with multiple features (i.e., a multivariate data set). The data set contains 50 samples from each of three species of Iris (*Iris setosa*, *Iris virginica*, and *Iris versicolor*). The data set contains four features for each sample: the length and the width of the sepals and petals. Based on the combination of these four features, Fisher developed a model to distinguish the species from each other. Fisher knew that the number of species was three. But, suppose he did not know the number of species and was trying to identify how many species of Iris there are based on the length of the sepals and petals. Similar examples exist in medicine, such as classifying illnesses based on symptoms. If the number of classes, or *k*, is unknown, the question is usually answered by first trying several values for *k*, for example, varying *k* from 2 to 20. Then, there are two techniques for determining which *k* is the best: reconstruction error (Alpaydin 2004) and peakedness (Johnson 2008).

2.7.3 Reconstruction Error

There are two steps to determining the best *k* with reconstruction error. The first step is to calculate the reconstruction error for each value of *k* as the sum of the mean squared error between all points ($\boldsymbol{x}^t$) in the cluster and the centroid of the cluster ($\boldsymbol{m}_i$) they were assigned to, or

$$E\left(\{\boldsymbol{m}_i\}_{i=1}^{k} \mid X\right) = \sum_t \sum_i b_i^t \left\|\boldsymbol{x}^t - \boldsymbol{m}_i\right\|^2$$

where

$$b_i^t = 1, \text{ if } \left\|\boldsymbol{x}^t - \boldsymbol{m}_i\right\| = \min_j \left\|\boldsymbol{x}^t - \boldsymbol{m}_i\right\|$$

$$0, \text{ otherwise.}$$

The second step is to plot the reconstruction error as a function of *k* and look for the "elbow" of the curve, as illustrated in Figure 2.18.

2.7.4 Peakedness

The second technique for determining the best *k* is to find the one where the largest cluster is the one with the highest peak in comparison to the other clusters. For example, if we plot the number of points in each cluster for a specific *k*, we might get a plot like that shown in Figure 2.19.

In Figure 2.19, there is a clear peak at cluster no. 7, with 6 points in it. We can find the *k* with the highest peak by calculating the "peakedness" for each *k* using the formula below and then using the one with the highest "peakedness" value:

$$\text{peakedness}(k) = \max(\text{cluster size for this } k)/(\text{total number of points}/k)$$

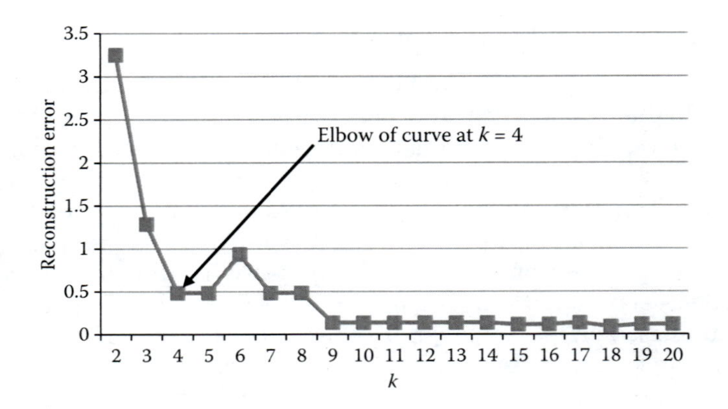

FIGURE 2.18
Example of using reconstruction error to determine the best value for k.

The total number of points remains constant for all k. It is the number of samples being clustered. In the example shown in Figure 2.19, if we assume that the number of samples being clustered is 20 and k is 10,

$$\text{peakedness}(k) = \max(\text{cluster size for this } k)/(\text{total number of points}/k)$$

$$\text{peakedness}(k) = 6/(20/10) = 3.0$$

For the case when the peakedness for two or more k is the same, the smallest k is used. For example, if the peakedness for $k = 3$, 4, and 6 is 4.5, and 4.5 is the largest peakedness value, then we would pick 3 as the best number of clusters for the data.

Figure 2.20 illustrates how the best number of clusters can be determined using peakedness.

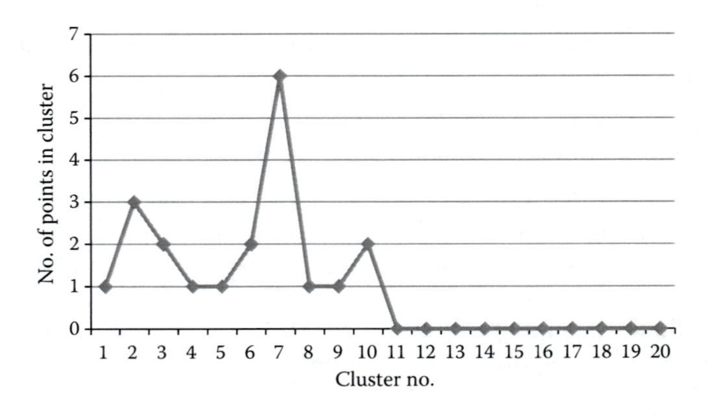

FIGURE 2.19
Plot of the number of points in each cluster for a specific k.

		k																		
		2	3	4	5	6	7	8	9	10	11	12	13	14	15	16	17	18	19	20
Cluster no.	1	11	4	5	7	1	2	2	1	1	1	1	2	1	1	2	2	1	1	1
	2	9	8	7	5	4	1	2	1	3	1	1	1	1	2	1	1	2	1	1
	3		8	5	3	6	3	3	1	2	1	2	1	3	1	2	1	2	1	1
	4			3	4	3	3	2	2	1	3	1	1	1	1	1	1	1	1	1
	5				1	5	2	3	6	1	1	1	2	1	1	1	2	1	1	1
	6					1	6	1	2	2	3	1	2	2	2	1	1	1	1	1
	7						3	6	2	6	2	2	1	2	1	1	1	1	1	1
	8							1	3	1	2	3	2	1	1	1	1	1	1	1
	9								2	1	2	3	1	2	2	1	1	1	1	1
	10									2	1	3	2	1	1	1	1	1	2	1
	11										3	1	1	1	1	2	1	1	1	1
	12											1	2	1	1	1	1	1	1	1
	13												2	2	2	1	1	1	1	1
	14													1	1	2	1	1	1	1
	15														2	1	1	1	1	1
	16															1	1	1	1	1
	17																2	1	1	1
	18																	1	1	1
	19																		1	1
	20																			1
Max		11	8	7	7	6	6	6	6	6	3	3	2	3	2	2	2	2	2	1
Number of vectors/*k*		10	7	5	4	3	3	3	2	2	2	2	2	1	1	1	1	1	1	1
Peakedness $p(k)$		1.1	1.2	1.4	1.8	1.8	2.1	2.4	2.7	3.0	1.7	1.8	1.3	2.1	1.5	1.6	1.7	1.8	1.9	1.0

Number of points in cluster

FIGURE 2.20
Example of determining the best number of clusters (k) using peakedness.

In this case, 10 clusters give the highest peakedness value (3.0). So, if we were trying to cluster diseases based on symptoms, then we would say there are 10 diseases. Then, we could use one of the other supervised machine learning techniques to classify a disease based on the value of the symptoms.

2.8 Principal Component Analysis

Principal component analysis (PCA) is another popular unsupervised machine learning technique that is used to identify the features that are the most important to use in classifying something (Pearson 1901). For example, suppose we are classifying US paper currency. We can measure the number on the bill, the width of the bill, and the length of the bill. Clearly, for US currency, the number on the bill is much more important than the width or length. Of course, this is not true with other countries' currencies where the sizes of the bills are different for different denominations. Returning to our example of classifying compounds by how well they kill cancer cells, we may have many features we can determine about a compound (e.g., atomic weight, acidity, color, elasticity). Some of those may not be as important as other features. PCA is a tool we can use to determine the most important features to use in our classification.

PCA begins by organizing the data into a centered matrix where each row represents a sample of the data (e.g., one compound) and each column represents the value of a feature (e.g., atomic weight). A centered matrix is one where the mean of the column has been subtracted from the values in each column. This is done so that the mean of each column is zero. Table 2.3 gives an example of the raw data, with the mean for each column calculated in the last row.

Table 2.4 shows the data in a centered PCA matrix.

Next, the covariance of the centered PCA matrix is calculated. The result is an n-by-n matrix, where n is the number of features (e.g., four). The covariance matrix for the centered PCA matrix in Table 2.4 is shown in Table 2.5.

The covariance matrix is a matrix whose cells give the covariance between the two features in the row and column. For example, in Table 2.5, the covariance between features A and B is −0.022. The diagonal of the covariance matrix is the variance of each feature. Covariance is a measure of how much two random variables change together, or in our case, how much two features change together.

The next step is to compute the eigenvector matrix and eigenvalue matrix of the covariance matrix with a computer-based algorithm for computing eigenvectors and eigenvalues. The resulting eigenvector and eigenvalue matrices are n-by-n matrices, where n once again is the number of features. The eigenvalues and eigenvectors are ordered and paired. The next step is to sort the columns of the eigenvector matrix and eigenvalue matrix in order of decreasing eigenvalue, while maintaining the correct pairings between the columns in each matrix. The resulting sorted matrices are then analyzed to determine the subset of features that cause the most variability in the data. The features that vary the most are the ones that make it easier to classify the data using one of the standard machine learning tools we learned about earlier. The resulting matrices also provide a transformation

TABLE 2.3

Example of PCA Matrix with the Mean of Each Column Shown in the Last Row

		Feature			
		A	B	C	D
Sample	1	0.29	0.76	0.08	0.02
	2	0.46	0.26	0.57	0.11
	3	0.38	0.32	0.01	0.80
	Mean	0.38	0.45	0.22	0.31

TABLE 2.4

Example of a Centered PCA Matrix with the Mean of Each Column Shown in the Last Row

		Feature			
		A	B	C	D
Sample	1	−0.09	0.31	−0.14	−0.29
	2	0.08	−0.18	0.35	−0.20
	3	0.01	−0.13	−0.21	0.49
	Mean	0.00	0.00	0.00	0.00

TABLE 2.5

Covariance Matrix for Centered PCA Matrix in Table 2.4

	A	B	C	D
A	0.008	−0.022	0.021	0.006
B	−0.022	0.073	−0.041	−0.057
C	0.021	−0.041	0.095	−0.067
D	0.006	−0.057	−0.067	0.095

TABLE 2.6

PCA-Transformed Data from Table 2.3

		Feature	
		A	C
Sample	1	0.27	0.09
	2	0.48	0.55
	3	0.26	0.20

matrix for adding the variability in the discarded features to the ones that are ignored. So, in essence, PCA concentrates the variability in a large set of features into a smaller set. For example, after selecting the appropriate features and transforming the data in Table 2.3, we might end up with a set of data to classify that looks like Table 2.6.

Note that the values of the features for each sample in Table 2.6 are different than they are in Table 2.3. But, that does not make a difference to whichever classifier we use since it is the variability of the data that is important, not the actual values.

PCA can also be used with *k*-means clustering to find better clusters.

References

Alpaydin, E. (2004). *Introduction to Machine Learning*. Cambridge, MA: MIT Press.

Baum, L. E. (1972). An inequality and associated maximization technique in statistical estimation of probabilistic functions of a Markov process. *Inequalities* 3: 1–8.

Baum, L. E. and Eagon, J. A. (1967). An inequality with applications to statistical estimation for probabilistic functions of Markov processes and to a model for ecology. *Bulletin of the American Mathematical Society* 73 (3): 360.

Baum, L. E. and Petrie, T. (1966). Statistical inference for probabilistic functions of finite state Markov chains. *The Annals of Mathematical Statistics* 37 (6): 1554–1563.

Baum, L. E. and Sell, G. R. (1968). Growth transformations for functions on manifolds. *Pacific Journal of Mathematics* 27 (2): 211–227.

Baum, L. E., Petrie, T., Soules, G., and Weiss, N. (1970). A maximization technique occurring in the statistical analysis of probabilistic functions of Markov chains. *The Annals of Mathematical Statistics* 41: 164.

Cortes, C. and Vapnik, V. (1995). Support-vector networks. *Machine Learning* 20 (3): 273–297.

Fisher, R. A. (1936). The use of multiple measurements in taxonomic problems. *Annals of Eugenics* 7 (2): 179–188.

Forgy, E. W. (1965). Cluster analysis of multivariate data: efficiency versus interpretability of classifications. *Biometrics* 21: 768–769.

Hopfield, J. J. (1982). Neural networks and physical systems with emergent collective computational abilities. *Proceedings of the National Academy of Sciences USA* 79, 2554–2558.

Johnson, D. O. (2008). Human robot interaction through semantic integration of multiple modalities, dialog management, and contexts. PhD Dissertation, Department of Electrical Engineering and Computer Science, University of Kansas, Lawrence, KS.

Johnson, D. O. and Agah, A. (2011). Recognition of marker-less human actions in videos using hidden Markov models. In *Proceedings of the ICAI'11 – The 2011 International Conference on Artificial Intelligence*, July 18–21, 2011, Las Vegas, NV, 95–100.

Lloyd, S. P. (1982). Least squares quantization in PCM. *IEEE Transactions on Information Theory* 28 (2): 129–137.

MacQueen, J. B. (1967). Some methods for classification and analysis of multivariate observations. In *Proceedings of 5th Berkeley Symposium on Mathematical Statistics and Probability*, Berkeley, CA: University of California Press, pp. 281–297.

McCarthy, J. (2007). *What is Artificial Intelligence?* Computer Science Department, Stanford University.

Pearson, K. (1901). On lines and planes of closest fit to systems of points in space. *Philosophical Magazine* 2 (11): 559–572.

Steinhaus, H. (1957). Sur la division des corps matériels en parties (in French). *Bulletin L'Académie Polonaise des Science* 4 (12): 801–804.

Watson, J. D., Baker, T. A., Bell, S. P., Gann, A., Levine, M., and Oosick, R. (2008). *Molecular Biology of the Gene*. San Francisco: Pearson/Benjamin Cummings.

3

Overview of Prominent Machine Learning and Data Mining Methods with Example Applications to the Medical Domain

Christopher M. Gifford

CONTENTS

3.1 Introduction

Modern medicine, health informatics, and bioinformatics systems create and leverage massive volumes of rich, complex, and highly coupled data sets. Hospitals, treatment centers, and doctors' offices capture a diverse collection of personal health data, health history, symptoms and diagnosis data, disease and treatment data, blood work and test results, and imaging products. Medical research institutions produce even further experimental data, metadata, and results at various resolutions, from cell-level tissue samples, to disease progression and propagation behaviors, to the sequencing of the human genome. These data are produced by a variety of devices, sensors, and information systems all over the world every day.

The speed and scale at which these data are produced frame a highly nonlinear, hyper-dimensional feature space that is far too large and complex for doctors and scientists to fully understand. Humans can no longer perform exhaustive tasks or ask even simple questions of these data in an efficient manner. This creates the need for increased automation, decision support mechanisms, and nontrivial algorithm development to relieve data overwhelm and enable practitioners to focus on higher-level tasks, analytical reasoning, and making informed decisions. Automated, algorithmic solutions are attractive for several important reasons:

- Enable sensemaking of high-volume, complex data sets
- Offer efficient summarization, modeling, and analysis of an enormous amount of data
- Provide compact, intuitive, and in many cases human-readable outputs
- Assist in the identification of abnormalities
- Provide greater automation, speed, and effectiveness of doctors, scientists, and officials

Artificial intelligence, primarily machine learning and data mining, has played and will continue to play a key role in the advancement of technologies, research, understanding, and treatment in the medical domain. This chapter is aimed at introducing the reader to many prominent artificial intelligence techniques being employed to improve modern medicine and practice. Section 3.2 provides an overview of machine learning and data mining, including descriptions of classes of algorithms and secondary decision methods utilized in the broader community. Section 3.3 discusses popular software suites and resources to support development for various applications and domains. Section 3.4 discusses several relevant and illustrative examples of medical applications of artificial intelligence. Section 3.5 concludes the chapter with closing remarks, leading the reader into the research chapters of the book.

3.2 Overview of Machine Learning and Data Mining

Machine learning and data mining are major fields within artificial intelligence that are heavily used in many domains for a variety of applications. This section describes some of the prominent methods and classes of algorithms. In many cases, the data set properties, expected output usage, and/or application domain drive the algorithm(s) to use for any particular problem.

3.2.1 Knowledge Discovery and Data Mining

The goal of data mining is to discover inherent and previously unknown information from data to represent as knowledge. In many cases, such methods are used to summarize and model very large data sets, capturing what salient (high support or occurring frequently) and interesting patterns reside in the data that may not have otherwise been discovered. Many of these methods were motivated by the growing demand to mine consumer transaction data to determine which items were purchased together, so that retail stores could better position their products within the store and track buying patterns. These methods therefore extend to online transactions and have therefore been heavily used by the Web community. As such, many algorithms leverage the transaction model for the core data structure and operating algorithmic assumptions. Two such methods are Association Rule Mining and Sequential Pattern Discovery.

3.2.1.1 *Association Rule Mining*

The methodology of association rule mining produces knowledge in rule form, called an *association rule*. An association rule describes a relationship among different attributes. Association rule mining was introduced by Agrawal et al. [1] as a way to discover interesting co-occurrences in supermarket data (the market basket analysis problem). It finds frequent sets of items (i.e., combinations of items that are purchased together in at least N database transactions) and generates from the frequent itemsets (such as $\{X, Y\}$) association rules of the form $X \rightarrow Y$ and/or $Y \rightarrow X$.

More formally, let $D = \{t_1, t_2,..., t_n\}$ be the transaction database, and let t_i represent the ith transaction in D. Let $I = \{i_1, i_2,..., i_m\}$ be the universe of items. A set $X \sqsubseteq I$ of items is called an itemset. When X has k elements, it is called a k-itemset. An association rule is an implication of the form $X \rightarrow Y$, where $X \subset I$, $Y \subset I$ and $X \cap Y = \varnothing$. Various statistical measures exist that offer metadata about the rule. *Support*, *confidence*, and *lift* are three such measures.

The *support* of an itemset X is defined as

$$\text{Support}(X) = \frac{\text{number records with } X}{\text{number records in } D}$$

The *support* of a rule $(X \rightarrow Y)$ is defined as

$$\text{Support}(X \rightarrow Y) = \frac{\text{number records with } X \text{ and } Y}{\text{number records in } D}$$

The *confidence* of a rule $(X \rightarrow Y)$ is defined as

$$\text{Confidence}(X \rightarrow Y) = \frac{\text{number records with } X \text{ and } Y}{\text{number records with } X}$$

Confidence can be treated as the conditional probability ($P(Y|X)$) of a transaction containing X and also containing Y. A high confidence value suggests a strong association rule. However, this can be deceptive. For example, if the antecedent (X) or consequent (Y) has a high support, it could have a high confidence even if it was independent. This is why the measure of lift was suggested as a useful metric.

The lift of a rule ($X \rightarrow Y$) measures the deviation from independence of X and Y. A lift greater than 1.0 indicates that transactions containing the antecedent (X) tend to contain the consequent (Y) more often than transactions that do not contain the antecedent (X). The higher the lift, the more likely that the existence of X and Y together is not just a random occurrence but, rather, due to the relationship between them.

$$\text{Lift}(X \rightarrow Y) = \frac{\text{Confidence}(X \rightarrow Y)}{\text{Support}(Y)}$$

A limitation of traditional association rule mining is that it only works on binary transaction data [e.g., an item is either purchased in a transaction (1) or not (0)]. In many real-world applications, data are either categorical (e.g., blue, red, green) or quantitative (e.g., number of molecules). For numerical and categorical attributes, Boolean rules are unsatisfactory. Extensions have been proposed to operate on these data, such as quantitative association rule mining [2] and fuzzy association rule mining [3,4].

Apriori [1] is the most widely used algorithm for finding frequent k-itemsets and association rules. It exploits the downward closure property, which states that if any k-itemset is frequent, all of its subsets must be frequent as well. The Apriori algorithm proceeds as follows:

Calculate the support of all *1*-itemsets and prune (i.e., remove) any that fall below the minimum support, specified by the user.

Loop:

1. Form candidate k-itemsets by taking each pair (p, q) of (k-1)-itemsets, where all but one item match. Form each new k-itemset by adding the last item of q onto the items of p.
2. Prune the candidate k-itemsets by eliminating any itemset that contains a subset not in the frequent (k-1)-itemsets.
3. Calculate the supports of the remaining candidate k-itemsets and eliminate any that fall below the specified minimum support threshold. The result is the frequent k-itemsets.

The extension of apriori to fuzzy apriori was proposed by Kuok et al. [3]. Both apriori and fuzzy apriori need to repeatedly scan the database when they construct the k-itemsets and calculate their support. This requirement caused these methods to be criticized by many researchers for being inefficient. The frequent pattern (FP)-growth algorithm was proposed by Han et al. [5] as a fast method for generating frequent itemsets. When generating frequent itemsets, FP-growth avoids generating a significant amount of candidates that are generated by apriori. FP-growth builds and operates on a data structure called the FP-tree and passes over the data set only twice. The FP-growth algorithm operates as follows:

FP-tree construction

1. Calculate the support of all 1-itemsets and prune any that fall below the minimum support threshold, specified by the user.
2. Sort the remaining frequent 1-itemsets in descending global frequency.
3. Build the FP-tree, tuple by tuple, from the first transaction to the last.

Deriving frequent itemsets from the FP-tree

1. A conditional FP-tree is generated for each frequent itemset, and from that tree, the frequent itemsets with the processed item can be recursively derived.

FP-growth mines the complete set of frequent patterns regardless of the length of the longest pattern. Lin et al. [6] proposed a partial fuzzification of FP-growth, including a fuzzy FP-tree (a tree structure for frequent fuzzy regions). The first step in their method is to transform quantitative values into fuzzy membership function values (as in fuzzy apriori). Their extension is only partially fuzzy, as later in the process, it only uses the fuzzy region (i.e., membership function) with maximum cardinality among all the transformed fuzzy regions of a variable.

3.2.1.2 Sequential Pattern Discovery

Traditional association rule mining was designed for mining salient patterns within a set of transactions, ignoring the temporal order of the transactions. Thus, association rule mining becomes inefficient for temporal and sequential applications that require ordered matching rather than simple subset testing. Although comparatively less mature, sequential pattern discovery methods have been developed specifically for these types of problems.

Here, we define a *sequential pattern* as a temporally ordered set of items or events. As with association rule mining, combinatorial explosion and the number of data passes affect speed and memory requirements of these algorithms. Further challenges are pattern summarization (i.e., providing an expressive, compact result set), the ability to naturally handle and analyze temporal data (e.g., activity sequences), and achieving acceptable performance for long sequences and reduced support (low frequency). Many early sequence mining algorithms were able to achieve good performance for short frequent sequence applications but suffered degraded performance for long sequences and low support thresholds.

Examples of existing methods in this space are generalized sequential pattern (GSP) [7], PrefixSpan [8], bi-directional extension (BIDE) [9], mining top-K closed sequential patterns (TSP) [10], and CloSpan [11]. In many cases, the full set of subsequences is not necessary to capture the expressive power of the data set. CloSpan, in particular, leverages an important concept to enable the efficient mining of long sequences with reduced support: *closed subsequences* [11]. A subsequence S is considered *closed* if no superset (extension) of S exists with the same support. By removing these superset sequences from consideration, redundant subsequences with the same support are eliminated. This dramatically reduces the result set, improves algorithm efficiency, and allows the mining of long sequences to be a tractable process.

Result set summarization is necessary for usability of the outputs of such methods. High-support thresholds are typically used to reduce the result set to a manageable size. Interestingness measures have also been developed to aid in post-pruning (i.e., selective filtering) the result set. Closed sequences provide a more compact result set to reduce analyst overwhelm. Sequential patterns can be efficiently summarized by leveraging *partial orders* [12]. Partial orders are created by collapsing and combining overlapping sequences into a directed graphical representation, viewable as a temporal "left-to-right" network. Thus, partial orders provide both an intuitive visualization of the result set to facilitate data understanding and a representative data model. By representing the result set in this manner, new incoming data can be evaluated against the sequential model as a basis for a variety of novel purposes, including the following:

- Anomaly detection (e.g., if an observed sequence does not fit the network, or events are observed out of order, an anomaly indicator can be triggered)
- Intervention/prediction (e.g., probable/expected future observations can be suggested using the temporal nature of the sequential network)
- Possible causes (e.g., probable prior observations can be suggested that lead to the currently observed event by exploiting the temporal nature of the sequential network)
- Further analytics and statistical measures

Potential medical applications of sequential pattern discovery include, but are not limited to, the following:

- DNA sequence alignment
- Motifs and tandem repeats in DNA sequences
- Modeling clinical visit patterns
- Analyzing infectious disease and micro-level disease observations
- Studying the evolution of a disease and agent/host interaction patterns
- Analysis of multi-omics experiment data involving temporal treatments

3.2.2 Machine Learning

The field of machine learning is concerned with algorithms and mechanisms to allow computers to learn (or model) data, behavior, or an environment. Learning is induced by automatically extracting information from data, typically in a statistical manner, and adjusting/adapting the underlying computational data structure, model, or resulting action/output over time. As a subfield of artificial intelligence, machine learning is highly related to other subfields of data mining, pattern recognition, classification, and optimization.

Enabling computers to model and provide automated decision support can eliminate the need for extensive human intervention or *a priori* assignment. It also provides a way to quickly and intelligently search through or summarize data to learn underlying patterns that are not evident or easily observable. The manner in which learning takes place, as well as the output of the learning process, depends on the underlying algorithm(s). These algorithms typically fall into one of several categories, depending on the task, underlying knowledge structure and extraction procedure, and desired output.

Many techniques exist within each category (e.g., unsupervised versus supervised). Some learning algorithms are *unstable*, meaning they experience major output changes with small input changes. Examples of unstable algorithms include artificial neural networks, decision trees, and rule-based methods. An example of a *stable* algorithm is a fuzzy system that makes use of non-crisp feature and decision boundaries. Machine learning algorithms are typically trained on known data and tested on (applied to) unknown/

unseen data. The following sections introduce a variety of machine learning algorithms, as well as multi-classifier decision fusion and meta-learning secondary decision methods.

3.2.2.1 Logistic Regression

Logistic regression is a statistical model that predicts the probability of an event occurring by fitting a logistic curve to the data. Several predictor variables are typically used to describe the relationship between risk factors and outcomes. Logistic regression has seen use in medical and social fields, including customer trending.

Logistic regression is based on the logistic function

$$f(Z) = \frac{1}{1+e^{-Z}}$$

where Z represents the input (set of risk factors) and the returned value represents the output (probability of a particular outcome). Z normally takes on the following form:

$$Z = a + b_1x_1 + b_2x_2 + \ldots + b_nx_n$$

where a is the intercept (where all risk factors are zero) and b_i represent the regression coefficients (risk factor contributions). Positive coefficients translate to that specific risk factor increasing the probability of an outcome, while negative coefficients translate to a decrease in probability of an outcome. Similarly, larger values correspond to higher influence. A multinomial logistic regression model with a ridge estimator is described in the work of le Cessie and van Houwelingen [13], which iterates until convergence.

3.2.2.2 Naïve Bayes

Statistical and probabilistic approaches to machine learning have been used for many years, achieving notable success in the medical domain. Several such approaches have become popular, such as Bayesian networks and the naïve Bayesian classifier, due to their simplicity and clear probabilistic semantics [14]. The naïve Bayes classifier operates on the assumption that numeric attributes result from a single Gaussian (normal) distribution. For real-world problems, this approximation is unlikely to be accurate but offers comparable performance to decision trees and other induction methods. It also assumes that the predictive attributes are conditionally independent given the class, working on the assumption that no concealed attributes influence prediction.

Naïve Bayes prediction typically takes place using the following process, where C represents a random variable indicating an instance's class, A represents a vector of random variables indicating the attribute values that are observed, c represents the label of a class, and a represents a vector of observed attribute values:

1. Use Bayes' rule to compute the probability of each class given a:

$$p(C = c \mid A = a) = \frac{p(C = c)\prod p(A_i = a_i \mid C = c)}{Z}$$

2. Predict the most probable class

The product in the numerator is due to the conditional independence assumption, and the denominator is a normalization factor, so the sum of $p(C = c|A = a)$ overall classes is 1. Normally, discrete attributes are modeled using a single probability value, while numeric/continuous attributes are modeled using a continuous probability distribution. A popular density estimation method is a single Gaussian. Kernel density estimation has been proposed so that continuous variables can be better estimated [14], especially in the case where the distribution is multimodal rather than normal.

3.2.2.3 Instance-Based Learners: k*-Nearest Neighbor and* K*

Some algorithms make use of specific instances rather than precompiled abstractions (trees, networks, etc.) during prediction. These methods are categorically termed instance-based learning or example-based learning. They have a strong underpinning as a nearest neighbor method and generally use some form of similarity measure to determine "distance" or "match" between instances [15]. One of the more popular instance-based learning algorithms is the k-nearest neighbor classifier, which utilizes the k most similar (nearest) training examples in various ways for prediction. In general, these algorithms work on the assumption that similar instances belong to similar classes. An advantage of such methods is their simplicity, which makes them applicable to many domains and problems. For instance, one could utilize $k = 5$ nearest neighbors for prediction and weigh the neighbors by the inverse of their distance when voting on class membership.

K^* is another flavor of instance-based learners that utilizes entropy (motivated by information theory) as a distance measure [16]. This allows for a more consistent approach to handling missing values and symbolic and real attribute values. The K^* distance method's general approach is similar to work done in comparing DNA sequences, where the distance between two instances is computed as a measure of the sum of all possible transforming paths (represented by transformation probabilities). The selected number of instances is termed the "sphere of influence," which dictates how the instances are weighed. During classification, the category corresponding to the highest probability is selected.

3.2.2.4 Decision Tree (Pruned and Unpruned)

Decision trees are one of the most well-known and widely used machine learning methods and have seen successful use in a variety of real-world problem domains. Their utility as a simple and fast learning approach is widely accepted. Many decision tree algorithms and variations exist, the most popular of which are Quinlan's ID3 and successor C4.5 [17]. Decision trees produced by these algorithms are typically small/shallow and accurate, making for fast and reliable classification.

The process of building a decision tree begins with a set of example cases or instances, each containing a series of numerical or symbolic attributes and a membership class. Each internal node of the tree represents a test that determines the branch to travel down. For example, if an internal node's test is "$x > 42$" and x is 30, the test returns false and proceeds down the right branch of the tree at that node (tests returning true proceed down the left branch). The tree's leaf nodes represent the possible classes that instances can belong to, which is the prediction produced by the tree for test instances.

C4.5 utilizes formulas based on information theory (information gain and gain ratio) to measure the "goodness" of tests for nodes. Thus, tests are chosen that extract the maximum information given a single attribute test and the set of cases. Overfitting is reduced by estimating the error rate of each subtree and replacing it with a leaf node if its estimated error is smaller, a process termed "pruning" [17]. Finally, the decision tree can be transformed into a set of rules by traversing the tree paths from the leaf nodes. The resulting set of rules can then be simplified in a variety of ways (and some rules completely removed) to arrive at a final set of rules for classification.

3.2.2.5 Random Forest

Rather than using a single decision tree for classification, many of them can be used in conjunction. Such a method is termed a random forest [18]. Random forests have their roots in work involving feature subset selection, random split selection, and ensemble classifiers. A random forest is constructed by selecting a random feature vector that is independent of the previous chosen feature vectors (but identically distributed) and growing a tree-structured classifier using this vector and the training set. This is done T times to create a random forest of T decision trees, each of which independently votes on a class. The most popular class from all trees is used to label the test instance.

Two primary measures are used to determine the accuracy and interdependence of the trees both individually and as a whole. The goal is to minimize tree correlation while maintaining individual tree strength. The accuracy of random forests, which has been found to be comparable to and sometimes better than AdaBoost, depends on the strength of the individual tree classifiers and the correlation/dependence between them [18]. Typically, choosing one or two random features to grow on each tree can provide near-optimum results. Random forests allow for any number of trees, which in some cases can be in excess of 100, considering one of a variety of feature subsets (e.g., $\log M + 1$ features, where M is the number of inputs).

3.2.2.6 Rule Learners: Repeated Incremental Pruning to Produce Error Reduction and PART

Rule-learning systems offer several desirable properties, including their human-understandable format and general efficiency. However, they are known to not scale well in relation to data size. One dominant rule-learning variant is the repeated incremental pruning to produce error reduction (RIPPER) propositional rule learner [19]. RIPPER is based on iterative reduced-error pruning (IREP), which, for rules, involves the training data being split into a growing set and a pruning set. A rule set is constructed, one rule at a time, in a greedy manner. When a rule is constructed, all examples that are covered by the rule are deleted via a pruning method.

The uncovered examples are randomly partitioned into two subsets: a growing set and a pruning set. A rule is grown by adding conditions that maximize an information gain criterion until the rule covers no negative examples. The rule is then pruned by deleting condition(s) that maximize a pruning function until the function's value cannot be improved by any deletion. This process is repeated to generate, prune, and optimize the rule set until there are no positive examples remaining, or a rule is found with a large error rate. RIPPER is generally considered to be competitive with C4.5 rules in terms of accuracy and is able to efficiently operate on very large (hundreds of thousands of examples) and noisy data sets. More details are provided on RIPPER's performance and comparison to C4.5 in the work of Cohen [19].

Another successful rule-learning algorithm is PART [20]. PART is a decision-list algorithm based on partial decision trees, combining the advantages of both C4.5 and RIPPER while eliminating some of their disadvantages. It is considered a separate-and-conquer algorithm, meaning that it generates one rule at a time, removes covered instances, and repeats. PART differs in the general construction of rules by creating a pruned decision tree for all current instances, building a rule corresponding to the leaf node with the largest coverage, then discarding the tree and continuing. Partial decision trees are constructed in a manner similar to C4.5. This is said to help avoid the problem of generalizing without knowing its implications [20]. PART avoids global optimization that is present in C4.5 and RIPPER but still produces accurate rule sets. In comparison, PART compares favorably to C4.5 in accuracy and outperforms RIPPER at the cost of a larger rule set.

3.2.2.7 Decision Table

Another rule-learning method is the decision table classifier, with a default rule mapping to the majority class [21]. A decision table has two main components: a schema (set of features) and a body (set of instances with feature values and class labels—the training set). An optimal feature set is determined by transforming the problem into a state space (feature subset) search, using best-first search with the heuristic being k-fold cross-validation to estimate the future prediction accuracy. Incremental cross-validation can be used to provide a speedup for algorithms that support incremental addition and deletion of rules [21].

Prediction for a test instance takes place by finding all labeled training examples that exactly match the features of the test instance. If this set is empty, the majority class of the entire training set is used for labeling the test instance; otherwise, the majority class of the set of matching training examples is used for labeling. Decision tables are more ideal for discrete attributes, as they attempt to find exact matches of the feature set for an instance for class assignment. A perfect match would be highly unlikely for continuous attributes in practice but may be more likely if the number of values is low. Also, as the accuracy prediction method is heuristic based, the best feature set may not be found. For some domains, the decision table was found to perform comparably with the C4.5 decision tree algorithm [17].

3.2.2.8 Artificial Neural Network

Artificial neural networks are computational models consisting of a network of interconnected neurons. The original concept for this approach was inspired by attempts to analyze and model the human brain, focusing on its nodes/neurons, dendrites, and axons. They are typically used to model, discover, or learn patterns in complex relationships between inputs and outputs. They operate on a connectionist approach, where neurons are fired, and their output is adjusted by weighted connections en route through the network via other neuron interfaces and layers.

Neurons represent simple processing entities, which, when connected in concert with other neurons, can exhibit complex global behavior. Each neuron utilizes a transfer function to adjust input for forwarding through connections to other neurons in subsequent neuron layers. A neural network is typically broken down into layers of neurons that behave similarly in terms of transfer function or connection interfaces. There is typically an input layer representing the nodes for inputs into the neural network, some hidden

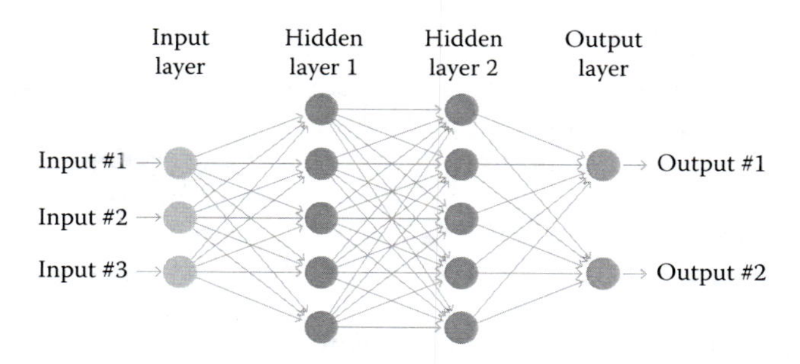

FIGURE 3.1
Fully connected neural network configuration with three input neurons, two hidden layers with five neurons each, and two output neurons.

intermediate layer(s) containing one or more neurons each, and an output layer, which produces the outputs from passing the input through the network. Figure 3.1 illustrates a fully connected neural network configuration, consisting of three input neurons, two hidden layers with five neurons each, and two output neurons.

Neural networks are trained by repeatedly providing the network with inputs and iterating until the error goal for network output has been achieved or a certain iteration threshold has been met. This occurs by providing the network with outputs corresponding to the inputs that it should attempt to learn. The network learns the association between inputs and outputs by passing each input through the network and calculating the error the network produces with its current connection weights. This error is then typically back-propagated to adjust connection weights in reverse, from output nodes to input nodes. By adjusting connection weights, the network can then adjust its output for given input. This process is repeated until a convergence goal is achieved. The connection weights of the network therefore represent the neural network's knowledge, given the instances it has seen and the output it has been provided to train toward. Neural networks exhibit primary disadvantages of being computationally slow in training and application phases, offering nondeterministic convergence (which can also be advantageous), and having difficulty in determining how/why the model made a particular decision. This affects both the use of a neural network model as well as its trustworthiness and acceptance of model outputs.

3.2.2.9 Support Vector Machine

Support vector machines are kernel-based learning machines that make no assumptions on the data distribution. Support vector machines are backed by strong mathematical rigor, are robust to noise, and are currently popular in the literature. They are designed for binary (two-class) problems but can be extended for multiclass classification data sets. The support vector Machine's goal is to find separators in the form of a hyperplane. The data are transformed into another feature space (including nonlinear features), followed by hyperplane construction. The best hyperplane is the one that represents the largest separation, or margin, between the two classes. Support vectors are samples on that margin. When new data are provided, a prediction is made based upon which portion of the margin the data point lays on. Various kernels have been developed for this method, one of which is the popular polynomial kernel.

3.2.3 Multi-Classifier Decision Fusion

Each machine learning paradigm has its own strengths and weaknesses stemming from its underlying theory, mathematical underpinning, and assumptions about the data and/or decision space. Therefore, the accuracy of each algorithm is dependent upon how well those assumptions apply to the input data. Different algorithms achieve different levels of accuracy on the same data, but specific classifiers can develop levels of expertise on portions of the decision space. These aspects provide support for investigating decision fusion and meta-learning approaches.

Since different learning methods typically converge to different solutions, combining decisions from different learning paradigms can be leveraged for improved accuracy [22,23]. Classifier combination is one such area in machine learning that has offered advances in classification accuracy for complex data sets. It has been termed differently in the literature, namely, classifier fusion, mixture of experts, committees, ensembles, teams, collective recognition, composite systems, and so forth. Generally, when predictions from multiple classifiers are combined, they are said to form an ensemble that is then used to classify new examples. Figure 3.2 illustrates the concept of combining decisions from N classifiers to arrive at a fused, consensus decision and confidence.

When developing a multi-classifier system, its members can be a mixture of weak (i.e., high-error-rate) and strong (i.e., low-error-rate) classifiers. Weak classifiers are typically simple to create, at the expense of their accuracy on complex data sets. Strong classifiers are typically time consuming and expensive to create, as their parameters are fine-tuned and tweaked for optimal performance. Combining weak/strong or homogeneous/heterogeneous classifiers offers the benefit of encompassing different levels of expertise and knowledge bases [23].

An important aspect related to differences in learners is their level of error correlation relative to one another, and as a unit. The more correlated (i.e., less disjoint) individual learners are, the less complementary they may be. If they are uncorrelated, they likely will misclassify different instances, and combining them better enables the system to correctly classify more instances. A significant improvement over a single classifier can only happen if the individual classifier theories are substantially different. It is desired to obtain a balance between high performance and complementarity in a team where decisions are combined. If one learner does not predict correctly, the other learners should be able to do so. Diverse models are therefore more likely

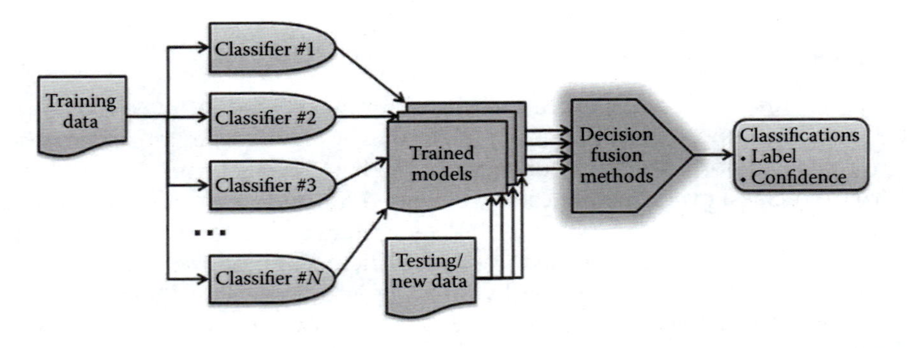

FIGURE 3.2
Decision fusion process for a multi-classifier system, where decisions are collected from a set of classifiers and combined in some way to arrive at a single classification label and confidence for each instance.

to make errors in different ways. Methods to accomplish this typically involve introducing diversity in terms of learning paradigms, feature subsets, or training sets. As modern data volumes grow in size, attempts at distributed processing and parallel machine learning are gaining popularity.

Various approaches exist for combining decisions from multiple classifiers, some of which perform better under certain circumstances. Pure voting methods are among the simplest ways to combine decisions from multiple classifiers. Each entity classifies (votes) that an input instance belongs to one of multiple classes. These votes are tallied in one of a number of ways to arrive at the final decision of the classifiers being fused in the form of a single class label. Examples of pure voting methods are majority vote, maximum confidence, average confidence, and product of confidences.

More intelligent voting methods are accuracy driven. Accuracy-driven voting methods are those that take into account the accuracy, confidence, or probability of each learner offering its vote. This moves a step beyond simple voting, where votes from classifiers that have exhibited strong classification accuracies are weighted more, and weak classifier votes are either discarded or weakly weighted. A popular accuracy-driven voting method is weighted voting, where each classifier is assigned a voting weight (usually proportional to its operational accuracy), and the class with the highest weighted vote is used for classification.

Data manipulation methods, which manipulate the training and testing data to attempt to achieve optimal classification accuracy, are by far the most popular means to train and combine multiple classifiers. They have seen widespread use due to their simplicity and mathematical background and have produced some of the best accuracies on challenging data sets. Some of the popular data manipulation methods include the following:

- **Input decimation:** Classifier correlation can be reduced by purposefully withholding some parts of each pattern (i.e., only using a subset of the features for certain classifiers). Feature inputs can be "pruned" by measuring how each affects the classifier output. Those features that have the least effect on the output can be selected for removal without compromising overall classifier performance. One example is to have one classifier per class, where each classifier only uses the features with high correlation to that class.
- **Boosting:** Boosting adaptively changes the training set distribution based on performance of previous classifiers, attempting to reduce both bias and variance. Based on results of previous classifiers, diverse training samples are collected such that instances that were incorrectly predicted play a more important role in training (i.e., further learning focuses on difficult examples). This method relies on multiple learning iterations. At each iteration, instances incorrectly classified are given greater weight in the next iteration. Thus, the classifier in each iteration is forced to concentrate on instances it was unable to correctly classify in previous iterations. All classifiers are then combined after all iterations have been processed, or a threshold is met.
- **Bagging:** Bagging creates a family of classifiers by training on stochastically different portions of the training set. *N* training "bags" are initially created, each obtained by taking a training set of size *S* and sampling the training set *S* times with replacement. Some instances could occur multiple times, while others may not appear at all. Each bag is then used to train a classifier, and classifiers are then combined using an equal weight for each.

3.2.4 Ensemble Learning, Meta-Learning, and Other Abstract Methods

One of the most active areas in supervised learning is constructing successful ensembles of classifiers. Ensemble methods construct a set of classifiers and then classify new data points by taking a (weighted) vote of their predictions [24]. Ensembles are well established as a method for obtaining highly accurate classifiers by combining less accurate ones. They can often outperform any single classifier. However, to be more accurate than a single classifier, the ensemble must be composed of both accurate and diverse classifiers [24].

Similarly, meta-learning uses a machine learning algorithm to model the decision patterns of a set of classifiers to construct a model that could yield increased accuracy. Figure 3.3 illustrates the meta-learning process for a multi-classifier system, where decisions are collected from a set of classifiers, used to create a model of the decision pattern of those classifiers, and that model is tested to arrive at a single classification for each instance. The goal of the meta-classifier is to learn the mapping of classifier decisions that provides the most correct labels based on the training data. The meta-classifier will then label input instances based on the decisions from the underlying individual classifiers. This process is also known as "stacking."

3.2.5 Evolutionary Algorithms

Evolutionary algorithms are a class of stochastic search and optimization methods that mimic natural biological evolution. The common underlying premise is that given a population of individuals, the environmental pressure causes natural selection, which further causes a rise in the quality of the population [25]. Given a quality function to be maximized (i.e., the *fitness function*), a random set of candidate solutions can be evolved that maximize or improve upon previous fitness. This fitness measure ensures that the better (more fit) candidates are selected to seed the next generation as baseline candidates, further evolved by applying reproduction operators such as *recombination* and *mutation* to them. Recombination is an operator that applies to two or more selected candidates (the parents) and results in one or more new candidates (the children). Mutation is applied to one candidate and results in one new candidate. Taken together, the operations of recombination and mutation produce a set of new candidate solutions (the offspring) that

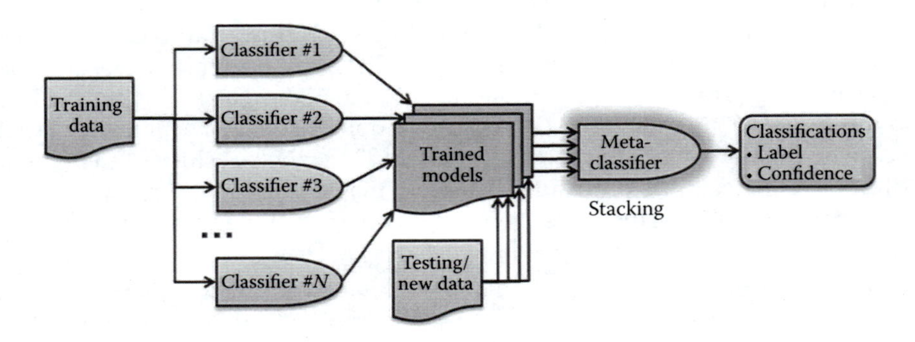

FIGURE 3.3
Meta-learning process for a multi-classifier system, where decisions are collected from a set of classifiers, used to create a model of the decision pattern of those classifiers, and that model is tested to arrive at a single classification label and confidence for each instance. This process is also known as "stacking."

compete—based on their fitness—with other candidates for inclusion in the next generation. This process, called a *generation*, is iterated until a termination condition is met. The termination condition can be the creation of a solution with sufficient quality, a time limit, or some computational limit [25]. The iterative process of computing and evolving a population is illustrated in Figure 3.4.

Evolutionary algorithms differ substantially from more traditional search and optimization methods [25]. The most significant differences are as follows:

- Use stochastic transition rules, not deterministic ones.
- Search multiple points in the solution space, not just a single point.
- Do not require problem-specific knowledge; only the fitness levels influence the directions of search [26].
- Are usually more straightforward to apply, as no restrictions for the definition of the objective function exist [26].
- Can provide a number of potential solutions to a given problem. The selection of the solution to utilize is left to the user. Thus, in cases where the particular problem does not have one individual/unique solution, as in the case of multiobjective optimization and scheduling problems, the evolutionary algorithm can be leveraged for identifying several high-quality alternative solutions.

The general pseudocode for an evolutionary algorithm can be summarized by the following:

```
BEGIN
      INITIALIZE Population with random Candidate solutions
      EVALUATE each Candidate;
REPEAT WHILE (TERMINATION CONDITION is not satisfied)
      SELECT Parents;
      RECOMBINE pairs of Parents;
      MUTATE resulting Offspring;
      EVALUATE new Candidates;
      SELECT Individuals for the next Generation;
END
```

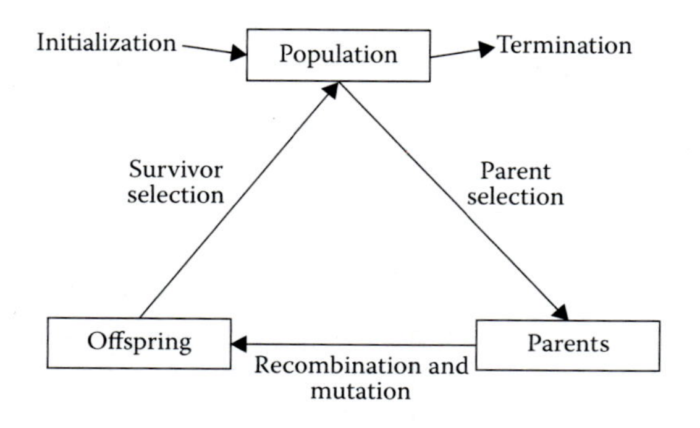

FIGURE 3.4
Iterative evolutionary process by which a population is created and evolved to arrive at a set of candidate solutions that maximize a fitness function.

3.2.5.1 Representation

The first step in defining any evolutionary algorithm is relating the paradigm to the problem at hand. The representation of individual candidate solutions is the manner by which the problem context is mapped to the problem-solving space where evolution can take place. Objects forming possible solutions within the problem context are referred to as *phenotypes*. The encodings of individuals in the population are referred to as *genomes* [25]. Thus, the design of any evolutionary algorithm must begin with specifying a representation, which is a mapping from the phenotypes onto a set of genotypes that represent candidate solutions. It is important to note that the phenotype space is often very different from genotype space [26]. For example, a binary string can be used as the genotype to describe a phenotype, which is a letter of the English alphabet. The entire evolutionary process occurs in the genotype space. The final solution is the phenotype that is obtained from decoding the fittest genotype after the termination condition.

3.2.5.2 Fitness Function

The fitness function represents a measure by which individuals of a population are evaluated and compared, with the goal of maximizing the overall fitness of a population through the evolutionary process. This function therefore represents the basis for parent and survivor selection and must be able to distinguish between individuals that are worthless, average, good, and excellent. The fitness function, together with the representation, provides the only view into the true problem space that the evolutionary algorithm computationally operates in [25]. The fitness function should generally be linear in its relationship with the problem objective function. It is possible that for a given problem, the fitness function is not the same as the problem objective function, as is the case with a minimization problem. In these cases, the fitness function may be the inverted objective function. In most cases, converting an objective function to a fitness function is trivial. Without a good fitness function, an evolutionary algorithm is directionless.

3.2.5.3 Population

The collection of individuals on which the evolutionary algorithm operates is called the "population." The population is a set of genotypes where multiple copies of the same genotype are allowed. Individual genotypes are static and do not evolve, so the unit of evolution is the whole population itself [25]. In many traditional implementations, defining a population is as simple as specifying the number of genomes held within it. In more sophisticated evolutionary algorithms, multiple populations can be simultaneously represented and evolved. Multipopulation algorithms include additional operators such as migration and competition.

3.2.5.4 Genetic Algorithms

Genetic algorithms are a special case of the more general class of evolutionary algorithms [25]. Genetic algorithms are the most widely known type of evolutionary algorithm. The primary construct that distinguishes genetic algorithms from other evolutionary algorithm variations is the representation of the candidate solutions. The representation of an individual in a genetic algorithm is defined as a string over a limited alphabet. Various applications fit this representation and general optimization strategy, including automated drug design, sequence alignment, molecular structure/folding/docking optimization, and general computational biology.

3.3 Machine Learning and Data Mining Resources

Many resources, most of which are open source, are available to support the development and application of these techniques across various domains and development environments. Toolboxes are available for most modern languages (Java, C#, Python) and scientific computing platforms (MATLAB®, IDL, R). For instance, MATLAB offers a neural network toolbox and a global optimization toolbox (contains genetic algorithm methods), among others, that cover many of the algorithms discussed in this chapter. Many other custom toolboxes can be downloaded that offer a range of techniques, variations, and support tools.

The Waikato Environment for Knowledge Analysis (Weka), developed and managed by the University of Waikato, is a comprehensive open-source set of object-oriented machine learning and data mining algorithms written in Java [27]. In addition to classification, regression, clustering, and association rule algorithms, it also provides methods for data filtering, preprocessing, and visualization. The set of packages provides interfaces to preprocessing routines including feature selection, categorical and numeric learning tasks, performance enhancement of classifiers, evaluation according to different criteria such as root mean squared error, and experimental support for verifying the robustness of models. Functionalities such as training, testing, and using classifiers are provided for all algorithms. Additionally, all algorithms are implemented using the same software architecture, so comparing, combining, and cross-validating algorithms is straightforward. Weka has been extensively used in machine learning research to develop new techniques and compare to the state of the art. RapidMiner [28] is another free machine learning suite similar to Weka that has a large user community.

3.4 Example/Illustrative Medical Applications

Artificial intelligence is widely used in the medical domain, several representative examples of which are included in other chapters of this book. Machine learning and data mining are currently popular mechanisms for supporting medical research and creating models to better understand complex data sets, relationships, and associations. This section includes several relevant examples that illustrate how these methods can be used for medical applications.

3.4.1 Multiagent Infectious Disease Propagation and Outbreak Prediction

Anticipating, predicting, and monitoring disease outbreaks can assist public health officials in their response efforts. To this end, the following capabilities are desirable: (1) to predict when and where an outbreak will occur; (2) should an outbreak occur in a certain location, to predict how it will spread; and (3) to monitor an outbreak as it is occurring. Various efforts are focused on (1) and (3); however, there is a need to develop the second capability (i.e., to predict how an outbreak will spread). Understanding the communicability circumstances (the how) as well as the spatial (the where) and temporal (the when, and how fast) spread characteristics of an infectious disease can aid in preparation and active intervention efforts. Often, this capability is called "disease propagation modeling."

Historically, models based on differential equations have been used to predict how a disease will spread. These models emphasize the numbers of people who are susceptible to contracting the disease, are infected/infectious with the disease, and have recovered from the disease. However, they typically do not describe how the disease will spread spatially. Another technique for disease propagation modeling that has gained recent prominence is agent-based modeling, which is based on the multiagent system paradigm. Rather than consider broad groups of people, this technique represents people individually and attempts to characterize their behavioral patterns. It is hoped that first-hand encounters between people can then also be predicted, therein directly representing the opportunities for the disease to be transmitted from one person to another.

The power of such a system is that new diseases or disease variations can be specified, and many simulations can be run to understand how those diseases spread under certain population, travel, and social constraints. It also enables the evaluation of vaccination/intervention mechanisms by simulating the intervention and measuring the corresponding decrease in infection, death, or spread as a surrogate variable of the intervention policy's utility. Similarly, such systems can be leveraged for sensor placement optimization, in cases where a limited number of sensors are available and need to be optimally placed for sufficient impact.

Efforts to model disease propagation are extensive, for many years focusing on *SEIR* differential equations models [29], where S is the number of susceptible individuals, E the number exposed, I the number infectious, and R the number recovered. Within the past decade, greater attention within the field has been given to agent-based models, and Connell et al. [30] offer a comparison of the two methods. Such agent-based models, of course, are conducive to computer simulation development, as has been documented in [31]. Incidentally, agent-based models have also been applied to many other areas of study, including containment of terrorist bioattacks [32], the spread of fear [33] or civil unrest [34], and catastrophic-event evacuations [35].

Infectious disease outbreak prediction represents an area of current global importance and research interest. Data mining can be leveraged to reveal patterns of disease outbreaks and forecast the location and time frame of an outbreak before its emergence. Dengue fever, for example, is an endemic without a vaccine that has tens of millions of cases each year. Predictive modeling efforts are currently being explored, such as the use of logistic regression and fuzzy association rule mining, to discover complex relationships between confirmed disease outbreaks and environmental, biological, ecological, and sociopolitical variables [36]. Predicting dengue fever outbreak in a timely manner could enable active public health interventions that can aid in limiting, localizing, or eliminating the outbreak from occurring.

3.4.2 Automated Amblyopia (Lazy Eye) Screening System

Amblyopia, commonly referred to as lazy eye, is a neurological vision disorder that studies show affects 2%–5% of the population. Current methods of treatment produce the best visual outcome if the condition is identified early in the patient's life. Several early screening procedures are aimed at finding the condition while the patient is a child, including an automated vision screening system (AVVDA) [37] that uses artificial intelligence algorithms to identify patients who are at risk for developing the amblyopic condition and should be referred to a specialist. AVVDA uses case-based reasoning, C4.5 decision tree, and artificial neural network classifiers to assist in making the decision. Various features are used by this system as input to the machine learning algorithms, including color density, Hirschberg reflex, and iris and pupil color slopes. Continued efforts based on multi-classifier decision fusion have improved upon the existing AVVDA performance [38].

3.4.3 Anticancer Drug Design and High-Throughput Screening

Taxol is a drug that has been shown to offer benefits in treating breast and ovarian cancers, and represents one of a small set of anticancer drugs that has been discovered, designed, and tested. Thus, Taxol becomes a candidate of further study to determine what features of Taxol make it an anticancer drug and to design other (potentially new) drugs that may offer improved performance or broader applicability. Machine learning algorithms (e.g., artificial neural networks) can be used to learn a model of Taxol's composite features and measured performance on breast and ovarian cancer cases. The model can then be used to evaluate existing drugs as anticancer drugs to consider for testing (i.e., classifier output is "test" versus "don't test" based on candidate drug composite features). The model could also be used to suggest (or design) composite feature sets that are highly similar to, or better than, that of Taxol and its anticancer performance metrics. A primary challenge of this problem area is limited and/or highly imbalanced training data.

Given recent advancements in high-performance computing and sequencing technologies, high-throughput screening has become commonplace in the medical community. The premise of high-throughput screening is simple: distinguish between active and nonactive cancer drugs (e.g., cobalt) given a set of experimentally determined and known features for each drug. The goal of such systems is to maximize the number of correctly identified active drugs (AA) while minimizing the number of nonactive drugs identified as active (NA). If a drug researcher plans on purchasing and testing all of the drugs the screening system classifies as active, it is desired to maximize the percentage (AA/NA) of those purchased drugs that are actually active. By raising this percentage, we lower the number of drugs it is necessary to buy to achieve the desired amount of active compounds, which saves a significant amount of time and money. If researchers simply buy all of the drugs they encounter, they will find a very low percentage of them to be active. Bayes classifiers and artificial neural network systems have been created for this task that were able to ensure that, on average, 6% were active, some leveraging more advanced methods such as multi-nets (combining multiple artificial neural networks with different characteristics).

3.4.4 Genetic Sequence Classification

Many biological classification problems exist that require sophisticated methods to learn the differences between large, highly overlapping data classes. One example is genetic sequence alignment, where a class of sequences with biologically significant properties is leveraged for developing a system and model capable of automatically identifying significant sequences. This represents a two-class problem: significant versus insignificant genetic sequences. Inputs are generally of the genetic sequence form (A,C,G,T). Within genetic sequence alignment, one focus area could be the identification of sequences that are found in an exonic splicing enhancers (ESE) motif recognized by the human serine rich protein SC35. Selected sequences for artificial neural network model creation, for instance, could be functional and specific by being those that promote splicing in a nuclear extract complemented by SC35 but not by SF2/ASF. Input test sequences could then be identified as potentially sharing one or two matches to a short and highly degenerate octamer consensus. Once a model capable of distinguishing between significant and insignificant sequences is established, it can be used to test new sequences for significance or to perform large-scale simulations to discover (potentially new) sequences that are deemed significant for further investigation.

3.5 Conclusions

This chapter has introduced many prominent machine learning and data mining methods and discussed considerations of their application and impact on the problem domain. Example applications to the medical domain were provided to illustrate how such methods can be applied to current complex problems. The field of artificial intelligence has and will continue to play an important role in the advancement of scientific research, modern health and human information systems, and the broader medical community. Upcoming chapters leverage many of these methods to further improve the state of the art and offer solutions to some of today's most challenging medical applications.

References

1. R. Agrawal, T. Imielinski, and A. Swami, Mining association rules between sets of items in large databases, In *Proceedings of the International Conference on Management of Data*, Washington, DC, pp. 207–216, 1993.
2. R. Srikant and R. Agrawal, Mining quantitative association rules in large relational tables, In *Proceedings of the International Conference on Management of Data*, Montreal, Quebec, Canada, pp. 1–12, 1996.
3. C.M. Kuok, A. Fu, and M.H. Wong, Mining fuzzy association rules in databases, In *ACM SIGMOD Record*, 27(1), New York, pp. 41–46, 1998.
4. A.L. Buczak and C.M. Gifford, Fuzzy association rule mining for community crime pattern discovery, In *Proceedings of the ACM SIGKDD Conference on Knowledge Discovery and Data Mining: Workshop on Intelligence and Security Informatics*, Washington, DC, 2010.
5. J. Han, J. Pei, and Y. Yin, Mining frequent patterns without candidate generation, In *Proceedings of the ACM International Conference on Management of Data*, New York, pp. 1–12, 2000.
6. C.-W. Lin, T.-P. Hong, and W.-H. Lu, Linguistic data mining with fuzzy FP-trees, *Expert Systems With Applications*, 37, pp. 4560–4567, 2010.
7. R. Srikant and R. Agrawal, Mining sequential patterns: generalizations and performance improvements, In *Proceedings of the International Conference on Extending Database Technology: Advances in Database Technology*, London, pp. 3–17, 1996.
8. J. Pei, J. Han, B. Mortazavi-Asl, J. Wang, H. Pinto, Q. Chen, U. Dayal, and M. Hsu, Mining sequential patterns by pattern-growth: the PrefixSpan approach, *IEEE Transactions on Knowledge and Data Engineering*, 16(11), pp. 1424–1440, 2004.
9. J. Wang and J. Han, BIDE: efficient mining of frequent closed sequences, In *Proceedings of the International Conference on Data Engineering*, Washington, DC, pp. 79–90, 2004.
10. P. Tzvetkov, X. Yan, and Y. Han, TSP: mining top-k closed sequential patterns, In *Proceedings of the IEEE International Conference on Data Mining*, Melbourne, FL, pp. 347–354, 2003.
11. X. Yan, J. Han, and R. Afshar, CloSpan: mining closed sequential patterns in large datasets, In *Proceedings of the International Conference on Data Mining*, San Francisco, pp. 166–177, 2003.

12. G. Casas-Garriga, Summarizing sequential data with closed partial orders, In *Proceedings of the International Conference on Data Mining*, Newport Beach, CA, pp. 380–391, 2005.
13. S. le Cessie and J. van Houwelingen, Ridge estimators in logistic regression, *Applied Statistics*, 41(1), pp. 191–201, 1992.
14. G.H. John and P. Langley, Estimating continuous distributions in Bayesian classifiers, In *Proceedings of the Conference on Uncertainty in Artificial Intelligence*, Morgan Kaufmann, San Mateo, CA, pp. 338–345, 1995.
15. D.W. Aha, D. Kibler, and M.K. Albert, Instance-based learning algorithms, *Machine Learning*, 6, pp. 37–66, 1991.
16. J.G. Cleary and L.E. Trigg, K*: an instance-based learner using an entropic distance measure, In *Proceedings of the International Conference on Machine Learning*, pp. 108–114, 1995.
17. J.R. Quinlan, *C4.5: Programs for Machine Learning*, Morgan Kaufmann, San Mateo, CA, 1993.
18. L. Breiman, Random forests, *Machine Learning*, 45(1), pp. 5–32, 2001.
19. W.W. Cohen, Fast effective rule induction, In *Proceedings of the International Conference on Machine Learning*, pp. 115–123, 1995.
20. E. Frank and I.H. Witten, Generating accurate rule sets without global optimization, In *Proceedings of the International Conference on Machine Learning*, Morgan Kaufmann, San Francisco, pp. 144–151, 1998.
21. R. Kohavi, The power of decision tables, In *Proceedings of the European Conference on Machine Learning*, pp. 174–189, 1995.
22. E. Alpaydin, Techniques for combining multiple learners, In *Proceedings of the Engineering of Intelligent Systems Conference*, Vol. 2, Tenerife, Spain, pp. 6–12, 1998.
23. C.M. Gifford, Collective machine learning: team learning and classification in multi-agent systems, PhD Dissertation, Department of Electrical Engineering and Computer Science, University of Kansas, Nov. 2009.
24. T.G. Dietterich, Ensemble methods in machine learning, *Lecture Notes in Computer Science: Multiple Classifier Systems*, 1857, pp. 1–15, 2000.
25. A.E. Eiben and J.E. Smith, *Introduction to Evolutionary Computing*, Springer-Verlag, 2003.
26. D. Goldberg, *Genetic Algorithms in Search, Optimization, and Machine Learning*, Addison-Wesley, Reading, MA, 1989.
27. I.H. Witten and E. Frank, *Data Mining: Practical Machine Learning Tools and Techniques*, 2nd Edition, Morgan Kaufmann, San Francisco, CA, 2005.
28. I. Mierswa, M. Wurst, R. Klinkenberg, M. Scholz, and T. Euler, YALE: rapid prototyping for complex data mining tasks, In *Proceedings of the ACM SIGKDD International Conference on Knowledge Discovery and Data Mining*, Philadelphia, PA, 2006.
29. P. Costa, J. Dunyak, and M. Mohtashemi, Models, prediction, and estimation of outbreaks of infectious disease, In *Proceedings of the IEEE SoutheastCon*, Fort Lauderdale, FL, 2005.
30. R. Connell, P. Dawson, and A. Skvortsov. Comparison of an agent-based model of disease propagation with the generalised SIR epidemic model, Australian Dept. of Defense, Aug. 2009.
31. J. Parker and J. Epstein, A distributed platform for global-scale agent-based models of disease transmission, *ACM Transactions on Modeling and Computer Simulation*, 22(1), Dec. 2011.
32. I.M. Longini Jr., M.E. Halloran, A. Nizam, Y. Yang, S. Xu, D.S. Burke, D.A. Cummings, and J.M. Epstein. Containing a large bioterrorist smallpox attack: a computer simulation approach, *International Journal of Infectious Diseases*, 11(2), 98-108, 2006.
33. J. Epstein, J. Parker, D. Cummings, and R. Hammond, Coupled contagion dynamics of fear and disease: mathematical and computational explorations, *The Electronic Journal of Differential Equations*, Spring, 3(12), 1–11, 2008.
34. J. Epstein, Modeling civil violence: an agent-based computational approach, *PNAS*, 99, pp. 7243–7250, May 2002.
35. J. Epstein, R. Pankajakshan, and R. Hammond, Combining computational fluid dynamics and agent-based modeling: a new approach to evacuation planning, *PLoS One*, 6(5), May 2011.

36. P. Koshute, A. Buczak, S. Babin, B. Feighner, C. Sanchez, E. Omar Napanga, and S. Lewis, Dengue fever outbreak prediction, *Emerging Health Threats Journal*, 4, 11054, 2011.
37. G.W. Cibis, Video vision development assessment in diagnosis and documentation of microtropia, *Binocular Vision and Strabismus Quarterly*, 20, pp. 151–158, 2005.
38. P.G. Clark, C.M. Gifford, J. Van Eenwyk, A. Agah, and G.W. Cibis, Applied machine learning and decision combination for identifying the lazy eye vision disorder, In *Proceedings of the International Conference on Artificial Intelligence*, Las Vegas, NV, 2012.

4

Introduction to Computational Intelligence Techniques and Areas of Their Applications in Medicine

Ali Niknejad and Dobrila Petrovic

CONTENTS

4.1 Introduction

Artificial intelligence (AI) is an adaptation of the human mind's way of decision making using computer systems. Since many of the important problems we face in everyday life cannot be solved using conventional mathematical and analytical approaches, computer algorithms have been developed that provide solutions to tackle these problems. These algorithms are often inspired by the decision-making processes in natural systems such as neural networks, evolutionary systems, and so forth; specifically, many problems in medical sciences require intelligence systems comparable to the human mind, and hence, AI has received considerable attention from researchers in this field. We refer interested readers to Pandey and Mishra (2009) for a detailed survey of AI applications in medicine.

AI can be divided into soft and hard AI. Hard AI includes methods such as expert systems, formal logic, Bayesian networks, and so forth, which look for neat, clear, and provably correct solutions. In contrast, soft AI provides solutions to many problems that are simply too complicated to be addressed by hard AI approaches and require more flexible methods. The soft AI methods often provide heuristic solutions that cannot be guaranteed to be correct or optimal but are useful solutions to problems that would have not been solved otherwise by hard AI methods within the provided time constraint. *Computational intelligence (CI)* is a subset of soft AI that mainly deals with particular methods, including *artificial neural networks (ANNs), fuzzy logic,* and *evolutionary computing (EC)* (Eberhart and Shi 2007).

In various problems, an enormous input space might be encountered that needs to be *classified* into two or more known classes. These problems are called *classification problems* and are very common in medical applications, where several inputs such as medical images, clinical data, or other medical inputs should be used for diagnosis, prognosis, or other purposes. As an example, in a typical medical application, classes can simply represent a positive or negative diagnosis of cancer or the possibility of existence of a tumor in a certain part of a produced medical image. To obtain a classifier, often, a set of *observations* needs to be provided (as a *training set*) for which the correct class is already known. Based on the provided training set, an AI method should be able to train itself so that the correct class for new observations can be determined. On the other hand, it is also possible that the training set data includes a set of observations without the correct class, and the algorithm is assumed to determine classes by classifying the nearest observations in one class; this is known as *clustering*. In clustering problems, the definition of each class (or cluster) is unknown initially and is assumed to be determined by the clustering algorithm. Clustering algorithms use a *distance function* to measure the similarity of data points with each other and then use this information to cluster them into a fixed number of clusters. The objective of these algorithms is to minimize the distances between observations in each cluster, while maximizing the distance of observations in one cluster with observations in another one.

CI methods have been researched extensively in medical applications, which can be explained by the complicated and vague nature of medical problems that CI methods are designed to tackle. In this chapter, the three CI methods are considered, and important areas in medicine where these methods have been applied are discussed. *ANNs* imitate the human's neural system by emulating a network of artificial *neurons,* which act similarly to the biological *neurons* in terms of decision-making and training capabilities. On the other hand, *genetic algorithm (GA)* is the most researched topic in *EC*, which is based on the concept of biological evolution. GA considers a set of fittest solutions, which reproduce and mutate to produce even stronger survivors. *Fuzzy logic* extends the traditional logic by considering vagueness of concepts as perceived by the human mind in contrast with the crisp interpretation of truth in traditional logic. These methods and their applications to medical sciences will be discussed in more detail in next sections.

4.2 Fuzzy Logic

Fuzzy logic is an extension of the traditional logic that addresses the uncertainty and imprecision that exist in real world but is not considered in the traditional logic. While in the traditional logic, every proposition has a value of either 0 (false) or 1 (true), in fuzzy logic, any value between 0 and 1 is acceptable as the *degree of truth* of a particular proposition. These degrees of truth represent ambiguity and vagueness in real-world information

(Pedrycz and Gomide 1998). As an example, fuzzy logic can provide a more realistic model of the truth by describing propositions to be "half true," "nearly false," or "quite true" as opposed to a crisp and binary description of either true or false.

Also, based on the same principle, fuzzy sets are extensions of classical sets in which the elements can have any *membership degree* between 0 and 1, as opposed to the classical set theory in which the element can be either a member (with a membership degree of one) or a nonmember (with a membership degree of 0). The function that assigns membership degrees to elements in the set's domain is referred to as the *membership function*. For example, when dealing with propositions such as "temperature is high" or "glucose level is low," classical set theory provides little flexibility in definition of vague terms such as *low*, *medium*, and *high*. In the temperature example, using the classical set theory, each term should be defined as a range of temperatures that belong to the particular term; as a partial membership is not possible in the classical set theory, there is no choice other than to use some thresholds to define each term, for example, any temperature below 5°C is low. But in reality, there is no particular "hard edge" between these terms, and any number determined to be the definition will be artificial and in contradiction to common sense. The concept of a partial membership, which is the core of fuzzy set theory, helps in addressing this problem by providing soft edges in the definition of the sets; for example, any temperature below 5°C is definitely low, but the degree of membership of temperatures above 5°C decreases gradually from 1 until it reaches 0 for 15°C, as can be seen in Figure 4.1.

The definition of fuzzy sets gives rise to the concept of *fuzzy linguistic variables*. In conventional mathematics, variables are typically numeric, but in fuzzy logic, variables can get linguistic terms as values; these terms describe vague concepts such as *cold*, *mild*, and *hot*. This proves to be particularly valuable when a human expert needs to understand and manipulate a problem model since humans are generally more familiar with linguistic descriptions than with numerical values.

Fuzzy logic has various applications in control, data clustering and classification, image processing, arithmetic, optimization, and so on. In this section, fuzzy clustering and fuzzy control will be discussed further. Fuzzy clustering has been extensively applied in medical image processing applications and is useful in many diagnosis and prognosis scenarios. Fuzzy control is also important in treatment planning problems such as anesthesia control, blood glucose level control, and control of respiratory systems.

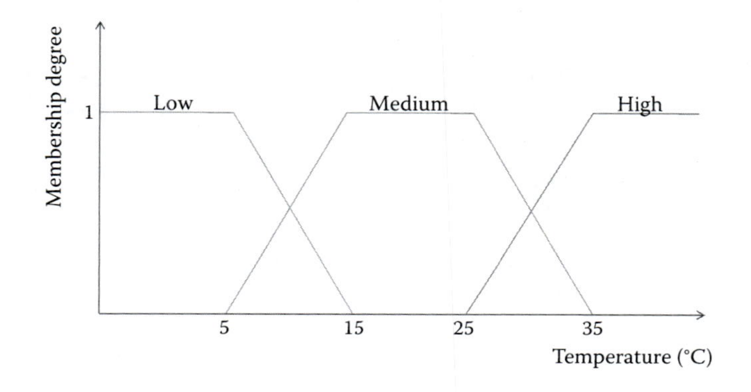

FIGURE 4.1
Membership functions of fuzzy sets low, medium, and high with regard to temperature.

4.2.1 Fuzzy Clustering

Classification of data into separate classes is among the most important applications of CI in medicine. If the number of classes is *a priori* known, but the definition of each class is unknown, *clustering* techniques can be used to classify the data. One of the well-researched and applied classical clustering techniques, *k-means algorithm*, defines an initial set of cluster center points and adjusts these center points repetitively in order to minimize the overall distance of observations to their corresponding cluster center points (Nock and Nielsen 2006). In each iteration, the distance of each observation to all cluster center points is calculated, and observations are assigned to the center point closest to them; sequentially, the center point of each cluster is recalculated as the average coordinates of the observations assigned to the particular cluster. The algorithm terminates when no change is observed in the assignment of center points. Figure 4.2 illustrates the changes that can happen in the center points of two clusters, in one iteration. As it is shown, new center points will yield to better clustering of observations.

Fuzzy clustering techniques, such as *fuzzy c-means*, use the same principle to cluster observations, but unlike hard clustering techniques, the fuzzy clustering technique can assign each observation to more than one cluster, with different degrees of membership (Nock and Nielsen 2006). The membership degrees to each cluster are usually initialized randomly. In each iteration, new center points are calculated as average coordinates of observations weighted by their membership degree to the center point as follows:

$$c_k = \frac{\sum_x \mu_k(x)x}{\sum_x \mu_k(x)}$$

where $\mu_k(x)$ is the membership degree of observation x to the cluster k and c_k is the center point of cluster k (both x and c_k are vectors). After updating the center points, the membership degrees of observations need to be updated as well, because they are inversely related to their distances from the corresponding center points. This is usually done using the following formula:

$$\mu_k(x) = \frac{1}{\sum_{c_i \in C} \left(\frac{d(c_k, x)}{d(c_i, x)} \right)^{2/(m-1)}}$$

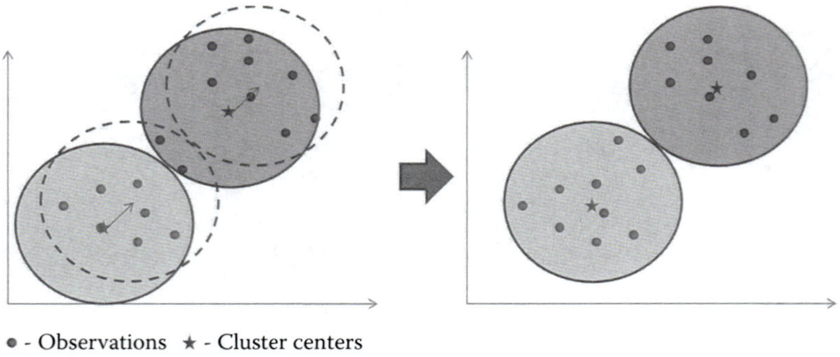

• - Observations ★ - Cluster centers

FIGURE 4.2
(See color insert.) Changes in clusters' center points between iterations.

where C is the set of cluster center points, *d* is the distance function, and *m* is the fuzziness parameter; higher values of *m* will increase the fuzziness of clusters, while for the extreme value of 1, the clustering will become crisp and similar to the *k*-means. The algorithm is repeated until the overall change in assignments of center points is not greater than an arbitrary small value—ε.

4.2.2 Fuzzy Control

Fuzzy control refers to the application of fuzzy logic to control theory. A fuzzy controller is a set of fuzzy if–then rules, defined using fuzzy linguistic variables, to control a system by utilizing analog input data and producing analog outputs. As the fuzzy if–then rules that constitute the core of the controller are specified in the form of natural language expressions, the main advantages of fuzzy controllers over traditional controllers are the ease of specifying, understanding, and manipulating the controller by the human expert. This is particularly useful when the system is not or cannot be identified mathematically, but a linguistic description of the system can be obtained from the expert based on the expert experience, intuition, or heuristics.

Fuzzy controllers consist of four main components: *fuzzification, fuzzy inference, fuzzy rule base,* and *defuzzification.* In the fuzzification component, input is translated into its equivalent fuzzy descriptions, which essentially are the membership degrees of the input to the corresponding fuzzy linguistic terms. Fuzzy inference is responsible for determining the outputs based on the fuzzy descriptions of inputs by using the fuzzy if–then rules, which are stored in the fuzzy rule base. Two main types of fuzzy inference are proposed: the *Mamdani type of inference* and *Takagi–Sugeno (TS) type of inference.* The main difference between the two types of inference is that TS type has crisp outputs typically modeled as functions of crisp input variables, while Mamdani type has fuzzy outputs (Zeng et al. 2000). Finally, for Mamdani-type inference, in the defuzzification component, the fuzzy outputs of the inference engine are converted to the representative scalar outputs for controlling the system. The overall schema is represented in Figure 4.3.

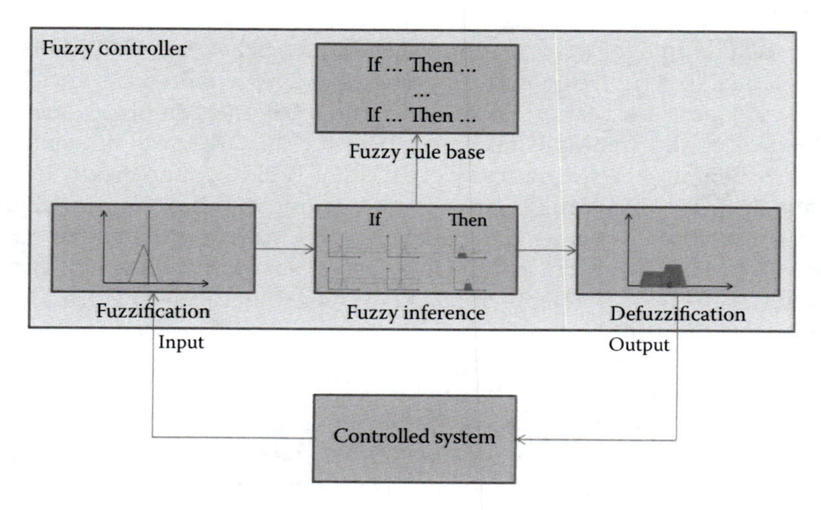

FIGURE 4.3
Schema of a typical fuzzy controller.

While fuzzy controllers allow for extracting knowledge represented by natural language expressions from experts, this approach might not be feasible or desirable in many cases; for example, in complicated systems where experts cannot understand the whole system or where the system properties change frequently and the rule base needs to be updated. To address this issue, several methods have been introduced with the purpose of *automatic identification* of the fuzzy controller's structure and parameters. Fuzzy clustering can be used to determine fuzzy rules and membership functions of linguistic terms used in fuzzy rules. Each rule represents a region in an input space that is related to a region in an output space; hence, clusters in the input/output space are equivalent to rules in the controller, and each cluster yields to one rule and the corresponding membership functions of the input/output variables. The basis for this approach has been introduced by Takagi and Sugeno (1985). Further on, other identification methods that use different AI techniques such as GA, ANNs, reinforcement learning methods, and so forth have been used to identify a fuzzy controller (Berenji and Khedkar 1992; Jang and Sun 1993; Halgamuge and Glesner 1994; Herrera et al. 1995; Pham and Karaboga 2000). One approach focused on combining ANNs with fuzzy inference, which is the basis for the adaptive network-based fuzzy inference system (ANFIS) (Jang 1993), has been proposed and applied to different application domains, such as classification (Nauck and Kruse 1999; Chikh et al. 2012; Wang et al. 2012), diagnosis (Übeyli 2009; Akgundogdu et al. 2010), monitoring and control (Belal et al. 2002; Kwok et al. 2003; Roy et al. 2009), and data mining (Ghazavi and Liao 2008). ANFIS learns parameters of a fuzzy controller by emulating the fuzzy inference in an adaptive neural network and using the conventional ANN learning methods.

4.2.3 Medical Applications

Fuzzy *c*-means and its variants have been proven to be very useful for image processing applications in various medical imaging scenarios such as magnetic resonance imaging (MRI) (Brandt et al. 1994; Ahmed et al. 2002; Zhang and Chen 2004; Wang et al. 2008), positron emission tomography (PET) images (Boudraa et al. 1996; Kim et al. 2007), retinal images (Tolias and Panas 1998; Muramatsu et al. 2011), near-infrared images (Mansfield et al. 1997, 1998), or other medical imaging purposes (Li et al. 2011; Plissiti et al. 2011). These applications include image segmentation (Masulli and Schenone 1999; Fletcher-Heath et al. 2001; Ahmed et al. 2002; He et al. 2008; Wang et al. 2008; Giannakeas and Fotiadis 2009; Ji et al. 2011), image classification (Wang and Fei 2009), image enhancement (Brandt et al. 1994; Boudraa et al. 1996; Mansfield et al. 1997; Ji et al. 2011), and image interpretation (Masulli and Schenone 1999; Muramatsu et al. 2011). Typically, in *c*-means image processing applications, image pixels and their values constitute the observations (volumetric pixels or *voxels*), and the goal is to group the relevant pixels or voxels to find the right segment, class, and so forth; for example, in an MRI image segmentation application, three-weighted magnetic resonance intensity images are used to create the feature space, and then the pixels are clustered to find similar tissues in the brain, which will be used to identify the brain tumor; a high-percentage match ranging from 53% to 91% is reported (Fletcher-Heath et al. 2001). It is worth noting that fuzzy *c*-means is helpful in these applications since a single pixel can represent more than one tissue (for example, a tumor-affected tissue and nonaffected tissue) and hence has partial memberships to its respective clusters. Furthermore, other applications of fuzzy clustering include DNA microarray classification and profiling (Tang et al. 2008; Avogadri and Valentini 2009; Giannakeas and Fotiadis 2009; Tari et al. 2009).

Fuzzy control has been widely applied to the anesthesia control problem. Various fuzzy controllers have been introduced to control the volume and timing of anesthesia infusion based on vital signals' inputs from the patient (Mason et al. 1994; Oshita et al. 1994; Nebot et al. 1996; Mason et al. 1997; Allen and Smith 2001; Mahfouf et al. 2005; Nunes et al. 2005; Denaï et al. 2009; Chou et al. 2010). Other applications include the respiratory system (Noshiro et al. 1994; Adlassnig 2001; Lin et al. 2001), psychological models (Hwang et al. 2009), medical image processing (Rafiee et al. 2004; Lin et al. 2005; Ciofolo and Barillot 2009), and blood glucose regulation (Ibbini and Masadeh 2005; Ibbini 2006; Ting and Quek 2009), among many others. A typical application of fuzzy control to anesthesia uses AEP signals that represent the brain's response to audio signals to control the administration of anesthetic drugs (Allen and Smith 2001). Features of the AEP signals, such as latencies, are extracted by an ANN and used as the inputs to the controller, and the infusion rate is obtained as the output of the controller. The results were found to be comparable to the human anesthetist. Also, in this scenario, 10 fuzzy rules are used, which were determined by the expert anesthetist. For more detailed description of fuzzy control applications in medicine, we refer the interested reader to Mahfouf et al. (2001).

4.3 Genetic Algorithm

GA belongs to a general class of methods called evolutionary computing, which is based on the concept of natural evolution. EC methods often simulate biological phenomena such as population, natural selection, genetic inheritance, mutation, survival of the fittest, and so forth, usually in a stochastic manner, to perform a guided random search (Eiben and Smith 2008). Particularly, GA is a *metaheuristic algorithm* that is used to provide useful solutions for optimization problems. Unlike optimization techniques, metaheuristics such as GA cannot guarantee an optimal or even a near-optimal solution, but metaheuristics perform much faster than the traditional optimization techniques and can often provide an acceptable solution. One particular benefit of GA is the flexibility in modeling the problem, which allows high nonlinearity, unlike the optimization techniques, which are usually limited to linear or integer problems.

GA is an iterative algorithm that keeps a list of good solutions—the *population*. In each iteration, new solutions are generated from the current population and evaluated using a *fitness function*. New solutions alongside the current population are compared with each other to select the most suitable solutions for the new population. The selection process needs to keep the best solutions (elite solutions) that are found so far to the next iteration, while it should also include other (though even worse) solutions to avoid getting trapped in local optima. GA iterations continue until one of the termination criteria, such as a maximum number of iterations reached or lack of a change in the best solution recorded in a certain number of iterations, is satisfied.

A problem is described in GA by an appropriate *fitness function* and *chromosome*. The purpose of the fitness function is to evaluate desirability of each solution. This is important in selecting solutions that should be kept or used to generate new solutions. Fitness functions can be of nonlinear form, but since the function evaluation happens frequently, they typically need to be easy to calculate. A solution in the solution space is represented as an encoding of the problem decision variables, which is usually done as an array of binary or real variables. By analogy to the genetics, the structure of this encoding is called the *chromosome,* while each particular solution (i.e., a member of the population) is usually called an *individual.* An element of a chromosome is called a *gene,* while the value that is represented by a gene is known as *allele.* It is worth mentioning that the encoding of chromosomes has a direct impact on the definition of genetic operations used in the algorithm and how the algorithm can create the next generation's population.

An important part of GA is to create new solutions from the previous generation's population. The new solutions should ideally preserve the good features of previous solutions, while new areas in the solution space must be explored as well. *Genetic operations* are used to generate new solutions from the previous ones. They combine some solutions from the previous generation (usually selected randomly) to create *offsprings* for the next generation. Two most important operations are *crossover* and *mutation.* Crossover combines two individuals to create two offsprings, which both take parts of their chromosome from either of the two parents. In contrast, mutation uses one individual and randomly changes one or more genes in the selected individual to create a new one. Crossover makes sure that good characteristics are carried on to the next generation, while mutation is necessary to create high diversity among chromosomes. Examples of the two operations are shown in Figures 4.4 and 4.5, respectively.

In summary, GA comprises six stages: initialization, evaluation, selection, genetic operations (including crossover and mutation), replacement, and termination. Initialization generates a few random solutions to create the initial population. In each iteration, evaluation is carried out to calculate the value of fitness function for each solution. It is used in the selection stage to determine if the solution will influence the next generation and how frequently it will be used. Different genetic operations are sequentially applied to the selected solutions to generate offsprings, and evaluation is repeated for the newly obtained solutions. In the replacement stage, some of the solutions in the current population are replaced with the offsprings based on their fitness values. To terminate the algorithm, either a set

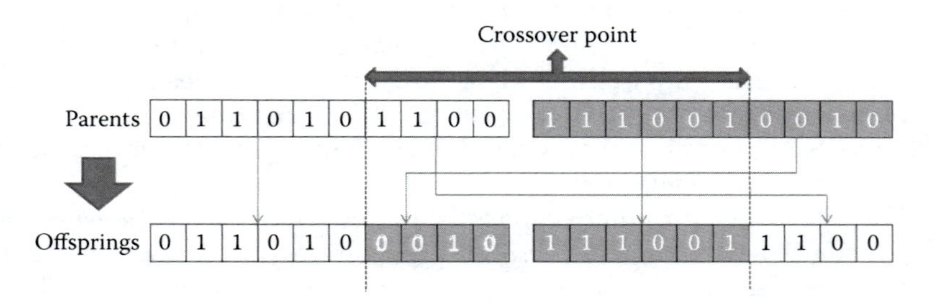

FIGURE 4.4
Example of a single-point crossover operation.

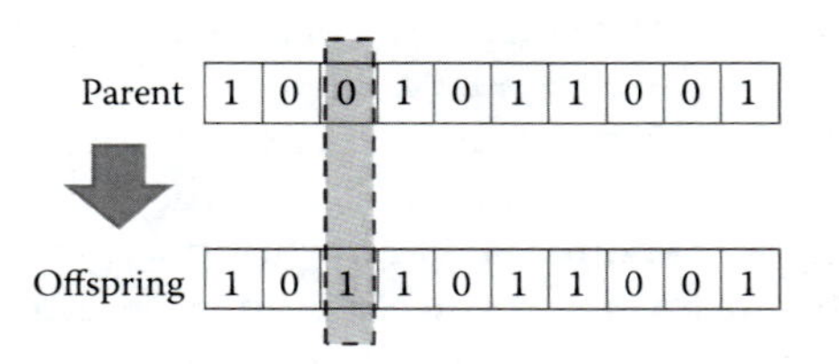

FIGURE 4.5
Example of a mutation operation.

number of iterations are used, or small changes in the best chromosomes' fitness function values between generations indicate the end of algorithm.

4.3.1 Multiobjective Genetic Algorithms

Typical optimization problems involve one objective function only, such as economic cost, survival rate, and so forth. Even in cases when more than one objective functions are necessary, they might be combined to form a single objective function. Often, it is the sum of weighted objective functions, where weights represent objectives preferences. However, in some situations, *a priori* knowledge about preferences between objectives might not be available. Further on, a single objective function does not preserve the multiobjective nature of a problem and can, for example, enable compensation among objectives. In such situations, it is desirable to obtain a set of *nondominated solutions* (or *Pareto optimal solutions*). A solution is nondominated if there is no other solution that is better or equal for all objective functions and is at least better for one objective function. This means that no improvement can be gained in any objective function without sacrificing another one. Otherwise, the solution is considered to be *dominated,* which means that another solution exists that is at least better for one objective function, while it is not worse for any objective function. Figure 4.6 shows a two-objective solution space with nondominated solutions and a dominated solution subspace.

Multiobjective genetic algorithms (MOGAs) are a variant of GAs that evaluate the solutions in a multiobjective manner and provide a set of nondominated solutions as the result. The fitness functions of MOGAs are usually based on *Pareto ranking* method in which the solution's rank is based on the number of solutions that dominate that particular solution; for nondominated solutions, the rank is one (only the solution dominates itself), and higher ranks are given to worse solutions. One of the most popular MOGAs is the *nondominated sorting genetic algorithm II (NSGA-II)* (Deb et al. 2002). NSGA-II performs a nondominated sorting (like many other MOGAs), but the difference is in the use of a complementary crowding distance sorting. Solutions are first chosen by their nondominated rank, and crowding distance is used when a few solutions need to be chosen among equally ranked solutions. In NSGA-II, the crowding distance is calculated as the sum of the solutions' distances to the nearest neighbors with regard to each objective function divided by the maximum range of that objective function. This feature leads to the higher diversity among the solutions, which helps the GA in the discovery of the solution space.

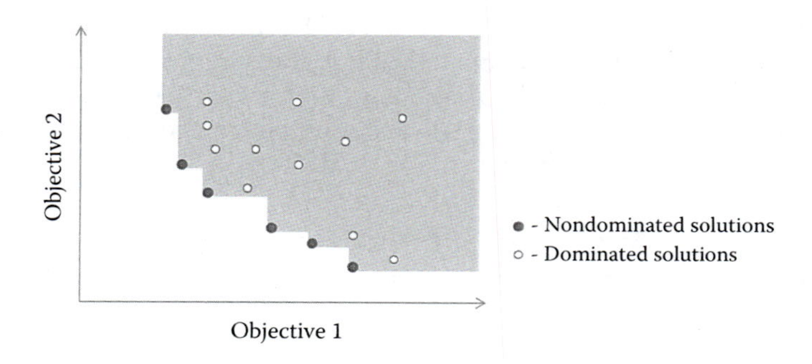

FIGURE 4.6
Pareto optimal (nondominated) solutions in a two-objective-function solution space.

4.3.2 Medical Applications

Naturally, one area of GA applications is genetic research. GAs have been applied to selection, mapping, profiling, and classification of genetic data and DNA microarray (Shah and Kusiak 2004; Tsai et al. 2004; Margolin et al. 2006; Lee 2008; Othman et al. 2008; Zacharia and Maroulis 2008; Liu et al. 2009; Paul and Iba 2009). As an example of these applications, in a microarray analysis problem with large number of features and small number of samples, a GA is used to determine the near-optimal selection of an *independent component (IC)* ensemble (Liu et al. 2009). In this work, each individual represents a mask for IC selection, and the number of correct classifications is used as the fitness function. The resulting classifier has been compared to several conventional methods, and significant improvements have been reported by using the GA-based method.

Another fruitful area of GA applications in medical domain has been related to cancer research; GAs have had applications in lung cancer diagnosis (Lee et al. 2001; Böröczky et al. 2006; Dehmeshki et al. 2007), breast cancer diagnosis (Pe and Sipper 1999; Jiang et al. 2007; Lambrou et al. 2011), breast cancer knowledge discovery (Tan et al. 2003; Peng et al. 2006; Mazurowski et al. 2008), liver cancer diagnosis (Gletsos et al. 2003), and pap smear diagnosis (Marinakis et al. 2009). In typical diagnosis problems, medical images are the diagnostic inputs, and GA is used to extract the appropriate image features that can be used to detect the existence of cancerous cells, while the fitness function is usually the percentage of correct classifications. High accuracy rates, in comparison with other methods, have been reported using this approach for data sets with two or more classes (Marinakis et al. 2009).

Image processing and computer vision has also been an area of interest for GA applications, including image registration (Mandava et al. 1989; Matsopoulos et al. 1999; Rouet et al. 2000), edge detection (Gudmundsson et al. 1998), MRI segmentation (Schroeter et al. 1998; Fan et al. 2002; Tohka et al. 2007), prostate boundary segmentation (Cosío 2008), liver PET segmentation (Hsu et al. 2008), and cell image segmentation (Yang and Jiang 2001). Rouet et al. (2000) use GA for image registration. First, a GA is used to find an initial rigid registration; the GA applies encoding with real parameters representing the angles of 3-D rotation. Subsequently, a second GA is used to determine a robust point matching; in this GA, the chromosome consists of eight pairs of points, which define the mapping between the images. On the other hand, in some applications, a gray-level correlation between images is used to determine fitness function. In these GAs, a feature-based fitness function is used, which improves the speed of algorithm. Maulik (2009) provides a detailed review regarding applications of GA in medical image segmentation.

GA has been applied to a great variety of other medical problems too. Some of these applications include radiation treatment planning (Haas et al. 1998; Yu et al. 2000; Wu and Zhu 2001; Cotrutz and Xing 2003; Li et al. 2003, 2004), radiotherapy scheduling (Petrovic et al. 2011), HIV treatment planning (Ying et al. 2006), computer-assisted surgery (Arámbula Cosío and Davies 1999), multidisorder diagnosis (Vinterbo and Ohno-Machado 2000), predicting pneumonia (Heckerling et al. 2004), detecting pharmacokinetic properties of drugs (Yang et al. 2009), and drug scheduling (Liang et al. 2006).

4.4 ANNs

ANNs have been one of the most researched and applied methods in AI since the 1940s (McCulloch and Pitts 1943; Dybowski and Gant 2007). ANNs mimic the natural neural network mechanism to simulate the way the human mind learns and makes decisions based on its sensory inputs. The main feature of the ANNs, in contrast with other techniques, is that it consists of several simple processing nodes, which carry on calculations independently. Hence, the ANNs are, by principle, highly parallelizable algorithms. Each of these nodes is called a *neuron*; the neuron receives several inputs either directly from the ANN's inputs or from other neurons, aggregates the captured inputs, and generates an output based on an aggregated input using an *activation function*. The neuron takes into account the influence of each input weight on the aggregated value, which needs to be adjusted to produce the desirable result; this process enables *learning* of the ANN.

4.4.1 Perceptron

Various network structures have been proposed in the literature, but one of the most applied options is the *perceptron*. Neurons in a perceptron network commonly act as binary classifiers; a binary value (either 0 or 1) is calculated based on the aggregated inputs. Each neuron in the perceptron network can be considered to be a linear classifier, because it divides the input space linearly into two subspaces: negative and positive. Since it is a linear classifier, the perceptron neuron can be represented by a dot product of weights and inputs ($w.x$) and a bias (b), where weights determine the influence of each input on the output and bias is a constant term that determines the threshold at which the output switches between the two classes.

A *single-layer perceptron* is the simplest form of the perceptron, which is formed by a number of perceptron neurons arranged in a single layer, in parallel. However, the most useful structure is the *multilayer perceptron (MLP)*, which has one or more *hidden layers*, that is, layers of neurons that are not directly producing outputs. Unlike a single-layer perceptron, which is a linear classifier, multilayer perceptron can classify nonlinear data as well. Each layer is fully connected to the following layer, while the influence of each output from the preceding layer on the particular neuron can vary based on the weights used. These weights (and the bias value) need to be learned, which is usually done through a training technique referred to as *backpropagation*. Backpropagation is described in more detail in the subsequent section. A three-layer perceptron network is illustrated in Figure 4.7. The input layer is responsible for distributing input values to the next hidden layer. The hidden layer is a layer of an arbitrary number of neurons that do not directly have input or output data but perform intermediate classification of data, which is ultimately necessary to enable nonlinear classification. The output layer is the last layer of neurons that produce outputs, similar to the layer of neurons in a single-layer perceptron.

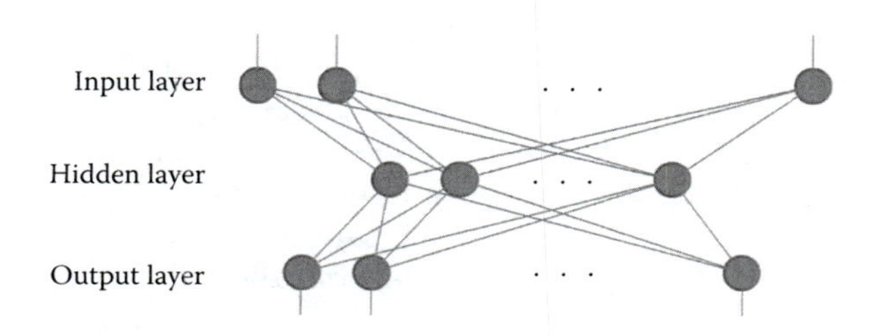

FIGURE 4.7
Layout of a three-layer perceptron.

4.4.2 Backpropagation

The backpropagation method is designed to propagate the error in output values back throughout the network and then adjust the parameters to correct the error. Training data (i.e., input data points with known output values) are fed into the network, and the values are propagated through the network until it reaches the output layer. The *perceptron equation*, which describes the forward calculations done at each neuron, is as follows:

$$y = g\left(\sum_i w_i x_i\right)$$

where y is the calculated output, g is the neuron's activation function, w_i is the weight for the ith input, and x_i is the ith input value. Based on the known output values and the obtained values, an error is calculated for each output in the output layer. Hence, the error represents misclassifications of training data by the network, which needs to be corrected. In the backpropagation process, each neuron's share in the error is calculated, and then, based on the error rate, weights are adjusted in such a way as to correct the error. In order to calculate the share of each neuron in the error, it is required that the activation function is differentiable. The formula used to calculate the changes in the neuron's weight is as follows:

$$\Delta w_i = \beta \sum_j (t_j - y_j) g'\left(\sum_i w_i x_{i,j}\right) x_{i,j}$$

where Δw_i represents the change in weight of the ith input; β is the *learning rate*, which is a scalar between 0 and 1 and determines how fast the weights change according to the error rates; j is the index for the set of training data samples; t_j is the correct known output value for sample j; y_j is the calculated value for sample j; g' is the derivative of the activation function; and $x_{i,j}$ is the ith input for sample j. The backpropagation process starts with neurons of the output layer. For the hidden layer, the error is calculated as the sum of the neuron's outputs, which are weighted by the changes in the output weights. Subsequently, changes in input weights are calculated in the same way. This process needs to be repeated several times (known as *epochs*) until the weights converge to certain values.

4.4.3 Medical Applications

Relative ease of training ANNs for classification purposes has led to many applications of ANNs in medicine. One of the most popular areas has been cancer diagnosis and prognosis. ANNs have been applied to skin cancer diagnosis (Wallace et al. 2000; Dreiseitl et al. 2001; Tomatis et al. 2005), breast cancer diagnosis (Setiono 1996, 2000; Chan et al. 1997; Kim and Park 1999; Yu and Guan 2000; Abbass 2002; Joo et al. 2004), breast cancer prognosis (Lisboa et al. 2003; Delen et al. 2005), lung cancer diagnosis (Penedo et al. 1998; Zhou et al. 2002; Suzuki et al. 2005, 2006), lung cancer prognosis (Santos-García et al. 2004), bladder cancer prognosis (Tasoulis et al. 2006), prostate cancer treatment (Rowbottom et al. 1999), brain cancer diagnosis (Ye et al. 2002), and ovarian cancer diagnosis (Tan et al. 2008). In the work of Chan et al. (1997), backpropagation ANN is used for texture classification of mammographic images by using a few extracted features as inputs, with the aim to determine malignant or benign tumors. Results show a 100% *sensitivity* (correct classification of malignant tumors) and 39% *specificity* (correct classification of benign tumors). In another study for detecting benign or malignant nodules in Computed Tomography (CT) images, *massive training ANNs* are used, which are capable of using images directly as training data; this resulted in high specificity of 48% and 100% sensitivity (Suzuki et al. 2005).

Other applications of ANNs in medicine have included radiotherapy treatment planning (Hosseini-Ashrafi et al. 1999; Rowbottom et al. 1999; Wu et al. 2000; Sharp et al. 2004), image segmentation and processing (Reddick et al. 1997; Heneghan et al. 2002; Lee et al. 2003), protein secondary structure prediction (Geourjon and Deleage 1995; Ruan et al. 2005; Babaei et al. 2010), epilepsy detection (Brasil et al. 2001; Argoud et al. 2006; Srinivasan et al. 2007), and so on. In an application to radiotherapy treatment planning, an ANN for prediction of tumor position for a moving tumor and latency in imaging is proposed (Sharp et al. 2004). Treatment accuracy when using an ANN's prediction is compared with no prediction and other prediction methods. Improvement is observed in comparison with no prediction, while in some cases, ANNs perform better than other prediction methods.

4.5 Conclusion

The complexity of many medical problems requires the use of CI techniques to provide timely and useful solutions. In this chapter, three CI techniques, namely, fuzzy logic, GAs, and ANNs, have been introduced, and their applications in medicine have been outlined. Fuzzy logic provides a framework for describing uncertainty and vagueness. This is particularly useful since many values and concepts in medical problems are vague and can be appropriately represented in the fuzzy framework. GA is a powerful optimization method that is able to deal with highly nonlinear problems, yet it can be applied in a variety of domains. ANNs are highly parallelizable algorithms that can be easily trained by a large volume of training data and are capable of classifying linear and nonlinear data.

Diagnosis and prognosis for various types of cancer; medical images processing including segmentation, classification, interpretation, enhancement, edge detection, and registration; treatment planning (radiation, HIV medicine, etc.); knowledge discovery; analysis of genetic data; and scheduling are among many current medical areas where the CI techniques have been successfully applied. We introduced some of these areas and applications in this chapter, but the use of CI in medicine is by no means limited to these. New applications of CI in medicine are constantly pursued by researchers, such as diagnosis of diabetes, Alzheimer's disease, heart diseases, rheumatic fever, psychosis, epilepsy, pneumonia, autism and multi diseases; treatment of blood infections and heart transplant, psychiatric, diabetic, and intensive care unit (ICU) patients; treatment scheduling; and dietary planning.

References

Abbass, H. A. (2002). An evolutionary artificial neural networks approach for breast cancer diagnosis. *Artificial Intelligence in Medicine* **25**(3): 265–281.

Adlassnig, K. P. (2001). The Section on Medical Expert and Knowledge-Based Systems at the Department of Medical Computer Sciences of the University of Vienna Medical School. *Artificial Intelligence in Medicine* **21**(1–3): 139–146.

Ahmed, M. N., S. M. Yamany, N. Mohamed, A. A. Farag and T. Moriarty (2002). A modified fuzzy C-means algorithm for bias field estimation and segmentation of MRI data. *IEEE Transactions on Medical Imaging* **21**(3): 193–199.

Akgundogdu, A., S. Kurt, N. Kilic, O. N. Ucan and N. Akalin (2010). Diagnosis of renal failure disease using adaptive neuro-fuzzy inference system. *Journal of Medical Systems* **34**(6): 1003–1009.

Allen, R. and D. Smith (2001). Neuro-fuzzy closed-loop control of depth of anaesthesia. *Artificial Intelligence in Medicine* **21**(1–3): 185–191.

Arámbula Cosío, F. and B. L. Davies (1999). Automated prostate recognition: a key process for clinically effective robotic prostatectomy. *Medical and Biological Engineering and Computing* **37**(2): 236–243.

Argoud, F. I. M., F. M. Azevedo, J. M. Neto and E. Grillo (2006). SADE3: an effective system for automated detection of epileptiform events in long-term EEG based on context information. *Medical and Biological Engineering and Computing* **44**(6): 459–470.

Avogadri, R. and G. Valentini (2009). Fuzzy ensemble clustering based on random projections for DNA microarray data analysis. *Artificial Intelligence in Medicine* **45**(2–3): 173–183.

Babaei, S., A. Geranmayeh and S. A. Seyyedsalehi (2010). Protein secondary structure prediction using modular reciprocal bidirectional recurrent neural networks. *Computer Methods and Programs in Biomedicine* **100**(3): 237–247.

Belal, S. Y., A. F. G. Taktak, A. J. Nevill, S. A. Spencer, D. Roden and S. Bevan (2002). Automatic detection of distorted plethysmogram pulses in neonates and paediatric patients using an adaptive-network-based fuzzy inference system. *Artificial Intelligence in Medicine* **24**(2): 149–165.

Berenji, H. R. and P. Khedkar (1992). Learning and tuning fuzzy logic controllers through reinforcements. *IEEE Transactions on Neural Networks* **3**(5): 724–740.

Böröczky, L., L. Zhao and K. P. Lee (2006). Feature subset selection for improving the performance of false positive reduction in lung nodule CAD. *IEEE Transactions on Information Technology in Biomedicine* **10**(3): 504–511.

Boudraa, A. E. O., J. Champier, L. Cinotti, J. C. Bordet, F. Lavenne and J. J. Mallet (1996). Delineation and quantitation of brain lesions by fuzzy clustering in positron emission tomography. *Computerized Medical Imaging and Graphics* **20**(1): 31–41.

Brandt, M. E., T. P. Bohan, L. A. Kramer and J. M. Fletcher (1994). Estimation of CSF, white and gray matter volumes in hydrocephalic children using fuzzy clustering of MR images. *Computerized Medical Imaging and Graphics* **18**(1): 25–34.

Brasil, L. M., F. M. De Azevedo and J. M. Barreto (2001). A hybrid expert system for the diagnosis of epileptic crisis. *Artificial Intelligence in Medicine* **21**(1–3): 227–233.

Chan, H. P., B. Sahiner, N. Patrick, M. A. Helvie, K. L. Lam, D. D. Adler and M. M. Goodsitt (1997). Computerized classification of malignant and benign microcalcifications on mammograms: texture analysis using an artificial neural network. *Physics in Medicine and Biology* **42**(3): 549–567.

Chikh, M. A., M. Ammar and R. Marouf (2012). A neuro-fuzzy identification of ECG beats. *Journal of Medical Systems* **36**(2): 903–914.

Chou, Y. C., M. F. Abbod, J. S. Shieh and C. Y. Hsu (2010). Multivariable fuzzy logic/self-organizing for anesthesia control. *Journal of Medical and Biological Engineering* **30**(5): 297–306.

Ciofolo, C. and C. Barillot (2009). Atlas-based segmentation of 3D cerebral structures with competitive level sets and fuzzy control. *Medical Image Analysis* **13**(3): 456–470.

Cosío, F. A. (2008). Automatic initialization of an active shape model of the prostate. *Medical Image Analysis* **12**(4): 469–483.

Cotrutz, C. and L. Xing (2003). Segment-based dose optimization using a genetic algorithm. *Physics in Medicine and Biology* **48**(18): 2987–2998.

Deb, K., A. Pratap, S. Agarwal and T. Meyarivan (2002). A fast and elitist multiobjective genetic algorithm: NSGA-II. *IEEE Transactions on Evolutionary Computation* **6**(2): 182–197.

Dehmeshki, J., X. Ye, X. Lin, M. Valdivieso and H. Amin (2007). Automated detection of lung nodules in CT images using shape-based genetic algorithm. *Computerized Medical Imaging and Graphics* **31**(6): 408–417.

Delen, D., G. Walker and A. Kadam (2005). Predicting breast cancer survivability: a comparison of three data mining methods. *Artificial Intelligence in Medicine* **34**(2): 113–127.

Denaï, M. A., M. Mahfouf and J. J. Ross (2009). A hybrid hierarchical decision support system for cardiac surgical intensive care patients. Part I: Physiological modelling and decision support system design. *Artificial Intelligence in Medicine* **45**(1): 35–52.

Dreiseitl, S., L. Ohno-Machado, H. Kittler, S. Vinterbo, H. Billhardt and M. Binder (2001). A comparison of machine learning methods for the diagnosis of pigmented skin lesions. *Journal of Biomedical Informatics* **34**(1): 28–36.

Dybowski, R. and V. Gant (2007). *Clinical Applications of Artificial Neural Networks*, Cambridge: Cambridge University Press.

Eberhart, R. C. and Y. Shi (2007). *Computational Intelligence: Concepts to Implementations*, San Francisco: Morgan Kaufmann Publishers.

Eiben, A. E. and J. E. Smith (2008). *Introduction to evolutionary computing (natural computing series)*, Berlin: Springer.

Fan, Y., T. Jiang and D. J. Evans (2002). Volumetric segmentation of brain images using parallel genetic algorithms. *IEEE Transactions on Medical Imaging* **21**(8): 904–909.

Fletcher-Heath, L. M., L. O. Hall, D. B. Goldgof and F. R. Murtagh (2001). Automatic segmentation of non-enhancing brain tumors in magnetic resonance images. *Artificial Intelligence in Medicine* **21**(1–3): 43–63.

Geourjon, C. and G. Deleage (1995). SOPMA: significant improvements in protein secondary structure prediction by consensus prediction from multiple alignments. *Computer Applications in the Biosciences* **11**(6): 681–684.

Ghazavi, S. N. and T. W. Liao (2008). Medical data mining by fuzzy modeling with selected features. *Artificial Intelligence in Medicine* **43**(3): 195–206.

Giannakeas, N. and D. I. Fotiadis (2009). An automated method for gridding and clustering-based segmentation of cDNA microarray images. *Computerized Medical Imaging and Graphics* **33**(1): 40–49.

Gletsos, M., S. G. Mougiakakou, G. K. Matsopoulos, K. S. Nikita, A. S. Nikita and D. Kelekis (2003). A computer-aided diagnostic system to characterize CT focal liver lesions: design and optimization of a neural network classifier. *IEEE Transactions on Information Technology in Biomedicine* **7**(3): 153–162.

Gudmundsson, M., E. A. El-Kwae and M. R. Kabuka (1998). Edge detection in medical images using a genetic algorithm. *IEEE Transactions on Medical Imaging* **17**(3): 469–474.

Haas, O. C. L., K. J. Burnham and J. A. Mills (1998). Optimization of beam orientation in radiotherapy using planar geometry. *Physics in Medicine and Biology* **43**(8): 2179–2193.

Halgamuge, S. K. and M. Glesner (1994). Neural networks in designing fuzzy systems for real world applications. *Fuzzy Sets and Systems* **65**(1): 1–12.

He, R., S. Datta, B. R. Sajja and P. A. Narayana (2008). Generalized fuzzy clustering for segmentation of multi-spectral magnetic resonance images. *Computerized Medical Imaging and Graphics* **32**(5): 353–366.

Heckerling, P. S., B. S. Gerber, T. G. Tape and R. S. Wigton (2004). Use of genetic algorithms for neural networks to predict community-acquired pneumonia. *Artificial Intelligence in Medicine* **30**(1): 71–84.

Heneghan, C., J. Flynn, M. O'Keefe and M. Cahill (2002). Characterization of changes in blood vessel width and tortuosity in retinopathy of prematurity using image analysis. *Medical Image Analysis* **6**(4): 407–429.

Herrera, F., M. Lozano and J. L. Verdegay (1995). Tuning fuzzy logic controllers by genetic algorithms. *International Journal of Approximate Reasoning* **12**(3): 299–315.

Hosseini-Ashrafi, M. E., H. Bagherebadian and E. Yahaqi (1999). Pre-optimization of radiotherapy treatment planning: an artificial neural network classification aided technique. *Physics in Medicine and Biology* **44**(6): 1513–1528.

Hsu, C. Y., C. Y. Liu and C. M. Chen (2008). Automatic segmentation of liver PET images. *Computerized Medical Imaging and Graphics* **32**(7): 601–610.

Hwang, S. L., G. F. Liang, J. T. Lin, Y. J. Yau, T. C. Yenn, C. C. Hsu and C. F. Chuang (2009). A real-time warning model for teamwork performance and system safety in nuclear power plants. *Safety Science* **47**(3): 425–435.

Ibbini, M. (2006). A PI-fuzzy logic controller for the regulation of blood glucose level in diabetic patients. *Journal of Medical Engineering and Technology* **30**(2): 83–92.

Ibbini, M. S. and M. A. Masadeh (2005). A fuzzy logic based closed-loop control system for blood glucose level regulation in diabetics. *Journal of Medical Engineering and Technology* **29**(2): 64–69.

Jang, J.-S. R. (1993). ANFIS: adaptive-network-based fuzzy inference system. *IEEE Transactions on Systems, Man and Cybernetics* **23**(3): 665–685.

Jang, J. S. R. and C. T. Sun (1993). Functional equivalence between radial basis function networks and fuzzy inference systems. *IEEE Transactions on Neural Networks* **4**(1): 156–159.

Ji, Z. X., Q. S. Sun and D. S. Xia (2011). A modified possibilistic fuzzy c-means clustering algorithm for bias field estimation and segmentation of brain MR image. *Computerized Medical Imaging and Graphics* **35**(5): 383–397.

Jiang, J., B. Yao and A. M. Wason (2007). A genetic algorithm design for microcalcification detection and classification in digital mammograms. *Computerized Medical Imaging and Graphics* **31**(1): 49–61.

Joo, S., Y. S. Yang, W. K. Moon and H. C. Kim (2004). Computer-aided diagnosis of solid breast nodules: use of an artificial neural network based on multiple sonographic features. *IEEE Transactions on Medical Imaging* **23**(10): 1292–1300.

Kim, J., W. Cai, S. Eberl and D. Feng (2007). Real-time volume rendering visualization of dual-modality PET/CT images with interactive fuzzy thresholding segmentation. *IEEE Transactions on Information Technology in Biomedicine* **11**(2): 161–169.

Kim, J. K. and H. W. Park (1999). Statistical textural features for detection of microcalcifications in digitized mammograms. *IEEE Transactions on Medical Imaging* **18**(3): 231–238.

Kwok, H. F., D. A. Linkens, M. Mahfouf and G. H. Mills (2003). Rule-base derivation for intensive care ventilator control using ANFIS. *Artificial Intelligence in Medicine* **29**(3): 185–201.

Lambrou, A., H. Papadopoulos and A. Gammerman (2011). Reliable confidence measures for medical diagnosis with evolutionary algorithms. *IEEE Transactions on Information Technology in Biomedicine* **15**(1): 93–99.

Lee, C. C., P. C. Chung and H. M. Tsai (2003). Identifying multiple abdominal organs from CT image series using a multimodule contextual neural network and spatial fuzzy rules. *IEEE Transactions on Information Technology in Biomedicine* **7**(3): 208–217.

Lee, Y., T. Hara, H. Fujita, S. Itoh and T. Ishigaki (2001). Automated detection of pulmonary nodules in helical CT images based on an improved template-matching technique. *IEEE Transactions on Medical Imaging* **20**(7): 595–604.

Lee, Z. J. (2008). An integrated algorithm for gene selection and classification applied to microarray data of ovarian cancer. *Artificial Intelligence in Medicine* **42**(1): 81–93.

Li, B. N., C. K. Chui, S. Chang and S. H. Ong (2011). Integrating spatial fuzzy clustering with level set methods for automated medical image segmentation. *Computers in Biology and Medicine* **41**(1): 1–10.

Li, Y., J. Yao and D. Yao (2003). Genetic algorithm based deliverable segments optimization for static intensity-modulated radiotherapy. *Physics in Medicine and Biology* **48**(20): 3353–3374.

Li, Y., J. Yao and D. Yao (2004). Automatic beam angle selection in IMRT planning using genetic algorithm. *Physics in Medicine and Biology* **49**(10): 1915–1932.

Liang, Y., K. S. Leung and T. S. K. Mok (2006). A novel evolutionary drug scheduling model in cancer chemotherapy. *IEEE Transactions on Information Technology in Biomedicine* **10**(2): 237–245.

Lin, D. T., C. R. Yan and W. T. Chen (2005). Autonomous detection of pulmonary nodules on CT images with a neural network-based fuzzy system. *Computerized Medical Imaging and Graphics* **29**(6): 447–458.

Lin, S. C., C. H. Luo and T. F. Yeh (2001). Fuzzy oxygen control system for the indirect calorimeter of premature infants. *Journal of Medical Engineering and Technology* **25**(4): 149–155.

Lisboa, P. J. G., H. Wong, P. Harris and R. Swindell (2003). A Bayesian neural network approach for modelling censored data with an application to prognosis after surgery for breast cancer. *Artificial Intelligence in Medicine* **28**(1): 1–25.

Liu, K. H., B. Li, Q. Q. Wu, J. Zhang, J. X. Du and G. Y. Liu (2009). Microarray data classification based on ensemble independent component selection. *Computers in Biology and Medicine* **39**(11): 953–960.

Mahfouf, M., M. F. Abbod and D. A. Linkens (2001). A survey of fuzzy logic monitoring and control utilisation in medicine. *Artificial Intelligence in Medicine* **21**(1–3): 27–42.

Mahfouf, M., C. S. Nunes, D. A. Linkens and J. E. Peacock (2005). Modelling and multivariable control in anaesthesia using neural-fuzzy paradigms: Part II. Closed-loop control of simultaneous administration of propofol and remifentanil. *Artificial Intelligence in Medicine* **35**(3): 207–213.

Mandava, V. R., J. M. Fitzpatrick and D. R. Pickens (1989). Adaptive search space scaling in digital image registration. *IEEE Transactions on Medical Imaging* **8**(3): 251–262.

Mansfield, J. R., M. G. Sowa, G. B. Scarth, R. L. Somorjai and H. H. Mantsch (1997). Fuzzy C-means clustering and principal component analysis of time series from near-infrared imaging of forearm ischemia. *Computerized Medical Imaging and Graphics* **21**(5): 299–308.

Mansfield, J. R., M. G. Sowa, J. R. Payette, B. Abdulrauf, M. F. Stranc and H. H. Manisch (1998). Tissue viability by multispectral near infrared imaging: a fuzzy C-means clustering analysis. *IEEE Transactions on Medical Imaging* **17**(6): 1011–1018.

Margolin, A. A., I. Nemenman, K. Basso, C. Wiggins, G. Stolovitzky, R. D. Favera and A. Califano (2006). ARACNE: an algorithm for the reconstruction of gene regulatory networks in a mammalian cellular context. *BMC Bioinformatics* **7**: 1–15 (SUPPL.1).

Marinakis, Y., G. Dounias and J. Jantzen (2009). Pap smear diagnosis using a hybrid intelligent scheme focusing on genetic algorithm based feature selection and nearest neighbor classification. *Computers in Biology and Medicine* **39**(1): 69–78.

Mason, D. G., D. A. Linkens, M. F. Abbod, N. D. Edwards and C. S. Reilly (1994). Automated delivery of muscle relaxants using fuzzy logic control. *IEEE Engineering in Medicine and Biology Magazine* **13**(5): 678–686.

Mason, D. G., J. J. Ross, N. D. Edwards, D. A. Linkens and C. S. Reilly (1997). Self-learning fuzzy control of atracurium-induced neuromuscular block during surgery. *Medical and Biological Engineering and Computing* **35**(5): 498–503.

Masulli, F. and A. Schenone (1999). A fuzzy clustering based segmentation system as support to diagnosis in medical imaging. *Artificial Intelligence in Medicine* **16**(2): 129–147.

Matsopoulos, G. K., N. A. Mouravliansky, K. K. Delibasis and K. S. Nikita (1999). Automatic retinal image registration scheme using global optimization techniques. *IEEE Transactions on Information Technology in Biomedicine* **3**(1): 47–60.

Maulik, U. (2009). Medical image segmentation using genetic algorithms. *IEEE Transactions on Information Technology in Biomedicine* **13**(2): 166–173.

Mazurowski, M. A., P. A. Habas, J. M. Zurada and G. D. Tourassi (2008). Decision optimization of case-based computer-aided decision systems using genetic algorithms with application to mammography. *Physics in Medicine and Biology* **53**(4): 895–908.

McCulloch, W. and W. Pitts (1943). A logical calculus of the ideas immanent in nervous activity. *Bulletin of Mathematical Biology* **5**(4): 115–133.

Muramatsu, C., T. Nakagawa, A. Sawada, Y. Hatanaka, T. Hara, T. Yamamoto and H. Fujita (2011). Automated segmentation of optic disc region on retinal fundus photographs: comparison of contour modeling and pixel classification methods. *Computer Methods and Programs in Biomedicine* **101**(1): 23–32.

Nauck, D. and R. Kruse (1999). Obtaining interpretable fuzzy classification rules from medical data. *Artificial Intelligence in Medicine* **16**(2): 149–169.

Nebot, A., F. E. Cellier and D. A. Linkens (1996). Synthesis of an anaesthetic agent administration system using fuzzy inductive reasoning. *Artificial Intelligence in Medicine* **8**(2): 147–166.

Nock, R. and F. Nielsen (2006). On weighting clustering. *IEEE Transactions on Pattern Analysis and Machine Intelligence* **28**(8): 1223–1235.

Noshiro, M., T. Matsunami, K. Takakuda, S. Ryumae, T. Kagawa, M. Shimizu and T. Fujino (1994). Fuzzy and conventional control of high-frequency ventilation. *Medical and Biological Engineering and Computing* **32**(4): 377–383.

Nunes, C. S., M. Mahfouf, D. A. Linkens and J. E. Peacock (2005). Modelling and multivariable control in anaesthesia using neural-fuzzy paradigms: Part I. Classification of depth of anaesthesia and development of a patient model. *Artificial Intelligence in Medicine* **35**(3): 195–206.

Oshita, S., K. Nakakimura and T. Sakabe (1994). Hypertension control during anesthesia. *IEEE Engineering in Medicine and Biology Magazine* **13**(5): 667–670.

Othman, R. M., S. Deris and R. M. Illias (2008). A genetic similarity algorithm for searching the Gene Ontology terms and annotating anonymous protein sequences. *Journal of Biomedical Informatics* **41**(1): 65–81.

Pandey, B. and R. B. Mishra (2009). Knowledge and intelligent computing system in medicine. *Computers in Biology and Medicine* **39**(3): 215–230.

Paul, T. K. and H. Iba (2009). Prediction of cancer class with majority voting genetic programming classifier using gene expression data. *IEEE/ACM Transactions on Computational Biology and Bioinformatics* **6**(2): 353–367.

Pe, C. A. and M. Sipper (1999). A fuzzy-genetic approach to breast cancer diagnosis. *Artificial Intelligence in Medicine* **17**(2): 131–155.

Pedrycz, W. and F. Gomide (1998). *An Introduction to Fuzzy Sets: Analysis and Design*, Cambridge, MA: MIT Press.

Penedo, M. G., M. J. Carreira, A. Mosquera and D. Cabello (1998). Computer-aided diagnosis: a neural-network-based approach to lung nodule detection. *IEEE Transactions on Medical Imaging* **17**(6): 872–880.

Peng, Y., B. Yao and J. Jiang (2006). Knowledge-discovery incorporated evolutionary search for microcalcification detection in breast cancer diagnosis. *Artificial Intelligence in Medicine* **37**(1): 43–53.

Petrovic, D., M. Morshed and S. Petrovic (2011). Multi-objective genetic algorithms for scheduling of radiotherapy treatments for categorised cancer patients. *Expert Systems with Applications* **38**(6): 6994–7002.

Pham, D. T. and D. Karaboga (2000). Intelligent optimisation techniques. In *Genetic Algorithms, Tabu Search, Simulated Annealing and Neural Networks*, New York: Springer.

Plissiti, M. E., C. Nikou and A. Charchanti (2011). Automated detection of cell nuclei in Pap smear images using morphological reconstruction and clustering. *IEEE Transactions on Information Technology in Biomedicine* **15**(2): 233–241.

Rafiee, A., M. H. Moradi and M. R. Farzaneh (2004). Novel genetic-neuro-fuzzy filter for speckle reduction from sonography images. *Journal of Digital Imaging* **17**(4): 292–300.

Reddick, W. E., J. O. Glass, E. N. Cook, T. David Elkin and R. J. Deaton (1997). Automated segmentation and classification of multispectral magnetic resonance images of brain using artificial neural networks. *IEEE Transactions on Medical Imaging* **16**(6): 911–918.

Rouet, J. M., J. J. Jacq and C. Roux (2000). Genetic algorithms for a robust 3-D MR-CT registration. *IEEE Transactions on Information Technology in Biomedicine* **4**(2): 126–136.

Rowbottom, C. G., S. Webb and M. Oldham (1999). Beam-orientation customization using an artificial neural network. *Physics in Medicine and Biology* **44**(9): 2251–2262.

Roy, S. H., M. S. Cheng, S. S. Chang, J. Moore, G. De Luca, S. H. Nawab and C. J. De Luca (2009). A combined sEMG and accelerometer system for monitoring functional activity in stroke. *IEEE Transactions on Neural Systems and Rehabilitation Engineering* **17**(6): 585–594.

Ruan, J., K. Wang, J. Yang, L. A. Kurgan and K. Cios (2005). Highly accurate and consistent method for prediction of helix and strand content from primary protein sequences. *Artificial Intelligence in Medicine* **35**(1–2): 19–35.

Santos-García, G., G. Varela, N. Novoa and M. F. Jiménez (2004). Prediction of postoperative morbidity after lung resection using an artificial neural network ensemble. *Artificial Intelligence in Medicine* **30**(1): 61–69.

Schroeter, P., J. M. Vesin, T. Langenberger and R. Meuli (1998). Robust parameter estimation of intensity distributions for brain magnetic resonance images. *IEEE Transactions on Medical Imaging* **17**(2): 172–186.

Setiono, R. (1996). Extracting rules from pruned neural networks for breast cancer diagnosis. *Artificial Intelligence in Medicine* **8**(1): 37–51.

Setiono, R. (2000). Generating concise and accurate classification rules for breast cancer diagnosis. *Artificial Intelligence in Medicine* **18**(3): 205–219.

Shah, S. C. and A. Kusiak (2004). Data mining and genetic algorithm based gene/SNP selection. *Artificial Intelligence in Medicine* **31**(3): 183–196.

Sharp, G. S., S. B. Jiang, S. Shimizu and H. Shirato (2004). Prediction of respiratory tumour motion for real-time image-guided radiotherapy. *Physics in Medicine and Biology* **49**(3): 425–440.

Srinivasan, V., C. Eswaran and N. Sriraam (2007). Approximate entropy-based epileptic EEG detection using artificial neural networks. *IEEE Transactions on Information Technology in Biomedicine* **11**(3): 288–295.

Suzuki, K., F. Li, S. Sone and K. Doi (2005). Computer-aided diagnostic scheme for distinction between benign and malignant nodules in thoracic low-dose CT by use of massive training artificial neural network. *IEEE Transactions on Medical Imaging* **24**(9): 1138–1150.

Suzuki, K., H. Abe, H. MacMahon and K. Doi (2006). Image-processing technique for suppressing ribs in chest radiographs by means of massive training artificial neural network (MTANN). *IEEE Transactions on Medical Imaging* **25**(4): 406–416.

Takagi, T. and M. Sugeno (1985). Fuzzy identification of systems and its applications to modeling and control. *IEEE Transactions on Systems, Man and Cybernetics* **15**(1): 116–132.

Tan, K. C., Q. Yu, C. M. Heng and T. H. Lee (2003). Evolutionary computing for knowledge discovery in medical diagnosis. *Artificial Intelligence in Medicine* **27**(2): 129–154.

Tan, T. Z., C. Quek, G. S. Ng and K. Razvi (2008). Ovarian cancer diagnosis with complementary learning fuzzy neural network. *Artificial Intelligence in Medicine* **43**(3): 207–222.

Tang, Y., Y. Q. Zhang, Z. Huang, X. Hu and Y. Zhao (2008). Recursive fuzzy granulation for gene subsets extraction and cancer classification. *IEEE Transactions on Information Technology in Biomedicine* **12**(6): 723–730.

Tari, L., C. Baral and S. Kim (2009). Fuzzy c-means clustering with prior biological knowledge. *Journal of Biomedical Informatics* **42**(1): 74–81.

Tasoulis, D. K., P. Spyridonos, N. G. Pavlidis, V. P. Plagianakos, P. Ravazoula, G. Nikiforidis and M. N. Vrahatis (2006). Cell-nuclear data reduction and prognostic model selection in bladder tumor recurrence. *Artificial Intelligence in Medicine* **38**(3): 291–303.

Ting, C. W. and C. Quek (2009). A novel blood glucose regulation using TSK0-FCMAC: a fuzzy CMAC based on the zero-ordered TSK fuzzy inference scheme. *IEEE Transactions on Neural Networks* **20**(5): 856–871.

Tohka, J., E. Krestyannikov, I. D. Dinov, A. MacKenzie Graham, D. W. Shattuck, U. Ruotsalainen and A. W. Toga (2007). Genetic algorithms for finite mixture model based voxel classification in neuroimaging. *IEEE Transactions on Medical Imaging* **26**(5): 696–711.

Tolias, Y. A. and S. M. Panas (1998). A fuzzy vessel tracking algorithm for retinal images based on fuzzy clustering. *IEEE Transactions on Medical Imaging* **17**(2): 263–273.

Tomatis, S., M. Carrara, A. Bono, C. Bartoli, M. Lualdi, G. Tragni, A. Colombo and R. Marchesini (2005). Automated melanoma detection with a novel multispectral imaging system: results of a prospective study. *Physics in Medicine and Biology* **50**(8): 1675–1687.

Tsai, H. K., J. M. Yang, Y. F. Tsai and C. Y. Kao (2004). An evolutionary approach for gene expression patterns. *IEEE Transactions on Information Technology in Biomedicine* **8**(2): 69–78.

Übeyli, E. D. (2009). Adaptive neuro-fuzzy inference systems for automatic detection of breast cancer. *Journal of Medical Systems* **33**(5): 353–358.

Vinterbo, S. and L. Ohno-Machado (2000). A genetic algorithm approach to multi-disorder diagnosis. *Artificial Intelligence in Medicine* **18**(2): 117–132.

Wallace, V. P., J. C. Bamber, D. C. Crawford, R. J. Ott and P. S. Mortimer (2000). Classification of reflectance spectra from pigmented skin lesions, a comparison of multivariate discriminant analysis and artificial neural networks. *Physics in Medicine and Biology* **45**(10): 2859–2871.

Wang, C. H., B. J. Liu and L. S. H. Wu (2012). The association forecasting of 13 variants within seven asthma susceptibility genes on 3 serum IgE groups in Taiwanese population by integrating of adaptive neuro-fuzzy inference system (ANFIS) and classification analysis methods. *Journal of Medical Systems* **36**(1): 175–185.

Wang, H. and B. Fei (2009). A modified fuzzy C-means classification method using a multiscale diffusion filtering scheme. *Medical Image Analysis* **13**(2): 193–202.

Wang, J., J. Kong, Y. Lu, M. Qi and B. Zhang (2008). A modified FCM algorithm for MRI brain image segmentation using both local and non-local spatial constraints. *Computerized Medical Imaging and Graphics* **32**(8): 685–698.

Wu, X. and Y. Zhu (2001). An optimization method for importance factors and beam weights based on genetic algorithms for radiotherapy treatment planning. *Physics in Medicine and Biology* **46**(4): 1085–1099.

Wu, X., Y. Zhu and L. Luo (2000). Linear programming based on neural networks for radiotherapy treatment planning. *Physics in Medicine and Biology* **45**(3): 719–728.

Yang, F. and T. Jiang (2001). Cell image segmentation with kernel-based dynamic clustering and an ellipsoidal cell shape model. *Journal of Biomedical Informatics* **34**(2): 67–73.

Yang, S. Y., Q. Huang, L. L. Li, C. Y. Ma, H. Zhang, R. Bai, Q. Z. Teng, M. L. Xiang and Y. Q. Wei (2009). An integrated scheme for feature selection and parameter setting in the support vector machine modeling and its application to the prediction of pharmacokinetic properties of drugs. *Artificial Intelligence in Medicine* **46**(2): 155–163.

Ye, C. Z., J. Yang, D. Y. Geng, Y. Zhou and N. Y. Chen (2002). Fuzzy rules to predict degree of malignancy in brain glioma. *Medical and Biological Engineering and Computing* **40**(2): 145–152.

Ying, H., F. Lin, R. D. MacArthur, J. A. Cohn, D. C. Barth-Jones, H. Ye and L. R. Crane (2006). A fuzzy discrete event system approach to determining optimal HIV/AIDS treatment regimens. *IEEE Transactions on Information Technology in Biomedicine* **10**(4): 663–676.

Yu, S. and L. Guan (2000). A CAD system for the automatic detection of clustered microcalcifications in digitized mammogram films. *IEEE Transactions on Medical Imaging* **19**(2): 115–126.

Yu, Y., J. B. Zhang, G. Cheng, M. C. Schell and P. Okunieff (2000). Multi-objective optimization in radiotherapy: applications to stereotactic radiosurgery and prostate brachytherapy. *Artificial Intelligence in Medicine* **19**(1): 39–51.

Zacharia, E. and D. Maroulis (2008). An original genetic approach to the fully automatic gridding of microarray images. *IEEE Transactions on Medical Imaging* **27**(6): 805–813.

Zeng, K., N. Y. Zhang and W. L. Xu (2000). A comparative study on sufficient conditions for Takagi-Sugeno fuzzy systems as universal approximators. *IEEE Transactions on Fuzzy Systems* **8**(6): 773–780.

Zhang, D. Q. and S. C. Chen (2004). A novel kernelized fuzzy C-means algorithm with application in medical image segmentation. *Artificial Intelligence in Medicine* **32**(1): 37–50.

Zhou, Z. H., Y. Jiang, Y. B. Yang and S. F. Chen (2002). Lung cancer cell identification based on artificial neural network ensembles. *Artificial Intelligence in Medicine* **24**(1): 25–36.

5

Satisficing or the Right Information at the Right Time: Artificial Intelligence and Information Retrieval, a Comparative Study in Medicine and Law

Paul Thompson

Information retrieval can be regarded as a process through which a gap in an individual's cognitive map is filled by information, or knowledge, retrieved from an external source of information, or knowledge. Viewed in this way, it is clear that artificial intelligence (AI) approaches could be used to fill this gap in the individual's cognitive map, and various approaches along these lines have been attempted. During the 1980s, the expert-system paradigm dominated AI research, particularly in medicine, but in many other fields as well. Despite this research interest, the main information retrieval systems used operationally to retrieve information for biomedical researchers have not incorporated AI approaches, nor is ranked document retrieval used by these systems, for example, PubMed. In the past few years, there has been much interest in Semantic Web technologies in the biomedical community, and in 2009, the National Institutes of Health (NIH) funded two companion stimulus grant projects for semantic repositories and search engines: eagle-i and VIVO [eagle-i 2012; VIVO 2012]. One research area that emerged from expert-system research was uncertainty and AI [Association 2012]. Expert systems are generally written using a rule-based formalism that captures the reasoning process of human experts in a given narrow domain. Originally, such rules were written either deterministically or with *ad hoc* confidence factors. The field of uncertainty and AI emerged as researchers became interested in providing better formalisms for representing uncertainty. More will be said about this later in the chapter. One formalism that emerged was the Bayesian belief network.

In the legal field, the situation was very different. In 1989, the Congressional Quarterly (CQ) Service and Personal Library Software (PLS) introduced the commercial online information retrieval world to ranked document retrieval [Pritchard-Schoch 1993] with their ranking of CQ documents in response to queries. The PLS software was based on the Syracuse Information Retrieval Experiment (SIRE) system from Syracuse University. In 1992, West Publishing Company, the provider of the legal information retrieval system Westlaw, deployed a version of the Inquery information retrieval system developed at the Center for Intelligent Information Retrieval at the University of Massachusetts, Amherst, which had been tailored for the legal domain, as its new ranked information retrieval search mode. Inquery was based on a Bayesian belief network approach to probabilistic information retrieval [Turtle and Croft 1991]. This was the first large-scale commercial ranked information retrieval system. Comparing the adoption of ranked information retrieval in the medical and legal domains, and how ranked document retrieval intersects with AI, is the focus of this chapter.

It is our hypothesis that users of document retrieval systems satisfice. Herbert Simon, the economist and AI pioneer, coined the concept of satisficing [Simon 1956]. People are not rational actors who optimize their activities to produce the best result but, rather, will accept results that are good enough to get by. Medical and legal information retrieval systems, such as PubMed, Westlaw, and LexisNexis, provide good enough retrieval, so that searchers continue to use suboptimal exact match retrieval methods even when, in the cases of Westlaw or LexisNexis, they could be using ranked document retrieval. On the other hand, when the quality of document retrieval becomes too poor, as with early Web search engines or with early approaches to e-Discovery in the legal domain, improved ranked document retrieval algorithms often provide a better solution [Baron and Thompson 2007]. In 2006, the federal rules of evidence were modified, giving electronic documents the same status as evidence in litigation as paper documents. Accordingly, the process of discovery, wherein parties to litigation request the discovery of documents from the other side(s), is now called e-Discovery, when electronic documents are involved. Often, these electronic documents are e-mail documents. Collections searched can have billions of documents. The best ranked retrieval results can only be obtained through the application of AI techniques. Eventually, these techniques may make it possible to achieve the often-stated, but difficult-to-reach, information retrieval goal of providing the right information to the right person at the right time.

AI research has had a long history. The term "artificial intelligence" is said to have first been used by the late John McCarthy, in a call for participation for the Dartmouth AI workshop of 1956 [Dartmouth 2012]. Over the years, many different types of research have been considered to be AI research. Rather than attempting to define AI, this chapter will consider AI research to include all of the various strands of research that have been considered to be AI research by the researchers involved. During the 1950s and early 1960s, there was an attempt to build systems, such as Newell and Simon's General Problem Solver [Newell et al. 1960], that would be able to solve problems as a human would from first principles. For example, researchers attempted to build systems that could play chess. Beginning in the 1960s and throughout the 1970s and 1980s, much attention was given to expert systems. The Dendritic Algorithm (DENDRAL) system created at Stanford by Edward Feigenbaum, which incorporated the expertise of Carl Djerassi and other mass spectrometry experts, is often considered the first expert system [Lindsay et al. 1993]. This system, developed beginning in 1965, used rules provided by Djerassi and the other experts to perform automated mass spectroscopy analysis. By the 1970s, various expert systems had been developed, especially at Stanford Medical School. One of the most well known of these systems was Mycin [Buchanan and Shortliffe 1984]. Since that time, Stanford Medical School has continued to develop many medical applications of AI. Mycin and other expert systems represented human expertise in the form of *if, then* rules, which could be chained together to reach conclusions. It was recognized that rules might only provide some evidence, or support, for their conclusions, so it was desirable to reason with confidence levels. Initially, these levels were *ad hoc*, and research turned to more formal ways of representing uncertainty. In 1985, the first Conference on Uncertainty and Artificial Intelligence was held [Association 2012]. This conference has met every year since then to the present. In the early years of this conference, there were many competing formalisms proposed to represent uncertainty in AI, for example, fuzzy set theory, Dempster–Shafer theory, or probability theory. Over time, the Bayesian belief network approach [Pearl 1988] became dominant.

Natural language understanding has been one of the major research areas for AI researchers. Through the 1980s, most approaches to natural language understanding were based on rule-based modeling of linguistic principles. Typically, approaches were based on syntactic, semantic, or pragmatic analysis, or combinations of two or more of these areas. Beginning with work on statistical machine translation by IBM [Brown et al. 1990], natural language understanding research moved away from linguistics-informed approaches towards brute-force statistical models. Similarly with clustering or classification research, the field of machine learning became more prominent around this time. Whether or not machine learning researchers consider themselves to be AI researchers, information retrieval research has become dominated by machine learning approaches, since at least the publication of Ponte's thesis in 1998 on the language modeling approach to document retrieval.

Used broadly, the term *information retrieval* can refer to a range of information access technologies including database retrieval, document retrieval, or question answering. Typically, the term is used more narrowly to mean document retrieval, where documents are usually publications, but can be any textual content, for example, e-mail. In this chapter, information retrieval is taken to mean document retrieval, though question answering will be considered, as well. Question answering has long been an area of AI research [Woods et al. 1972; Lehnert 1978; Strzalkowski and Harabagiu 2006]. Information retrieval was one of the early nonnumerical applications of computing, but many approaches to information retrieval have not involved AI. Although academic information retrieval researchers have always primarily focused on the development of algorithms for ranked document retrieval, commercial or other operational document retrieval systems did not use ranked document retrieval until the early 1990s. Instead, searchers formulated their queries as Boolean logic expressions. A document was retrieved if the document's descriptors, or words in its text, satisfied the logic of the search expression. It was necessary to list documents in some way for the user. One common way of listing documents was in reverse chronological order. This is often a good default approach, because users tend to be more interested in recent documents, other things being equal. Ranked document retrieval, on the other hand, attempts to rank documents according to some metric, such as similarity to the query or the probability that the document will satisfy the searcher's information need. In the United States, the vector space model, going back to the 1960s, was the dominant approach to document retrieval [Salton 1983]. With the vector space, model documents and queries are represented as *n*-dimensional vectors, where each component of the vector stands for one stemmed unique word from the document collection. Thus, if a document collection contained 10,000 unique word stems, each document and query would be represented by a 10,000-dimensional vector, where the value of each component of the vector is a weight corresponding to the word stem. The weights might be as simple as a 0 or a 1, depending on whether or not the given word stem appeared in the document, or the query. More typically, the weights are more complicated. One widely used weighting scheme is referred to as *tf*idf* weighting, which stands for term frequency * inverse document frequency. The weight is a combination of two factors: (1) the frequency of the term in the document, or query, and (2) the proportion of the documents in the collection that contain at least one occurrence of the term. The assumption behind this weighting scheme is that a document containing many occurrences of a term that is also in the query is more likely to be relevant, while a term appearing in only a small fraction of the collection is a better predictor of relevance than a term that occurs more frequently throughout the collection. For example, a word that appears in most documents in the collection is not likely to be a good term. In addition, common words, such as "the" and "and," are considered stop words and are not used to calculate scores to rank documents. With this or a similar representation scheme, the vector space model then calculates the cosine correlation coefficient between the query vector and the document vector, or some other similarity metric, to produce a score. These scores are used to rank documents [Salton and Buckley 1988].

The alternative ranking model to the vector space model in academic research is the probabilistic retrieval model [Maron and Kuhns 1960; Robertson and Sparck Jones 1976; van Rijsbergen 1979]. With the probabilistic model, an attempt is made to combine all evidence to calculate the probability that a document will be relevant, rather than measuring the similarity of the query and document vectors.

Although ranked document retrieval has been an academic research topic since the 1960s, commercial information retrieval systems until the 1990s were based on another model of retrieval: Boolean logic. Boolean logic is still a widely used approach to information retrieval in the medical and legal domains. With the Boolean model, a searcher combines query terms with Boolean operators, that is, AND, OR, and NOT. A document is retrieved if the Boolean logic expression is satisfied by the words, or descriptors, of the document. For example, a query might be (computer OR automation) AND economics. To be retrieved, a document would have to contain either the word *computer* or the word *automation*, as well as the word *economics*. As with the vector space and probabilistic models that used word stems, the Boolean model typically supports truncation operators. Thus, the search just mentioned might be, instead, (Comput! OR automat!) AND econom!, which would allow matches to take place on words such as *computing*, *computed*, *automatic*, or *economy*. Many implementations of Boolean logic also deviate from a pure logical formalism and allow many other operators, for example, that two terms must appear adjacent to each other in the text or within *n* characters of each other. While many experienced searchers, such as law librarians, prefer Boolean searching to ranked retrieval searching, there is empirical evidence, as well as theoretical arguments, in favor of ranked retrieval [Cooper 1988].

In 1992, West Publishing Company introduced a ranked retrieval search mode to its Westlaw legal information retrieval system. Shortly afterwards, other commercial search systems, such as LexisNexis in the legal domain and Dialog, widely used in libraries, also introduced ranked retrieval modes. Around the same time, Web search engines emerged for the newly developed World Wide Web. These search engines also ranked documents. Thus, by the mid-1990s, legal-domain searchers were able to use ranked document retrieval to find documents using either of the two most widely used commercial retrieval systems for legal information. It is an interesting fact that only a small percentage of users of Westlaw or LexisNexis use the ranked retrieval search mode of these two systems. The traditional Boolean search mode has remained available for these systems, and most users have stayed with the Boolean search mode.

The situation in the medical domain has been different. The US National Library of Medicine (NLM) maintains the Medical Literature Analysis and Retrieval System Online (MEDLINE) database, which is the most widely used source for biomedical document retrieval, particularly using its current online interface, PubMed. While academic biomedical researchers have explored ranked retrieval [Hersh and Voorhees 2009], including those at the NLM [Lu et al. 2009], PubMed continues to provide only Boolean retrieval.

The information retrieval systems developed by the medical and legal communities were somewhat similar up until the early 1990s. The NLM was a pioneer in online information retrieval, offering a service since 1964 [Rogers 1964]. The legal domain did not have comparable online retrieval services until the mid-1970s, when first LexisNexis and then West Publishing Company introduced legal information retrieval systems. In the biomedical field and in law documents are valuable resources, and there is extensive manual indexing and abstracting, or curation. In the legal domain, at West Publishing Company, each case law document, a judge's written opinion from an appellate court case, was examined by an attorney/editor, who was a specialist in the area of law of the case law document, say criminal law. The attorney/editor identified every point of law mentioned in the case and wrote a headnote, or abstract, for the point of law. This headnote was also classified according to the Key Number System [Westlaw 2012], a proprietary classification scheme with approximately 100,000 classes with up to eight levels of hierarchy. These case law documents with their headnotes and various indices to the documents are all published as print publications, as are the statutes for various states and other jurisdictions with their back-of-the-book indices created by attorney/editors. Once these case law and other legal documents became available online through the Westlaw legal information retrieval system, all of the editorial enhancements provided for the print products were also available as metadata for the online documents.

While there are a large number of appellate court cases in the various state and federal jurisdictions, there are far fewer such documents than there are biomedical journal articles and conference proceedings. The NLM also has biomedical librarians that provide indexing and abstracting for these articles and proceeding papers, but it is not possible with manual curation to provide the fine-grained level of metadata in the biomedical domain, such as the headnote in the legal domain. The NLM has, however, developed extensive thesauri and classification schemes, which have been merged together as the Unified Medical Language System (UMLS) [McCray et al. 1999].

The use of citation indexes was also pioneered in the legal domain. Starting in the 1890s, the legal publisher Shepards had paralegals manually categorize each citation in an appellate case to another case according to a taxonomy of reasons that one judge would cite another judge's opinion, for example, to overturn a case. Having this information allows online legal searchers to determine if a case is still considered to represent good law. While Science Citation Index, which was modeled on Shepards, and the open-source CiteSeer, allow a searcher to see which scientific articles cite a given article, these citation links are not categorized.

All of these editorial enhancements, or curation, in the legal and medical domains provide a rich environment for search by an experienced searcher. In the legal domain, the former West Publishing Company, now a part of Thomson–Reuters, has continued to develop these enhancements. Still, it has been difficult to effectively integrate editorial enhancements, or manual curation, with ranked document retrieval, either in the legal domain or in biomedicine. The fact that a document has been categorized with a certain UMLS descriptor or Key Number classification should be a feature of the document that could be taken into account when ranking documents. In the medical field, systems such as PubMed still do not rank documents. In the legal domain, documents are ranked but only based on free text features, such as *tf*idf* weighting, as discussed above.

While trained biomedical and legal searchers were able to find relevant documents without too much difficulty, thanks in part to the extensive editorial enhancements, or curation, provided by human editors and indexers, the search needs of the intelligence community were not as easily met. In the early 1990s, the US government's Tipster program funded several contractor teams to develop revolutionary new algorithms for information extraction and for information retrieval [Tipster 2000]. The University of Massachusetts, Amherst, was one of the Tipster contractors for both information retrieval and information extraction. Its Inquery search engine software, developed through the Tipster program, was adapted by West Publishing Company for its ranked retrieval search mode. Later, as search engines began to appear on the World Wide Web, one of the early Web search engines, Infoseek, also, for a short time, used a version of Inquery as its search engine. While Inquery worked well for Tipster and for Westlaw, it did not work as well for Web search. Nor did any of the early Web search engines rank Web content well. Yahoo! made an attempt to manually categorize documents on the Web, much as medical or legal documents are curated, but this approach soon broke down as the number of documents on the Web grew exponentially. A typical Web search from the late 1990s might have one, two, or three query terms, which retrieved, say, 5 million documents. These documents were ranked, but often, the quality of the ranking was poor. New algorithms for Web search were developed, such as Hyperlink-Induced Topic Search (HITS) [Kleinberg 1998] and PageRank [Page et al. 1999]. These algorithms ranked documents based on the graph structure of the Web. Although there have been many proprietary refinements made to these algorithms by companies such as Google, the basic idea behind such ranking schemes is that a document that meets the minimum standards for retrieval, that, contains variants of the one, two, or three search terms, should be ranked by its popularity as determined by the number of other Web documents that link to it and the popularity, in turn, of those other Web documents.

Although the Tipster program funded only a handful of contractors, many of the data sets developed for the program were made available to the wider research community through the Message Understanding Conferences (MUCs) [Message Understanding Conference 2001] and the Text REtrieval Conference (TREC) [TREC 2012]. MUC was an annual benchmark evaluation for information extraction technology, while TREC is still a similar evaluation for information retrieval.

US government-funded research on information retrieval and information extraction that began with Tipster has continued with programs such as Translingual Information Detection Extraction and Summarization (TIDES) [TIDES 2001], Automatic Content Extraction (ACE) [ACE 2009], Advanced Question Answering for Intelligence (AQUAINT) [AQUAINT 2010], and the Defense Advanced Research Projects Agency (DARPA) Machine Reading program [Machine Reading 2012].

Much biomedical AI research focuses on automating, or semiautomating, the curation process. It is widely acknowledged that manual curation cannot keep up with the pace of the growing number of publications and other biomedical knowledge sources [Baumgartner et al. 2007]. Although this research can be expected to help manual curation, it is not being applied to ranked biomedical document retrieval systems such as PubMed. There have been other biomedical information retrieval systems developed with NIH funding in recent years, such as eagle-i and VIVO, that use AI knowledge representations and ranked retrieval, but even with these systems, the focus of the research has been on the development of knowledge representations, such as ontologies, rather than on improved ranked retrieval algorithms [Vasilevsky et al. 2012]. The widely used open source vector space model search engine Lucene [2012] is used by both eagle-i and VIVO.

Just as medical expert systems of the 1970s and 1980s initially tried to avoid formal modeling of uncertainty with the *ad hoc* confidence factors attached to their rules, so these Semantic Web biomedical retrieval systems attempt to use Lucene and its vector space weighting approaches without recourse to probabilistic information retrieval modeling. Still, as with Web search and e-Discovery, the need for ranked information retrieval has been recognized. It may also be expected that the early *ad hoc* approaches to ranking will give way to more suitable formalisms. Approaches are being developed that combine Semantic Web knowledge representations with probabilistic retrieval principles [Elbassouni 2011].

At present, there is also much interest in question answering in the biomedical community. IBM developed Watson, the question-answering system that defeated two human *Jeopardy* champions in 2011 [Ferucci et al. 2010]. *Jeopardy* is a popular television game show based on answering questions. Watson's *Jeopardy* performance is seen by IBM as being a proof of concept for a system that can assist with medical decision making [Silobrcic 2012]. Question-answering approaches typically include a stage of ranked document retrieval. Just as West's ranked retrieval search mode so far has not been heavily used by end-user searchers, the technology of ranked retrieval, nevertheless, has been useful behind the scenes in supporting more automated curation [Jackson and Al-Kofahi 2011]. Similarly, ranked retrieval can support more automated curation in the medical domain. Moreover, as with e-Discovery in the legal domain or with Web search, ranked retrieval approaches informed by AI techniques may ultimately lead to improved ranked retrieval for biomedical searchers.

References

ACE. 2009. Automatic Content Extraction (ACE) evaluation, http://tides.nist.gov/, accessed November 15, 2012.

AQUAINT. 2010. Advanced Question Answering for Intelligence: AQUAINT, http://www-nlpir.nist.gov/projects/aquaint/, accessed November 15, 2012.

Association. 2012. Association for Uncertainty in Artificial Intelligence, http://www.auai.org/, accessed November 13, 2012.

Baron, J. and Thompson, P. 2007. The search problem posed by large heterogeneous data sets in litigation: possible future approaches to research, In *International Conference on Artificial Intelligence and Law 2007 (ICAIL 2007)*, Palo Alto, CA, June 4–8, 2007.

Baumgartner, W. A.; Cohen, K. B.; Fox, L. M.; Acquaah-Mensah, G.; and Hunter, L. 2007. Manual curation is not sufficient for annotation of genomic databases. *Bioinformatics*, Jul. 23:13, pp. 141–148.

Brown, P. F.; Cocke, J.; Della Pietra, S. A.; Della Pietra, V. J.; Jelinek, R.; Lafferty, J. D.; Mercer, R. L.; and Roossin, P. S. 1990. A statistical approach to machine translation. *Computational Linguistics*, vol. 16, no. 2, pp. 79–85.

Buchanan, B. G. and Shortliffe, E. H. 1984. *Rule Based Expert Systems: The Mycin Experiments of the Stanford Heuristic Programming Project*, Reading, MA: Addison-Wesley.

Cooper, W. S. 1988. Getting beyond Boole. *Information Processing and Management*, vol. 24, no. 3, pp. 243–248.

Dartmouth. 2012. The Dartmouth Artificial Intelligence Conference: The next 50 years, http://www.dartmouth.edu/~ai50/homepage.html, accessed November 13, 2012.

eagle-i. 2012. eagle-i, https://open.med.harvard.edu/display/eaglei/Welcome, accessed November 13, 2012.

Elbassuoni, S. 2011. Effective searching of RDF knowledge bases, Ph.D. Thesis, Max-Planck-Institut für Informatik.

Ferucci, D.; Brown, E.; Chu-Carroll, J.; Fan, J.; Gondek, D.; Kalyanpur, A. A.; Lally, A.; Murdock, W.; Nyberg, E.; Prager, J.; Schlaefer, N.; and Welty, C. 2010. Building Watson: an overview of the DeepQA Project. *AI Magazine*, vol. 31, no. 3, pp. 59–79.

Hersh W. and Voorhees E. 2009. TREC genomics special issue overview. *Information Retrieval*, vol. 12, pp. 1–15.

Jackson, P. and Al-Kofahi, K. 2011. Human expertise and artificial intelligence in legal search. In Geist, A.; Brunschwig, C.R.; Lachmeyer, F.; and Schefbeck, G., *Strukturierung der Juristischen Semantik—Structuring Legal Semantics*, Bern: Editions Weblaw, pp. 417–427.

Kleinberg, J. M. 1998. Authoritative sources in a hyperlinked environment. In *Proceedings of the ACM-SIAM Symposium on Discrete Algorithms*, San Francisco.

Lehnert, W. 1978. *The Process of Question Answering: A Computer Simulation of Cognition*, Hillsdale, NJ: Erlbaum Associates.

Lindsay, R. K.; Buchanan, B. G.; Feigenbaum, E. A.; and Lederberg, J. 1993. DENDRAL: a case study of the first expert system for scientific hypothesis formation, *Artificial Intelligence*, vol. 61, pp. 209–261.

Lu, Z.; Kim, W.; and Wilbur, J. W. 2009. Evaluating relevance ranking strategies for MEDLINE retrieval. *Journal of the American Medical Informatics Association*, vol. 16, pp. 32–36.

Lucene. 2012. Lucene, http://lucene.apache.org/, accessed November 13, 2012.

Machine Reading. 2012. Machine Reading, http://www.darpa.mil/Our_Work/I2O/Programs/Machine_Reading.aspx, accessed November 15, 2012.

Maron, M. E. and Kuhns, J. 1960. On relevance, probabilistic indexing and information retrieval. *Journal of the Association for Computing Machinery*, vol. 7, no. 3, pp. 216–244.

McCray, A. T.; Loane, R. F.; Browne, A. C.; and Bangalore, A. K. 1999. Terminology issues in user access to Web-based medical information, In *Proceedings of the American Medical Informatics Symposium*, Marriott Wardman Park, Washington, DC, pp. 107–111.

Message Understanding Conference. 2001. Introduction to information extraction, http://www.itl.nist.gov/iaui/894.02/related_projects/muc/, accessed November 15, 2012.

Newell, A.; Shaw, J. C.; and Simon, H. A. 1960. Report on a general problem-solving program, *Proceedings of the International Conference on Information Processing*, UNESCO, Paris, June 15–20, 1959, pp. 256–264.

Page, L.; Brin, S.; Motawani, R.; and Winograd, T. 1999. The PageRank citation ranking: bringing order to the Web, Technical Report, Stanford InfoLab.

Pearl, J. 1988. *Probabilistic Reasoning in Intelligent Systems: Networks of Plausible Inference*, San Mateo, CA: Morgan Kaufmann.

Ponte, J. 1998. A language modeling approach to information retrieval, Ph.D. Thesis, University of Massachusetts, Amherst.

Pritchard-Schoch, T. 1993. Natural language comes of age, *Online*, vol. 17, no. 3.

Robertson, S. E. and Sparck Jones, K. 1976. Relevance weighting of search terms. *Journal of the American Society for Information Science* 27, pp. 129–146 Reprinted In: Willett, P. (ed.), *Document Retrieval Systems*. Taylor Graham, 1988, pp. 143–160.

Rogers, F. B. 1964. The development of MEDLARS, *Bulletin of the Medical Library Association*, vol. 52, no. 1, pp. 150–151.

Salton, G. 1983. *Introduction to Modern Information Retrieval*, New York: McGraw-Hill.

Salton, G. and Buckley, C. 1988. Term-weighting approaches in automatic text retrieval, *Information Processing and Management*, vol. 24, no. 5, pp. 513–523.

Silobrcic, J. 2012. Innovation in clinical decision support—a new role for Watson in health care, presentation at the *National Quality Colloquium*.

Simon, H. A. 1956. Rational choice and the structure of the environment, *Psychological Reviews*, vol. 63, no. 2, pp. 129–138.

Strzalkowski, T. and Harabagiu, S. (eds.). 2006. *Advances in Open Domain Question Answering*, Dordrecht, The Netherlands: Kluwer Academic Publishers (Text, Speech and Language Technology, vol. 32).

TIDES. 2001. Translingual Information Detection, Extraction, and Summarization Evaluation Site, http://tides.nist.gov/, accessed November 15, 2012.

Tipster. 2000. The TIPSTER Text Program: a multi-agency, multi-contractor program, http://www.itl.nist.gov/iaui/894.02/related_projects/tipster/, accessed November 15, 2012.

TREC. 2012. Text REtrieval Conference (TREC), http://trec.nist.gov/, accessed November 15, 2012.

Turtle, H. and Croft, W. B. 1991. Evaluation of an inference network-based retrieval model. *ACM Transactions on Information Systems (TOIS)*, vol. 9, no. 3, pp. 187–222.

van Rijsbergen, C. J. 1979. *Information Retrieval*, 2nd ed., London: Butterworth.

Vasilevsky, N.; Johnson, T.; Corday, K.; Torniai, C.; Brush, M.; Segerdell, E.; Wilson, M.; Shaffer, C.; Robinson, D.; and Haendel, M. 2012. Research resources: curating the new eagle-i discovery system, *Database*, vol. 2012, doi:10.1093/database/bar067.

VIVO. 2012. VIVO, http://vivoweb.org/, accessed November 13, 2012.

Westlaw. 2012. The Keynumber System: yesterday and today, http://lawschool.westlaw.com/knumbers/history.asp?mainpage=16&subpage=4, accessed November 15, 2012.

Woods, W.; Kaplan, R. M.; and Nash-Webber, B. L. 1972. The Lunar Sciences Natural Language Information System: Final Report, BBN Report No. 2378, Cambridge, MA: Bolt Beranek and Newman Inc. (Available from NTIS as N72-28984.)

6

Soft Tissue Characterization Using Genetic Algorithm

Yongmin Zhong, Yashar Madjidi, Bijan Shirinzadeh, Julian Smith, and Chengfan Gu

CONTENTS

6.1 Introduction

Soft tissue properties are important to many modern applications of technology to medicine, such as robotic surgery, soft tissue modeling, and surgical simulation with force feedback. However, realistic acquisition of soft tissue properties is extremely challenging, not only because of the nonlinearity, anisotropy, nonhomogeneity, rate dependence, and time dependence of soft tissues but also due to the layered and nonhomogeneous structures of soft tissues [Samur et al. 2007; Kim et al. 2008; Zhong et al. 2010, 2012]. It is understood that soft tissue properties may dynamically change during the surgical process according to different patients, different organs, different functional regions and layers crossed by the surgical tools, and different physiological conditions. Therefore, it requires that mechanical properties of soft tissues be acquired and studied through a real-time intraoperative measurement process. Currently, soft tissue properties are commonly acquired by conducting standard mechanical testing such as tensile and indentation tests on material samples extracted from organs under well-defined loading and boundary conditions [Yamada and Evans 1970; Fung 1981; Woo et al. 1993; Ahn and Kim 2009; Ahn et al. 2012; Koo et al. 2012]. Since the thickness and transversal area of organs are not easy to measure, strain and stress values cannot be obtained directly from the measured displacement and force data [Samur 2007; Chawla et al. 2009]. As such experimental conditions are impossible to achieve, the parameter estimation methods may not be representative when they are determined based on *in vitro* experiments [Kohandel et al. 2008]. Further, material properties of soft tissues change in time, and the results obtained from the *in vitro* experiments can be misleading [Samur et al. 2007]. Therefore, it is necessary to study an analytical solution for accurately predicting the mechanical behaviors of soft tissues and satisfying the time-saving requirement of intraoperative measurement.

The quasilinear viscoelastic (QLV) model has been widely used for soft tissue modeling [DeFrate et al. 2006; DeFrate and Li 2007]. It has the capability of modeling materials with time-dependent viscoelastic behavior. Various estimation methods were reported for Fung's QLV parameters [Woo 1982; Dortmans et al. 1984; Nigul and Nigual 1987; Myers et al. 1991; Kwan et al. 1993]. Fitting the QLV model for stress relaxation tests were previously complicated by the need to have a very fast ramp to the applied strain. Abramowitch and Woo reported an improved method to analyze the stress relaxation of ligaments following a finite ramp time based on the QLV theory to obtain the constants of the QLV theory for goat femur–medial collateral ligament–tibia complexes [Abramowitch and Woo 2004]. It was found that the regression algorithm converged to a solution that was stable and validated by predicting the stress response of a separate experiment. The ramping and relaxation portions of the data were simultaneously fitted to the constitutive equation based on the strain history approach, and the related equations were minimized using the nonlinear Levenberg–Marquardt optimization algorithm. The finite element method was also reported for soft tissue characterization, in which a finite element model of soft tissue is constructed and used with an optimization method to match the experimental data with the numerical solution through iterations [Liu 2004; Samur et al. 2007; Sangpradit et al. 2011; Hollenstein et al. 2012]. However, the finite element method is computationally expensive and is generally conducted off-line.

In general, the existing studies are mainly based on regression algorithms to find a gradient-based solution, rather than gradient-free optimization algorithms, for characterization of soft tissue properties. The gradient-based algorithm has a fast convergence by using the derivatives with respect to the objective function. However, it is difficult to handle discontinuous and nondifferentiable problems and also involves the variability of constants [Mitchell 1999; Chawla et al. 2009]. Alternatively, the gradient-free method is typically designed to solve optimization or approximation problems whose objective function is computed by a black box. It is particularly favored in the case when the gradient computation is unavailable or expensive in terms of speed/time and accuracy, as estimating derivatives by the finite-difference method may be prohibitively costly. Among gradient-free algorithms, a genetic algorithm (GA) can be applied to solve a variety of optimization problems that are specifically not well suited for standard optimization algorithms, including problems in which the objective function is discontinuous, nondifferentiable, stochastic, or highly nonlinear [Asthana 2000]. GA is also suitable for a large number of quantized parameters and is less susceptible to getting stuck at local optima.

There has been limited research focusing on using GA for soft tissue characterization. Based on the approach reported in the work of Abramowitch and Woo [2004], Kohandel et al. [2008] reported a GA to obtain the QLV parameters. However, this study was at a preliminary stage, only providing simple results for verifying the curve matching ability. Chawla et al. [2009] studied GA for parameter estimation in the finite element method to identify soft tissue properties. However, the GA is only used to minimize the differences between experimental and finite element force responses, rather than directly identify the parameters of soft tissues.

This chapter focuses on establishing a gradient-free solution for estimation of the QLV model parameters, rather than a gradient-based approach, as in the previously existing studies. It investigates the estimation ability of a direct search tool to calculate and predict the parameters of soft tissues for achieving optimum intraoperative characterization. An improved time-saving GA is developed as a gradient-free direct search tool to estimate and predict the QLV parameters explicitly without approximating their derivatives. Experiments and analysis have been conducted to comprehensively evaluate the performance of the proposed method in comparison with the exponential formulation model [Pioletti et al. 1998] and the Mooney–Rivlin (MR) model [Holzapfel 2000]. Since extensive theories and experiments demonstrate that the robustness of gradient-free optimization methods does not suffer from the cases with a moderate level of noise, it is reasonable that the experimental analysis is conducted under an assumption that no noise is existed.

6.2 Biomechanical Models

6.2.1 QLV Model

The QLV model has become the most widely used theory in soft tissue biomechanics [DeFrate and Li 2007]. The QLV theory combines elastic and time dependence components of a tissue's mechanical response using an integral formulation. The basic concept of this theory is that (1) the stress at a given time can be described by a convolution integral form, separating the elastic response and the relaxation function, and (2) the relaxation function has a specific continuous spectrum. According to the QLV theory, the complete stress in a tissue subjected to a step strain can be expressed by the following convolution formula

$$\sigma(t) = G(t) * \sigma^{e}(\lambda) \tag{6.1}$$

where $G(t)$ is the reduced relaxation function, $\sigma^{e}(\lambda)$ is the nonlinear elastic response, and $\lambda(t)$ is the stretch ratio. In general, $G(t)$ is a fourth-order tensor to describe the direction-dependent relaxation phenomenon.

Using the Boltzmann superposition principle and representing the strain history as a series of infinitesimal step strains, the overall stress relaxation function can be expressed as the sum of all individual relaxations. For a general strain history, the stress at time t is represented by the convolution integral of $G(t)$ over time:

$$\sigma(t) = \int_{-\infty}^{t} G(t-\tau) \cdot \frac{\partial \sigma^{e}(\lambda)}{\partial \lambda} \cdot \frac{\partial \lambda}{\partial \tau} \cdot d\tau \tag{6.2}$$

where $\partial\sigma^e(\lambda)/\partial\lambda$ represents the instantaneous elastic response and $\partial\lambda/\partial\tau$ is the stretch history. For biological soft tissues, it is commonly assumed that the relaxation function is the same and continuous in all directions, and thus, $G(t)$ can be simplified as a scalar:

$$G(t) = \frac{1 + c \cdot \left[E_1(t/\tau_1) - E_1(t/\tau_2)\right]}{1 + c \cdot \ln(\tau_2/\tau_1)} \tag{6.3}$$

where $E_1(y) = \int_y^\infty (e^{-z}/z) \cdot \mathrm{d}z$ is the exponential integral function, τ_1 and τ_2 are the time constants that bind the lower and upper limits of the constant damping range for the relaxation function, and c is a dimensionless constant that scales the degree such that viscous effects are occurred. These three viscoelastic material coefficients can be determined from the stress relaxation experimental analysis.

For the purpose of integrability and time-saving calculation, the temporal behaviors of the relaxation function is described by a decaying exponential equation [Toms et al. 2002]:

$$G(t) = ae^{-bt} + ce^{-dt} + ge^{-ht} \tag{6.4}$$

where coefficients a, c, and g and exponents b, d, and h are the constants to be either determined experimentally or estimated numerically.

6.2.2 MR Model

The MR model has been widely considered to estimate the elastic stress response of soft tissues in biomechanics [Johnson 1996; Holzapfel 2000]. In this model, the strain energy function may be expressed in terms of the principal invariants I_1 and I_2 of the right Cauchy–Green tensor C as

$$W = C_1(I_1 - 3) + C_2(I_2 - 3) \tag{6.5}$$

$$I_1 = Tr(\mathbf{C}) = \lambda_1^2 + \lambda_2^2 + \lambda_3^2 \tag{6.6}$$

$$I_2 = \frac{1}{2}\left[Tr(\mathbf{C})^2 - Tr(\mathbf{C}^2)\right] = \lambda_1^2 \cdot \lambda_2^2 + \lambda_1^2 \cdot \lambda_3^2 + \lambda_2^2 \cdot \lambda_3^2 \tag{6.7}$$

where λ_1, λ_2, and λ_3 represent the principal stretches and C_1 and C_2 are the material constants. The principal Cauchy stresses for an incompressible and isotropic material in the case of uniaxial tension can be calculated by

$$\sigma^e = 2(\lambda^2 - 1/\lambda) \cdot (C_1 + C_2/\lambda); \quad C_1 + C_2 > 0 \tag{6.8}$$

where σ^e and λ are the stress and stretch in the axial direction, respectively. The axial stretch can be converted to the engineering strain by

$$\varepsilon = \frac{\Delta L}{L} = \lambda - 1 \tag{6.9}$$

where L is the length of the specimen and ΔL is the change in the length.

6.2.3 Exponential Formulation

Exponential formulation has been widely used in soft tissue models for describing the tensile behaviors of ligaments and tendons. One of the most widely used exponential expressions is the empirical equation, which was firstly proposed for the tensile behavior of skin [Kenedi et al. 1975]:

$$\sigma^e = A(e^{B\varepsilon} - 1); \quad A \,\&\, B > 0 \tag{6.10}$$

where A and B are the material constants, which are determined by fitting the model to the experimental data. However, this model does not consider three-dimensional stress states and is not generally expressed in terms of a strain energy function.

6.3 GA

GA is a global search and optimization method based on Darwin's natural evolution theory with the underlying principle "survival of the fittest." It uses three main types of rules (selection, crossover, and mutation) at each step to create the next generation from the current population: (1) selection rules choosing the individuals, called parents, that contribute to the population at the next generation; (2) crossover rules combining two parents to form children for the next generation; and (3) mutation rules applying random changes to individual parents to form children.

6.3.1 Fitness Function

The population members are ranked on the basis of fitness function, and then their ranks are typically divided by the number of individuals to provide a probability threshold for the selection. Scaling is another factor used to consider the recent history of the population and assign fitness values based on the comparison of individual performance to the recent average performance of the population. Fitness values can be assigned based on their actual distances from the floor value or can be equally spaced for the purpose of simplicity.

In this chapter, the relaxation function (Equation 6.4) and the elastic model (Equation 6.8) are used in the QLV model (Equation 6.2). Considering engineering strain over the ramping period $0 < t < t_0$, the stress resulted from a ramp phase with a constant strain rate γ may be written as

$$\sigma(t : 0 < t < t_0, \theta) = 2\gamma \int_0^t [ae^{-b(t-\tau)} + ce^{-d(t-\tau)} + ge^{-h(t-\tau)}] \left[2C_1(1+\gamma\tau) + \frac{C_1(1+\gamma\tau)+2C_2}{(1+\gamma\tau)^3} + C_2 \right] d\tau \tag{6.11}$$

and the stress response from t_0 onwards, where the strain rate is zero, is simply left with

$$\sigma(t : t \geq t_0, \theta) = 2\gamma \int_0^{t_0} [ae^{-b(t-\tau)} + ce^{-d(t-\tau)} + ge^{-h(t-\tau)}] \left[2C_1(1+\gamma\tau) + \frac{C_1(1+\gamma\tau)+2C_2}{(1+\gamma\tau)^3} + C_2 \right] d\tau \tag{6.12}$$

where $\theta = \{C_1, C_2, a, b, c, d, g, h\}$ is the vector of constants to be estimated by using the proposed GA.

For a set of experimental data, the ramping portion of the data is defined as $(t_i, \mathbf{R}_i)$ for $0 < t < t_0$ and the relaxation data as $(t_i, \mathbf{S}_i)$ for $t_0 < t < \infty$. Thus, by summing the squares of the differences between the experimental and theoretical data, the following equations may be written:

$$f(\theta) = \sum_i \left[\mathbf{R}_i - \sigma(t_i : 0 < t_i < t_0, \theta)\right]^2 \tag{6.13}$$

and

$$g(\theta) = \sum_i \left[\mathbf{S}_i - \sigma(t_i : 0 < t_i < t_0, \theta)\right]^2 \tag{6.14}$$

The objective is to minimize the two functions represented by Equations 6.13 and 6.14 simultaneously by using the GA. Thus, the sum of the two functions is considered as the fitness function:

$$\textit{fitness function} = f(\theta) + g(\theta) \tag{6.15}$$

6.3.2 Population

The initialization of the population is usually performed stochastically. It requires that the population represent a wide assortment of individuals. The main trade-off on the size of populations is obvious, that is, a large population searches the space more completely but at a higher computational cost. Further, although the population size tends to increase with the individual string length linearly rather than exponentially, the optimal population size also depends on a specific problem. In this study, an initial size of population n = 200 was firstly chosen, and the best individuals did have the highest fitness of all possibilities when the population convergence was occurred. To satisfy the time-saving requirement of intraoperative soft tissue measurement, the initial size of population n = 100 was finalized according to the required precision for the fitness function (see Figure 6.1).

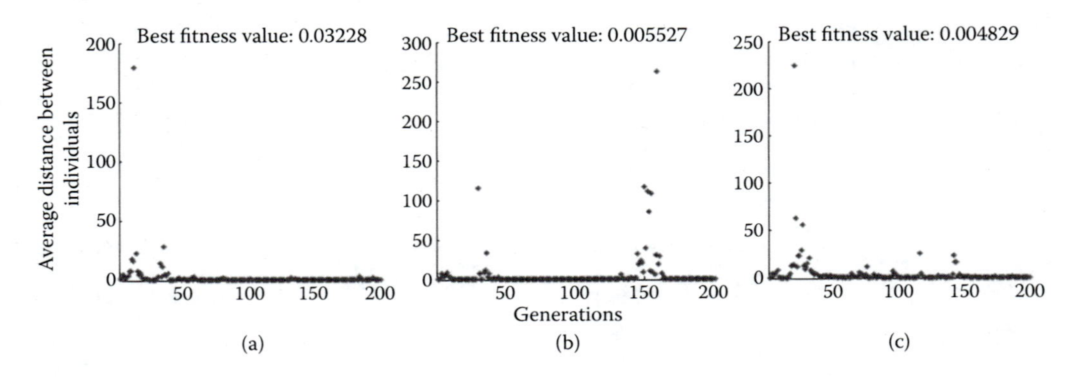

FIGURE 6.1
The GA's behavior where the initial sizes of populations are 50 (a), 100 (b), and 200 (c).

6.3.3 Selection

The selection function chooses parents for the next generation based on their scaled values from the fitness scaling function. The main three widely used selection schemes are roulette wheel implementation, tournament selection, and elitism. With the standard roulette wheel selection, the individual with the highest fitness value in a given generation may not survive reproduction, crossover, and mutation, thus being unaltered in the new generation. Tournament selection selects each parent by randomly considering the size of tournament players (individuals) to obtain the best individual (the one with higher fitness). This study chooses tournament selection as the selection rule, and the size of tournament players for choosing individuals is set to two.

6.3.4 Crossover

The most effective operator in GA is crossover based on the metaphor of sexual combination. Crossover has two main attributes, which can be varied. One is the probability of occurrences, and the other is the type of crossover to be implemented. The most basic crossover type is one-point crossover, which involves selecting a single crossover point at random and exchanging the portions of the individual strings to the right of the crossover point. Two-point crossover with a probability of 0.60–0.80 is a relatively common choice for another type of crossover. Heuristic crossover uses the fitness of the parents to determine the search direction. Uniform crossover is also a useful crossover type, in which a random decision is made at each bit position in the string according to whether the bits between parent strings are exchanged or not. Uniform crossover sometimes works better with a slightly lower crossover probability. It is also common to start out running the GA with a relatively higher value for crossover, then taper off the value linearly to the end of the computational trial, ending with a value of one-half or two-thirds of the initial value [Asthana 2000].

In this study, considering the significant influence of crossover function and its fraction value on the convergence performance of GA, experiments were conducted with different functions (Figure 6.2) and fractions (Figure 6.3). Consequently, single-point function with less than 60% of the fraction value was chosen for the proposed GA.

With the conventional genetic operators, there is no guarantee that the offspring are better than their parents. Therefore, the direction-based crossover has been adopted by the

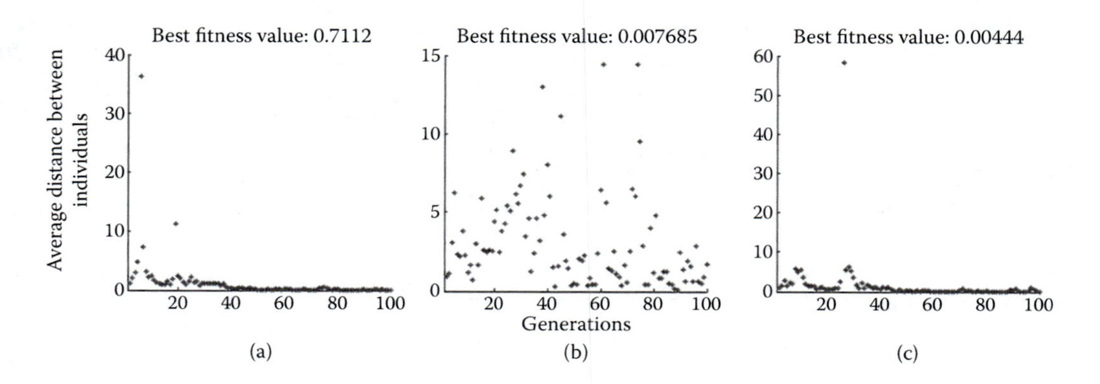

FIGURE 6.2
The GA's behavior where the crossover fraction is 0.6 and the crossover type is heuristic (a), two-point (b), and one-point (c).

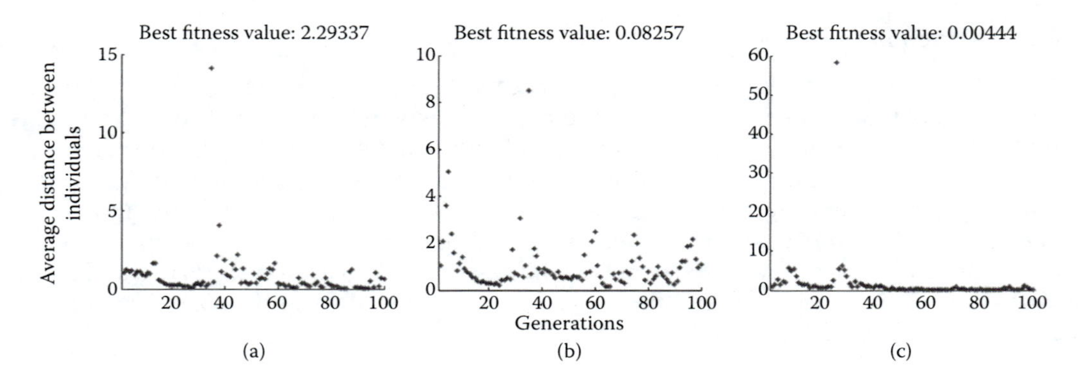

FIGURE 6.3
The GA's behavior where the crossover function is single point with the value of 0.8 (a), 0.7 (b), and 0.6 (c).

proposed GA. This operator uses the values of the objective function for determining the direction of genetic search. The operator generates a single offspring x' from two parents x_1 and x_2 according to the following rule:

$$x' = r(x_2 - x_1) + x_2 \tag{6.16}$$

where r is a random number between 0 and 1. It also assumes that parent x_2 is not worse than x_1, that is,

$$f(x_2) \leq f(x_1) \tag{6.17}$$

6.3.5 Mutation

Mutation is the stochastic flipping of bits occurring at each generation, which is often held constant for the entire computational trial of the GA. In this study, the fixed rate of 0.01 was defined for the mutation process, and the change of the mutation rate during the computational trials was not hypothesized as an influential operator.

As a matter of fact, mutation is not an especially important operator in GA and is usually set at a very low rate, or sometimes omitted altogether. Traditional crossover and mutation operators are based on a randomization mechanism. However, improved GAs generally include new adaptive penalty schemes and adaptive mutations as well as adaptive crossover operators. In this study, a penalty factor higher than common values was used. In particular, setting a higher value for the initial penalty after randomization of the initial population was observed as a noticeable factor in the GA convergence performance.

6.4 Performance Analysis

A prototype system has been implemented with the proposed method for soft tissue characterization. Experiments have been conducted to investigate the elastic stress–strain and stress relaxation responses to verify the capability of the proposed GA to fit the QLV model. Experiments have also been conducted to examine the ability of the proposed QLV model for predicting the elastic stress relaxation behavior. Comparison analysis of the proposed method with the existing methods using the MR model and the exponential formulation for prediction and analysis of soft tissue responses is also discussed in this section.

Among soft biological tissues, tendons and ligaments play an important role in the mechanical integrity of the body by transferring loads between bones for the ligaments, or between muscles and bones for the tendons. The anterior cruciate ligament (ACL) is one of the major ligaments of the knee, which has been the subject of a great numbers of both experimental and computational studies. Given the fact that a number of QLV models based on the same experimental data are available in the literature [Defrate and Li 2007] and ACL is one of the most investigated tissues, the ACL is chosen as the test subject, and the same data available in the literature are used to evaluate and predict the proposed method.

6.4.1 Modeling of Stress–Strain Response

Two elastic models (Equations 6.8 and 6.10) are used to fit the experimental data through a gradient-based solution and a gradient-free solution. The experimental data are derived from the previous studies on uniaxial tensile tests of ACL [DeFrate and Li 2007; Johnson 1996]. The experimental data are estimated by digitizing images of the graphs through a raster-to-vector converter software package. For the gradient-based solution, each model is fitted to the stress–strain curve up to 4% strain based on a least-squares fitting method using a mathematical software package (Mathematica, Wolfram Research). The elastic models are fitted under the assumption that the load is applied instantaneously. For the gradient-free solution, GA is used to fit the stress–strain curve up to 4% strain using a technical computing software package (MATLAB®, The MathWorks). All of the models are then used to model stress–strain behavior at higher strain levels. The estimating ability of each model is evaluated as the models were fitted up to different strain levels (Figure 6.4). The exponential and MR formulations using the gradient-based method are termed "Exp." and "MR" respectively, and the MR model using GA is termed as "GA" on the diagrams.

As shown in Figure 6.4, below the 3% strain level, all the models closely fit the experimental data. Beyond this level, the ability of the models to fit the experimental data is limited. However, the MR formulation (Equation 6.8) models the stress response more accurately than the exponential formulation (Equation 6.10). Considering the models to fit the experimental data up to 4% strain level (Figure 6.4b), at 7% strain level, the exponential formulation overestimates the stress by 88% and the MR formulation by 15%. Even by fitting the models to the experimental data up to 6% strain level (Figure 6.4c), at 7% strain level, the exponential formulation overestimates the stress by 18% and the MR formulation by approximately 11%.

The results for the gradient-based solution used with these two elastic models were previously reported [DeFrate and Li 2007]. It can be seen from Figure 6.4 that the proposed GA method has the same calculation precision as the gradient-based solution for the MR model, indicating the same results in terms of parameter estimation. Generally speaking, the GA is more practical when finding a solution to a multidimensional discrete problem is unavailable or time consuming.

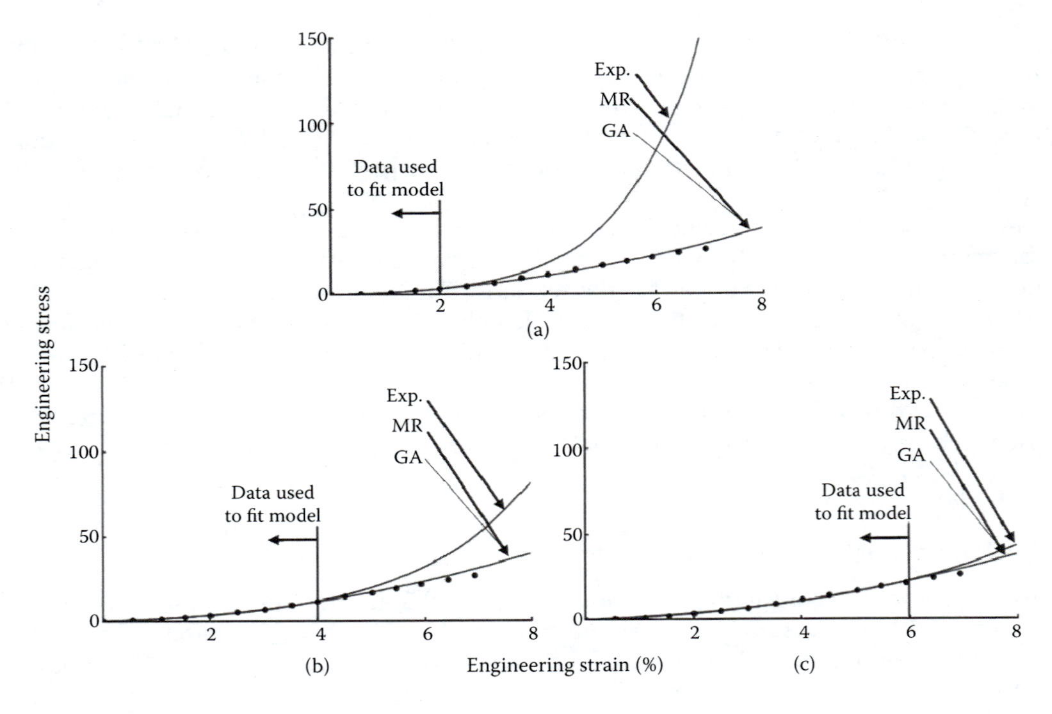

FIGURE 6.4
Engineering stress versus strain for the *ACL*. The experimental data (dots) derived from the previous study [Woo et al. 1993] are used to fit the models to three different levels of strain: 2% (a), 4% (b), and 6% (c). (From Woo, S. L. et al., *Journal of Biomechanical Engineering*, vol. 115, 1993, pp. 468–473.)

6.4.2 Modeling of Stress Relaxation Response

In the derivation of the constitutive equation, as the first modeling step, choosing appropriate models for the instantaneous elastic response and the relaxation function is required. As mentioned previously, the exponential and MR formulations are chosen to be used in the QLV model to describe the stress relaxation response of the ACL. The related experimental data are derived from the previously investigated data sets [Pioletti et al. 1998; Pioletti and Rakotomanana 2000; Defrate and Li 2007]. By using the gradient-based and GA approaches, three different QLV models are investigated in this study. The first QLV model consists of the elastic model based on the exponential formulation and the exponential relaxation function. The second model is made up of the MR elastic model and the exponential relaxation function. The parameters of these two models are estimated by a gradient-based solution to fit the elastic models and the relaxation function separately, assuming that the load is applied instantaneously. The third QLV model consists of the MR elastic model and the exponential relaxation function, but it uses the GA to estimate QLV parameters. Similar to previous studies [Abramowitch and Woo 2004; Kohandel et al. 2008], the ramping and relaxation portions are simultaneously fitted to the constitutive equation in order to remove the assumption of a step change in strain. All the models are fitted to the stress relaxation data measured at 2% strain level to determine the QLV parameters. Subsequently, these models are used to predict the stress relaxation responses at higher strain levels, which will be discussed in the next section.

6.4.3 Prediction of Elastic Stress Relaxation

After verifying the QLV models with the stress relaxation data measured at the 2% strain level to determine the material constants, these models are used to predict the stress relaxation response of ACL at 4% and 6% strain levels for the comparisons with the experimental data. Although all the models closely match the experimental data at 2% strain level (Figure 6.5a), the prediction ability of these models at higher strain levels is limited. Figure 6.5b and c generally indicates that the higher strain levels, in comparison with the strain level at which the QLV parameters are determined by fitting the experimental data, cause more overestimated stress response, especially after a long relaxation time.

During the prediction of the stress relaxation response at 4% strain level (Figure 6.5b), the MR model overestimates the experimental data by less than 2% and 6% at 100 s and 1600 s, respectively, and the exponential model overestimates the stress response by 36% and 45% at 100 s and 1600 s, respectively. The GA-based model shows a similar behavior to the MR model at 100 s but indicates more overestimation with up to 7% at 1600 s.

In the case of predicting the stress relaxation at 6% strain (Figure 6.5c), the prediction of the overall stress response undergoes a greater difference from the experimental data. On the change point of phases from ramping to relaxation (the peak stress point at 100 s), the GA-based model overestimates by 1% in comparison with the overestimation of 3% obtained by the MR model. However, during the relaxation period of up to 1600 s, the MR model predicts the stress response with a more decreasing rate than the GA-based model. The slow decrement of the GA-based model during the relaxation portion is monitored

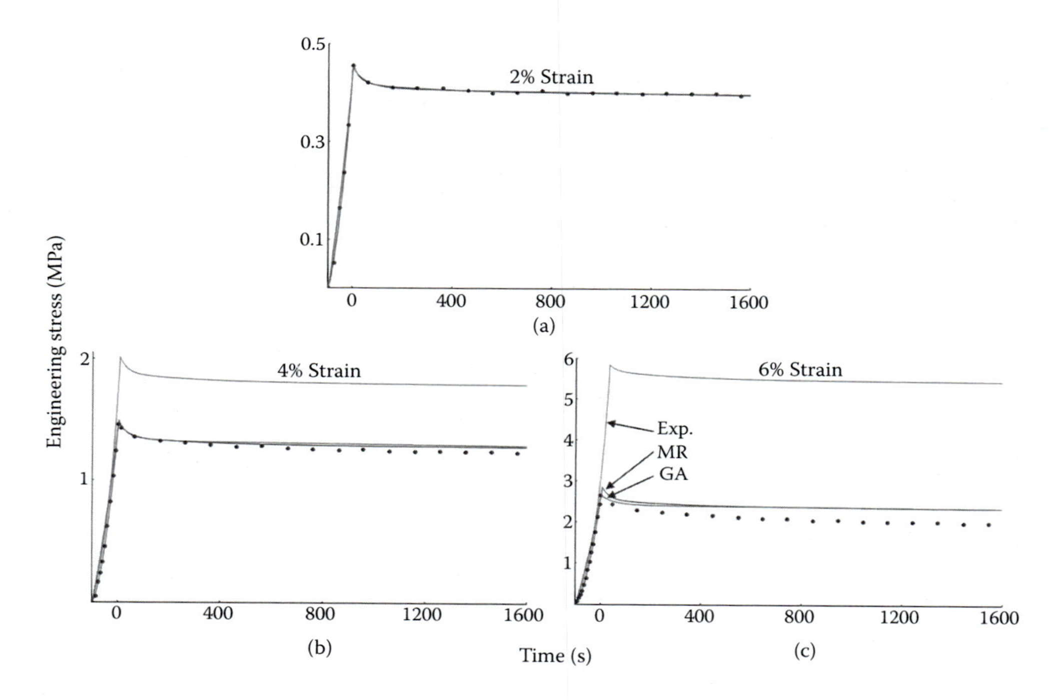

FIGURE 6.5
Engineering stress versus time for the *ACL*. The experimental data are depicted by the dots [Pioletti et al. 1998, 2000]. The validated models at 2% strain (a), and the predicted models at 4% strain (b), and 6% strain (c), are indicated by the lines. (From Pioletti, D. P. et al., *Journal of Biomechanics*, vol. 31, 1998, pp. 753–757; Pioletti, D. P. and L. R. Rakotomanana, *Journal of Biomechanics*, vol. 33, 2000, pp. 1729–1732.)

during the computational trials. This trend is not unexpected and can be relatively adjusted when the specific part of experimental data is considered. At 6% strain level, the prediction of the exponential model is very poor, and an overestimation of more than 100% is observed.

6.4.4 Time-Saving Factor

When the models are fitted to the stress relaxation data for ACL, all of the models closely match the experimental data (Figure 6.5a). During the approximation process based on the GA, the trend of curve matching can be tracked through generations by determining its behavior based on the fitness function value, average distance between individuals, and the curve matching quality analysis. Two former factors have been evaluated in section 6.3 to determine the different rates of GA operators. In the following, we will discuss the timing factor for the matching ability of the proposed GA for the stress relaxation data.

Experiments have also been conducted to evaluate the time-saving factor in terms of curve matching quality. One investigation is focused on defining the weight coefficient around the change point of data portion from the ramping phase to the relaxation phase of the diagram (see Figure 6.6b). The other is focused on defining the appropriate slope for the relaxation phase in accordance with the time-dependent reduction of stress at the end of the time period (1600 s) (see Figure 6.6a). Experiments show that these two techniques can get the desired curve matching with fewer generations. However, the latter solution (Figure 6.6a) tends to obtain the optimally estimated QLV parameters with fewer generations in comparison with the former solution (Figure 6.6b). In addition to the above two techniques, estimation of the upper and lower bounds for starting the search algorithm is also an effective way to achieve a faster convergence. The convergence ability of this method can be adjusted according to the required precision by determining the fitness function limit as one of the stopping criteria. With the ability provided by these techniques, the proposed GA is able to satisfy the real-time computational requirement of intraoperative soft tissue measurement to determine the QLV parameters.

Apart from the time factor in terms of curve matching, this proposed GA method has also been evaluated by using two different crossover functions, i.e. the conventional single-point operator with less than 60% of the fraction value and the direction-based single-point

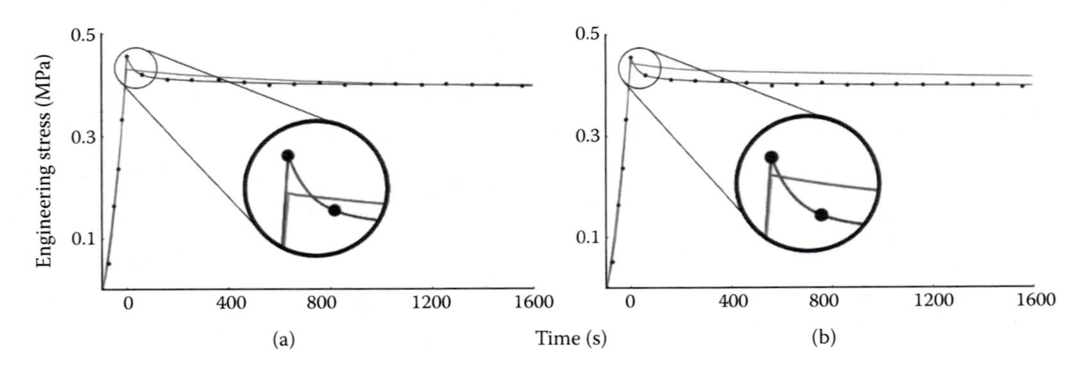

FIGURE 6.6
(See color insert.) Engineering stress versus time for the *ACL*. The experimental data are depicted by the dots [Pioletti et al. 1998, 2000]. The fitted models at 2% strain with the focus on change point (peak stress point) (a) and relaxation portion (b) are indicated by the lines. (From Pioletti, D. P. et al., *Journal of Biomechanics*, vol. 31, 1998, pp. 753–757; Pioletti, D. P. and L. R. Rakotomanana, *Journal of Biomechanics*, vol. 33, 2000, pp. 1729–1732.)

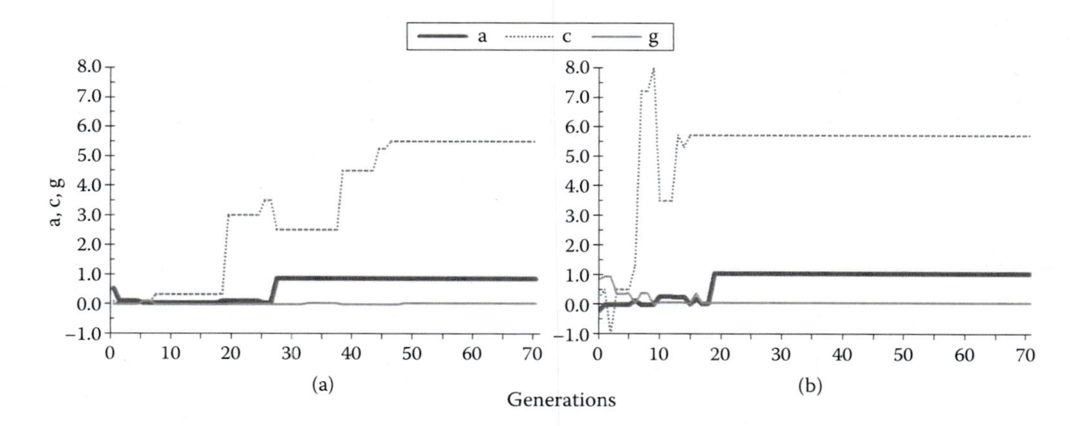

FIGURE 6.7
Evolution of magnitudes of three QLV model parameters through generations using GA based on the conventional single-point crossover operator (a) and the direction-based crossover function (b).

crossover. Experiments have been conducted to estimate the relaxation function parameters of the QLV model by the proposed GA using the conventional single-point crossover function and direction-based single-point crossover, respectively. As shown in Figure 6.7, the desired fitness function value using the conventional single-point crossover operator is obtained after an average of 50 generations, while the direction-based crossover function obtains the same accuracy with less than 20 generations. Considering the history of the QLV model parameters through generations, the QLV model parameters are observed to be optimized with a higher convergence rate of GA using the direction-based crossover function. This also demonstrates that the proposed GA method using the direction-based crossover function is able to determine the QLV model parameters within a reasonable time period.

6.5 Conclusions

This chapter presents a gradient-free method to intraoperative soft tissue characterization by developing an improved GA to estimate and predict the QLV model parameters. Three different constitutive laws are used to describe the elastic stress response through both a gradient-based solution and a gradient-free direct search solution. Experiments and comparison analysis have been conducted to validate the approximation and prediction performance of the improved GA for estimation of QLV model parameters. Experimental results demonstrate that the proposed GA method can accurately estimate soft tissue parameters in a time-saving process and thus is suitable for intraoperative soft tissue characterization.

Future research will focus on establishment of an inverse solution for characterization of soft tissue properties from the experimental data obtained by *in situ* robot-assisted measurements to predict the stress–strain behavior in response to different loading conditions such as cyclic stress relaxation and creep. Measurement of *in situ* stress and strain poses extreme experimental difficulties, and these data cannot be obtained directly from the measured displacements and forces. A minimally invasive robotic indenter and associated analytical algorithms will be established for measurement of *in situ* stress and strain.

Acknowledgments

This research is supported by the Australian Research Council (ARC) Discovery Grants (DP0986814 and DP110104970).

References

Abramowitch S. D., S. L. Woo, An Improved method to analyse the stress relaxation of ligaments following a finite ramp time based on the quasi-linear viscoelastic theory, *Journal of Biomechanical Engineering*, vol. 126, no. 1, 2004, pp. 92–97.

Ahn B., J. Kim, An efficient soft tissue characterization under large deformations in medical simulations, *International Journal of Precision Engineering and Manufacturing*, vol. 10, no. 4, 2009, pp. 115–121.

Ahn B., Y. Kim, C. K. Oh, J. Kim, Robotic palpation and mechanical property characterization for abnormal tissue localization, *Medical and Biological Engineering and Computing*, vol. 50, no. 9, 2012, pp. 961–971.

Asthana R. G. S., *Evolutionary Algorithms and Neural Networks, Soft Computing and Intelligent Systems: Theory and Applications*, N. K. Sinha, L. A. Zadeh, M. M. Gupta (eds.), San Diego, CA: Academic Press, 2000.

Chawla A., S. Mukherjee, B. Karthikeyan, Characterization of human passive muscles for impact loads using genetic algorithm and inverse finite element methods, *Biomechanics and Modeling in Mechanobiology*, vol. 8, no. 1, 2009, pp. 67–76.

DeFrate L. E., G. Li, The prediction of stress-relaxation of ligaments and tendons using the quasi-linear viscoelastic model, *Biomechanical and Modeling in Mechanobiology*, vol. 6, no. 4, 2007, pp. 245–251.

DeFrate L. E., A. van der Ven, P. J. Boyer, T. J. Gill, G. Li, The measurement of the variation in the surface strains of Achilles tendon grafts using imaging techniques, *Journal of Biomechanical Engineering*, vol. 39, no. 3, 2006, pp. 399–405.

Dortmans L. J., A. A. Sauren, E. P. Rousseau, Parameter estimation using the quasi-linear viscoelastic model proposed by Fung, *Journal of Biomechanical Engineering*, vol. 106, no. 3, 1984, pp. 198–203.

Fung Y. C., *Biomechanics: Mechanical Properties of Living Tissues*. Berlin, Heidelberg: Springer, 1981.

Hollenstein M., M. Bajka, B. Röhrnbauer, S. Badir, E. Mazza, Measuring the in vivo behavior of soft tissue and organs using the aspiration device, In *Soft Tissue Biomechanical Modeling for Computer Assisted Surgery*, Y. Payan (ed.), Springer-Verlag: Berlin, Heidelberg, 2012, pp. 201–228.

Holzapfel G. A., *Nonlinear Solid Mechanics: A Continuum Approach for Engineering*, New York: Wiley, 2000.

Johnson G. A., G. A. Livesay, S. L. Y. Woo, K. R. Rajagopal, A single integral finite strain viscoelastic model of ligaments and tendons, *Journal of Biomechanical Engineering*, vol. 118, no. 2, 1996, pp. 221–226.

Kenedi R. M., T. Gibson, J. H. Evans, J. C. Barbenel, Tissue mechanics, *Physics in Medicine and Biology*, vol. 20, no. 5, 1975, pp. 699–717.

Kim J., B. Ahn, S. De, M. A. Srinivasan, An efficient soft tissue characterization algorithm from in-vivo indentation experiments for medical simulation, *International Journal of Medical Robotics and Computer Assisted Surgery*, vol. 4, no. 3, 2008, pp. 277–285.

Kohandel M., S. Sivaloganathan, G. Tenti, Estimation of the quasi-linear viscoelastic parameters using a genetic algorithm, *Mathematical and Computer Modelling*, vol. 47, no. 3–4, 2008, pp. 266–270.

Koo T. K., J. H. Cohen, Y. Zheng, A mechano-acoustic indentor system for in vivo measurement of nonlinear elastic properties of soft tissue, *Journal of Manipulative and Physiological Therapeutics*, vol. 34, no. 9, 2012, pp. 584–593.

Kwan M. K., T. H. Lin, S. L. Woo, On the viscoelastic properties of the anteromedial bundle of the anterior cruciate ligament, *Journal of Biomechanics*, vol. 26, no. 4–5, 1993, pp. 447–452.

Liu Y., A nonlinear finite element model of soft tissue indentation, *Lecture Notes in Computer Science*, vol. 3078, 2004, pp. 67–76.

Mitchell M., *An Introduction to Genetic Algorithms*, Cambridge, MA: MIT Press, 1999.

Myers B. S., J. H. McElhaney, B. J. Doherty, The viscoelastic responses of the human cervical spine in torsion: experimental limitations of quasi-linear theory, and a method for reducing these effects, *Journal of Biomechanics*, vol. 24, no. 9, 1991, pp. 811–817.

Nigul I., U. Nigul, On algorithms of evaluation of Fung's relaxation function parameters, *Journal of Biomechanics*, vol. 20, no. 4, 1987, pp. 343–352.

Pioletti D. P., L. R. Rakotomanana, On the independence of time and strain effects in the stress relaxation of ligaments and tendons, *Journal of Biomechanics*, vol. 33, no. 12, 2000, pp. 1729–1732.

Pioletti D. P., L. R. Rakotomanana, J. F. Benvenuti, P. F. Leyvraz, Viscoelastic constitutive law in large deformations: application to human knee ligaments and tendons, *Journal of Biomechanics*, vol. 31, no. 8, 1998, pp. 753–757.

Samur E., M. Sedef, C. Basdogan, L. Avtan, O. Duzgun, A robotic indenter for minimally invasive measurement and characterization of soft tissue response, *Medical Image Analysis*, vol. 11, no. 4, 2007, pp. 361–373.

Sangpradit K., H. Liu, P. Dasgupta, K. Althoefer, L. D. Seneviratne, Finite-element modeling of soft tissue rolling indentation, *IEEE Transactions on Biomedical Engineering*, vol. 58, no. 12, 2011, pp. 3319–3327.

Toms S. R., G. J. Dakin, J. E. Lemons, A. W. Eberhardt, Quasi-linear viscoelastic behavior of the human periodontal ligament, *Journal of Biomechanics*, vol. 35, no. 10, 2002, pp. 1411–1415.

Woo S. L., Mechanical properties of tendons and ligaments. I. Quasi-static and nonlinear viscoelastic properties, *Biorheology*, vol. 19, no. 3, 1982, pp. 385–396.

Woo S. L., G. A. Johnson, B. A. Smith, Mathematical modeling of ligaments and tendons, *Journal of Biomechanical Engineering*, vol. 115, no. 4B, 1993, pp. 468–473.

Yamada H., F. G. Evans, *Strength of Biological Materials*. Baltimore, MD: Williams & Wilkins, 1970.

Zhong Y., B. Shirinzadeh, J. Smith, C. Gu, Thermal-mechanical based soft tissue deformation for surgery simulation, *International Journal of Advanced Robotics*, vol. 24, no. 12, 2010, pp. 1719–1739.

Zhong Y., B. Shirinzadeh, J. Smith, C. Gu, Soft tissue deformation with reaction-diffusion process for surgery simulation, *Journal of Visual Languages and Computing*, vol. 23, no. 1, 2012, pp. 1–12.

7

An Investigation on Support Vector Machines and Wavelet Transform in Electroencephalogram Signal Classification

Clodoaldo A. M. Lima, Renata C. B. Madeo, Sarajane Marques Peres, and Marcio Eisencraft

CONTENTS

7.1 Introduction

Time series have been intensively studied in recent years in different areas, such as medicine, the financial market, and climatology. Many techniques have been used to extract the information encoded in time series, aiming to make some estimation (prediction) or to classify a current situation, comparing it with past situations.

Transitory electrical disturbances of the brain are a kind of time series that can cause epileptic seizures. Sometimes seizures may go unnoticed, depending on their presentation, and sometimes they may be confused with other events, such as a stroke, which can also cause falls or migraines. Unfortunately, the occurrence of an epileptic seizure seems unpredictable, and its course of action is not completely understood. Research is needed for a better understanding of the mechanisms causing epileptic disorders. Careful analysis of the records of electroencephalogram (EEG) signals can provide valuable insights into this widespread brain disorder. For example, the detection of epileptiform discharges occurring in the EEG between seizures is an important component in the diagnosis of epilepsy (Subasi 2007).

EEG recordings continue to play an important role in both the diagnosis of neurological diseases and understanding of the psychophysiological processes. In order to extract relevant information from recordings of brain electrical activity, a variety of computerized analysis methods have been developed. Due to the nature of nonstationary EEG signals, this is a challenging task. To improve the performance of the EEG signal analysis, both the temporal and spatial aspects must be considered.

Most of the techniques for extracting features are based on Fast Fourier Transform (FFT), Short-Time Fourier Transform (STFT), and wavelet transform [Discrete Wavelet Transform (DWT)]. Representations based on FFT have been the most commonly applied. This approach is based on earlier observations that the EEG spectrum contains some characteristic waveforms that fall primarily within four bands—delta (<4 Hz), theta (4–8 Hz), alpha (8–13 Hz), and beta (13–30 Hz). However, the FFT suffers from large noise sensitivity. Parametric power spectrum estimation methods such as autoregressive methods reduce the spectral loss problems and give better frequency resolution. Since the EEG signals are nonstationary, the parametric methods are not suitable for frequency decomposition of these signals (Guler et al. 2001; Adeli et al. 2003). Furthermore, the FFT cannot capture the transient features in a given signal, and the time–frequency information is not readily seen in the FFT coefficients. Already, STFT is a time–frequency analysis method; information regarding time and frequency is localized by a uniform time window for all frequency ranges (Xu et al. 2009). Wavelet decomposition overcomes the shortcomings of the STFT for the analysis of nonstationary signals, permitting higher time resolution of higher frequencies, as well as temporal localization of nonstationary signals. In this chapter, the following types of mother wavelet were considered: Haar, Daubechies of order 2 (Db2), and Daubechies of order 4 (Db4). Wavelets of the Daubechies family were emphasized because of their smoothing feature, which, according to Subasi (2007), makes them more adequate to detect changes in EEG signals.

Moreover, many nonlinear classification methods have been proposed. Among them, we can mention artificial neural networks (Revett et al. 2006) and Support Vector Machines (SVMs) for EEG signal binary classification (Liao et al. 2006) and also for the multiclass classification problem (Güler and Übeyli 2007; Übeyli 2008a).

Nowadays, there is much interest in the study of SVM for both regression (Cortes and Vapnik 1995) and classification (Gunn 1998) problems. The SVM approach is based on the minimization of the structural risk (Vapnik 1995), which asserts that the generalization error is delimited by the sum of the training error and a parcel that depends on the Vapnik–Chervonenkis dimension. By minimizing this summation, high generalization performance may be obtained. Besides, the number of free parameters in SVM does not explicitly depend upon the input dimensionality of the problem at hand. Another important feature of the support vector learning approach is that the underlying optimization problems are inherently convex and have no local minima, which comes as the result of applying Mercer's conditions on the characterization of kernels (Cristianini and Shawe-Taylor 2000).

SVM was originally proposed for binary classification. Different methods have been proposed in order to adapt them to be applied in multiclass classification problems. There are two ways to achieve this goal: the first is by combining several binary classifiers [one vs one (OvsO), one vs all (OvsA), error-correcting output coding (ECOC), and minimal output coding (MOC), among others]; the second considers all the classes in the optimization problem formulation. In general, the first form is preferred, because it is simpler to solve several binary classification problems than a single problem with all classes. In our experiments performed with multiple classes, only this first approach is considered.

In previous work (Lima and Coelho 2011), statistical measures derived from DWT coefficients and chaotic dynamic measures, like Lyapunov exponents, from the EEG were investigated with respect to their influence on the performance achieved by the different types of kernel machines. However, all analyses were restricted to the binary discrimination between normal and epileptic conditions (sets A and E). In this chapter, this analysis was extended to binary discrimination between other sets (B, C, and D) and also for multiclass discrimination, considering only a kernel machine and statistical measures derived from DWT.

Thus, in this study, for each mother wavelet, a comparative study involving different values as kernel parameters is presented, in order to indicate which produces greater gain in performance when SVM is adopted to solve EEG signal binary and multiclass classification problems. In the next section, we describe the data analyzed in the experiments and the techniques used to preprocess them; in Section 7.3, we present some theory about the SVM method used for implementing classifiers; in Section 7.4, we describe the different ways of applying binary classifiers to multiclass problems. Finally, in Section 7.5, we show the results obtained and their accuracy rate, and then in Section 7.6, we conclude the study.

7.2 Data Analysis

In order to better organize the explanation about the data used in the experiments, this section is divided into three parts: data selection, data preprocessing, and feature extraction.

7.2.1 Data Selection

In this study, we used the publicly available EEG data described in Elger et al. (2001). The complete data set consists of five sets (denoted A to E, see Figure 7.1), each containing 100 single-channel EEG segments. These segments were selected and cut out from continuous multichannel EEG recordings after visual inspection for artifacts, for example, due to muscle activity or eye movements.

Sets A and B consist of segments taken from surface EEG recordings that were carried out on five healthy volunteers using a standardized electrode placement scheme. Volunteers were relaxed in an awake state with eyes open (set A) and eyes closed (set B). Sets C, D, and E originated from EEG archives of presurgical diagnosis. EEGs from five patients were selected, all of whom had achieved complete seizure control after resection of one of the hippocampal formations, which was therefore correctly diagnosed to be the epileptogenic zone. Segments in set D were recorded from within the epileptogenic zone, and those in set C were recorded from the hippocampal formation of the opposite hemisphere of the brain. While sets C and D contain only activity measured during seizure-free intervals, set E only contains seizure activity. Here, segments were selected from all recording sites exhibiting ictal activity. All EEG signals were recorded with the same 128-channel amplifier system, using an average common reference. The data were digitized at 173.61 Hz using 12-bit resolution. Band-pass filter settings were 0.53–40 Hz (12 dB/oct). In Lima and Coelho (2011), Subasi (2007), and Revett et al. (2006), only two sets (A and E) were used.

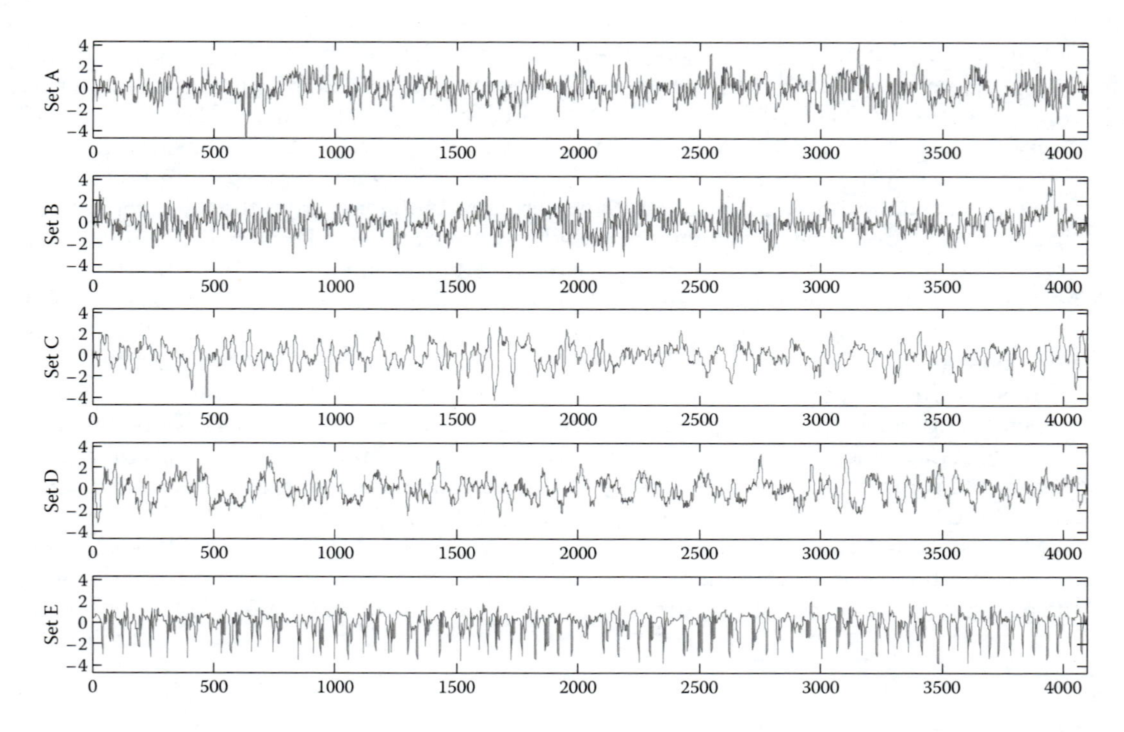

FIGURE 7.1
Examples of EEG signals from sets A, B, C, D, and E.

7.2.2 Data Preprocessing

Wavelet transform is a spectral estimation technique in which any general function can be expressed as an infinite series of wavelets. The basic idea underlying wavelet analysis consists of expressing a signal as a linear combination of a particular set of functions (DWT), obtained by shifting and dilating one single function, called a mother wavelet. The decomposition of the signal leads to a set of coefficients called wavelet coefficients. Therefore, the signal can be reconstructed as a linear combination of the wavelet functions weighted by the wavelet coefficients. In order to obtain an exact reconstruction of the signal, an adequate number of coefficients must be computed (Guler and Übeyli 2005; Übeyli 2008b).

The key feature of wavelets is the time–frequency localization. Time–frequency localization means that most of the energy of the wavelet is restricted to a finite time interval. When compared to STFT, the advantage of time–frequency localization is that wavelet analysis varies the time–frequency aspect ratio, producing good frequency localization at low frequencies (long time windows) and good time localization at high frequencies (short time windows). This produces a segmentation or tiling of the time–frequency plane that is appropriate for most physical signals, especially those of a transient nature. The wavelet technique applied to the EEG signal reveals features related to the transient nature of the signal that are not obvious with the use of Fourier transform (Okandan and Kara 2007).

Selection of a suitable wavelet and number of decomposition levels is important in analysis of signals using the DWT (Subasi 2007). The number of decomposition levels is chosen based on the dominant frequency components of the signal. The levels are chosen such that those parts of the signal that correlate well with the frequencies necessary for classification of the signal are retained in the wavelet coefficients. In the present study, since

TABLE 7.1

Frequencies Corresponding to Different Levels of Decomposition, Considering a Signal with a Sampling Frequency of 173.61 Hz

Decomposed Signal	Frequency Range (Hz)
D1	43.40–86.80
D2	21.70–43.40
D3	10.85–21.70
D4	5.45–10.85
D5	2.72–5.45
A5	0–2.72

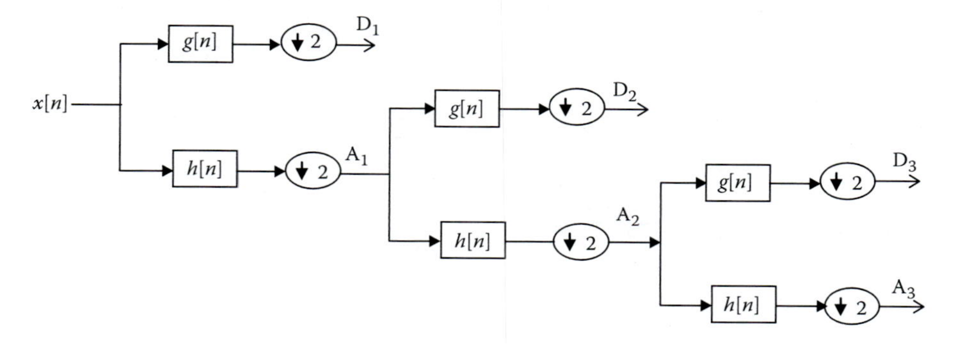

FIGURE 7.2
Subband decomposition of DWT implementation; $g[n]$ is the impulse response of a high-pass filter, and $h[n]$ is the impulse response of a low-pass filter.

the EEG signals do not have any useful frequency components above 30 Hz, the number of decomposition levels was chosen to be five (see Table 7.1). Thus, the EEG signals were decomposed into details D1–D5 and one final approximation, A5 (Subasi 2007). Figure 7.2 shows the procedure of multiresolution decomposition of a signal $x[n]$.

It can be seen from Table 7.1 that the components of A5 are within the delta range (1–4 Hz), D5 decomposition is within the theta range (4–8 Hz), D4 decomposition is within the alpha range (8–13 Hz), and D3 decomposition is within the beta range (13–30 Hz).

We have used a Db4. Its smoothing feature makes it more appropriate to detect changes in EEG signal, as discussed by Subasi (2007) and Revett et al. (2006).

7.2.3 Feature Extraction

In order to reduce the dimensionality of the extracted feature vectors (Güler and Übeyli 2005; Revett et al. 2006; Subasi 2007; Übeyli 2008b), some statistics measures for each sub-band wavelet coefficient were used to generate the input to the SVM:

- Average of wavelet coefficients (W_Avg)
- Standard deviation of wavelet coefficients (W_Std)
- Maximum of wavelet coefficients (W_Max)
- Minimum of wavelet coefficients (W_Min)

- Average power of the wavelet coefficients (W_PAv)
- Mean of the absolute values of the wavelet coefficients (W_MAb)
- Ratio of the absolute mean values of adjacent subbands (W_RAb)
- Combination of all statistical features extracted from the wavelet coefficients (W_ALL)

Lima and Coelho (2011) showed that the features extracted through the standard deviation of the wavelet coefficients are good sources of signal representation for purpose of discrimination. Figure 7.3 shows the distribution of standard deviation extracted over the

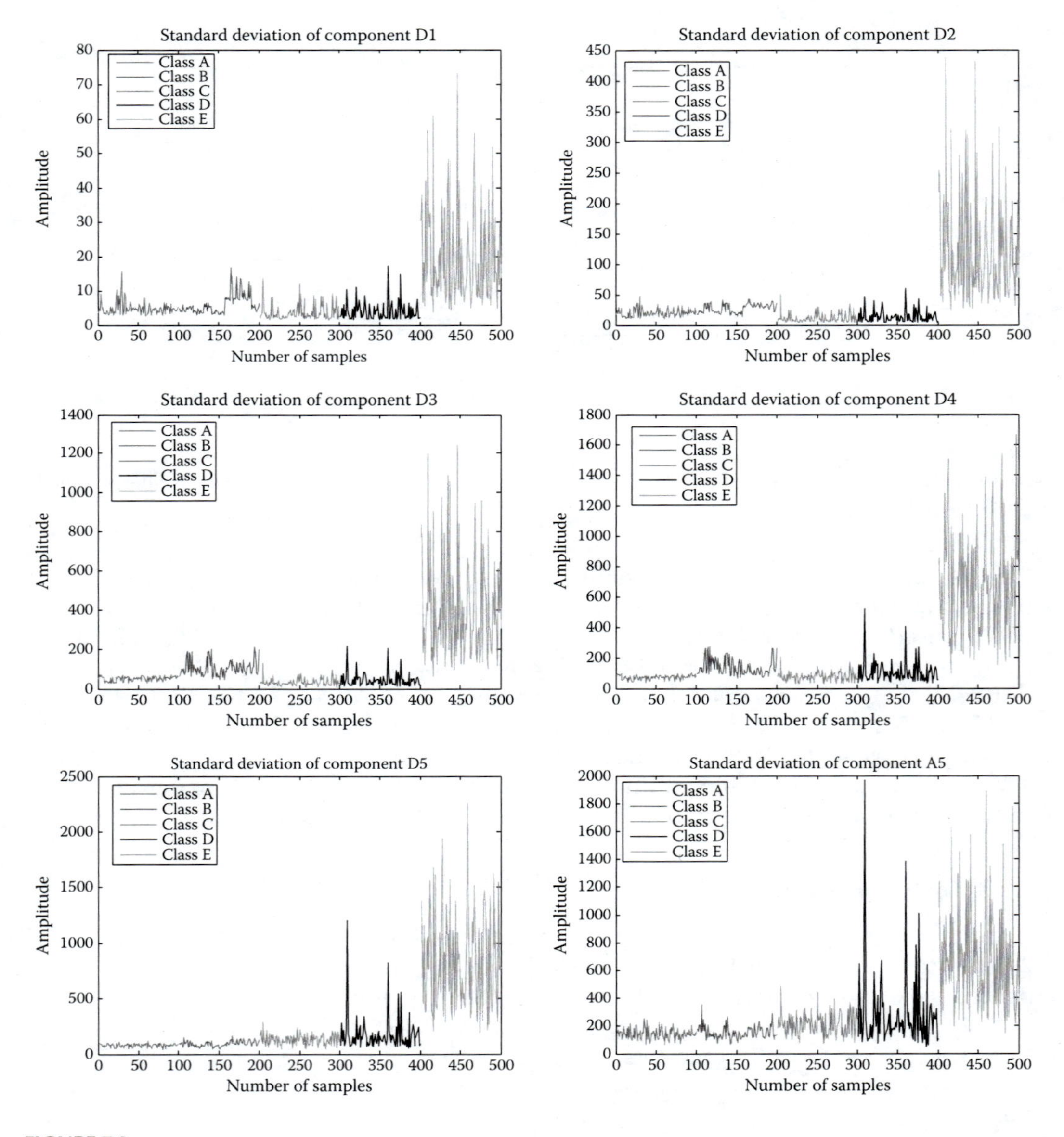

FIGURE 7.3
(See color insert.) Distribution of the standard deviation over the wavelet coefficients for Db4 within the data set.

wavelet coefficients to Db4. Indeed, note that W_Std for class E is quite different from the other classes. This indicates that this class can be easily separated from other classes.

7.3 Support Vector Machines

In this section, we review the use of SVM in classification problems. Let the training set be $\{(x_i, y_i)\}_{i=1}^{N}$, with input $x_i \in \mathfrak{R}^m$ and $y_i \in \{\pm 1\}$. The SVM first accomplishes a mapping $\varphi: \mathfrak{R}^m \rightarrow \mathfrak{R}^n$. Usually, n is much higher than m in such a way that the input vector is mapped into a high-dimensional space. For binary classification, when data are linearly separable, the SVM builds a hyperplane in $\mathfrak{R}^n$ $w^T\phi(x) + b$ in which the boundary between examples belonging to positive and negative classes is maximized. It can be shown that w, for this optimal hyperplane, may be defined as a linear combination of $\phi(x_i)$, that is $w = \sum_{i=1}^{N} \alpha_i y_i \phi(x_i)$.

The generalized optimal separating hyperplane is determined by the vector w, obtained by minimizing the functional:

$$\min_{w,b,\xi} J(w,b,\xi) = \frac{1}{2}(w^T w) + C\sum_{i=1}^{N} \xi_i, \tag{7.1}$$

where C is the usual regularization parameter that controls the trade-off between errors of the SVM on training data and margin maximization. Note that Equation 7.1 is subjected to the constraint

$$y_i[w^T\varphi(x_i) + b] > 1 - \xi_i,\ i = 1,\ldots,N.$$

Then, the resulting quadratic programming (QP) problem may be written as

$$\max_{\alpha} J(\alpha) = \max_{\alpha} \sum_{i=1}^{N} \alpha_i - \frac{1}{2}\sum_{i=1}^{N}\sum_{j=1}^{N} \alpha_i \alpha_j y_i y_j \phi(x_i)^T \phi(x_j)$$

subject to $\sum_{i=1}^{N} \alpha_i y_i = 0$ and $0 \le \alpha_i \le C$, for $i = 1,\ldots,N$. To obtain $\phi(x_i)^T\phi(x_j)$ in the QP problem, we do not need to calculate $\phi(x_i)$ and $\phi(x_j)$ explicitly. Instead, for some φ, we can design a kernel $K(.,.)$ such that $K(x_i, x_j) = \phi(x_i)^T\phi(x_j)$.

Then, the expression of QP problem becomes

$$\max_{\alpha} J(\alpha) = \max_{\alpha} \sum_{i=1}^{N} \alpha_i - \frac{1}{2}\sum_{i=1}^{N}\sum_{j=1}^{N} \alpha_i \alpha_j y_i y_j K(x_i, x_j).$$

For the training samples along the decision boundary, the corresponding α_is are greater than zero, as ascertained by the Kuhn–Tucker theorem (Cristianini and Shawe-Taylor 2000). These samples are known as support vectors. The number of support vectors is generally much smaller than N, being proportional to the generalization error of the classifier (Vapnik 1995). A test vector $x \in \Re^m$ is then assigned to a given class with respect to the expression $f(x) = \text{sign}(w^T \varphi(x) + b) = \text{sign}\left(\sum_{i=1}^{N} \alpha_i y_i K(x, x_i) + b\right)$.

7.4 Description of Multiclass Methods

In this section, we briefly review the strategies for multiclass classification implementations that are applied in this chapter. For a given multiclass problem, *nc* denotes the number of classes C_i, $i = 1,\ldots, nc$. For binary classification, we refer to the two classes as positive and negative; a binary classifier produces an output function that gives values greater than zero for examples from the positive class and values less than zero for examples belonging to the negative class.

7.4.1 OvsA

This method constructs *nc* binary classifiers. The *i*th classifier output function f_i is trained using the examples from C_i as positive and the examples from all other classes as negative. For a new example x, OvsA strategy assigns it to the class with the largest value of f_i.

The advantage of this method is the small number of classifiers, leading to memory saving and faster classification. However, the training phase is very time consuming, because each classifier must be trained with all training samples. Although the original problem can be balanced, this strategy produces an unbalanced data set.

7.4.2 OvsO

This method was introduced by Knerr et al. (1990), and later, the max-wins strategy was proposed by Friedman (1996). This method constructs one binary classifier for every pair of distinct classes. Altogether, $nc(nc-1)/2$ binary classifiers are constructed. The *ij*th binary classifier is trained to take the examples from C_i as positive and the examples from C_j as negative. For a new example x, if the *ij*th classifier says that x is in class C_i, then the vote for class C_i is increased by one. Otherwise, the vote for class C_j is increased by one. After each one of the $nc(nc-1)/2$ binary classifiers makes its vote, OvsO strategy assigns x to the class with the largest number of votes.

This method has the advantage of being easy to understand and to implement, in addition to providing a good performance. The disadvantage of this method is the huge number of binary classifiers, which means a great memory load. Furthermore, there is a great computational load in the test step because each unknown sample must be submitted to all of the $nc(nc-1)/2$ binary classifiers.

7.4.3 ECOC

This method constructs n binary classifiers to distinguish between the *nc* different classes. Each class is given a codeword of length n according to a binary matrix. Each row of this matrix corresponds to a specific class. Table 7.2 shows an example for $nc = 5$ classes and $n = 7$-bit codewords. Each class is given a row of the matrix, and each column is used to

TABLE 7.2
ECOC Design Example

	f_1	f_2	f_3	f_4	f_5	f_6	f_7
Class 1	−1	−1	−1	−1	−1	−1	−1
Class 2	−1	1	1	−1	−1	−1	1
Class 3	−1	1	1	1	1	1	−1
Class 4	1	−1	1	1	1	−1	−1
Class 5	1	1	−1	1	1	−1	1

train a distinct binary classifier. When testing an unseen example, the output codeword from the n classifiers is compared to the given nc codewords, and the one with minimum Hamming distance is considered the class label for that example.

In Allwein et al. (2000), two methods for generating the binary matrix were proposed. The first, called the dense method, generates codewords of length $10\log_2 nc$. Each element is generated randomly from {+1,−1}. A code matrix is built by randomly generating various matrices and choosing the one with the highest minimum Hamming distance among its rows.

The second method, called the sparse method, has codewords of length $15\log_2 nc$. Each entry in the matrix is built randomly to get the value of 0 with probability 1/2 and {−1,+1} with a probability of 1/4 each. As before, the matrix with the highest minimum Hamming distance is chosen among all randomly generated matrices.

7.4.4 Minimal Output Coding

This method is similar to ECOC, but it aims at finding the minimal output for a given number of classes. Thus, up to $\log_2 nc$ classifiers are trained, each of them aiming to separate a different combination of classes.

Considering $nc = 5$, we defined a coding with $n = 3$, generating a codebook represented by a matrix, where the rows represent the different classes and the columns indicate the results of the binary classifiers. The multiclass classifier output code for a pattern is a combination of targets of these three classifiers. In this example, the codebook in Table 7.3 would be generated.

When testing an unseen example, the output codeword from three classifiers is compared to the given five codewords, and the one with minimum Hamming distance is considered the class label for that example. The advantage of this method is the small number of classifiers. However, the training phase is very time consuming, because each classifier must be trained with all training samples. Although the original problem can be balanced, this strategy produces an unbalanced data set. Moreover, in general, the results obtained with this strategy may be lower when compared with the ones obtained with the other strategies.

TABLE 7.3
MOC Design Example

	f_1	f_2	f_3
Class 1	−1	−1	−1
Class 2	−1	−1	1
Class 3	−1	1	−1
Class 4	−1	1	1
Class 5	1	−1	−1

7.5 Experimental Results

In what follows, we provide details on how the experiments were conducted and present a comparative analysis of the SVM with respect to the Radial Basis Function (RBF) and Exponential Radial Basis Function (ERBF) kernel parameter variation for each mother wavelet.

$$\text{RBF kernel: } K(x_i, x_j) = \exp\left(-\frac{\|x_i - x_j\|^2}{2\sigma^2}\right) \qquad \text{ERBF kernel: } K(x_i, x_j) = \exp\left(-\frac{\|x_i - x_j\|}{2\sigma^2}\right)$$

Note that σ^2 determines the width in the case of the RBF and ERBF kernel.

7.5.1 Configuration of the Experiments

In the experiments accomplished, as mentioned before, we have assessed the performance of the SVM, for each mother wavelet, regarding the variation of kernel parameter, keeping the value of the regularization parameter C constant at 100. This value was achieved after some preliminary experiments and agrees with the fact that SVM models with low values of C tend in general to achieve better performance than those with high values of this parameter. Although we know that there are several rules of thumb to select values for the variance parameter (Cherkassky and Ma 2004) of the RBF kernel, we have opted to set the values of σ as 2^i, $i = -10, -9, \ldots, 14, 15$. For each of the 26 values in this range, a tenfold cross-validation was performed to better gauge the average performance of the models.

7.5.1.1 Experiment 1

The objective of this experiment was analyzing the impact of using different types of wavelets in the task of distinguishing between combinations of classes, namely, A versus B, C versus D, and A and B versus C and D. Tables 7.4, 7.5, and 7.6 present error rates obtained by SVM classification based on feature vectors extracted from wavelets Haar, Db2, and Db4, respectively.

Based on these results, it is also possible to note that, in most cases, best performance was achieved with feature vectors based on standard deviation and mean of the absolute values of wavelet coefficients. It is worth noting that the problem of separating samples of classes C and D has shown to be more difficult than separating samples of other classes.

Also, there was no significant difference between results obtained with RBF and ERBF kernel functions. The values 1, 2, and 4 for kernel parameter were also associated with best results in this experiment.

7.5.1.2 Experiment 2

The goal of this experiment was to verify the behavior of SVM in the task of multiclass classification (A, B, C, D, and E) of the EEG data set described in the beginning of this chapter. For performing the multiclass classification, OvsO, OvsA, MOC, and ECOC methods were used to separate samples of each class. These methods were already described in Section 7.4 and are very popular for multiclass classification. As in previous experiments, tenfold cross validation was used. Again, parameter C was fixed at 100, and kernel parameter varied in the range 2^i for $i \in [-10:15]$. Table 7.7 presents the average and standard

TABLE 7.4

Error Rate in Classification Using Feature Vectors Based on Haar Wavelet Features (Multiclass Classification)

		A versus B		C versus D		AB versus CD	
Kernel Type	Feature Vector	Kernel Parameter	Error Rate	Kernel Parameter	Error Rate	Kernel Parameter	Error Rate
RBF	W_Avg	4	0.365 ± 0.0280	32	0.455 ± 0.0345	4	0.285 ± 0.0308
	W_Std	4	0.060 ± 0.0194	0.5	0.240 ± 0.0287	2	0.014 ± 0.0042
	W_Max	16	0.115 ± 0.0248	16	0.345 ± 0.0263	4	0.080 ± 0.0186
	W_Min	4	0.115 ± 0.0183	4	0.320 ± 0.0429	4	0.072 ± 0.0120
	W_PAv	1	0.070 ± 0.0186	0.125	0.305 ± 0.0369	1	0.012 ± 0.0042
	W_MAb	1	0.050 ± 0.0167	0.5	0.240 ± 0.0420	2	0.017 ± 0.0053
	W_RAb	2	0.410 ± 0.0297	0.5	0.430 ± 0.0374	0.5	0.322 ± 0.0192
	W_ALL	4	0.100 ± 0.0258	8	0.300 ± 0.0333	16	0.037 ± 0.0077
ERBF	W_Avg	0.5	0.345 ± 0.0229	16	0.500 ± 0.0333	4	0.290 ± 0.0251
	W_Std	8	0.055 ± 0.0189	2	0.225 ± 0.0250	4	0.015 ± 0.0056
	W_Max	16	0.145 ± 0.0263	8	0.355 ± 0.0398	8	0.074 ± 0.0177
	W_Min	8	0.115 ± 0.0224	8	0.330 ± 0.0318	1	0.060 ± 0.0100
	W_PAv	4	0.070 ± 0.0200	0.5	0.335 ± 0.0395	1; 2	0.017 ± 0.0099
	W_MAb	4	0.050 ± 0.0149	2	0.250 ± 0.0279	2; 4	0.012 ± 0.0042
	W_RAb	8	0.410 ± 0.0379	0.25	0.415 ± 0.0373	8	0.300 ± 0.0171
	W_ALL	16	0.095 ± 0.0217	8	0.280 ± 0.0318	2; 4	0.032 ± 0.0084

TABLE 7.5

Error Rate in Classification Using Feature Vectors Based on Db2 Wavelet Features (Multiclass Classification)

		A versus B		C versus D		AB versus CD	
Kernel Type	Feature Vector	Kernel Parameter	Error Rate	Kernel Parameter	Error Rate	Kernel Parameter	Error Rate
RBF	W_Avg	4	0.345 ± 0.0229	32	0.455 ± 0.0293	2	0.290 ± 0.0292
	W_Std	2; 4; 8	0.050 ± 0.1972	0.5	0.225 ± 0.0227	2	0.007 ± 0.0038
	W_Max	4; 8; 16	0.155 ± 0.0273	16	0.415 ± 0.0365	4	0.075 ± 0.0110
	W_Min	16	0.160 ± 0.0256	8	0.305 ± 0.0293	4	0.082 ± 0.0106
	W_PAv	2	0.050 ± 0.0183	0.25	0.270 ± 0.0403	2	0.025 ± 0.0075
	W_MAb	2	0.050 ± 0.0197	0.5	0.255 ± 0.0321	4	0.017 ± 0.0065
	W_RAb	0.25	0.370 ± 0.0291	0.5	0.435 ± 0.0334	0.25	0.417 ± 0.0264
	W_ALL	32	0.080 ± 0.0170	8	0.265 ± 0.0289	8	0.030 ± 0.0082
ERBF	W_Avg	8	0.375 ± 0.0214	16	0.505 ± 0.0311	2	0.262 ± 0.0277
	W_Std	2; 4	0.045 ± 0.0203	2	0.220 ± 0.0281	0.5; 1; 2; 4	0.010 ± 0.0049
	W_Max	8	0.155 ± 0.0293	0.5	0.380 ± 0.0374	1	0.070 ± 0.0012
	W_Min	8	0.145 ± 0.0302	16	0.285 ± 0.0317	4	0.064 ± 0.0106
	W_PAv	1; 2; 4	0.055 ± 0.0138	0.5	0.230 ± 0.0327	2	0.004 ± 0.0038
	W_MAb	1	0.045 ± 0.0138	2	0.190 ± 0.0323	1; 2; 4	0.010 ± 0.0055
	W_RAb	2	0.390 ± 0.0306	4	0.435 ± 0.0342	4	0.402 ± 0.0206
	W_ALL	4; 8	0.085 ± 0.0183	4	0.250 ± 0.0279	2; 8	0.022 ± 0.0069

TABLE 7.6

Error Rate in Classification Using Feature Vectors Based on Db4 Wavelet Features (Multiclass Classification)

		A versus B		C versus D		AB versus CD	
Kernel Type	Feature Vector	Kernel Parameter	Error Rate	Kernel Parameter	Error Rate	Kernel Parameter	Error Rate
RBF	W_Avg	8	0.370 ± 0.0351	2	0.350 ± 0.0337	4	0.310 ± 0.0287
	W_Std	8	0.045 ± 0.0138	0.5	0.245 ± 0.0345	2	0.010 ± 0.0055
	W_Max	16	0.150 ± 0.0269	0.25	0.355 ± 0.0241	4	0.070 ± 0.0133
	W_Min	4	0.140 ± 0.0125	2	0.310 ± 0.0314	4	0.057 ± 0.0065
	W_PAv	2	0.050 ± 0.0129	0.25	0.255 ± 0.0450	2	0.025 ± 0.0053
	W_MAb	8	0.055 ± 0.0138	1	0.220 ± 0.0359	1	0.007 ± 0.0038
	W_RAb	2	0.405 ± 0.0320	0.03125	0.450 ± 0.0316	0.5	0.362 ± 0.0218
	W_ALL	32	0.080 ± 0.0200	2	0.275 ± 0.0214	16; 32	0.027 ± 0.0058
ERBF	W_Avg	0.5	0.370 ± 0.0396	2	0.475 ± 0.0271	1	0.265 ± 0.0261
	W_Std	2; 4	0.055 ± 0.0157	2	0.250 ± 0.0387	0.5; 1; 2; 4	0.012 ± 0.0056
	W_Max	16	0.145 ± 0.0189	0.25	0.375 ± 0.0423	8	0.062 ± 0.0107
	W_Min	1	0.170 ± 0.0170	4	0.270 ± 0.0238	1	0.052 ± 0.0120
	W_PAv	2; 4	0.055 ± 0.0189	2	0.215 ± 0.0350	8	0.010 ± 0.0041
	W_MAb	2; 4	0.050 ± 0.0149	4	0.200 ± 0.0269	8	0.007 ± 0.0038
	W_RAb	16	0.405 ± 0.0369	2	0.450 ± 0.0236	4	0.402 ± 0.0206
	W_ALL	4; 8	0.090 ± 0.0233	2	0.245 ± 0.0376	2; 8	0.022 ± 0.0069

deviation of the error rate for the best result obtained, with the respective value for the kernel parameter.

The results illustrated in Table 7.7 demonstrate sharply the influence of the type of feature vector on the classifier performance. It is possible to perceive that the standard deviation is an important feature to separate classes, improving significantly the classifier performance. This result has been already perceived for binary classification problems, such as that illustrated in the previously described experiment.

The following figures show the classifier performance for each combination strategy, that is, OvsO, OvsA, MOC, and ECOC. Figure 7.4 shows how accuracy for each class, using standard deviation as a feature, varies according to the value of the kernel parameter. The bars in these graphs represent the variance in accuracy (1 standard deviation from the mean) for each value of σ considered. Observe that there is a range of values for the kernel parameter that produces really interesting results. However, there is no value for the kernel parameter, within this range, that produces 100% correct classification for classes C and D. Classes A, B, and E are easier to classify correctly. Among employed strategies, OvsO and OvsA were those that produced the best results in terms of classification error.

Figure 7.5a and b presents the error rate for each strategy of multiclass classification according to the value of the kernel parameter. Due to space limitations, we have decided to present only the results obtained using features extracted through the standard deviation of the wavelet coefficients and without preprocessing (non-preprocessed signals served as input to the classifier). It is worth noting that there is a kernel parameter that produces similar error rates among all combination methods.

Tables 7.8, 7.9, 7.10, and 7.11 present the confusion matrix for some results illustrated in Table 7.5. Specifically, these matrices correspond to the best result achieved by SVM using

TABLE 7.7

Results Obtained with SVM Using RBF Kernel

Feature Vector	Wavelet Type	OvsO		OvsA		MOC		ECOC	
		σ^2	Error Rate	σ^2	Error Rate	σ^2	Error Rate	σ^2	Error Rate
Raw data	——-	32	0.258 ± 0.01749	32	0.302 ± 0.01698	16	0.34 ± 0.024766	16	0.386 ± 0.02191
W_Avg	Haar	2	0.540 ± 0.0208	1	0.554 ± 0.0238	0.5	0.646 ± 0.0438	1	0.546 ± 0.0169
	Db2	0.5	0.510 ± 0.0193	1	0.522 ± 0.0081	1	0.534 ± 0.0115	0.5	0.536 ± 0.0151
	Db4	1	0.533 ± 0.0133	2	0.515 ± 0.0358	1	0.595 ± 0.0468	1	0.465 ± 0.0353
W_Std	Haar	1	0.162 ± 0.0145	0.125	0.156 ± 0.0136	0.25	0.196 ± 0.0240	0.25	0.164 ± 0.0136
	Db2	0.5	0.150 ± 0.0108	0.25	0.144 ± 0.0145	0.125	0.150 ± 0.01238	0.125	0.160 ± 0.0133
	Db4	2	0.150 ± 0.0166	0.25	0.142 ± 0.0141	1	0.152 ± 0.0201	0.25	0.152 ± 0.0121

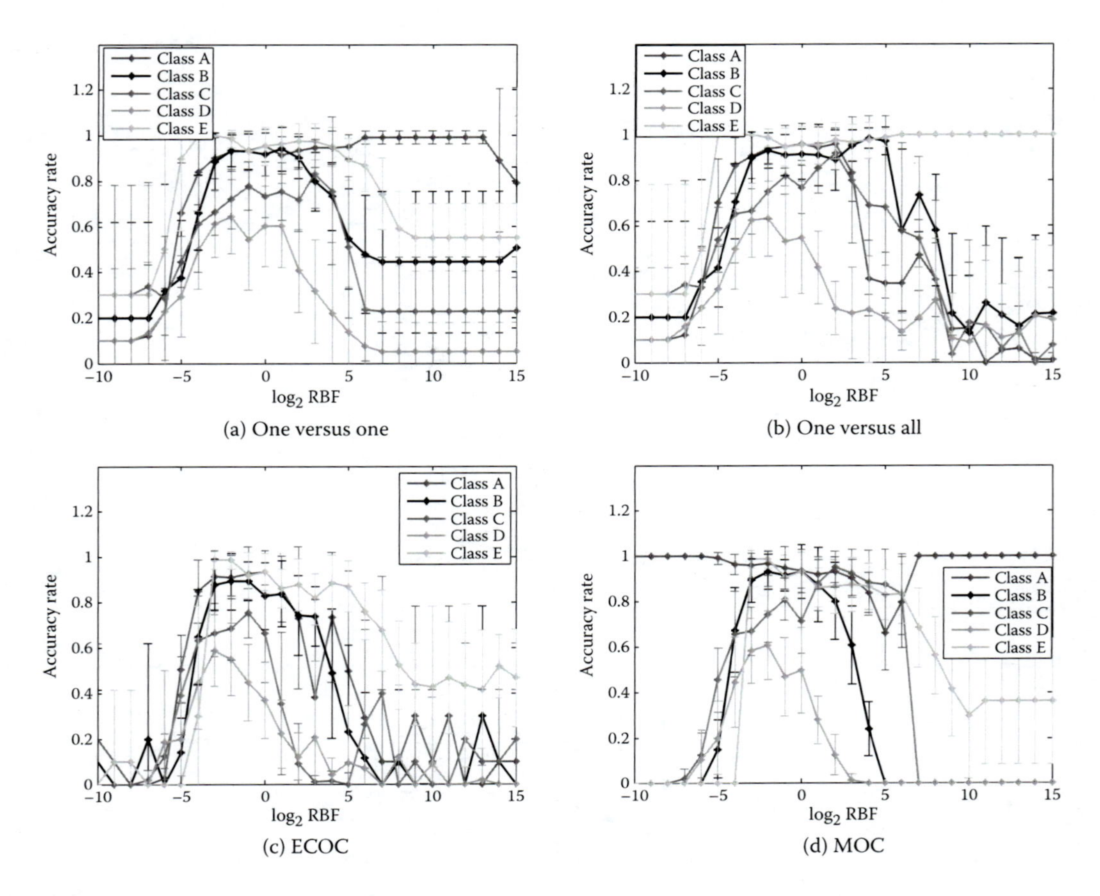

FIGURE 7.4
(See color insert.) Accuracy rate by class for different strategies of multiclass classification using the RBF kernel and feature vectors extracted through the standard deviation of the wavelet coefficients.

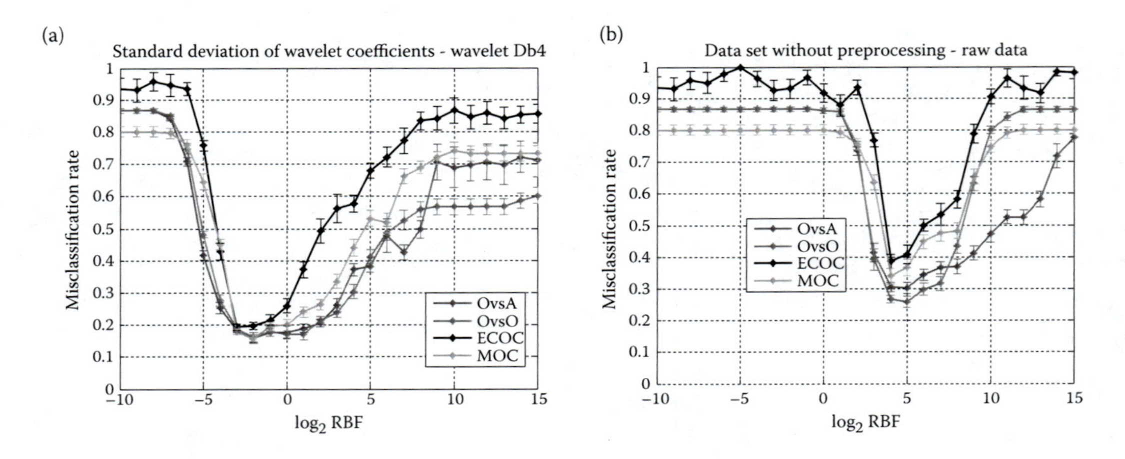

FIGURE 7.5
(See color insert.) Error rate for different strategies for multiclass classification using the RBF kernel and feature vectors extracted through the standard deviation of the wavelet coefficients.

TABLE 7.8

Confusion Matrix for SVM with RBF Kernel ($\sigma = 0.125$), Using Standard Deviation over Db2 Wavelet Coefficients and ECOC Strategy for Multiclass Classification

	Predicted Class				
Actual Class	**Class A**	**Class B**	**Class C**	**Class D**	**Class E**
Class A	0.891 ± 0.0428	0.043 ± 0.0228	0.000 ± 0.0100	0.010 ± 0.0100	0.009 ± 0.0091
Class B	0.056 ± 0.0270	0.916 ± 0.0267	0.000 ± 0.0000	0.000 ± 0.0000	0.000 ± 0.0000
Class C	0.018 ± 0.0182	0.000 ± 0.0000	0.779 ± 0.0490	0.145 ± 0.0490	0.017 ± 0.0116
Class D	0.009 ± 0.0091	0.000 ± 0.0000	0.245 ± 0.0581	0.616 ± 0.0581	0.100 ± 0.0310
Class E	0.000 ± 0.0000	0.000 ± 0.0000	0.000 ± 0.0000	0.012 ± 0.0125	0.987 ± 0.0125

TABLE 7.9

Confusion Matrix for SVM with RBF Kernel ($\sigma = 0.125$), Using Standard Deviation over Db2 Wavelet Coefficients and MOC Strategy for Multiclass Classification

	Predicted Class				
Actual Class	**Class A**	**Class B**	**Class C**	**Class D**	**Class E**
Class A	0.901 ± 0.0389	0.062 ± 0.0294	0.016 ± 0.0167	0.010 ± 0.0100	0.009 ± 0.0091
Class B	0.069 ± 0.0270	0.930 ± 0.0270	0.000 ± 0.0000	0.000 ± 0.0000	0.000 ± 0.0000
Class C	0.018 ± 0.0182	0.000 ± 0.0000	0.795 ± 0.0428	0.154 ± 0.0487	0.031 ± 0.0169
Class D	0.009 ± 0.0091	0.000 ± 0.0000	0.266 ± 0.0418	0.624 ± 0.0569	0.100 ± 0.0310
Class E	0.000 ± 0.0000	0.000 ± 0.0000	0.000 ± 0.0000	0.012 ± 0.0125	0.987 ± 0.0125

TABLE 7.10

Confusion Matrix for SVM with RBF Kernel ($\sigma = 0.125$), Using Standard Deviation over Db2 Wavelet Coefficients and One-versus-All Strategy for Multiclass Classification

	Predicted Class				
Actual Class	**Class A**	**Class B**	**Class C**	**Class D**	**Class E**
Class A	0.956 ± 0.0228	0.043 ± 0.0228	0.000 ± 0.0000	0.000 ± 0.0000	0.000 ± 0.0000
Class B	0.045 ± 0.0158	0.933 ± 0.0195	0.000 ± 0.0000	0.021 ± 0.0152	0.000 ± 0.0000
Class C	0.009 ± 0.0091	0.023 ± 0.0161	0.794 ± 0.0228	0.155 ± 0.0323	0.017 ± 0.0116
Class D	0.009 ± 0.0091	0.022 ± 0.0151	0.258 ± 0.0440	0.641 ± 0.0520	0.068 ± 0.0247
Class E	0.000 ± 0.0000	0.012 ± 0.0125	0.000 ± 0.0000	0.025 ± 0.0167	0.965 ± 0.0267

TABLE 7.11

Confusion Matrix for SVM with RBF Kernel ($\sigma = 0.125$), Using Standard Deviation over Db2 Wavelet Coefficients and One-versus-One Strategy for Multiclass Classification

	Predicted Class				
Actual Class	**Class A**	**Class B**	**Class C**	**Class D**	**Class E**
Class A	0.946 ± 0.0229	0.042 ± 0.0225	0.011 ± 0.0111	0.000 ± 0.0000	0.000 ± 0.0000
Class B	0.055 ± 0.0217	0.930 ± 0.0224	0.000 ± 0.0000	0.014 ± 0.0143	0.000 ± 0.0000
Class C	0.018 ± 0.0182	0.000 ± 0.0000	0.807 ± 0.0406	0.164 ± 0.0395	0.010 ± 0.0100
Class D	0.000 ± 0.0000	0.000 ± 0.0000	0.295 ± 0.0477	0.638 ± 0.0539	0.065 ± 0.0194
Class E	0.000 ± 0.0000	0.012 ± 0.0125	0.007 ± 0.0077	0.041 ± 0.0174	0.938 ± 0.0281

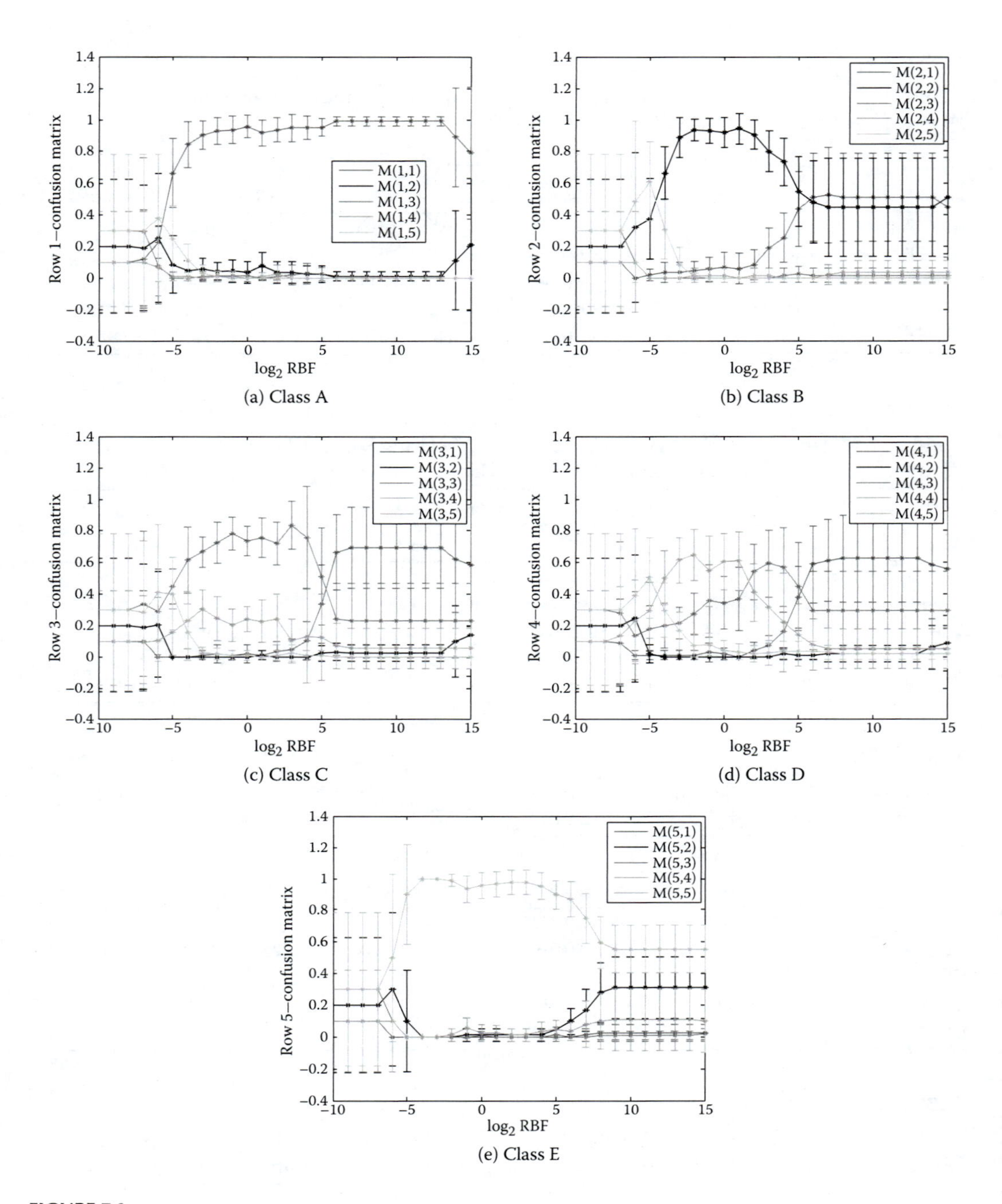

FIGURE 7.6
(See color insert.) Effect of varying the kernel parameter in the classification of each class using standard deviation of wavelet coefficients and one-versus-one strategy.

the RBF kernel and standard deviation of Db2 wavelet coefficients as a feature. Analyzing the elements of the diagonal, which corresponds to correct classification for each class, it is possible to observe that the precision for classes A, B, and E reaches more than 90%; for classes C and D, precision reaches no more than 80%.

Figure 7.6 shows the effect of varying the kernel parameter in a confusion matrix, where M(i, j) means the rate in which the class i (A, B, C, D, E) is labeled to each class j by the classifier. For instance, M(1, 1) means that the correct class is A, and the classifier assigns the label A; M(1, 2) means that the correct class is A, and the classifier assigns the label B; M(1, 3) means that the correct class is A, and the classifier assigns the label C; M(1, 3) means that the correct class is A, and the classifier assigns the label D; and M(1, 5) means that the correct class is A, and the classifier assigns the label E. Already in Figure 7.6b, it is possible to note that there is a region in which the classifier presents good performance. As already mentioned, analyzing Figure 7.6d and e, it is noted that there is no value for the kernel parameter that produces 100% correct classification for elements of classes C and D.

Generally, the methods for combining binary classifiers presented equivalent performances with each other, with OvsO and OvsA presenting slightly better results.

7.6 Conclusion

In this chapter, SVM is used to classify a set of EEG signals, using different feature vectors as input for different values to the kernel parameter. Overall, it has presented good performances for both binary and multiclass classification of EEG signals. Since the raw data are vectors containing all characteristics available in the data set with dimensions equal to 4096, it is important to consider more compact feature vectors. In this work, the feature vectors extracted using statistics over the series were processed using wavelet transform, producing vectors with dimensions equal to 6 and, even then, providing classifications with very low error rates. Therefore, these vectors provide a much more compact representation, containing characteristics that allow the classifier to achieve good results: Experiments have shown that vectors resulting from a feature extraction process using wavelets, especially features extracted through the standard deviation and mean of the absolute values, have allowed the classifier to obtain superior results, for binary and multiclass classification problems, using a much more compact data representation compared to the raw data. The reduction of the input vector dimension affects positively the computational cost to perform classification.

The results obtained show that the strategies of OvsO and OvsA all present competitive performance with each other. The ECOC strategy was the one that presented the worst performance compared to all the others, though. For the multiclass strategy addressed, the parameters of the kernel function had a significant impact on the classifier performance. However, the choice of the type of wavelet seems not to be so relevant.

From the experiments, it can be observed that the SVM using a feature vector based on the standard deviation of wavelet coefficients, with a kernel parameter equal to 2 and 4 for binary and 0.5 or 0.125 for multiclass classification problems, is a good starting point for classifying segments of the EEG data set analyzed in this chapter and potentially for other EEG data sets.

The ability to distinguish among different subsets of EEG signals presented by SVM using feature extraction based on statistics over wavelet coefficients extracted from the original data set can be useful for analyzing EEG signals associated with a specific disease, such as epilepsy or schizophrenia.

Acknowledgments

The authors wish to thank Andrzejak et al. (2001) for making the data publicly available at http://www.meb.uni-bonn.de/epile-ptologie/science/physik/eegdata.html. The second author thanks the São Paulo Research Foundation (FAPESP) for its support under grant #2011/04608-8. The first and last authors thank the National Council for Technological and Scientific Development (CNPq) for its support under grant #308816/2012-9 and #303926/2010-4.

References

Adeli, H.; Zhou, Z.; Dadmehr, N. Analysis of EEG records in an epileptic patient using wavelet transform, *Journal of Neuroscience Methods*, 123, pp. 69–87, 2003.

Allwein, E.; Shapire, R.; Singer, Y. Reducing multiclass to binary: A unifying approach for margin classifiers. *Journal of Machine Learning Research*, 1, pp. 113–141, 2000.

Andrzejak R. G.; Lehnertz K.; Mormann F.; Rieke C.; David P.; Elger C. E. Indications of nonlinear deterministic and finite dimensional structures in time series of brain electrical activity: dependence on recording region and brain state. *Physical Review E, Statistical Nonlinear and Soft Matter Physics*, 64, 2001.

Cherkassky, V.; Ma, Y. Practical selection of SVM parameters and noise estimation for SVM regression. *Neural Networks*, 17, pp. 113–126, 2004.

Cortes, C.; Vapnik, V. Support-vector networks. *Machine Learning*, 20, pp. 273–207, 1995.

Cristianini, N.; Shawe-Taylor, J. *An Introduction to Support Vector Machines*. London: Cambridge University Press, 2000.

Elger, C. E.; Rieke, C.; Mormann, F.; Lehnertz, K.; David, P.; Andrzejak, R. G. Indications of nonlinear deterministic and finite dimensional structures in time series of brain electrical activity: Dependence on recording region and brain state. *Physical Review E*, 64, n. 6, 8 pp., 2001, Article ID 061907.

Friedman, J. H. Another approach to polychotomous classification, Technical Report, Stanford University, Dept. of Statistics, 1996.

Guler, I.; Kiymik, M. K.; Akin, M.; Alkan, A. AR spectral analysis of EEG signals by using maximum likelihood estimation. *Computers in Biology and Medicine*, 31, pp. 441–450, 2001.

Güler, I.; Übeyli, E. D. Adaptive neuro-fuzzy inference system for classification of EEG signals using wavelet coefficients. *Journal of Neuroscience Methods*, 148, pp. 113–121, 2005.

Güler, I; Übeyli, E. D. Multiclass support vector machines for EEG-signals classification. *IEEE Transactions on Information Technology in Biomedicine*, 11, pp. 117–126, 2007.

Gunn, S. Support vector machine for classification and regression. Image Speech and Intelligent Systems Group, Technical Report ISIS-1-98, University of Southampton, Nov. 1998.

Knerr, S.; Personnaz, L.; Dreyfus, G. Single-layer learning revisited: A stepwise procedure for building and training a neural network. In F. Fogelman Soulié and J. Hérault (eds.), *Neurocomputing: Algorithms, Architectures and Applications*, volume F68 of NATO ASI Series, pp. 41–50. Berlin, Heidelberg: Springer, 1990.

Liao, X.; Yin, Y.; Li, C.; Yao, D. Application of SVM framework for classification of single trial EEG. In: J. Wang et al. (eds.), *Advances in Neural Networks – ISNN 2006, Lecture Notes in Computer Science*, vol. 3973, Berlin, Heidelberg: Springer, pp. 548–553, 2006.

Lima, C. A. M.; Coelho, A. L. V. Kernel machines for epilepsy diagnosis via EEG signal classification: A comparative study. *Artificial Intelligence in Medicine*, 53, pp. 83–95, 2011.

Okandan, M.; Kara, S. Atrial fibrillation classification with artificial neural networks. *Pattern Recognition*, 40, pp. 2967–2973, 2007.

Revett, K.; Jahankhani, P.; Kodogiannis, V. EEG Signal classification using wavelet feature extraction and neural networks. *IEEE John Vincent Atanasoff International Symposium*, pp. 120–124, 2006.

Subasi, A. EEG signal classification using wavelet feature extraction and a mixture of expert model. *Journal of Expert Systems With Applications*, 32, pp. 1084–1093, 2007.

Übeyli, E. D. Analysis of EEG signals by combining eigenvector methods and multiclass support vector machines. *Computers in Biology and Medicine*, 38, pp. 14–22, 2008a.

Übeyli, E. D. Wavelet/mixture of experts network structure for EEG signals classification. *Journal of Expert Systems With Applications*, 34, pp. 1954–1962, 2008b.

Vapnik, V. N. *Statistical Learning Theory*. New York: Wiley, 1995.

Xu, Qi; Zhou, H.; Wang, Y.; Huang, J. Fuzzy support vector machine for classification of EEG signals using wavelet-based features. *Medical Engineering and Physics*, 31, pp. 858–865, 2009.

8

Building Naïve Bayes Classifiers with High-Dimensional and Small-Sized Data Sets

Lin Liu and Jiuyong Li

CONTENTS

8.1 Introduction

Naïve Bayes classifiers (NBCs) [1] are well known for their simplicity, clear probabilistic semantics, as well as the naïve assumption of conditional independence among features, and they have shown surprisingly good performance in many domains [2].

With the rapid increase of high-dimensional data, such as microarray gene expression data, and the needs for analyzing them, many new classification algorithms and extensions to existing methods, including those to NBCs, have been proposed [3–5].

By its nature, an NBC should perform well with high-dimensional data, because the conditional independence assumption reduces a high-dimensional task to multiple single-dimensional tasks. The main objection to or concern with using NBCs, particularly in gene expression data classification, however, is also this assumption of conditional independence among features, because in practice, features are often related. With gene expression data, it is a consensus that genes (features) normally interact with each other.

However, research on the optimality of NBCs has provided us some theoretical support about the suitability of NBCs in classifying data whose features may be related. In the work of Kuncheva [6], it is found that feature selection, if used with an NBC, may remove dependent features. More interestingly, some explanations have been found for the good performance of NBCs in certain cases when features are dependent. For example, Zhang [7] shows that it is not just the individual dependence relationships that affect the classification; more importantly, it is the combined dependence that affects the classification. When the dependence relationships of all features are put together, they may cancel each other out; hence, their overall effect may no longer exist.

Such findings have encouraged us to investigate the performance of NBCs in high-dimensional data classification, particularly gene expression data classification, where features are believed to be highly dependent. Our aim is to evaluate how NBCs perform with real-world data comparing to the more complicated classifiers and, importantly, to study the factors affecting the performance of NBCs and the assessment of their performance.

An NBC can be used with discrete or continuous features. With continues features, an NBC typically assumes a normal distribution for the values of a feature given a class. John and Langley [2] propose to use kernel functions to approximate the distributions, because features are often not normally distributed. The other approach is to discretize the data before using them to train an NBC [8]. Bouckaert [9] has done a good comparison of the normal approximation, the kernel approximation, and a supervised discretization method, with data sets from the University of California, Irvine (UCI), Machine Learning Repository [10]. The result shows that no method can perform universally better. With gene expression data classification, data are continuous, so depending on the methods for dealing with data, different classification outcomes with NBCs may be obtained. Among the very little work on data approximation or discretization methods for microarray gene expression data, Lustgarten et al. [11] have found that a supervised discretization method improves the classification outcomes of support vector machines (SVMs), random forests (RFs), and NBCs. In this chapter, we will conduct further work to investigate the effect of both supervised and unsupervised discretization methods for NBCs.

With high-dimensional and small-sized data sets, feature selection has to play an important role in achieving good classification results [12,13]. Some classifiers, such as SVMs, have embedded feature selection processes. Some, like NBCs, do not. In this case, we often conduct feature selection before applying the classification algorithm. In this study, we will investigate the effect of feature selection on the classification outcomes of NBCs.

Cross-validation is the most common approach to assessing the performance of a classifier [14]. As supervised data filtering (e.g., feature selection and supervised discretization) relies on class information of samples, cross-validation must be carried out on a classifier itself, as well as on the supervised data filtering (if used). That is, for each run of a cross-validation, both data filtering and classifier training must be done with the portion of samples for training only, without accessing the testing samples for this cross-validation run. However, in microarray gene expression data classification literature, we often see work that firstly applies data filtering to the whole training data set [15,16], and then the filtered data are used for classifier training in each run of a cross-validation. Since the filtered data in this case are obtained with the knowledge of all samples, the cross-validation process has peeked through the portion of testing samples. The results of the cross-validation may still be valid if the purpose is to compare the performance of different classifiers. However, they give us overoptimistic estimations of the performance of individual classification methods. In this chapter, we will address the issue of misusing cross-validation, to reassure us of the results of performance assessment of NBCs when feature selection and/or supervised discretization are used.

The chapter is not aimed at proposing new classification methods. Rather, it is focused on the use of NBCs in classifying high-dimensional and small-sized data sets, to study the external factors that impact on the outcome of NBCs and the result of performance assessment of classifiers, and to provide guidelines for using NBCs with high-dimensional and small-sized data sets. Additionally with our experimental analysis of the consequences of misusing cross-validation, we would like to show the limitations of existing classification methods, in the context of classifying microarray gene expression data.

The rest of the chapter is organized as follows. Section 8.2 introduces NBCs and the discretization and feature selection methods used in this study. Section 8.3 presents a series of experiments on NBCs, with three parts (Sections 8.3.3, 8.3.4, and 8.3.5). Section 8.4 discusses the experiment results and the issues revealed by the results. Finally, Section 8.5 concludes the chapter.

8.2 Naïve Bayes Classifier

8.2.1 Definition

Considering a set of K feature variables, $A_1, A_2,\ldots,A_K$, and the class variable C with M possible values (class labels), given a sample, $\boldsymbol{a} = (a_1, a_2,\ldots,a_K)$, the observed values of the K features, a classifier is a function, f, that maps the sample to a class, that is, $c = f(\boldsymbol{a})$, where $c \in C$. The process of classification comprises two steps: (1) learning a classifier from the training data set, $\boldsymbol{D} = \{\boldsymbol{a}_1, \boldsymbol{a}_2,\ldots,\boldsymbol{a}_N\}$, and (2) using the classifier to predicate the class of a given data sample.

An Naïve Bayes Classifier (NBC) is a simple Bayesian classification method [1]. For a given sample, a Bayesian classifier calculates the posterior probability, $P(C = c \mid A_1 = a_1, A_2 = a_2,\ldots,A_K = a_K)$, for all $c \in C$, and assigns the class with the highest probability (called c_{BC} below) to the given sample:

$$c_{BC} = \arg\max_{c \in C} P(C = c \mid A_1 = a_1, A_2 = a_2, \ldots, A_K = a_K) \tag{8.1}$$

Using Bayes rule, we have

$$P(C = c \mid A_1 = a_1, A_2 = a_2, \ldots, A_K = a_K) = \frac{P(A_1 = a_1, A_2 = a_2, \ldots, A_K = a_K \mid C = c)P(C = c)}{P(A_1 = a_1, A_2 = a_2, \ldots, A_K = a_K)} \tag{8.2}$$

As the denominator in Equation 8.2 is the same for all classes, Equation 8.1 becomes

$$c_{BC} = \arg\max_{c \in C} P(A_1 = a_1, A_2 = a_2, \ldots, A_K = a_K \mid C = c)P(C = c) \tag{8.3}$$

To use the Bayesian classifier defined in Equation 8.3, firstly we need estimate the two probability items in the equation from a training set, which is infeasible in most cases. Taking the simple case when all features and the class variable have binary values, the number of parameters to be estimated is $2(2^K - 1)$. This means that, for example, with only 20 features, we would need a training set with over 2 million samples to see each instance once, not to mention getting an accurate estimation, for which we need to see each instance many times.

This is how the NBC comes in, which assumes that all features are independent given a class (the naïve assumption); thus we have

$$P(A_1 = a_1, A_2 = a_2, \ldots, A_K = a_K \mid C = c)P(C = c) = P(C = c)\prod_{i=1}^{K} P(A_i = a_i \mid C = c) \tag{8.4}$$

Then from Equation 8.3, we have the NBC:

$$c_{NBC} = \arg\max_{c \in C} P(C = c)\prod_{i=1}^{K} P(A_i = a_i \mid C = c) \tag{8.5}$$

Comparing with the classifier in Equation 8.3, the number of parameters to be estimated is only $2(K - 1)$ with binary features and the class variable, which dramatically reduces the complexity.

Additionally, from the definition of an NBC, under the naïve assumption, NBCs are optimal in terms of the posterior probability of the class. However, the assumption is only a sufficient condition of an NBC being optimal, instead of a sufficient and necessary condition. So it is possible to get optimal classification even if the assumption is not satisfied. This gives an explanation to why NBCs can perform well when features are not totally independent [6].

8.2.2 Parameter Estimation

8.2.2.1 Discrete Features

When features have discrete values, the maximum likelihood estimation of the parameters in Equation 8.5 is simply based on counting the cases [1,9]:

$$P(C = c) = \frac{1 + N_c}{M + N} \tag{8.6}$$

where N is the total number of samples in the training set $\boldsymbol{D} = \{\boldsymbol{a}_1, \boldsymbol{a}_2, \ldots, \boldsymbol{a}_N\}$, and N_c is the number of samples whose class labels are c. The numbers 1 and M (number of different classes) are a result of assuming a Dirichlet prior for the class variable. Using 1 and M also avoids the zero-count problem of the estimation. Similarly,

$$P(A_i = a_i \mid C = c) = \frac{1 + N_{a_i,c}}{|A_i| + N_c} \tag{8.7}$$

where $N_{a_i,c}$ is the number of samples of which the value of A_i is a_i and the class label is c, and $|A_i|$ is the number of possible values of the feature variable A_i.

8.2.2.2 Continuous Features and Data Discretization

With continuous features, the most commonly used method is the so-called normal method [9], which assumes a normal distribution for each feature given a class. Parameters of the normal distributions are estimated from the training data. That is, for feature A, assume $A|c \sim \mathcal{N}(\mu_c, \sigma_c^2)$, and

$$\mu_c = \frac{1}{N_c}\sum_{j=1}^{N_c} a_{j,c} \tag{8.8}$$

$$\sigma_c = \sqrt{\frac{1}{N_c - 1}\sum_{j=1}^{N_c}\left(a_{j,c}^2 - \mu_c^2\right)} \tag{8.9}$$

where $a_{j,c}$ is a value of feature A in a sample whose class labels are c.

The kernel method [2] attempts to improve situations where features are not normally distributed. With the method, $P(A|c)$ is approximated using the sum of a set of normal kernels, that is, $A|c \sim \sum_{j=1}^{N_c} N(\mu_c, \sigma_c'^2)$, where $\sigma_c' = 1/N_c$.

Another method for continuous features is to discretize the data and then use Equations 8.6 and 8.7 to learn the parameters with the discretized data. This can be useful for dealing with gene expression data, which often are not normally distributed [13]; thus, the normal method is not a good choice. Furthermore, normal kernels around sample points are not a good approximation of gene expression data either. This can be seen from Figure 8.1, where the histograms (drawn with the Free Statistics Software [17]) are shown for the values of randomly selected features from the data sets used in our experiments (see Table 8.1). From the histograms, we cannot see the bell-shaped peaks around sample values.

A discretization method can be supervised or unsupervised [14]. The former uses class information of samples, while the latter does not. In this chapter, we will look at both types of discretization methods, to see how they affect the classification outcome of NBCs.

In our experiments in Section 8.3, for supervised discretization, we use the entropy-based method with the minimum description length (MDL) stopping criterion by Fayyad and Irani [18]. It is regarded as one of the best supervised discretization methods [14]. With unsupervised discretization, we use a basic binary binning method, which is based on the common assumption that a gene has two states: on (expressed) and off (not expressed). We use 1 to represent the on state, 0 the off state, and the sample mean of a gene (i.e., the mean of the gene's expression values across all samples) as the dividing point for the on and off states.

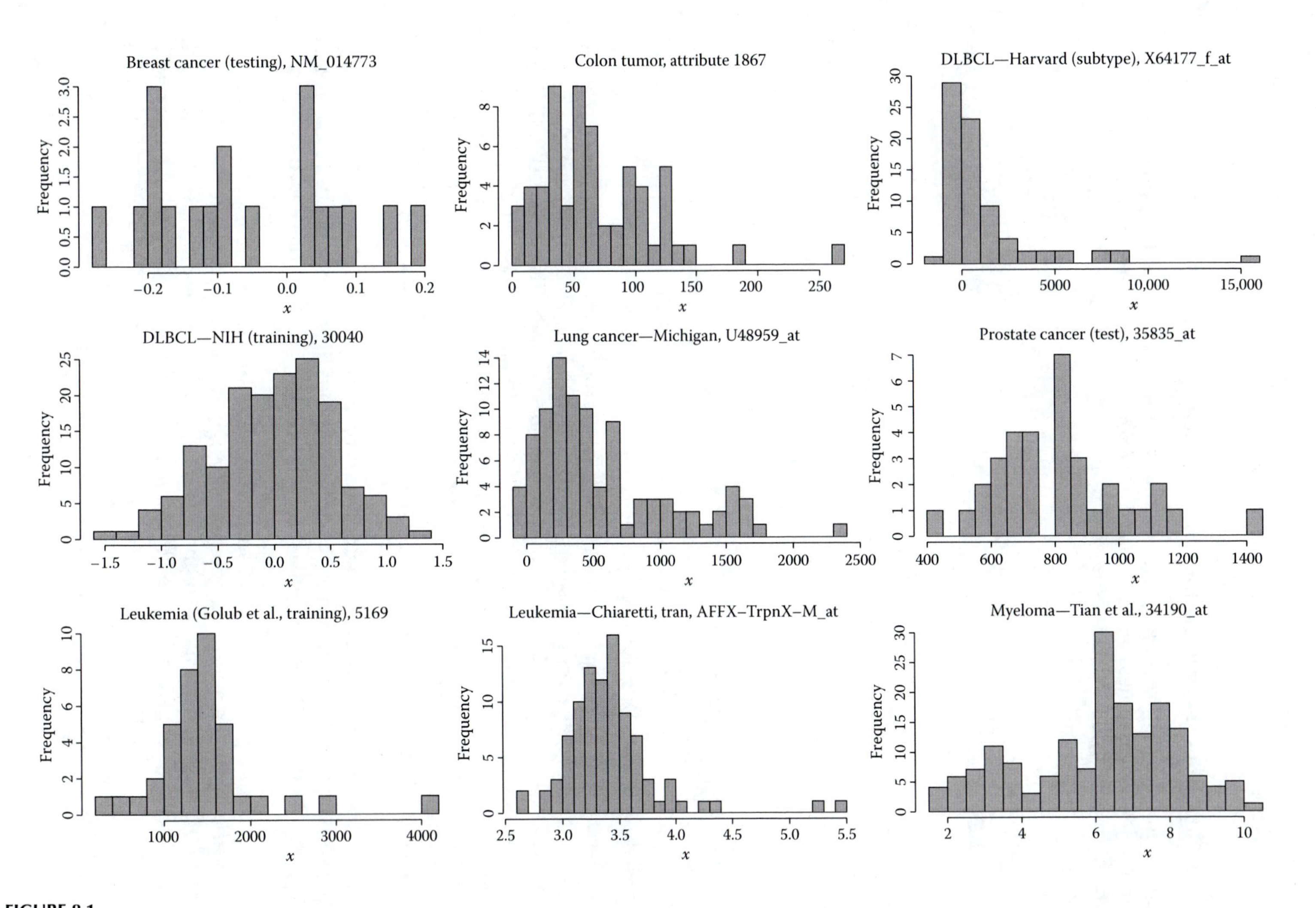

FIGURE 8.1
Histograms of features from the experiment data sets.

TABLE 8.1
Data Sets Used in Experiments

No.	Name	No. of Features	Class Labels and No. of Samples in a Class
1	Breast cancer (training)	24,481	34 relapse, 44 nonrelapse
2	Breast cancer (testing)	24,481	12 relapse, 7 nonrelapse
3	Colon tumor	2000	40 negative, 22 positive
4	DLBCL—Harvard (subtype)	6817	58 BLBCL, 19 FL
5	DLBCL—Harvard (outcome)	7129	32 cured, 26 fatal
6	DLBCL—NIH (training)	7399	72 alive, 88 dead
7	DLBCL—NIH (testing)	7399	30 alive, 50 dead
8	DLBCL—Stanford	4026	24 geminal, 23 activated
9	Leukemia—MLL (Armstrong et al. 2002)	12,582	20 ALL, 20 AML
10	Leukemia—Chiaretti et al. 2004 (MDR)	12,625	101 negative, 24 positive
11	Leukemia—Chiaretti et al. 2004 (origin)	12,625	95 B cell, 33 T cell
12	Leukemia—Chiaretti et al. 2004 (relapse)	12,625	35 false, 65 true
13	Leukemia—Chiaretti et al. 2004 (translocation)	12,625	26 true, 67 false
14	Leukemia—ALLAML, Golub et al. 1999 (training)	7129	27 ALL, 11 AML
15	Leukemia—ALLAML, Golub et al. 1999 (testing)	7129	20 ALL, 14 AML
16	Leukemia—subtype, St. Jude (training)	12,558	28 T-ALL, 52 TEL-AML1
17	Leukemia—subtype St. Jude (testing)	12,558	15 T-ALL, 27 TEL-AML1
18	Lung cancer—Harvard, Bhattacharjee et al. 2001 (aden vs. normal)	12,600	17 normal, 139 aden
19	Lung cancer—Harvard, Bhattacharjee et al. 2001 (normal vs. non-SCLC)	12,600	17 normal, 180 non-SCLC
20	Lung cancer—Harvard, Gordon et al. 2002 (training)	12,533	16 MPM, 16 ADCA
21	Lung cancer—Harvard, Gordon et al. 2002 (testing)	12,533	15 MPM, 134 ADCA
22	Lung cancer—Michigan	7129	86 tumor, 10 normal
23	Lung cancer—Ontario	2880	24 relapse, 15 nonrelapse
24	Myeloma—Tian et al. 2003	12,625	36 false, 137 true
25	Central nervous system	7129	21 class 1, 39 class 2
26	Prostate cancer (training)	12,600	50 normal, 52 tumor
27	Prostate cancer (testing)	12,600	9 normal, 25 tumor
28	Prostate cancer (outcome)	12,600	8 relapse, 13 nonrelapse

Source: Data sets are from I. B. Jeffery. Higgins Laboratory People: Comparison and Evaluation of Microarray Feature Selection Methods, [http://www.bioinf.ucd.ie/people/ian/], last accessed on February 15, 2013; and Kent Ridge Bio-medical Dataset. [http://datam.i2r.a-star.edu.sg/datasets/krbd/], last accessed on February 15, 2013.

8.2.3 Feature Selection

As described previously, due to the conditional independence assumption, the number of parameters has been reduced dramatically. However, with gene expression data, the number of features is often tens of thousands, while the number of samples is, at most, a few hundreds. An NBC trained using such data will inevitably have poor performance. On the other hand, it is believed that most genes (features) are irrelevant to the classification of samples. Therefore, a common way of alleviating the high-dimensionality problem is to do feature selection to remove irrelevant genes before using the data to train a classifier.

There are two categories of feature selection methods: wrapper and filter approaches. The former takes the classification algorithm to be used to select features, and the latter is independent of the classifier. As we are interested in how feature selection influences the outcome of NBCs, in this chapter, we only look at the filter approach. We use the information gain ratio method [13,14], a popular and widely implemented feature selection method.

8.3 Experiments

8.3.1 Overview

In this section, we carry out three sets of experiments with a large number of microarray gene expression data sets, and the experiments are described in three parts:

- Section 8.3.3: experiments with NBCs only. The aim of the experiments is to find out if the individual use or combined use of feature selection and data approximation or discretization can improve the results of NBCs.
- Section 8.3.4: experiments on comparing NBCs with the commonly used classifiers, including SVMs, decision trees (DTs), and RF, to see if NBCs can perform as effectively as these more sophisticated classifiers, when proper data filtering methods (feature selection and/or supervised discretization) are applied.
- Section 8.3.5: experiments of investigating the impact of misuse of cross-validation on the outcome of performance evaluation. In the misuse case, firstly a filtering method is applied to the whole data set (both training and testing portions of the data set), and cross-validation is done with the classifier only.

All experiments are conducted using the machine learning tool, the Waikato Environment for Knowledge Analysis (Weka) [19], including using the classifiers, feature selection, and data approximation and discretization methods implemented in the tool. To test classification results, the tenfold cross-validation is used. With the experiments in Sections 8.3.3 and 8.3.4, cross-validations are used correctly, that is, a data filtering method is always applied to the training data portion in each run of a cross-validation.

Two performance measures are used: classification accuracy and area under the receiver operating characteristic (ROC) curve (AUC). The latter is considered to be a good measure for comparing performance of classifiers [20]. For each single experiment with a specific method and data set, cross-validation is done 100 times with Weka Experimenter. Then we calculate and use the average performance measures in the analysis.

8.3.2 Data Sets

In total, 28 data sets are used in our experiments. The numbering, name, number of features (genes or probes), and number of samples of each data set are listed in Table 8.1.

The data sets are sourced from the Huggins Laboratory Web site (sets 10 to 13 and 24) [21] and the Kent Ridge data repository (the other 23 data sets) [22]. The five Huggins data matrices (data sets) are transposed to follow the data format of Weka. The 23 Kent Ridge data sets are already in the format required by Weka.

Details of the Huggins data sets, including their original source, can be found in the work of Jeffery et al. [23]. Among the five data sets, sets 10 to 13 are extracted from the leukemia data of Chiaretti et al. [24] based on the class labels. Specifically, the samples in [24] are divided into four groups in terms of (1) origin: B cell versus T cell; (2) status of multidrug resistance (MDR): with versus without; (3) patient outcome: relapse versus nonrelapse; and (4) the status of chromosome translocation: with versus without. The four groups of samples give us data sets 10 to 13, respectively.

Some original data sets contain two independent sets one for training, and one for testing. In this case, we treat the training and the testing sets as two separate data sets. For example, data sets 1 and 2 are generated from the breast cancer data set at the Kent Ridge data repository.

Similar to the leukemia data of Chiaretti et al. [24], some data sets from the Kent Ridge data repository have multiple ways of classification. In this case, we consider each way of the classification as a separate data set. For example, samples in the original DLBCL—Harvard data set are labeled in terms of the types (DLBCS vs. FL) and patient outcome (cured vs. fatal). So from this data set, we create two data sets (4 and 5) for our experiments.

Some data sets comprise multiclass samples. In this case, we create one or more binary class data sets from the original data set. For example, with the leukemia—MLL data, only two subtypes are considered, that is, we include training samples labeled ALL or AML in data set 9.

8.3.3 Experiments with NBCs Only

8.3.3.1 Data Approximation and Discretization

Corresponding to the different approaches to dealing with continuous features, we consider four methods of using NBCs: NBC with normal data approximation (NB-N), NBC with kernel data approximation (NB-K), NBC with supervised discretization (NB-SD), and NBC with unsupervised discretization (NB-UD) (Table 8.2). The first two are the NBCs with the normal and kernel data approximations, respectively, and the last two, the NBCs with the supervised and unsupervised discretization methods described in Section 8.2, respectively.

Cross-validation results of the four methods with the 28 data sets are shown in Figures 8.2, 8.3, and 8.4. In each figure, the result of NB-N is shown in dashed lines, as the baseline for comparisons. The top diagram of each figure shows classification accuracies, and the bottom diagram of each figure shows the AUCs.

Comparing NB-N with NB-K (Figure 8.2), none of the two methods has obvious advantage over or is consistently better than the other. However, NB-SD outperforms NB-N for nearly all data sets (Figure 8.3). With most data sets, NB-UD in general is slightly better than NB-N (Figure 8.4). The average accuracies of NB-N, NB-K, NB-SD, and NB-UD are

TABLE 8.2

Four Methods of Using NBCs

Name	Method
NB-N	NBC with normal data approximation
NB-K	NBC with kernel data approximation
NB-SD	NBC with supervised discretization
NB-UD	NBC with unsupervised discretization

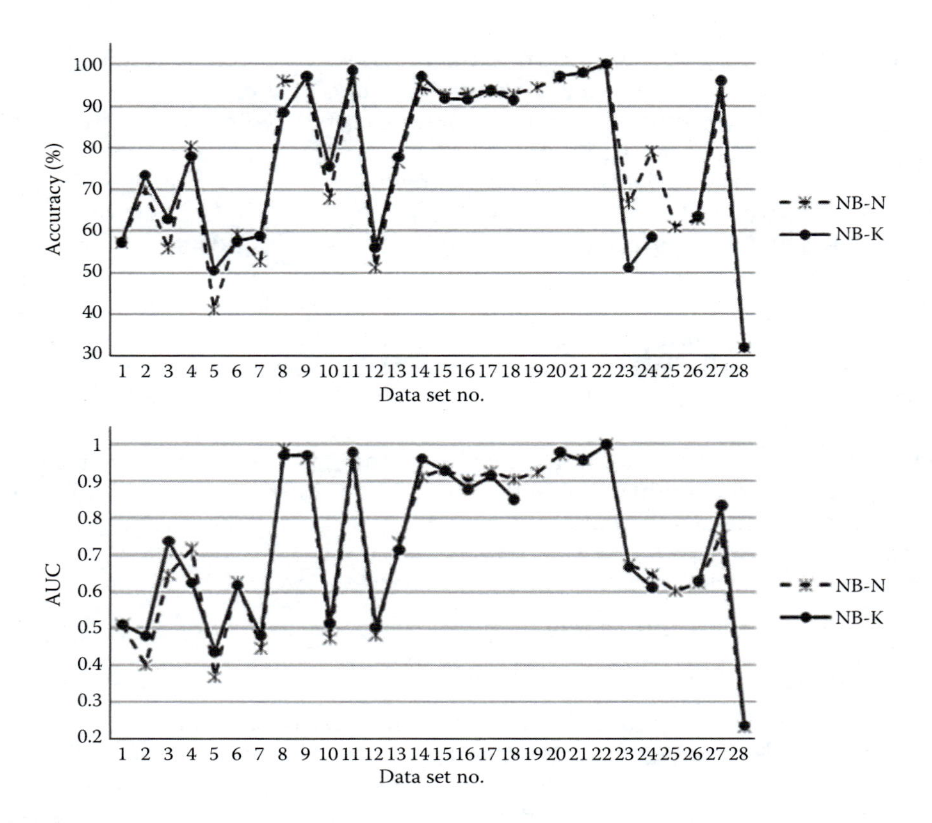

FIGURE 8.2
Accuracies and AUCs of NB-N and NB-K.

76.78%, 76.69%, 81.32%, and 77.06%, respectively (Table 8.3), which are very close to each other, except for that of NB-SD.

The average AUCs of NB-N, NB-K, NB-SD, and NB-UD are 0.724, 0.73, 0.79, and 0.76, respectively (Table 8.3), again indicating that NB-SD outperforms the other three methods, and the AUC of NB-UD is higher than those of NB-N and NB-K. Therefore, we can say that discretizing microarray data is more effective than approximating the data with normal distributions or kernel functions, and supervised discretization is more effective than unsupervised discretization.

Interestingly, a similar trend is shown for the four methods across the 28 data sets. Referring to Figures 8.2 through 8.4, we see that almost all methods achieve their best performance with the same data sets and their worst performance with the same data sets.

8.3.3.2 Feature Selection

To investigate the effects of feature selection, we filter each data set with the information gain ratio method. In terms of the number of features to use, there is no common

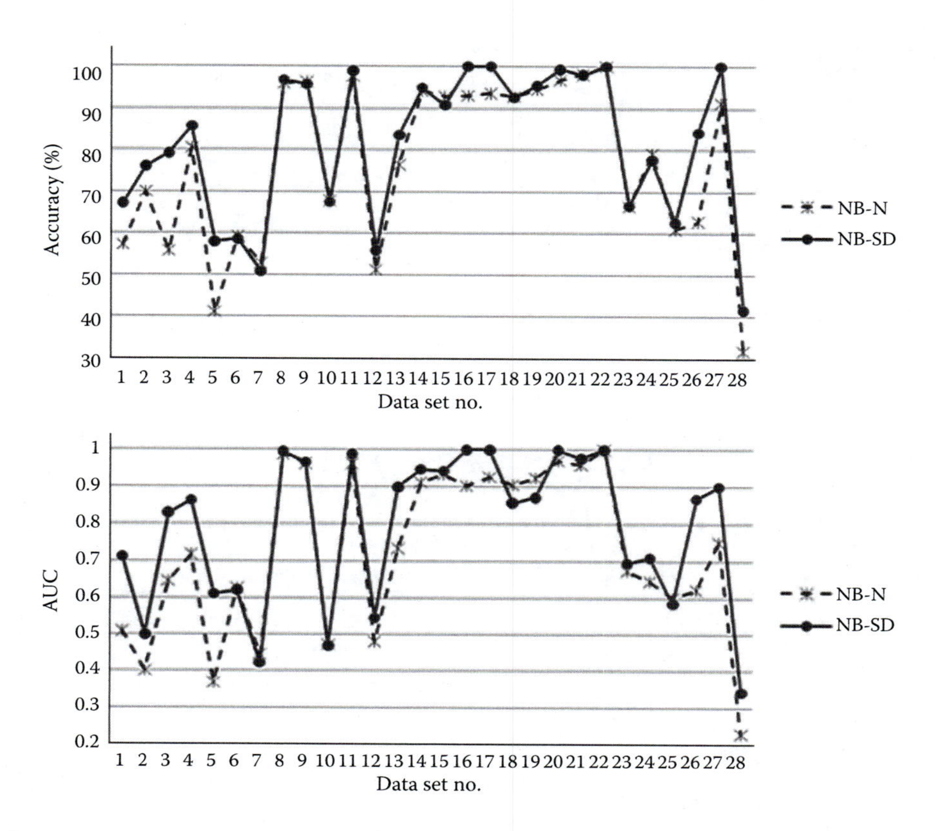

FIGURE 8.3
Accuracies and AUCs of NB-N and NB-SD.

agreement from the literature. From the work of Symons and Nieselt [25], 100 to 500 seems to be a preferable range for the number of features. We experimented with 50, 100, 150, 200, and 250 features with the data sets in Table 8.1 and found that the numbers 50 or 250 gave worse outcomes than the others. So in the experiments presented in the next section, the number of features to be selected is set to 100.

To find out the individual effects of feature selection, at this stage of experimenting, we consider two methods: NB-N with feature-selected data (NB-N-FS) and NB-K with feature-selected data (NB-K-FS). NB-N is again used as the baseline method for comparison.

From Figures 8.5 and 8.6 we see that feature selection improves the performance of NB-N and NB-K. For most data sets, NB-N-FS and NB-K-FS have higher or similar accuracies and AUCs comparing to NB-N. The average accuracies of NB-N-FS and NB-K-FS are 81.3% and 81.05% respectively, and the average AUCs are both 0.8 (recall that the average accuracy and AUC of NB-N are 76.78% and 0.724, respectively). Furthermore, NB-N and NB-K perform equally better on feature-selected data. Comparing to NB-SD (81.32% accuracy and 0.79 AUC), we see that feature selection and discretization have similar impact on the performance of NBCs.

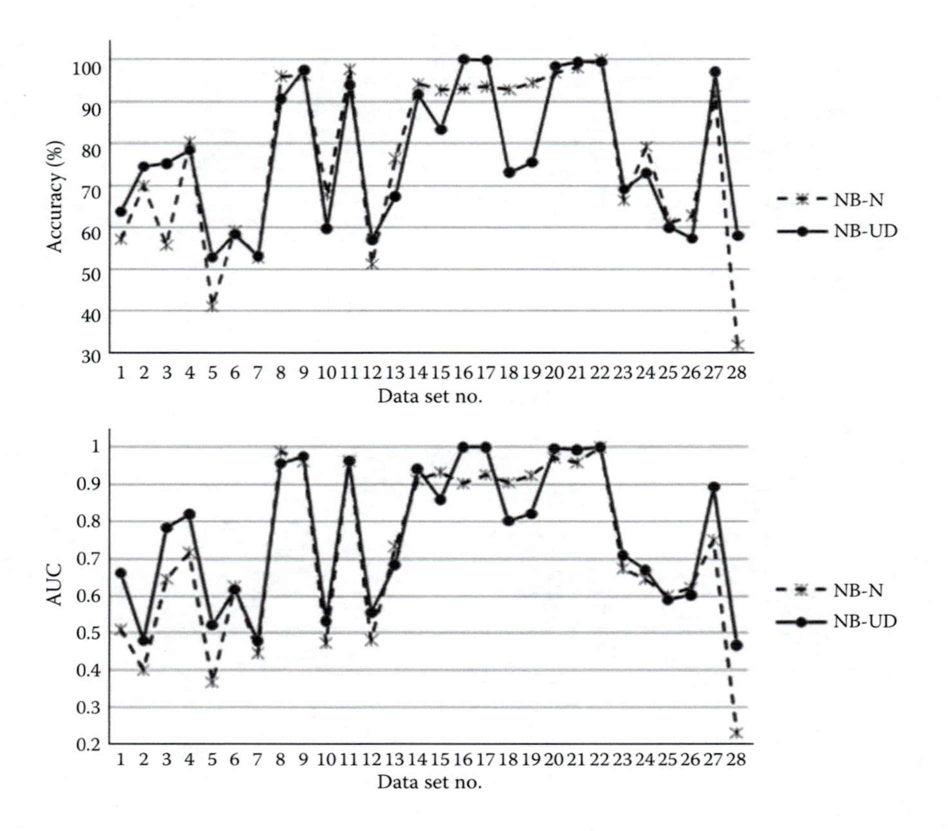

FIGURE 8.4
Accuracies and AUCs of NB-N and NB-UD.

TABLE 8.3
Summary of the Performance of NBCs

Method	Average Accuracy	Average AUC
NB-N	76.78%	0.72
NB-K	76.69%	0.73
NB-SD	81.32%	0.79
NB-UD	77.06%	0.76
NB-N-FS	81.30%	0.80
NB-K-FS	81.05%	0.80
NB-FS-SD	82.25%	0.81

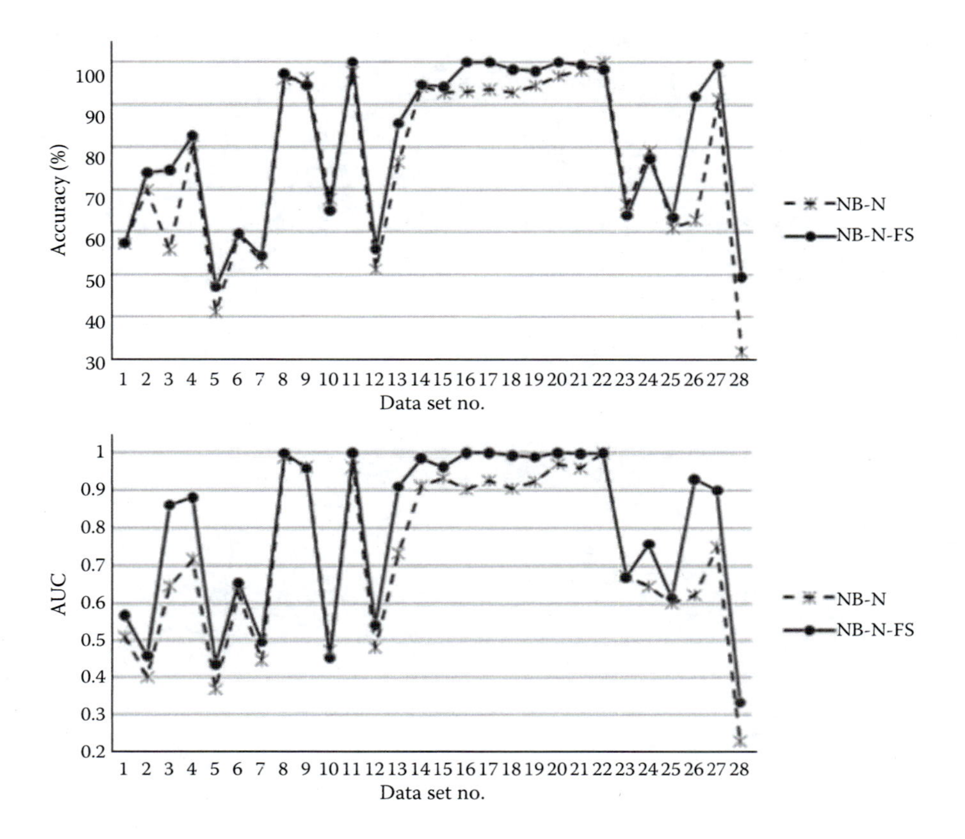

FIGURE 8.5
Accuracies and AUCs of NB-N and NB-N-FS.

8.3.3.3 Feature Selection and Data Discretization

The above results show that discretization or feature selection improves the outcomes of NBCs, and supervised discretization is better than the unsupervised method. To study the combined effect of feature selection and discretization, we do experiments with NBC with both feature-selected and discretized data (NB-FS-SD), the NBC with data filtered by the feature selection followed by the supervised discretization.

From Figure 8.7, the outcome of combining feature selection and discretization is better than NB-N for all data sets. The average accuracy and AUC of NB-FS-SD are 82.25% and 0.81, respectively. However, comparing this with NB-N-FS and NB-SD, the improvement is not significant. This is due to the fact that supervised discretization makes use of the correlation of features and class, which is similar to the basic idea of feature selection.

Table 8.3 summarizes the results of the experiments with NBCs. We see that data approximation methods [-N (normal) or -K (normal kernel)] are not as effective as supervised discretization (-SD), and feature selection (-FS) improves classification outcome. We can combine feature selection with supervised discretization to obtain slightly better outcomes.

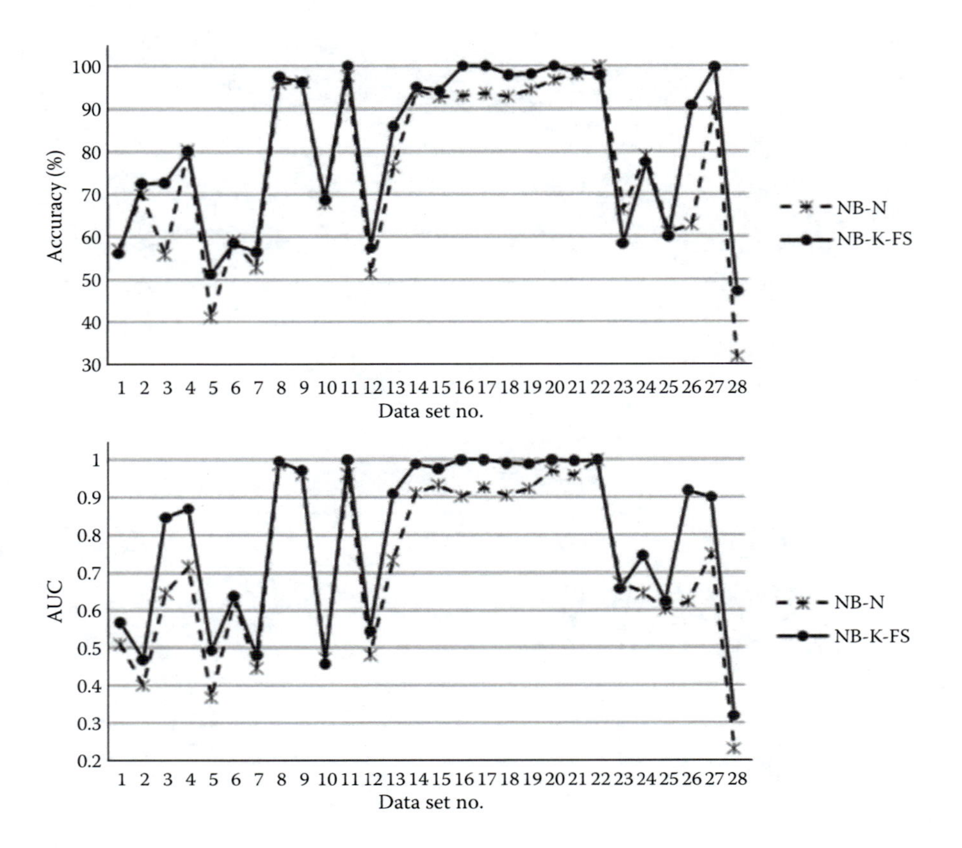

FIGURE 8.6
Accuracies and AUCs of NB-N and NB-K-FS.

8.3.4 Experiments Comparing NBCs with Other Classifiers

To investigate if the best-performing NBC (NB-FS-SD) has comparable performance to other classifiers, in this section, we compare it with the commonly used classifiers, including SVMs, DTs, and RFs, implemented in Weka.

With each of the three classification methods, we train the classifier using unfiltered data, feature-selected data (-FS), discretized data with supervised discretization (-SD), and both feature-selected and discretized data (-FS-SD). Table 8.4 lists the average performance of each method. From the table, NB-FS-SD has similar accuracy as SVMs with different types of data. Although NB-FS-SD has slightly lower accuracy than SVMs, it has a higher AUC than all the SVMs. NB-FS-SD outperforms all versions of DTs and slightly outperforms all versions of RFs. This confirms the previous report that NBCs can have comparative performance to some of the more modern and complicated classifiers.

Figure 8.8 shows the accuracies and AUCs of the best-performing version of each of the classifiers under consideration, NB-FS-SD, SVM, DT-SD, and RF-FS-SD, for all 28 data sets. We see the same trend of performance in regards to individual data sets among the

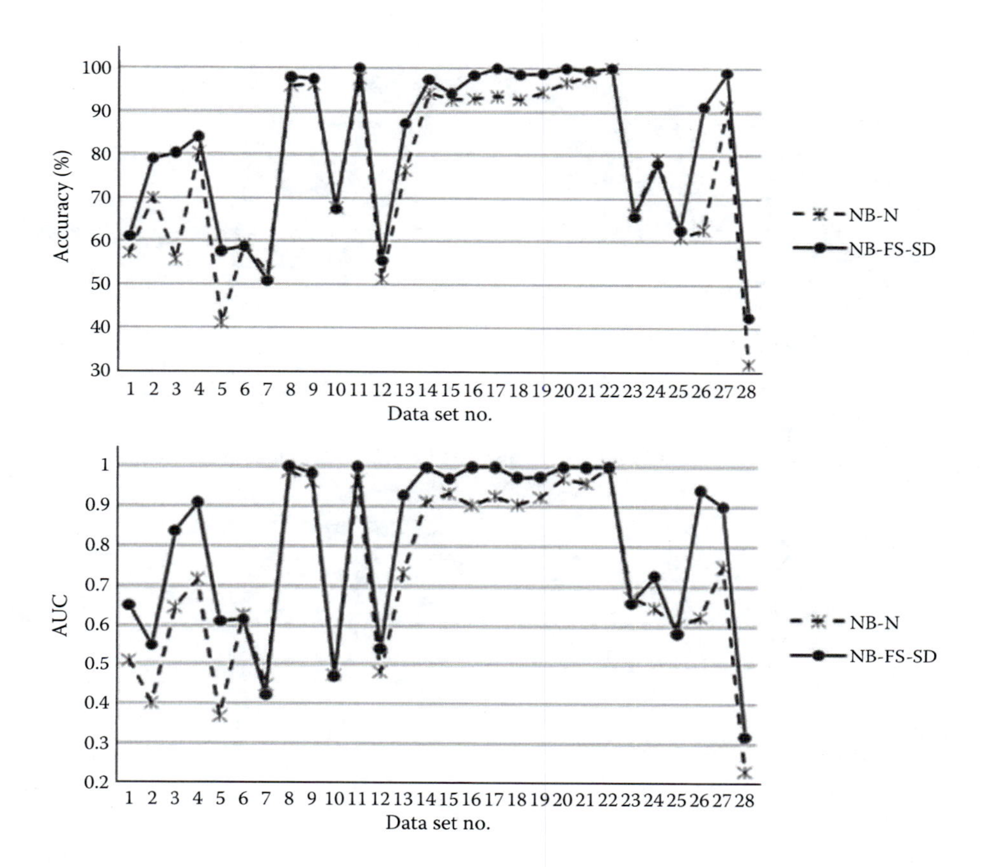

FIGURE 8.7
Accuracies and AUCs of NB-N and NB-FS-SD.

TABLE 8.4
Summary of Performance of NBC and Other Classifiers

Method	Average Accuracy	Average AUC
NB-FS-SD	82.25%	0.81
SVM	83.41%	0.78
SVM-SD	82.75%	0.77
SVM-FS	82.90%	0.78
SVM-FS-SD	82.48%	0.78
DT	76.02%	0.72
DT-SD	77.42%	0.73
DT-FS	75.65%	0.72
DT-FS-SD	77.08%	0.72
RF	77.95%	0.77
RF-SD	79.78%	0.79
RF-FS	81.33%	0.80
RF-FS-SD	81.38%	0.80

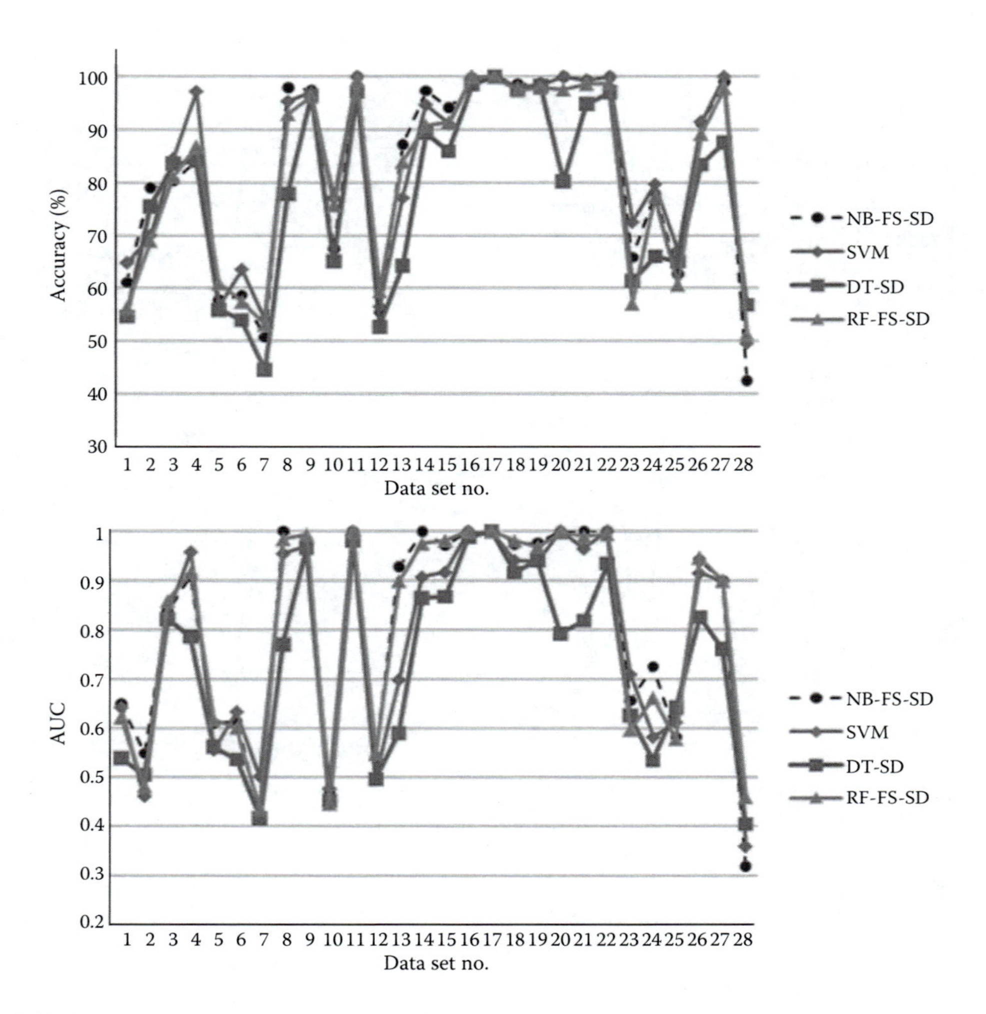

FIGURE 8.8
(See color insert.) Accuracies and AUCs of NB-FS-SD and the other classifiers.

classifiers, and none of them can handle some of the data sets well (the accuracies are very low, only around 40%–50%).

8.3.5 Experiments on the Effects of Misusing Cross-Validation

As described earlier, when supervised data filtering (feature selection and supervised discretization) is used, cross-validation must be applied not only to the classification part but also to the data filtering part. It is incorrect to do supervised data filtering using the full data sets, as class information of the test samples in a cross-validation run will be seen indeed; thus, the evaluation of the performance of the classification method could be overoptimistic.

The issue of misusing cross-validation has been addressed in microarray data studies [15,26]. In terms of the effect of the misuse on the evaluation of NBCs, we have not seen much work. Also, many publications do not describe explicitly how their cross-validation is carried out. So in this section, we compare the evaluation results of the NBCs in the case of correct use of cross-validation (i.e., all the results shown in Sections 8.3.3 and 8.3.4) to those in the misuse case.

The comparison results are provided in Table 8.5, which is the same as Table 8.3, except that the performance figures for the misuse cases (when applicable) are written inside parentheses.

From the table, we see that NB-FS-SD is still the best-performing method, with 97% accuracy and an AUC of 0.97. The differences (increases) are 14.75% and 0.16 in accuracy and AUC, respectively, comparing to the case when cross-validation is correctly used. Notably, for the data sets with which all the NBCs performed badly, in the misuse case, this is not indicated at all. For example, with data set 28, previously, the accuracy of NB-FS-SD was only 42.5% (see Figure 8.7), worse than a random guess, but from the result with the misuse of cross-validation, with the same data set, the accuracy of NB-FS-SD is 100%, and the AUC is 0.8 (previously, it was 0.32).

When comparing NB-FS-SD with other classifiers with feature-selected and/or discretized data (see Table 8.6), the same conclusion as that in Section 8.3.4 can be drawn on the relative performance of the classifiers. To see this more clearly, in Figure 8.9, we show

TABLE 8.5

Summary of Performance of NBCs for Both Cases of Correct and Incorrect Use of Cross-Validation

Method	Average Accuracy (Misuse)	Average AUC (Misuse)
NB-N	76.78%	0.72
NB-K	76.69%	0.73
NB-SD	81.32% (96.48%)	0.79 (0.96)
NB-UD	77.06%	0.76
NB-N-FS	81.30% (89.58%)	0.80 (0.91)
NB-K-FS	81.05% (89.20%)	0.80 (0.91)
NB-FS-SD	82.25% (97.00%)	0.81 (0.97)

TABLE 8.6

Summary of Performance of NBC and Other Classifiers in Correct and Incorrect Use of Cross-Validation

Method	Average Accuracy (Misuse)	Average AUC (Misuse)
NB-FS-SD	82.25% (97.00%)	0.81 (0.97)
SVM	83.41%	0.78
SVM-SD	82.75% (97.62%)	0.77 (0.95)
SVM-FS	82.90% (91.12%)	0.78 (0.87)
SVM-FS-SD	82.48% (97.35%)	0.78 (0.95)
DT	76.02%	0.72
DT-SD	77.42% (87.51%)	0.73 (0.84)
DT-FS	75.65% (83.43%)	0.72 (0.79)
DT-FS-SD	77.08% (91.82%)	0.72 (0.88)
RF	77.95%	0.77
RF-SD	79.78% (96.05%)	0.79 (0.96)
RF-FS	81.33% (89.93%)	0.80 (0.91)
RF-FS-SD	81.38% (95.87%)	0.80 (0.96)

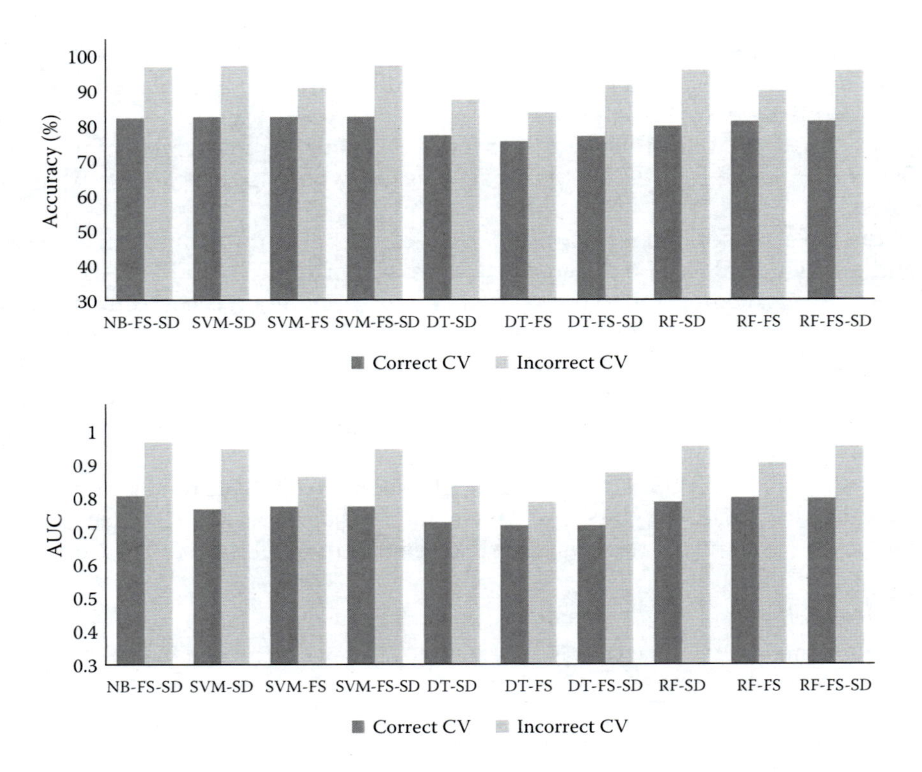

FIGURE 8.9
Average accuracies and average AUCs of classifiers in cases of correct and incorrect use of cross-validation (CV).

the bar chart diagrams for average accuracies and AUCs of all the classifiers with differently filtered data. We see that the differences among the classifiers in the case of correct use of cross-validation and the misuse case are similar, as indicated by the same trend of the bars for the two cases.

Therefore, when cross-validation is used incorrectly, the results of comparative study across different classification methods may still be valid. However, the assessment of the performance of individual classifiers can be extremely optimistic with the misuse of cross-validation. This finding is unlike the discussions in the work of Symons and Nieselt [25], where the difference between two cases is considered minor.

8.4 Discussions and Problems

From the experiments presented in Sections 8.3.3 and 8.3.4, we can reach the following conclusions:

1. The NBCs with normal or kernel data approximation have modest performance in classifying the gene expression data.
2. Data discretization, particularly the supervised discretization method, improves the overall performance of NBCs.

3. Feature selection improves classification outcome.
4. The NBC with feature-selected and discretized data is competitive with more popular classifiers, including SVMs, DTs, and RF.

At the same time, we can see that although data discretization and feature selection improve the performance of NBCs, the improvements are quite modest. As from Table 8.3, with supervised discretization, the average accuracy increases from 76.78% to 81.32%, only a 5% improvement. Similar improvement is obtained by using feature selection or using the combination of supervised discretization and feature selection. With some data sets, such as set 28, none of the data filtering approaches can improve the classification accuracy to an acceptable level. Low performance is also shown by other classifiers with those difficult data sets.

Also from the experiments in Section 8.3.5, we have noticed that in terms of the assessment of individual classifiers, the outcome of the misuse of cross-validation can be extremely overoptimistic and misleading. For example, the average accuracy of the NBC with supervised discretization increases from 76.78% to 96.48% (see Table 8.5), which is huge. With data set 28, in this case, the accuracy in fact is 100%, which seriously hides the problem.

Therefore, the results have raised a question: Are current classification methods indeed adequate for solving the problem of classifying high-dimensional data, especially gene expression data?

8.5 Conclusion and Future Work

In this chapter, we have studied NBCs in classifying high-dimensional data sets with small numbers of samples, using a large number of microarray gene expression data sets. We firstly reviewed NBCs and discussed the naïve assumption and its influence on classification outcome. Also, based on the characteristics of microarray data, we described the importance of data approximation, data discretization, and feature selection. Then we conducted a series of experiments with 28 microarray gene expression data sets related to different diseases.

Given the large number of existing studies of high-dimensional data classification and gene expression data classification, we consider the significance of having another look at NBCs (thus the contribution of this work) as twofold. Firstly, this work particularly focuses on NBCs in high-dimensional classification. It provides a comprehensive experimental study of NBCs and has identified the impact of approximation and discretization of continuous data and the impact of feature selection, through experiments with a large number of real-world microarray data sets. Secondly, it addresses the issue of misusing cross-validation in evaluating the performance of NBCs and the other classifiers, when supervised data filtering is used together with the classifiers. The experimental analysis brings to light the inadequacy of existing classification methods in dealing with difficult data sets. This reminds us that there are still challenges ahead in high-dimensional data classification, and we need to pursue new directions to advance this area.

This may include the study of integrative analysis of data sets from different sources that address the same problems and fusing other types of data. For gene expression analysis, many laboratories have produced gene expression data sets for the same purposes, and there is also a huge amount of different types of data, such as clinical data related to same problems. Integrative analysis of gene expression data from different sources and fusing gene expression data with other types of data hopefully not only resolve the performance issue of gene expression data classification, but also help with the poor transferability of classifiers trained with individual data sets. Research in this area has been very active in recent years, and many challenging issues remain to be tackled [27].

References

1. T. M. Mitchell. *Machine Learning*. Burr Ridge, Illinois: McGraw-Hill, 1997.
2. G. H. John and P. Langley. Estimating continuous distributions in Bayesian classifiers. In *Proceedings of the 11th Conference in Uncertainty in Artificial Intelligence*, Morgan Kaufmann, San Francisco, pp. 338–345, 1995.
3. J. W. Lee, J. B. Lee, M. Park, and S. H. Song. An extensive comparison of recent classification tools applied to microarray data. *Computational Statistics and Data Analysis*, vol. 48, no. 4, pp. 869–885, 2005.
4. L. Wang, J. Zhu, and H. Zou. Hybrid huberized support vector machine for microarray classification and gene selection. *Bioinformatics*, vol. 24, no. 8, pp. 412–419, 2008.
5. B. Chandra and M. Gupta. Robust approach for estimating probabilities in Naïve–Bayes classifier for gene expression data. *Expert Systems With Applications*, vol. 38, no. 3, pp. 1293–1298, 2011.
6. L. I. Kuncheva. On the optimality of naïve Bayes with dependent binary features. *Pattern Recognition Letters*, vol. 27, no. 7, pp. 830–837, 2006.
7. H. Zhang. The optimality of naïve Bayes. In *Proceedings of the 17th International Florida Artificial Intelligence Research Society Conference*, AAAI Press, 2004.
8. J. Dougherty, R. Kohavi, and M. Sahami. Supervised and unsupervised discretization of continuous features. In *Proceedings of the 12th International Conference on Machine Learning*, Morgan Kaufmann, San Francisco, pp. 194–202, 1999.
9. P. R. Bouckaert. Naïve Bayes classifiers that perform well with continuous variables. In *Proceedings of the 17th Australian Conference on AI, LNCS*, Springer, New York, vol. 3339, pp. 85–116, 2004. Extended report version [http:www.cs.waikato.ac.nz/~remco/disc.ps], last accessed on September 27, 2012.
10. A. Frank and A. Asuncion. UCI Machine Learning Repository [http://archive.ics.uci.edu/ml]. Irvine, CA: University of California, School of Information and Computer Science, 2010.
11. J. L. Lustgarten, V. Gopalakrishnan, and H. Grover. Improving classification performance with discretization on biomedical datasets. In *Proceedings of 2008 American Medical Informatics Association Symposium*, Curran Associates, Inc., pp. 445–449, 2008.
12. W. Dubitzky, M. Granzow, C. S. Downes, and D. Berrar. Introduction to microarray data analysis. Chapter 1 of *A Practical Approach to Microarray Data Analysis*, edited by D. P. Berrar, W. Dubitzky, and M. Granzow, Kluwer Academic Publishers, Dordrecht, the Netherlands, 2003.
13. E. P. Xing. Feature selection in microarray analysis. Chapter 6 of *A Practical Approach to Microarray Data Analysis*, edited by D. P. Berrar, W. Dubitzky, and M. Granzow, Kluwer Academic Publishers, Dordrecht, the Netherlands, 2003.
14. I. H. Witten and E. Frank. *Data Mining: Practical Machine Learning Tools and Techniques*. 2nd Edition, Elsevier, Amsterdam, 2005.
15. D. M. Rocke, T. Ideker, O. Troyanskaya, J. Quackenbush and J. Dopazo. Papers on normalization, variable selection, classification, or clustering of microarray data. The Editorial. *Bioinformatics*, vol. 25, no. 6, pp. 701–702, 2006.

16. S. Dudoit and R. Gentleman. Classification in microarray experiments. Presentation at the bioconductor short course: a short course on computational and statistical aspect of microarray analysis, 2003 [http://www.bioconductor.org/help/course-materials/2003/Milan/Lectures/classif.pdf], last accessed on February 15, 2013.
17. P. Wessa. Histogram (v1.0.11) in Free Statistics Software (v1.1.23-r6), Office for Research Development and Education [http://www.wessa.net/rwasp_histogram.wasp/].
18. U. M. Fayyad and K. B. Irani. Multi-interval discretization of continuous valued attributes for classification learning. In *Proceedings of the 13th International Joint Conference on Artificial Intelligence*, pp. 1022–1027, 1993.
19. M. Hall, E. Frank, G. Holmes, B. Pfahringer, P. Reutemann, and I. H. Witten. The WEKA data mining software: An update. *SIGKDD Explorations*, vol. 11, no. 1, pp. 10–18, 2009.
20. A. P. Bradley. The use of the area under the ROC curve in the evaluation of machine learning algorithms. *Pattern Recognition*, vol. 30, no. 7, pp. 1145–1159, 1997.
21. I. B. Jeffery. Higgins Laboratory people: comparison and evaluation of microarray feature selection methods [http://www.bioinf.ucd.ie/people/ian/], last accessed on February 15, 2013.
22. Kent Ridge Bio-medical Dataset [http://datam.i2r.a-star.edu.sg/datasets/krbd/], last accessed on February 15, 2013.
23. I. B. Jeffery, D. G. Higgins, and A. C. Culhane. Comparison and evaluation of methods for generating differentially expressed gene lists from microarray data. *BMC bioinformatics* vol. 7, p. 359, 2006.
24. S. Chiaretti, X. Li, R. Gentleman, A. Vitale, M. Vignetti, F. Mandelli, J. Ritz, and R. Foa. Gene expression profile of adult T-cell acute lymphocytic leukemia identifies distinct subsets of patients with different response to therapy and survival. *Blood*, vol. 103, no. 7, pp. 2771–2778, 2004.
25. S. Symons and K. Nieselt. Data mining microarray data—comprehensive benchmarking of feature selection and classification methods. Preprint from [http://it.inf.uni-tuebingen.de/public/SymonsNieselt.pdf], last accessed on February 15, 2013.
26. S. Dudoit and J. Fridlyand. Introduction to classification in microarray experiments. Chapter 7 of *A Practical Approach to Microarray Data Analysis*, edited by D. P. Berrar, W. Dubitzky, and M. Granzow, Kluwer Academic Publishers, Dordrecht, the Netherlands, 2003.
27. J. S. Hamid, P. Hu, N. M. Roslin, V. Ling, C. T. Greenwood, and J. Beyene. Data integration in genetics and genomics: methods and challenges. *Human Genomics and Proteomics*, 2009, Article ID: 869093.

9

Predicting Toxicity of Chemicals Computationally

Meenakshi Mishra, Jun Huan, and Brian Potetz

CONTENTS

9.1 Introduction

A huge gap is emerging between the new chemicals being produced and their toxicological information that is available to us. The Office of Toxic Substances (OTS) of the United States Environmental Protection Agency (EPA) had listed around 70,000 industrial chemicals in the 1990s, with around 1000 chemicals added each year for which there is little or no toxicological information available, as mentioned in an article by Polishchuk et al. [1]. Polishchuk et al. also went ahead and mentioned in the same article that more than 30,000 of these compounds are manufactured at the rate of more than a ton per year. Moreover, humans are regularly exposed to a lot of these chemicals (10,000–30,000) in the form of water contaminants, antimicrobials, pesticide active and inert ingredients, or other forms [2–4]. Hence, there is an urgent need to obtain the toxicological information for these chemicals.

The barrier that lies in the way of overcoming this information gap includes both time and money. The traditional toxicity methods involve giving varying doses of the chemicals to animal subjects and observing them for years. Each of these chronic bioassays for carcinogenicity can take up to several years and can cost millions of dollars [5]. Other drawbacks of the traditional toxicity testing methods include the guidelines to perform these tests being outdated, the tests possessing a black-box nature when the objective is to study the pathway associated with the development of toxic properties, the unavoidable extrapolation of results between species that take place, and the huge number of animals that are sacrificed and forced to a painful death each year [6,7]. We have to come up with new methods to predict toxicity that will save time and cost.

The efforts to find an alternative approach to obtain the toxicological profiles of the chemicals have been spread worldwide, and there has been a great deal of political support to deal with this problem as well. In 2007, the regulation on Registration, Evaluation, Authorization and Restriction of Chemicals (REACH) [8] came into force in Europe, and it produced a list of around 30,000 chemicals that needed urgent attention in terms of discovering their toxic behavior [9]. The United States of America has geared up as well, and the EPA produced a list of around 10,000 chemicals that demanded immediate attention [9]. Hence, upon the EPA's request to the National Research Council, a committee was formed that met to explore the long-range vision and strategy for toxicity, the results of the meeting being summarized in the National Research Council's report, "Toxicity Testing in the 21st Century: A Vision and a Strategy," published in 2007 [10]. The meeting revolved around finding alternative faster and less expensive methods that will speed up the toxicity computation process, get information about underlying biological pathways for each toxicological end point, and reduce animal suffering. More focus was given on methods such as *in vitro* testing and computational methods.

One way to determine the activity of drugs is Quantitative Structure Activity Relationship (QSAR). We can regard toxicity as an activity of the drug as well, even though it is undesirable activity. Hence, QSAR methods can be used for predicting toxicity as well [1,11–13]. Modest von Korff and Thomas Sander [14] used QSAR methods to discover the structural clues that might lead to toxicity. Stefan Rannar and Patrik L. Andersson [15] used hierarchical clustering that can help in predicting the impact of certain industrial chemicals on the environment. There are many more studies that have been done on similar grounds. However, the present status of toxicity prediction is very accurately summarized by the results of the Predictive Toxicology Challenge 2000–2001 [5,16]. A total of seventeen groups participated in this challenge to predict the carcinogenicity induced in mice by various chemicals. Surprisingly, out of the 111 models submitted, only five of them performed significantly better ($p < 0.05$) than random guessing. Efforts to develop an efficient computational model to predict toxicity are still pretty prevalent.

There has been considerable development in the technologies to culture cells, tissues, and also the 'omic' technologies like genomics, proteomics, metabolomics, and so forth having the potential to lead to better models for *in vitro* testing of chemicals. The clear advantage of *in vitro* testing is that these tests can be directly performed on the human cells, saving us from errors that might result from extrapolation between species [7]. Another area people are working on is to find a correlation between the *in vitro* biomarkers and the *in vivo* toxicity results [17–19]. But, there are still challenges associated with this kind of mapping [19]. For example, it is not possible to have assays for all the toxicity targets that exist. We still do not know the pathways that are associated with toxicity, and hence, we cannot just test each protein for chemical perturbations. Also, the possibility that toxicity is a function of a much more complex process rather than a single biochemical reaction is also a barrier in developing mappings from *in vitro* assay results to *in vivo* predictions [18,20].

The EPA has started a new project, ToxCast, whose aim is to find novel reliable methods of toxicity prediction [3,4]. The basic objectives of ToxCast are "(i) to identify the *in vitro* assays that can reliably output signs that correlate with the *in vivo* toxicity, (ii) develop signatures or prediction models that will achieve higher accuracy than any single assay or computational model alone and (iii) then be able to use these models to predict the toxicity profiles of previously untested chemicals" [3,4]. Basically, the idea is to bring together the lab methods and computational methods that will help in the mapping from *in vitro* to *in vivo* predictions. To achieve its objectives, ToxCast produced a list of chemicals, most of which are or have been commonly used pesticides and whose chemical profiles are already known, and performed many *in vitro* assays on these chemicals. The toxicity profiles of these chemicals were also collected. Both the *in vitro* test results and the toxicity profiles are easily available online.

In IEEE Conference on Bioinformatics and Biomaterials (BIBM) 2010, we published a paper [21] that showed that both random forests (RF) and naïve Bayes (NB) had a good performance on this data set. In the same paper, we further improved upon RFs. We considered the RF to be simply a fully connected weighted graph in which trees can be called the nodes and the measure of similarity between each pair of trees can be called the edge weight. We used a graph boosting algorithm to weigh the performance of some trees over others. This process did improve the performance slightly, but there was still room for considerable improvement. Hence, we decided to delve further into this area.

Since NB [22] had also done a decent classification on this data set (when compared to other classifiers used), we went ahead to explore Bayesian classifiers for chemical toxicity prediction [23]. A major drawback of NB algorithms is that they make an assumption about the conditional independencies of features given the labels. There are algorithms that relax these assumptions, like tree-augmented NB [22]. However, since the dimensionality of the given data is high, it would be difficult to make any kind of independence assumptions. We therefore decided to consider other kinds of Bayesian algorithms that use message propagation techniques to compute the posterior probabilities of a linear classifier [24]. In BIBM 2011, we published another paper in which we explored the performance of two such message propagation algorithms, namely, belief propagation (BP) [25] and expectation propagation (EP) [24,26], in the prediction of any kind of lesion that takes place in the mouse liver. In this paper, we showed that if, instead of approximating the posterior distribution of parameters as Gaussian, we allow the posterior distribution to take any shape, we can improve the performance of the Bayes machine. This chapter will discuss our findings in these two papers.

9.2 Data Set

The data set used in this work is provided by ToxCast. This data set consists of a total of 309 unique chemicals that are either currently or previously used pesticides. It was expected that a full toxicity profile would be available for these chemicals. The data also consisted of some industrially used chemicals. The data set consisted of various assay results as performed on these chemicals.

The assays used to screen these chemicals can be divided into nine different assay technologies, namely, multiplex transcription reporter assays [19], cell-based protein level assays [18], cytotoxicity assays, high-throughput genotoxicity screening, cell imaging assays, transcription assays, receptor binding and enzyme inhibitor assays, nuclear receptor assays, and real-time cell electronic sensing (RT-CES). The total number of *in vitro* assay end points measured 624. These 624 assays have been mapped to 331 genes; 231 assays can be mapped to human genes and 65 mapped to rodents. The results are reported in terms of either half maximal activity concentration (AC50) or lowest effective concentration (LEC) values. The unit used is micromolar units. If a chemical did not show any reaction to a particular assay at the maximum dosage level, the chemical was said to be nontoxic for that particular assay and given a value of 1,000,000.

Computed physical and chemical properties were also included in the data set. Two hundred and five different physical and chemical properties were computed using various software available in the market and provided with the data set. We also calculated molecular properties of these chemicals using software called DRAGON, version 5.5. This software allows different kinds of descriptors that can be created. For our purpose, we used the descriptors called "functional group count." This group of descriptors produces a 154-length vector associated with each chemical. Each element of this vector is a count of the particular functional group present in the chemical.

The feature set for our analysis was formed from concatenating the assay end points, computed physical chemical properties, and molecular descriptors.

There were some redundancies that were purposely included in the data set to ensure accuracy of experiments. Three of the chemicals had three samples taken from the same source in the data set, and five chemicals had two samples taken from different sources, making a total of 320 samples. The chemicals tested in the data set had purity of more than 90%, 80% of the chemicals having purity of more than 97%.

The *in vivo* toxicity profile for these chemicals was compiled by the EPA Toxicity Reference Database (ToxrefDB). This study includes rat and mouse 2-year cancer bioassays, rat multigenerational reproductive toxicity assays, and rat and rabbit prenatal developmental assays performed on multiple end points, reported in LEC. The ToxCast and ToxrefDB databases can be downloaded at http://www.epa.gov/ncct/toxcast/ and http://epa.gov/ncct/toxrefdb/, respectively.

9.3 Preprocessing and Computation

There were a large number of end points tested in the ToxrefDB data set. In our analysis, we only considered those end points that had at least 20% of positive cases.

We used the molecular descriptors and the assay results from ToxCast to predict the *in vivo* toxicity at a particular end point. The chemicals whose end point result was unknown were simply removed. All the features of the chemicals that were unknown were replaced by '-999'. We also tried replacing the missing value with the average of that particular feature but did not notice any significant difference in result.

The training algorithms that were used are support vector machines (SVMs), RFs, NB (NB), and *k*-nearest neighbor (KNN). SVM was implemented on MATLAB®, NB on the Waikato Environment for Knowledge Analysis (Weka), and RF on Spider. For model selection, a simple cross-validation method was used. First, we divided the data randomly into 80% and 20% for training and testing. Then, we used tenfold cross-validation on the 80% of the data set (using 70% for training and 30% for validation) to select the parameters for each algorithm. Error rate, or the ratio of the number of samples misclassified to the total number of samples in the validation data, was the model selection criteria used. The final model was tested on the remaining 20% of test data. This experiment was repeated 10 times, and the reported results are the mean accuracy, precision, and recall of the 10 testing results thus obtained.

$$\text{Accuracy} = \frac{\text{tp} + \text{tn}}{\text{tp} + \text{tn} + \text{fp} + \text{fn}}$$

$$\text{Precision} = \frac{\text{tp}}{\text{tp} + \text{fp}}$$

$$\text{Recall} = \frac{\text{tp}}{\text{tp} + \text{fn}}$$

where tp is true positive, tn is true negative, fp is false positive, and fn is false negative.

We compared the performance of different algorithms (shown in Figure 9.1) based on all three parameters, namely accuracy, precision, and recall, because relying on accuracy alone can be misleading. This is especially the case when the data set is biased towards one class.

As can be seen in Figure 9.1, RF gave the highest accuracy of 75% for the target mouse liver hypertrophy. But we cannot suggest that RF performed well for this end point as the recall was very low, only 0.23.

Similarly, we notice that the accuracies of all algorithms were about the same. However, SVMs had very poor values of precision and recall as mostly, they either were near zero or had a divide-by-zero error. This is a usual scenario if a classifier gives a biased result and classifies entire samples as either positive or negative. Even KNN had similar results and performed poorly if the metric was recall values.

NB classifier and RF were the best classifiers amongst the classifiers compared here. The accuracy of this algorithm was comparable with the results of other algorithms, but the precision and recall values were generally higher than other algorithms. The higher values of precision and recall show that these classifiers were more unbiased than others. Hence, we favored NB and RF more in our further studies.

We hereby continue the study using only the target, any kind of lesion shown on the mouse liver. This target was chosen because the number of positive samples was maximum at this particular end point, with around 42% of the samples being positive, which ensures unbiased data.

9.4 RF with Boosting Algorithm

An RF algorithm [27] can be thought of as an ensemble of decision tree classifiers. This algorithm first creates a number of decision trees and then assigns the class that is voted by the majority of the trees created [27]. It has been shown that RF is an effective tool for the QSAR tasks [1].

In this work, we wanted to show that if we use small and more relevant trees in RF, which contribute to right decision making, we can further improve RF's performance. For this purpose, we used the RF algorithm implemented by Leo Breiman [28]. This algorithm had a small feature selection attribute to it as well. There was a parameter called 'mtry,' which indicated

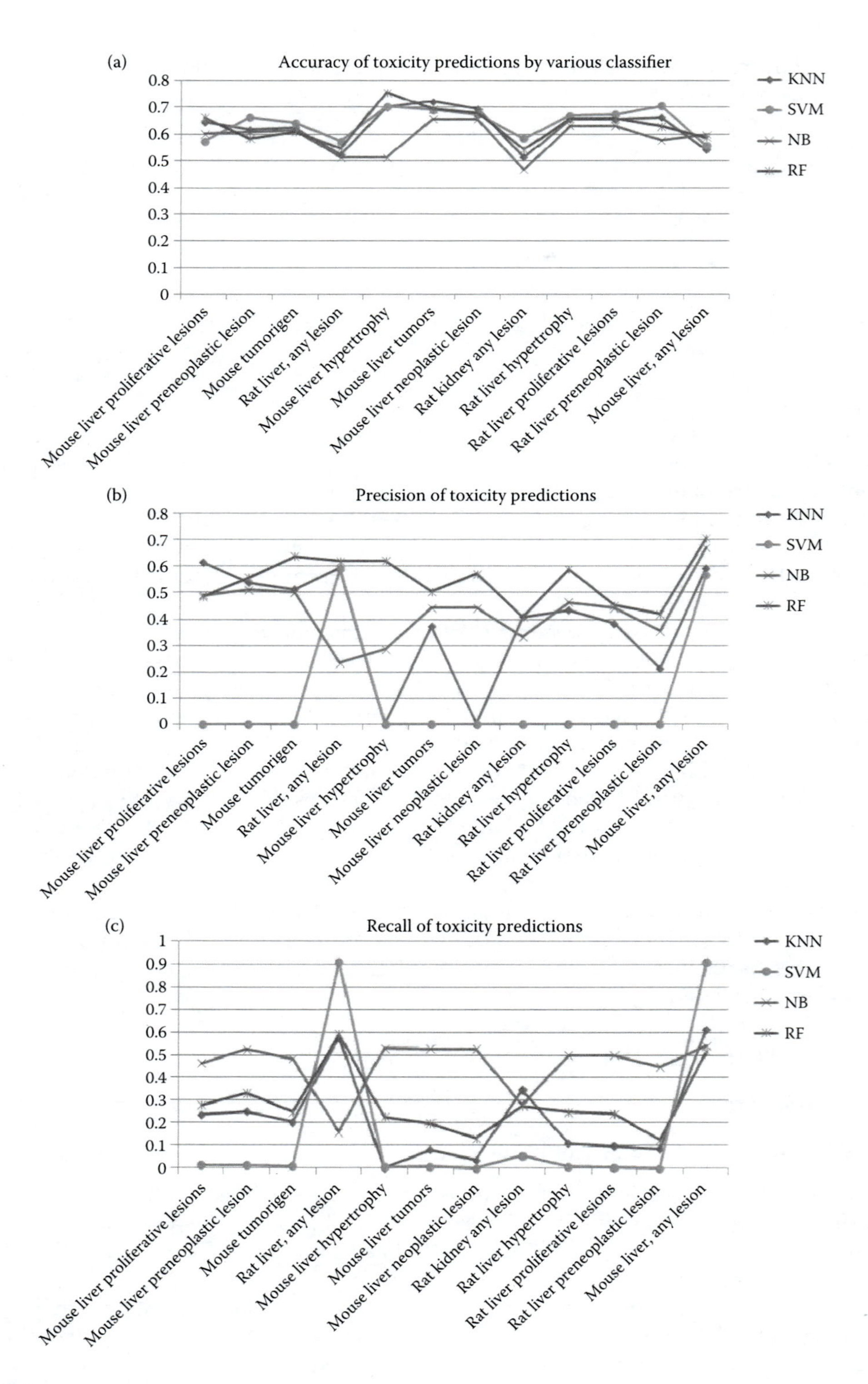

FIGURE 9.1
The (a) accuracy, (b) precision, and (c) recall values of the toxicity predictions made by different algorithms.

the number of features tried for each node to be split, and the feature that gave the best performance was used. Ideally, 'mtry' should be large so that we can find more relevant trees.

The RF thus created was treated as an undirected graph. Each tree in this graph served as a single node. The features that were picked by the trees in the RF were grouped together by the category of assays they belong to. The nodes or the trees in this undirected graph were connected by edges whose weights were computed based on the groups of features shared by the trees.

$$w_{i,j} = \frac{\sum_{k=r,s,t} \min\left(n_k^i, n_k^j\right)}{\sum_{k=r,s,t} \max\left(n_k^i, n_k^j\right)}$$

where n_k^i is the number of features in group k of tree i. For example, suppose a feature belonging to group r, two features belonging to group s, and one feature from group t were picked by group A, and one feature belonging to group r, none to group s, and one to group t were picked by tree B. In that case, the weight of the edge between trees A and B would be (1 + 0 + 1)/(1 + 2 + 1) = 0.5. Basically, the intersection between two trees is denoted by the sum of minimum values, and the union by the sum of maximum values. Then, we used a graph boosting algorithm [29] to enhance the performance of the more relevant trees in the RF.

In the boosting algorithm, each node or tree can be considered as base learner, and the weights of the edges between the nodes $w_{i,j}$ represent the structural relationship that exists between nodes. This boosting algorithm makes use of L1 norm regularization on the coefficients of the base learners for obtaining sparse representation. For utilizing structural information in boosting, the algorithm makes use of L2 regularized graph Laplacian, defined as $L = D - W$. Here W is the p-by-p (p being number of base learners or trees) edge weight matrix given as $W = (w_{i,j})_{i,j=1}^{p}$, and D is density matrix of W, given by $D = (d_{i,j})_{i,j=1}^{p}$, where

$$d_{i,j} = \begin{cases} \sum_{k=1}^{p} W_{i.k} & \text{If } i = j \\ 0 & \text{Otherwise} \end{cases}$$

The effects of L2 norm regularization on graph Laplacian are a smoothness and grouping effect. Using the L2 norm ensures that the coefficients of neighboring base learners are close to each other and to the feature graph Laplacian penalty term, resulting in smoothness. Also, if a node is selected, its neighbors have a higher chance of being selected, resulting in grouping effect. Hence, using this boosting algorithm, highly relevant trees were picked, which contributed more towards prediction of the end points. For more information about the boosting algorithm, refer to the work of Fei and Huan [29].

The accuracy, precision, and recall values obtained before applying the boosting algorithm and after applying the boosting algorithm to the RF algorithm were compared and are shown in Figure 9.2. As can be seen from Figure 9.2, we do improve the accuracy after using more relevant trees. The precision does go down slightly, by 1%, but the recall increases significantly, by around 12%. The results thus produced using boosting with RF proved to be better than both NB and RF in base form.

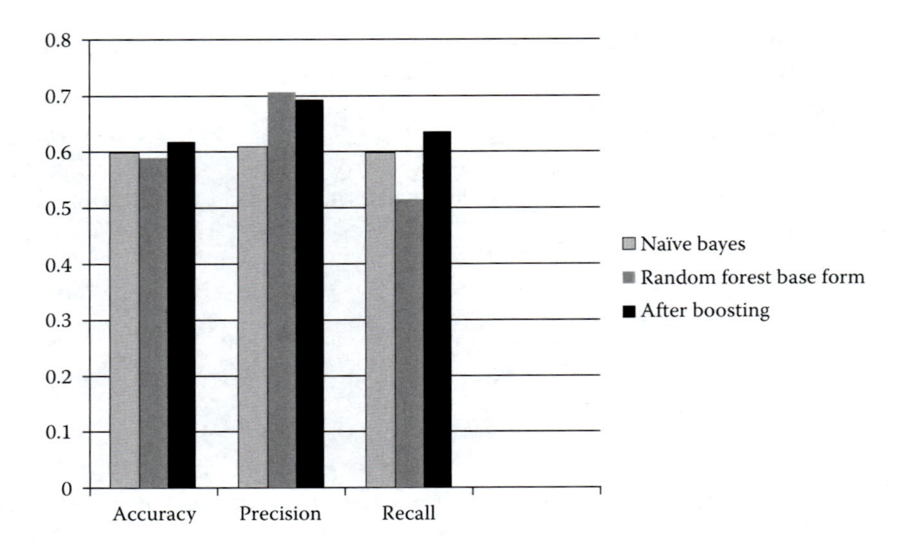

FIGURE 9.2
A summary of performance of random forest algorithms before and after boosting and its comparison to NB.

9.5 Toxicity Prediction Using Bayesian Classifiers

Since NB [22] also performed well in classification on this data set (when compared to other classifiers used), we decided to explore Bayesian classifiers for chemical toxicity prediction. The major drawback of NB is its assumption about the conditional independence of the input features given the class. Although classifiers like tree-augmented NB [22] relax this assumption, making any kind of conditional independency assumption is difficult for this data set, given its high dimensionality. Hence, we look at another family of Bayesian classifiers that makes use of message propagation algorithms [24].

9.5.1 Background on Bayesian Classifiers

Bayesian classifiers basically make use of probabilistic models to estimate the likelihood of the output class y. So, if x is an input vector whose class needs to be computed, y will be a hypothetical choice of output class, and Z will be the training data consisting of feature vectors x_i and the class of each feature vector, y_i. Bayesian classifiers will attempt to estimate $P(y|x,Z)$.

Generally, Bayesian classifiers come in two forms: generative and discriminative. Generative Bayesian classifiers are those that attempt to model the data distribution for each class, $P(x|y,Z)$. Once models are learned for each class y, classification can proceed using Bayes rule:

$$
\begin{aligned}
y^* &= \arg\max_y P(y \mid x, Z) \\
&= \arg\max_y P(x \mid y, Z) P(y \mid Z)
\end{aligned}
$$

Conversely, discriminative Bayesian classifiers forgo learning the distribution of data x and instead model $P(y|x,Z)$ directly, typically by modeling the likelihoods of different classifier parameters.

9.5.1.1 NB Classifiers

One popular and simple generative Bayesian classifier is NB. Like all generative classifiers, NB attempts to model $P(x|y,Z)$ for each output class y. This is a challenging task, especially given high-dimensional data x. An attempt to model these high-dimensional distributions accurately would require many parameters, and thus, large amounts of data. The NB classifier reduces the number of parameters by making a strong assumption: that given the class, all the features are conditionally independent. Thus,

$$P(y \mid x, Z) = cP(y \mid Z) \prod_i P(x_i \mid y, Z)$$

where y is the label, **X** is the feature vector with each feature given by x_i, and c is some constant. Z is the given data set composed of input feature vectors, X, and their labels y.

In most applications, it is not possible to verify whether the NB assumption holds. Other Bayesian network classifiers, like the tree-augmented NB classifier, relax this assumption and try to explore any kind of relationship that might exist between the features.

9.5.1.2 Bayes Optimal Classifiers

In order to model the distribution of data $P(x|y,Z)$, generative models typically must make sacrifices: They must either fit $P(x|y,Z)$ using a model with many parameters (thus requiring large amounts of data) or make strong assumptions about the distribution (such as the independence assumption of the NB classifier). Discriminative classifiers avoid this trade-off by modeling $P(y|x,Z)$ directly instead of $P(x|y,Z)$.

Discriminative Bayesian classifiers are typically approximations of the Bayes optimal classifier. In a Bayes optimal classifier,

$$\begin{aligned} y^* &= \arg\max_y P(y \mid x, Z) \\ &= \arg\max_y \int_H P(y, h \mid x, Z) \mathrm{d}h \\ &= \arg\max_y \int_H P(y \mid x, h) P(Z \mid h) P(h) \mathrm{d}h \end{aligned}$$

Here, H is the hypothesis space, or the space of all possible classifiers. Example hypotheses h might include different decision boundaries in the space of X. The integrand over H is generally intractable. However, the Bayes optimal classifier is theoretically interesting because it will, on average, outperform all other classifiers, given the same prior knowledge [30].

9.5.1.3 Bayes Point Machines

The high-dimensional integrand of the Bayes optimal classifier not only is large, but it must be reevaluated for every novel input vector x. Discriminative Bayesian classifiers typically often approximate the Bayes optimal classification process by selecting a single classifier h that achieves performance that is as close as possible to the Bayes optimal. Sometimes, the most likely, or *maximum a posteriori* (MAP), classifier is selected.

$$h_{\mathrm{MAP}} = \mathrm{argmax}_h \, P(h|Z)$$

However, it has been proved that the mean of the posterior distribution of classifiers

$$h_{\mathrm{BP}} = \mathbf{E}_{P(h|Z)}[h]$$

approximates the Bayes optimal classifier better and also leads to improved generalization performance [31,32]. Such a classifier, which estimates the mean of the posterior distribution $P(h|Z)$, is called the Bayes point classifier.

Intuitively, one reason for the improved generalization performance of the Bayes point classifier is that it achieves greater robustness by taking all possible classifiers into account. Either noise in the data or a coincidental property of the training data may cause a spurious peak in $P(h|Z)$, which would not occur if more data are present. In this case, h_{MAP} would perform well in the training data but would generalize poorly to test data. The mean of $P(h|Z)$ is robust towards spurious peaks in $P(h|Z)$.

Another intuitive way to understand the advantage of the Bayes point classifier over the MAP classifier is to recall that the MAP point estimate minimizes the 0-1 loss over all classifiers; that is, it maximizes the probability that h_{MAP} is exactly equal to the optimal classifier. However, in the event that h_{MAP} is not optimal (an event with a probability of 1, assuming H is continuous), h_{MAP} may be located very far from the optimal classifier and may have poor performance. The mean of $P(h|Z)$ is also known as the *minimum mean-squared* error point estimate (MMSE), because it minimizes the squared error between itself (h_{BP}) and the optimal classifier. Thus, test performance, on average, is approximately minimized.

We now address how $P(h|Z)$ is modeled. First, let us choose our hypothesis space H to be the set of all binary linear classifiers. Thus, every hypothesis h corresponds to a different linear weight vector w and bias b and classifies novel input vectors x by

$$y = h(x) = sign(w^T x + b)$$

Thus, h classifies x as +1 or −1 depending whether x lies above or below the hyperplane specified by w and b.

Here, we only used linear classifiers, because we had found that SVM with the linear kernel usually performs as well as or better than SVM with Gaussian or polynomial kernels on this particular data set. This is a typical scenario in problems where the number of features is far greater than the number of data samples. In data sets that do not have this property, we can imagine projecting the data into a higher dimension, such as x^2 or x^3, in which a linear decision boundary is a good approximation.

Our goal, then, is to locate the mean of the distribution $P(h|Z)$, where h is defined by vector w and scalar b. According to Bayes rule, $P(h|Z) \propto P(Z|h)P(h)$, where the term $P(Z)$ is treated as a normalizing constant, since it is constant with respect to h.

$P(Z|h)$ is the likelihood of the data given the parameters. Assuming that each training point is independent of one another, $P(Z|h)$ factorizes:

$$P(Z\,|\,h) \propto \prod_{i=1}^{S} \exp\left(C \cdot L(y_i \cdot (w^T x_i + b))\right)$$

where C is a hyperparameter, y_i and x_i are the class and input features for the ith training sample, and S is the number of training samples.

L is known as a *loss function*, and it describes the penalty incurred by a classifier for misclassifying a point. To maximize performance over unseen test data, the loss function should be chosen to resemble the method used to evaluate classifiers on the test data. For example, if classifiers are evaluated based on the total percentage of correctly classified test points, L should be chosen as the *0-1 loss function*, and $L_{01}(x)$ is 1 for negative x and 0 for positive x. Thus, L_{01} penalizes all mistakes equally, regardless of how confident the classifier was in its mistake or whether the mistake was a false positive or a false negative.

9.5.1.4 Message-Propagating Algorithms

While the goal of computing the Bayes point classifier h_{BP} is computationally more efficient than the Bayes optimal classifier (because it needs to be computed only once, during training), it still requires integrating over a large, high-dimensional integrand. Brute-force integration is impractical even for data sets with few features. Sampling methods have been proposed for estimating the Bayes point [31], but past methods have not been shown to outperform soft-margin SVM. Here, we will explore message-passing algorithms developed for probabilistic inference.

Gaussian EP [26] is one of the message propagation algorithms that have been used to estimate the Bayes point by approximating the posterior distribution $P(h|Z)$ as Gaussian. The mean of the posterior estimated is used for h_{BP}. The results of comparison of Gaussian EP and soft-margin SVM have been mixed [32].

9.5.1.5 BP Using Linear Multiple-Choice Knapsacks

Gaussian EP approximates the posterior as Gaussian; however, it is not necessary that Gaussian is the best approximation for the posterior $P(h|Z)$. The approximation of the posterior can have a significant impact on the performance. We have shown in our paper that data sets with a large number of features compared to the number of samples, like our toxicity data set, can have a non-Gaussian posterior.

In this work, we make use of another message-passing algorithm for computing the mean of $P(h|Z)$, known as belief propagation. BP uses a discrete, or histogram, representation for messages to compute $P(h|Z)$. It has been shown that EP using histograms is the same as traditional BP [26].

At this point, we must note that the traditional BP algorithm is intractable when solving for Bayes point. BP in a fully connected graphical representation, such as that of classification, is even slower than the brute-force computation of the integral to solve for h_{BP}. Traditional, the BP algorithm has a time complexity of $O(SDM^D)$ per iteration, where S is the number of training samples, D is the number of features, and M is the number of states each variable w_j can have.

Recently, we published a paper in which we developed a faster method to compute the Bayes point using BP efficiently, based on the classical linear knapsack problem [33]. The time complexity of each iteration of the new BP boils down to $O(RSD^2M)$, where R is the number of concave regions in each of the messages computed (typically less than 3). The downside of this algorithm is that the Bayes point is computed as the mean of the max-marginals of $P(h|Z)$ rather than using the true mean of $P(h|Z)$ (see the work of Potetz [33] for details). However, this approximation appears to be reasonable as this algorithm outperformed linear kernel soft-margin SVM and Gaussian EP in six of seven data sets from UC Irvine Machine Learning Repository.

9.5.2 Classification for Toxicity Profiles

We use the same data set as used for the boosted RF. The *in vitro* assay end points, the computed physical chemical properties of the chemicals, and the molecular descriptors of the chemicals were concatenated and used as feature vector for the chemicals. The feature vector for each chemical was the input to the classifier. The output of the classifier was the prediction of whether the particular chemical would produce any kind of lesion on the mouse liver. There were only 245 samples for which the output information was available; hence, we used only 245 samples out of the 320 in the ToxCast data set.

We have shown that SVM provided a highly biased result in this data set, computing every chemical in question as either toxic or nontoxic. We used a linear kernel for SVM as previous studies showed that the linear kernel had the best performance over Gaussian and polynomial kernels. The soft-margin constraint value C was also adjusted to give the best possible results using cross-validation. The accuracy obtained using SVM was 55.20% ± 1.60. We have shown that the NB classifier and RF classifier showed better results when compared with SVM or KNN. We also modified the RF algorithm, and we saw that we could improve the results but not significantly enough to really be able to use the method in a real-world application. Hence, we continued the study, this time looking into the Bayesian classifier. We also explored a commonly used classifier, EP. We continued using the linear kernel for the EP algorithm as well. However, we made a minor change in the original EP. We used a modified step function given by

$$\theta(x) = \begin{cases} \varepsilon_1 & \text{when } x < 0 \\ 1 - \varepsilon_1 & \text{when } x > 0 \end{cases}$$

as the basic loss function so that some error could be tolerated.

The value of ε_1 is zero in the original EP framework giving the basic 0-1 step function. We do investigate the effect of ε_1 in our paper [23]; however, here we just discuss the case when ε_1 is zero. The accuracy obtained from using EP was 59.39% ± 1.59. These results were better than the results from SVM. As discussed earlier, the EP approximates the posterior distribution as Gaussian, which might or might not be ideal. Also, even though the EP computes the covariance of the posterior, the covariance term has no effect in the prediction of the class for the given input feature vector. The weights used finally are just the mean of the posterior. Thus, we decided to delve into the BP framework. We hypothesized that if the marginals of the posterior can be allowed to take any shape instead of limiting to Gaussian, we can achieve better results. The BP algorithm allowed us to model the posterior marginals as a histogram. If the histogram bins are small enough, the posterior marginals can be given any shape. Since traditional BP is slow, we implemented a version of the BP algorithm utilizing the linear knapsack problem approach, to classify the chemicals into toxic and nontoxic [33]. The accuracy obtained from this method was 62.45% ± 1.22, which is higher than both EP and SVM.

All the results presented here are the average test results of 20 trials.

9.6 Concluding Remarks

As discussed above, there are a variety of algorithms that can be used to predict the toxicity profiles of chemicals. Given the fact that there are thousands of chemicals that need to be tested, an improvement of a couple of percent in accuracy can go a long way in saving cost and time for toxicity testing of the chemicals. We see here that Bayesian methods have an edge over other methods in toxicity prediction for the ToxCast data set. The reason is that the number of features greatly outnumbers the number of samples. For the same reason, linear kernels have an edge over other kernels as well. However, we still have a long way to go in order to develop a model that can be used in the commercial world for toxicity prediction.

References

1. P. G. Polishchuk, E. N. Muratov, A. G. Artemenko, O. G. Kolumbin, N. N. Muratov, and V. E. Kuz'min, Application of random forest approach to QSAR prediction of aquatic toxicity, *J Chem Inf Model*, vol. 49, pp. 2481–8, Nov 2009.
2. R. Judson, F. Elloumi, R. W. Setzer, Z. Li, and I. Shah, A comparison of machine learning algorithms for chemical toxicity classification using a simulated multi-scale data model, *BMC Bioinformatics*, vol. 9, p. 241, 2008.
3. R. S. Judson, K. A. Houck, R. J. Kavlock, T. B. Knudsen, M. T. Martin, H. M. Mortensen, D. M. Reif, D. M. Rotroff, I. Shah, A. M. Richard, and D. J. Dix, *In vitro* screening of environmental chemicals for targeted testing prioritization: the ToxCast project, *Environ Health Perspect*, vol. 118, pp. 485–92, Apr 2010.
4. D. J. Dix, K. A. Houck, M. T. Martin, A. M. Richard, R. W. Setzer, and R. J. Kavlock, The ToxCast program for prioritizing toxicity testing of environmental chemicals, *Toxicol Sci*, vol. 95, pp. 5–12, Jan 2007.
5. C. Helma and S. Kramer, A survey of the Predictive Toxicology Challenge 2000–2001, *Bioinformatics*, vol. 19, pp. 1179–82, Jul 1, 2003.
6. N. Bhogal, C. Grindon, R. Combes, and M. Balls, Toxicity testing: creating a revolution based on new technologies, *Trends Biotechnol*, vol. 23, pp. 299–307, Jun 2005.
7. R. J. Kavlock, G. Ankley, J. Blancato, M. Breen, R. Conolly, D. Dix, K. Houck, E. Hubal, R. Judson, J. Rabinowitz, A. Richard, R. W. Setzer, I. Shah, D. Villeneuve, and E. Weber, Computational toxicology—a state of the science mini review, *Toxicol Sci*, vol. 103, pp. 14–27, May 2008.
8. G. Schoeters, The REACH perspective: toward a new concept of toxicity testing, *J Toxicol Environ Health B Crit Rev*, vol. 13, pp. 232–41, Feb 2010.
9. T. B. Knudsen, K. A. Houck, N. S. Sipes, A. V. Singh, R. S. Judson, M. T. Martin, A. Weismann, N. C. Kleinstreuer, H. M. Mortensen, D. M. Reif, J. R. Rabinowitz, R. W. Setzer, A. M. Richard, D. J. Dix, and R. J. Kavlock, Activity profiles of 309 ToxCast chemicals evaluated across 292 biochemical targets, *Toxicology*, vol. 282, pp. 1–15, Mar 28, 2011.
10. D. Krewski, D. Acosta, Jr., M. Andersen, H. Anderson, J. C. Bailar, 3rd, K. Boekelheide, R. Brent, G. Charley, V. G. Cheung, Jr. Green S, K. T. Kelsey, N. I. Kerkvliet, A. A. Li, L. McCray, O. Meyer, R. D. Patterson, W. Pennie, R. A. Scala, G. M. Solomon, M. Stephens, J. Yager, and L. Zeise, Toxicity testing in the 21st century: a vision and a strategy, *J Toxicol Environ Health B Crit Rev*, vol. 13, pp. 51–138, Feb 2010.
11. P. L. Piotrowski, B. G. Sumpter, H. V. Malling, J. S. Wassom, P. Y. Lu, R. A. Brothers, G. A. Sega, S. A. Martin, and M. Parang, A toxicity evaluation and predictive system based on neural networks and wavelets, *J Chem Inf Model*, vol. 47, pp. 676–85, Mar–Apr 2007.
12. S. C. Basak, G. D. Grunwald, B. D. Gute, K. Balasubramanian, and D. Opitz, Use of statistical and neural net approaches in predicting toxicity of chemicals, *J Chem Inf Comput Sci*, vol. 40, pp. 885–90, Jul–Aug 2000.
13. I. Kahn, S. Sild, and U. Maran, Modeling the toxicity of chemicals to Tetrahymena pyriformis using heuristic multilinear regression and heuristic back-propagation neural networks, *J Chem Inf Model*, vol. 47, pp. 2271–9, Nov–Dec 2007.

14. M. von Korff and T. Sander, Toxicity-indicating structural patterns, *J Chem Inf Model*, vol. 46, pp. 536–44, Mar–Apr 2006.
15. S. Rannar and P. L. Andersson, A novel approach using hierarchical clustering to select industrial chemicals for environmental impact assessment, *J Chem Inf Model*, vol. 50, pp. 30–6, Jan 2010.
16. H. Toivonen, A. Srinivasan, R. D. King, S. Kramer, and C. Helma, Statistical evaluation of the Predictive Toxicology Challenge 2000–2001, *Bioinformatics*, vol. 19, pp. 1183–93, Jul 1, 2003.
17. R. Kikkawa, M. Fujikawa, T. Yamamoto, Y. Hamada, H. Yamada, and I. Horii, *In vivo* hepatotoxicity study of rats in comparison with *in vitro* hepatotoxicity screening system, *J Toxicol Sci*, vol. 31, pp. 23–34, Feb 2006.
18. K. A. Houck, D. J. Dix, R. S. Judson, R. J. Kavlock, J. Yang, and E. L. Berg, Profiling bioactivity of the ToxCast chemical library using BioMAP primary human cell systems, *J Biomol Screen*, vol. 14, pp. 1054–66, Oct 2009.
19. M. T. Martin, D. J. Dix, R. S. Judson, R. J. Kavlock, D. M. Reif, A. M. Richard, D. M. Rotroff, S. Romanov, A. Medvedev, N. Poltoratskaya, M. Gambarian, M. Moeser, S. S. Makarov, and K. A. Houck, Impact of environmental chemicals on key transcription regulators and correlation to toxicity end points within EPA's ToxCast program, *Chem Res Toxicol*, vol. 23, pp. 578–90, Mar 15, 2010.
20. U. S. Bhalla and R. Iyengar, Emergent properties of networks of biological signaling pathways, *Science*, vol. 283, pp. 381–7, Jan 15, 1999.
21. M. Mishra, H. Fei, and J. Huan, Computational prediction of toxicity, presented at the *IEEE International Conference on Bioinformatics and Biomedicine*, Hong Kong, 2010.
22. N. Friedman, D. Geiger, and M. Goldszmidt, Bayesian network classifiers, *Machine Learning*, vol. 29, pp. 131–63, 1997.
23. M. Mishra, B. Potetz, and J. Huan, Bayesian classifiers for chemical toxicity prediction, in *IEEE International Conference on Bioinformatics and Biomedicine*, Atlanta, USA, 2011.
24. T. Minka, A family of algorithms for approximate Bayesian inference, PhD, Electrical Engineering and Computer Science, MIT, 2001.
25. T. Heskes, K. Albers, and B. Kappen, Approximate inference and constrained optimization, presented at the *Uncertainty in Artificial Intelligence*, 2003.
26. T. Minka, Expectation propagation for approximate Bayesian inference, in *Uncertainty in Artificial Intelligence*, pp. 362–9, 2001.
27. L. Breiman, Random forests, *Machine Learning*, vol. 45, pp. 5–32, 2001.
28. L. Breiman and A. Cutler. Random forests. Available: http://www.stat.berkeley.edu/~breiman/RandomForests/
29. H. Fei and J. Huan, Graph boosting with spatial and overlapping regularization, presented at the *16th ACM SIGKDD Conference on Knowledge Discovery and Data Mining (SIGKDD'10)*, Washington, DC, 2010.
30. C. M. Bishop, *Pattern Recognition and Machine Learning*. New York: Springer, 2006.
31. R. Herbrich, T. Graepel, and C. Campbell, Bayes point machine, *Journal of Machine Learning Research*, pp. 245–79, 2001.
32. M. Opper and O. Winther, A Bayesian approach to on-line learning, *On-Line Learning in Neural Networks*, 1999, pp. 363–78.
33. B. Potetz, Estimating the Bayes point using linear knapsack problems, presented at the *International Conference in Machine Learning*, Bellevue, Washington, USA, 2011.

10

Cancer Prediction Methodology Using an Enhanced Artificial Neural Network–Based Classifier and Dominant Gene Expression

Manaswini Pradhan and Ranjit Kumar Sahu

CONTENTS

10.1 Introduction

Genomic signal processing [1] (GSP) is a comparatively novel area in bioinformatics. It deals with the utilization of traditional digital signal processing (DSP) techniques in the representation and analysis of genomic data. As the gene microarray becomes more accessible, microarray gene expression data are finding applications in diverse areas of cancer study, and they can be used to test and generate new hypotheses [2]. A fundamental step in the understanding of a genome is the computational recognition, and in the analysis of newly sequenced genomes, it is one of the challenges. Conventional and modern signal processing techniques play a vital part as an accurate and speedy tool in the analysis of genomic sequences [3,4].

DNA microarrays, which are influential tool in the study of collective gene reaction to changes in the environment, are presented by a gene expression microchip, and also offers indications about the structures of the involved gene networks [5]. Nowadays, in a solitary experiment employing microarrays, the expression levels of thousands of genes, possibly all genes in an organism, can be measured simultaneously [6]. In monitoring, genome-wide expression levels of gene DNA microarray technology has become a requisite tool [7]. The evaluation of the gene expression profiles in a variety of organs, using microarray technologies, discloses separate genes, gene ensembles, and the metabolic ways underlying the structural and functional organization of an organ and its physiological function [8]. By employing microarray technology, the diagnostic chore can be automated, and the precision of the conventional diagnostic techniques can be enhanced. Simultaneous examination of thousands of gene expressions is being facilitated by microarray technology [9].

DNA microarrays are being increasingly used for diagnostic classification of cancers [10]. The field of computational biology that is involved with algorithmically distinguishing the stretches of sequence, generally genomic DNA that are biologically functional, is known as gene finding. This in particular not only engrosses protein-coding genes but also includes added functional elements, for instance, RNA genes and regulatory regions [11]. Some of the research on gene prediction are works by Singhal et al. [12], Freudenberg and Propping [13], and Stanke and Waack [14].

In this chapter, we propose an effective gene prediction technique that predicts the dominant genes. Initially, the classified microarray gene data set [central nervous system (CNS) tumor, colon tumor, lung tumor and diffuse large B-cell lymphoma (DLBCL)], which is of high dimension, is reduced through the hybrid form of genetic algorithm (GA) with artificial neural network (ANN) to generate the training data set for the neural network (NN). Consequently, the designed data set is used to predict the dominant genes of the cancer. Subsequently, the gene that causes the cancer is predicted without analyzing the entire database. The rest of the chapter is organized as follows. In Section 10.2, a brief review of some of the existing works in gene prediction is presented. The proposed effective gene prediction is detailed in Section 10.3. Sections 10.4 and 10.5 describe the results and discussion. The conclusions are summed up in Section 10.6.

10.2 Review of Related Research

Some of the recent related research works are reviewed here.

Wong and Wang [15] have proposed the regulation-level method using microarray data of cancer classification that can be optimized utilizing GAs. The proposed symbolization decreases the dimensionality of microarray data to a greater extent, and several statistical machine learning methods have become usable and efficient in cancer classification. It has been confirmed that the three-regulation-level representation monotonically converges to a solution by experimental results on real-world microarray data sets. In addition to the improvement to cancer classification capability, the ternary regulation level promotes the visualization of microarray data.

Sarhan [8] has developed an ANN and the discrete cosine transform (DCT)-based stomach cancer detection system. Classification features are extracted by the proposed system from stomach microarrays utilizing DCT. ANN does the classification (tumor or no tumor) upon application of the features extracted from the DCT coefficients. Sarhan's study used the microarray images that were obtained from the Stanford Medical Database (SMD). The ability of the proposed system to produce a very high success rate has been confirmed by simulation results.

Hoff et al. [16] proposed a machine learning approach as a gene prediction algorithm for metagenomic fragments. Initially, linear discriminants were employed for mono-codon usage, di-codon usage, and translation initiation sites. Secondly, an open reading frame encodes a protein, and an ANN combines these characteristics with open reading frame length and fragment guanine cytosine (GC)-content. This probability was employed for categorizing and achieving the gene candidates. By means of extensive training, this technique formed fast single fragment predictions with fine-quality sensitivity and specificity on artificially fragmented genomic DNA. Additionally, with high consistency, this technique can precisely calculate translation initiation sites and distinguish complete genes from incomplete genes.

10.3 Dominant Gene Prediction Using ANN

Exploitation of a large gene data set for disease analysis increases the computation time and reduces the performance of the process. Therefore, a technique that requires less computational time to predict dominant genes is essential. So an efficient technique is proposed to predict the dominant genes of cancer in the form of CNS tumors, colon tumors, lung tumors, and DLBCL from a microarray gene data set.

10.3.1 Preprocess for Dominant Gene Prediction

The preprocessing steps for predicting dominant genes are explained in the steps shown in Figure 10.1.

In order to generate the training set for the ANN, the possible combinations of the gene data set are required. The two processes involved in the generation of a training data set are generation of possible combinational data and dimensionality reduction [17,18].

Possible combinational data are generated with the intention of simplifying the learning process for dominant gene prediction by classifying the microarray gene data set with a lot of combinations within the data set. Let M_{ij} be the microarray gene data set, where $0 \le i \le N_s - 1$ and $0 \le i \le N_g - 1$. Here, N_s represents the number of samples, and N_g represents the number of genes; the size of M_{ij} is given by $N_s \times N_g$. The number of possible combinational

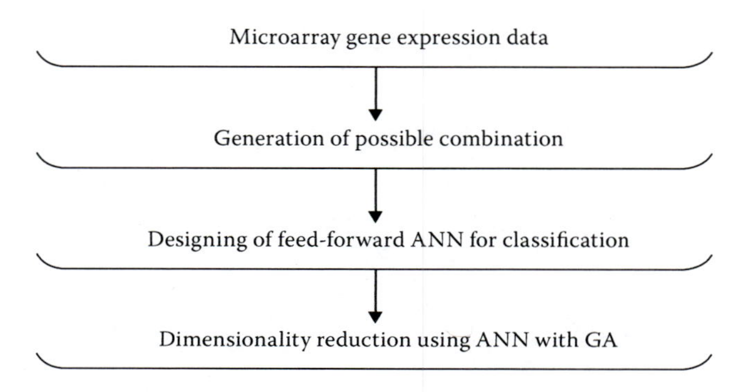

FIGURE 10.1
Preprocessing steps for dominant gene prediction.

data $M_{c_{ij}}$ has a high dimension of $N_s' \times N_g'$, which has to be reduced for diagnosis purposes, calculated as follows:

$$M_{cij} = \frac{(N_s \times N_g)!}{((N_s \times N_g) - k)!\ k\ !} \tag{10.1}$$

The dimensionality reduction is achieved using the hybrid-form GA with feed-forward ANN (FF-ANN), and the high-dimensional $M_{c_{ij}}$ was converted to low dimension using the single-point crossover and interchange method in mutation. The dimensionality-reduced data is utilized as the training data set for the NN. Thus, training data set $\hat{M}_{c_{ij}}$ for the ANN is generated with reduced dimension $N_s'' \times N_g''$.

10.3.2 Training through FF-ANN

The dimensionality-reduced microarray gene data set is utilized for training the feed-forward neutral network with backpropagation (BP). The proposed technique incorporates a multilayer FF-ANN with BP for predicting the dominant genes of the CNS tumor, colon tumor, lung tumor, and DLBCL. In this NN, N_s'' (dimensionality-reduced) input neurons and a bias neuron, N_g'' hidden neurons and a bias neuron, and an output neuron y_i are presented. A feed-forward network maps a set of input values to a set of output values, and the graphical representation of a parametric function is supposed. The NN to train the gene data set is shown in Figure 10.2. The steps to be followed in this process are given below.

Step 1: Set the input weights of every neuron network with N_g'' input layers, and N_g'' hidden layers and an output layer are designed.

Step 2: The designed NN is weighted and biased. The developed NN is shown in Figure 10.2.

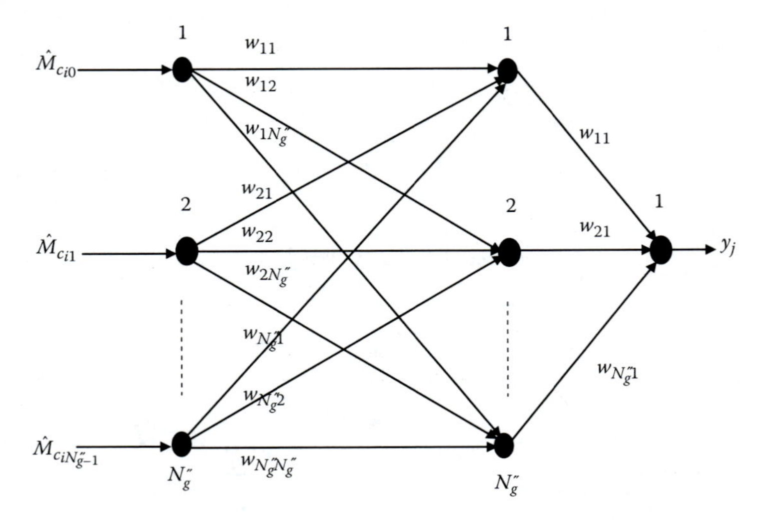

FIGURE 10.2
n-inputs one-output neural network to train the gene data set.

Step 3: The basis function and the activation function that are chosen for the designed NN are shown in Equations 10.2 and 10.4, respectively, whereas $\hat{M}_c$ is the dimensionality-reduced microarray gene data, w_{ij} is the weight of the neuron, α is the bias, and y refers to the ranges [0, 1].

$$Y_i = \alpha + \sum_{j=0}^{N_g''-1} w_{ij}\hat{M}_{c_{ij}}, \quad 0 \le i \le N_s'' - 1 \tag{10.2}$$

$$g(y) = \frac{1}{1+e^{-y}} \tag{10.3}$$

Step 4: The basis function given in Equation 10.1 is commonly used in all the remaining layers (hidden and output layer, but with the number of hidden and output neurons, respectively). The output of the ANN is determined is determined by giving it $\hat{M}_c$ as the input.

Step 5: The learning error is determined for the NN as follows:

$$E = \frac{1}{N_s''}\sum_{b=0}^{N_s''-1} D - Y_b \tag{10.4}$$

where E is the error in the FF-ANN, D is the desired output and Y_b is the actual output.

10.3.3 Minimization of Error by BP Algorithm

The BP algorithm is used to train the network and to minimize the total errors. The steps involved in backpropagation algorithm are discussed below:

Step 1: Assign random weights to the links range [0, 1] in the ANN above.

Step 2: In case of training, perform output calculation based on two functions, that is, basis function (the product of weights and inputs) and activation function (nonlinear).

Step 3: In order to determine the BP error using Equation 10.5, the training gene data sequence is given to the NN. Equations 10.2, 10.3, and 10.4 show the basis function and transfer function.

Step 4: The weights of all the neurons are adjusted when the BP error is determined as follows:

$$w_{ij} = w_{ij} + \Delta w_{ij} \tag{10.5}$$

The change in weight Δw_{ij} given in Equation 10.3 can be determined as $\Delta w_{ij} = \gamma \cdot y_{ij} \cdot E$, where E is the BP error and γ is the learning rate; normally, it ranges from 0.2 to 0.5.

Step 5: After adjusting the weights, steps 2 and 3 are repeated until the BP error is minimized. Normally, it is repeated till the criterion $E < 0.1$ is satisfied.

When the error is minimized, it is construed that the designed ANN is well trained for its further testing phase, and the BP algorithm is terminated. Thus, the NN is trained by using the samples. Then, to determine the dominant genes of the cancer, the GA with ANN is applied.

10.3.4 GA-Based Dominant Gene Prediction of Cancer Diagnosis

In order to train the NN, selected features were normalized; this normalization was necessary to prevent nonuniform learning, in which the weight associated with some features converges faster than that associated with others. After normalization, a randomly chosen sample was divided into training, cross-validation, and testing data sets. The training data set was presented to the network for learning. The cross-validation data set was used to measure the training performance during training or stop training if necessary. The GA was applied on the classified test sequence, and then this test sequence was evaluated, and the dominant genes were predicted. In this GA-based dominant gene prediction, initially, the random chromosomes are generated. The random chromosomes are the indices of the test sequence, which are classified as CNS tumor (or) colon tumor (or) lung tumor (or) DLBCL.

The process of GA to predict the dominant gene is discussed below, and the flowchart is shown in Figure 10.3.

Step 1: Generation of chromosomes in a random manner, which are the indices of the test sequences.

$$D^{(k)} = \left\{D_0^{(k)}, D_2^{(k)}, D_3^{(k)}, \ldots, D_{n-1}^{(k)}\right\} 0 \le k \le N_p - 1, 0 \le l \le n-1 \quad (10.6)$$

In Equation 10.6, $D_l^{(k)}$ represents the lth gene of the kth chromosome, and N_p is the number of random chromosomes; the number of genes in each chromosome relies on N_g'', that is, the number genes in the training data set.

Step 2: The fitness function is evaluated using Equation 10.7. In Equation 10.7, N_{out} is the network output obtained from the FF-ANN for the kth chromosome, and N_{fit} in Equation 10.8 is the fitness value of the initially generated chromosomes.

$$\mu_{net} = \frac{\sum_{k=0}^{N_p-1} N_{out}}{|k|} \quad (10.7)$$

$$N_{fit} = \frac{1}{(1-\mu_{net})^c} \quad (10.8)$$

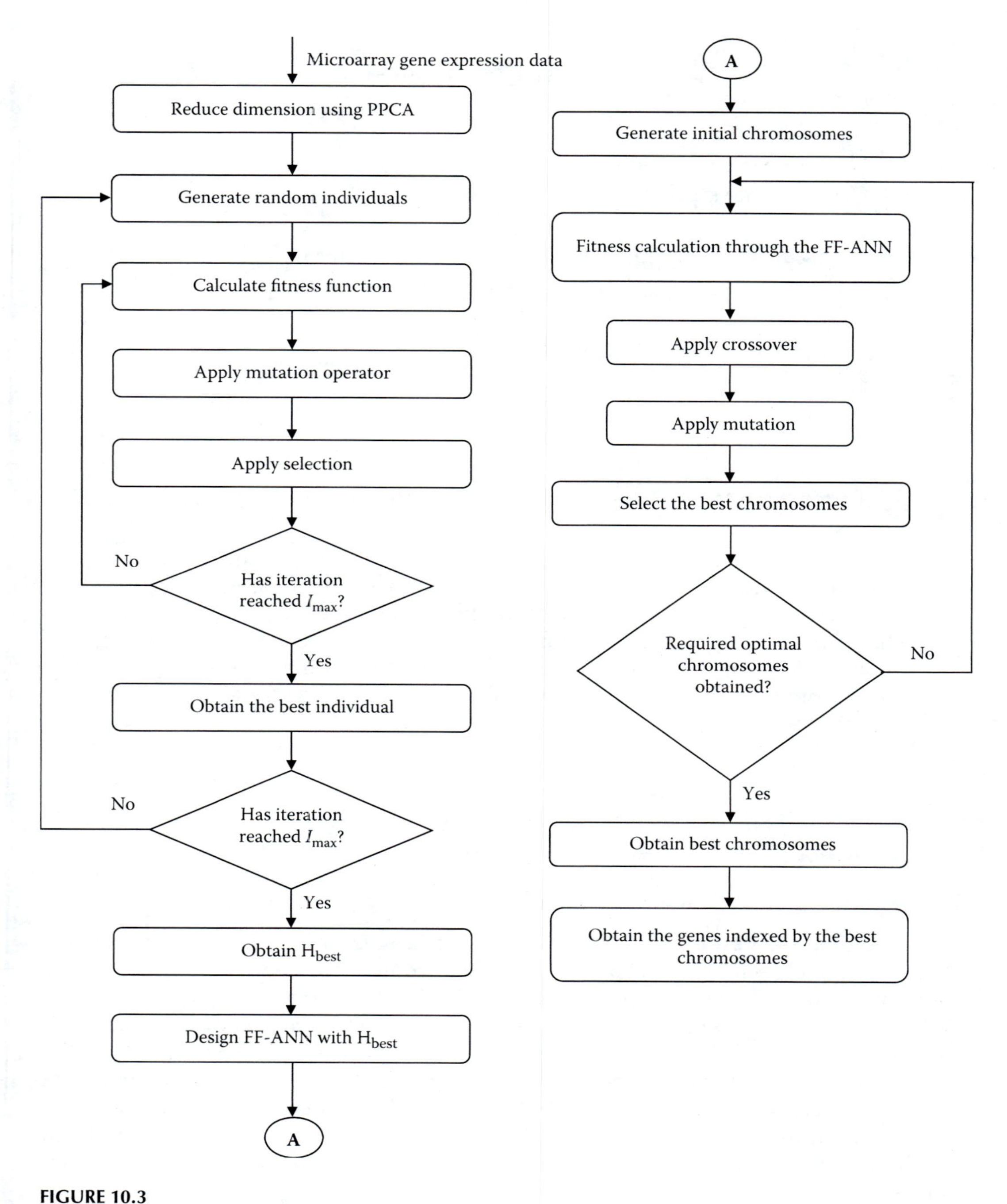

FIGURE 10.3
FF-ANN enhanced using evolutionary programming (EP) technique and dominant gene prediction based on genetic algorithm.

$$\left.\begin{cases} c = 0 \text{ if test sequence is class 1} \\ c = 1 \text{ if test sequence is class 2} \end{cases}\right\} \text{in case of CNS tumor}$$

$$\left.\begin{cases} c = 0 \text{ if test sequence is negative} \\ c = 1 \text{ if test sequence is positive} \end{cases}\right\} \text{in case of colon tumor}$$

$$\left.\begin{cases} c = 0 \text{ if test sequence is malignant pleural mesothelioma (MPM)} \\ c = 1 \text{ if test sequence is adenocarcinoma (ADCA)} \end{cases}\right\} \text{in case of lung cancer}$$

$$\left.\begin{cases} c = 0 \text{ if test sequence is germinal center B-like} \\ c = 1 \text{ if test sequence is activated B-like} \end{cases}\right\} \text{in case of DLBCL}$$

Step 3: Perform random crossover and mutation until a maximum iteration of I_{max} is reached.

Step 4: The best chromosomes are selected from the group of chromosomes that is obtained after the process is repeated I_{max} times. The obtained best chromosomes are used to retrieve the corresponding gene values from the test sequence. The gene values of the cancer represented by the indices, which are obtained from the genes of the best chromosomes, are the dominant genes of the CNS tumor, colon tumor, lung tumor, and DLBCL, and they are retrieved in an effective manner.

10.4 Results and Discussion

In this section, the performance of the proposed classifier technique is compared with the conventional technique, probability principal component analysis (PPCA), used in the previous work [17,18]. The proposed classification technique is implemented in the MATLAB® platform, and it is evaluated using the microarray gene expression data. The training data set is of dimension $N_g = 7129$ and $N_s = 60$ in case of CNS tumor, $N_g = 2000$ and $N_s = 62$ in case of colon tumor, $N_g = 12533$ and $N_s = 32$ in case of lung tumor, and $N_g = 4026$ and $N_s = 47$ in case of DLBCL [19]. This high-dimensional training data set is subjected to dimensionality reduction using ANN with GA, and so a data set of dimension 50×60, 50×62, 50×32, and 50×47 for CNS tumor, colon tumor, lung tumor, and DLBCL, respectively, is obtained.

This training data set is utilized to design the FF-ANN, and then the test input sequence is tested through the GA. After the completion of the crossover and mutation operations, based on the conditions given in Section 10.3, the optimal chromosomes are obtained. Testing is done under the evaluation of sensitivity and specificity values. These values are among the terms true positive (TP), true negative (TN), false positive (FP), and false negative (FN) [20], which are given in Table 10.1.

TABLE 10.1

TP, TN, FP, and FN Values for Four Data Sets for the Classifiers PPCA and GA with ANN

	Dimension Reduction Using PPCA				Dimension Reduction Using GA with ANN			
Data Sets	**TP**	**TN**	**FP**	**FN**	**TP**	**TN**	**FP**	**FN**
CNS tumor	3	2	3	2	4	3	2	1
Colon tumor	2	3	2	3	3	4	1	2
Lung tumor	2	2	3	3	4	2	1	3
DLBCL	3	3	2	2	2	4	3	1

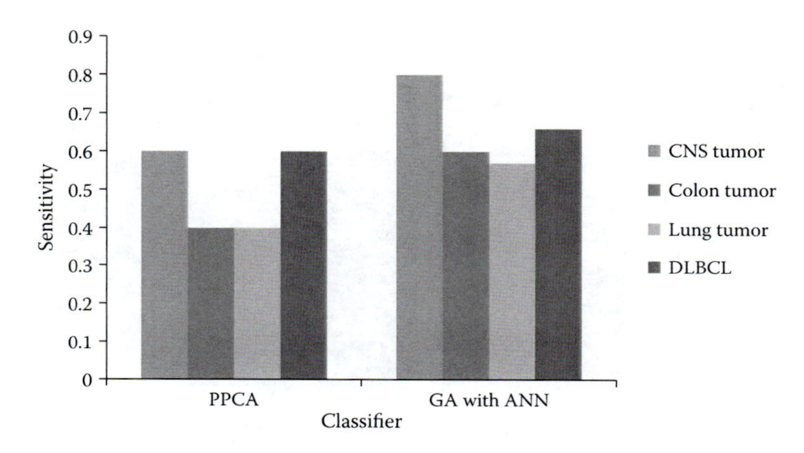

FIGURE 10.4
Graphical representation of sensitivity value for PPCA and GA with ANN classifiers under different data sets.

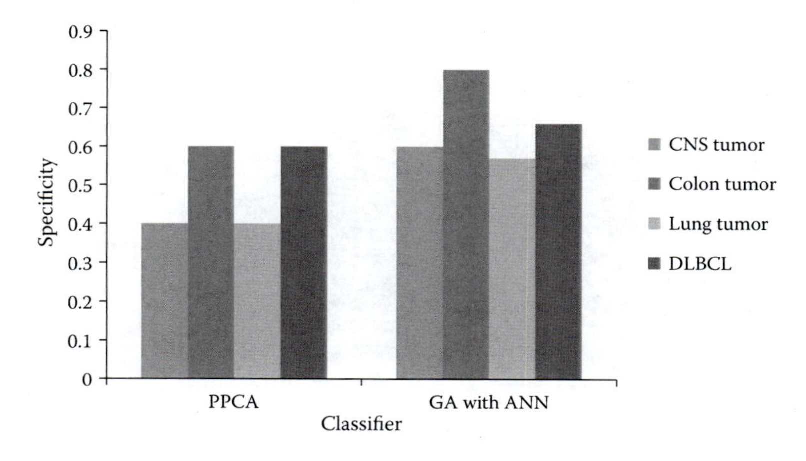

FIGURE 10.5
Graphical representation of specificity value for PPCA and GA with ANN classifiers under different data sets.

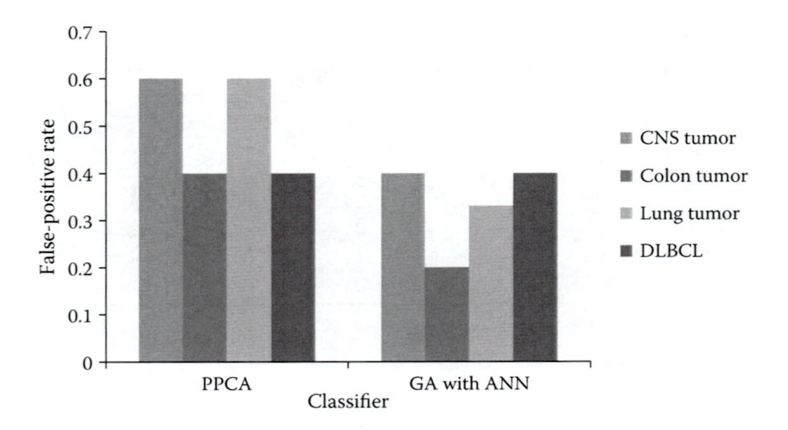

FIGURE 10.6
Graphical representation of false-positive rate for PPCA and GA with ANN classifiers under different data sets.

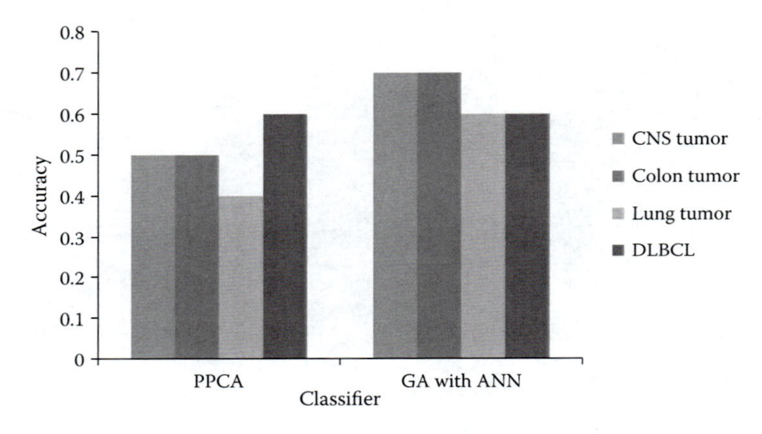

FIGURE 10.7
Graphical representation of accuracy of PPCA and GA with ANN classifiers under different data sets.

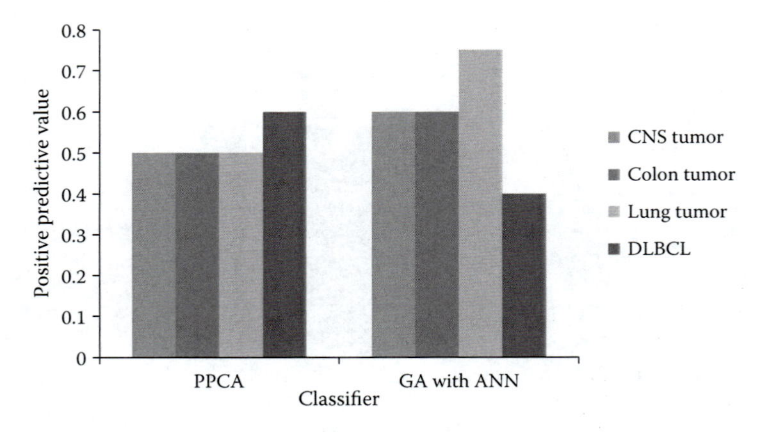

FIGURE 10.8
Graphical representation of positive prediction value for PPCA and GA with ANN classifiers under different data sets.

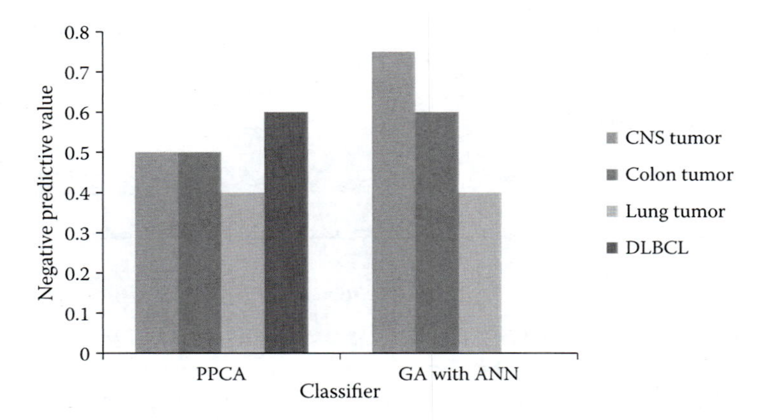

FIGURE 10.9
Graphical representation of negative prediction value for PPCA and GA with ANN classifiers under different data sets.

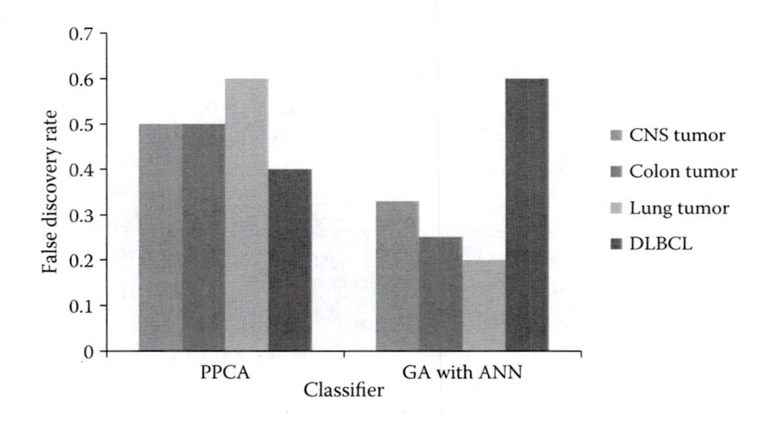

FIGURE 10.10
Graphical representation of false discovery rate for PPCA and GA with ANN classifiers under different data sets.

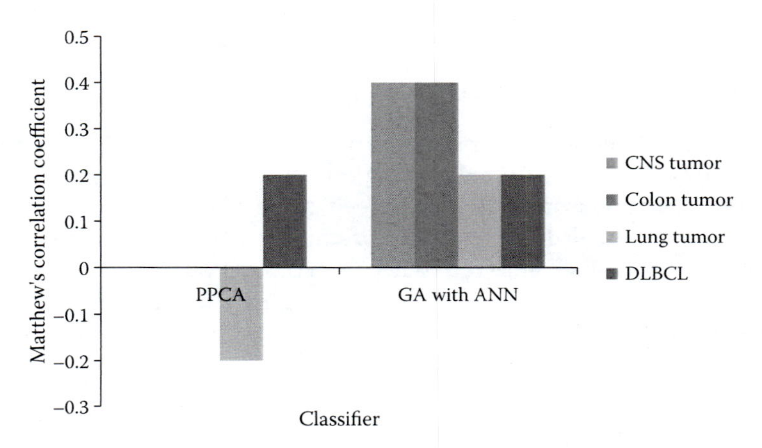

FIGURE 10.11
Graphical representation of Matthews correlation coefficient for PPCA and GA with ANN classifiers under different data sets.

The graphical representation for the sensitivity, specificity, false positive rate (FPR), accuracy, positive predictive value (PPV), negative predictive value (NPV), false discovery rate (FDR), and Matthew's correlation coefficient (MCC) for the PPCA and GA with ANN classifiers under different data sets are shown in Figures 10.4 through 10.11.

10.5 Discussion

Considering the values in Table 10.1, sensitivity, specificity, accuracy, positive prediction value, negative prediction value, false-positive rate, false discovery rate, and Matthew's correlation coefficient can be found out. The graphical representations of the testing values are plotted in the Figures 10.4 through 10.11.

In the case of sensitivity in Figure 10.4, the cancer prediction rate using the classifier GA with ANN is better than PPCA: GA with ANN performs at 57%–80%, but PPCA does so at only 40%–60%. The specificity in Figure 10.5 also yields a better outcome with 60%–80% in the case of GA with ANN as opposed to PPCA, with 40%–60%. The false-positive rate in Figure 10.6, which is recognized as an error, is observed to be a maximum of 60% in PPCA and a maximum of 40% in GA with ANN. The accuracy rate in Figure 10.7 is 40%–60% in PPCA and 60%–70% in GA with ANN. In Figure 10.8, positive predictive value, which is the proportion of the positive results, has been identified to be a maximum of 60% in PPCA and 75% in GA with ANN. In Figure 10.9, the negative prediction value, the proportion of negative results, is 40%–60% in PPCA and 40%–75% in GA with ANN. The false discovery rate in Figure 10.10, which is the identification of false results, is analyzed as a minimum of 40% in PPCA and 20% in GA with ANN. The Matthew's correlation coefficient has been used for the identification of the results, which hold its value from the range of 0 to 1. The value of 1 corresponds to a correct identification; while analyzing Figure 10.11, it is concluded that GA with ANN has better results than PPCA. So the cancer prediction can be done easily using the proposed classifier GA with ANN.

10.6 Conclusion

We have presented a mechanism to predict cancer patients using ANNs. The training and testing of the four data sets under consideration is undergone using the cross-validation and classification function (sensitivity and specificity measures), respectively. Results presented in this chapter show that the cancer prediction is easily done using the proposed GA-with-ANN mechanism compared with the conventional PPCA method. Even the accuracy level of the proposed classifier is encouraging. ANNs need some enhancement to improve their accuracy level. Some adaptive techniques with the ANN will be used for improvement in future works.

Acknowledgment

The author acknowledges the University Grants Commission (UGC), New Delhi, India, for research support for the ongoing study.

References

1. D. Anastassiou, Genomic signal processing, *IEEE Signal Processing Magazine*, vol. 18, pp. 8–20, 2001.
2. Z. Peng Qiu, J. Wang, and K.J. Ray Liu, Genomic processing for cancer classification and prediction, *IEEE Signal Processing Magazine*, pp. 100–110, January 2007.
3. A. Rodriguez, J. V. Lorenzo, and R. Grau Abalo, A new predictor of coding regions in genomic sequences using a combination of different approaches, *International Journal of Biological and Life Sciences*, vol. 3, no. 2, pp. 106–110, 2007.
4. M. Pradhan and R.K. Sahu, An extensive survey on gene prediction methodologies, *International Journal of Computer Science and Information Security*, vol. 8, no. 7, October 2010.
5. Y. Xu, V. Olman, and D. Xu, Minimum spanning trees for gene expression data clustering, *Genome Informatics*, vol. 12, pp. 24–33, 2001.
6. A. Gauthaman, Analysis of DNA microarray data using association rules: a selective study, *World Academy of Science, Engineering and Technology*, vol. 42, pp. 12–16, 2008.
7. C.K. Sarmah, S. Samarasinghe, D. Kulasiri, and D. Catchpoole, A simple affymetrix ratio-transformation method yields comparable expression level quantifications with cDNA data, *World Academy of Science, Engineering and Technology*, vol. 61, pp. 78–83, 2010.
8. N.S. Khlopova, V.I. Glazko, and T.T. Glazko, Differentiation of gene expression profiles data for liver and kidney of pigs, *World Academy of Science, Engineering and Technology*, vol. 55, pp. 267–270, 2009.
9. A.M. Sarhan, Cancer classification based on microarray gene expression data using DCT and ANN, *Journal of Theoretical and Applied Information Technology*, vol. 6, no. 2, pp. 207–216, 2009.
10. J. Khan, J.S. Wei, M. Ringner, et al. Classification and diagnostic prediction of cancers using gene expression profiling and artificial neural networks, *Nature Medicine*, vol. 7, pp. 673–679, 2001.
11. S. Lambrosa, I. Panosc, and L. Spiridona, Coding potential prediction in Wolbachia using artificial neural networks, *Silico Biology*, vol. 7, pp. 105–113, 2007.
12. P. Singhal, B. Jayaram, S.B. Dixit, and D.L. Beveridge, Prokaryotic gene finding based on physicochemical characteristics of codons calculated from molecular dynamics simulations, *Biophysical Journal*, vol. 94, pp. 4173–4183, June 2008.
13. J. Freudenberg and P. Propping, A similarity-based method for genome-wide prediction of disease-relevant human genes, *Bioinformatics*, vol. 18, no. 2, pp. 110–115, April 2002.
14. M. Stanke and S. Waack, Gene prediction with a hidden Markov model and a new intron submodel, *Bioinformatics*, vol. 19, no. 2, pp. 215–225, 2003.
15. H.-S. Wong and H.-Q. Wang, Constructing the gene regulation-level representation of microarray data for cancer classification, *Journal of Biomedical Informatics*, vol. 41, no. 1, pp. 95–105, February 2008.

16. K.J. Hoff, M. Tech, T. Lingner, R. Daniel, B. Morgenstern, and P. Meinicke, Gene prediction in metagenomic fragments: a large scale machine learning approach, *BMC Bioinformatics*, vol. 9, no. 217, pp. 1–14, April 2008.
17. M. Pradhan, S.S. Pattnaik, B. Mitra, and R.K. Sahu, GA-ANN based dominant gene prediction in microarray dataset, *International Journal of Computer Science and Information Security*, vol. 8, no. 8, November 2010.
18. M. Pradhan, S.S. Pattnaik, and B. Mitra, Effective classification technique by blending of PPCA and EP-enhanced supervised classifier: classifies microarray gene expression data, *American Journal of Scientific Research*, vol. 11, pp. 60–71, 2010.
19. CNS tumor, colon tumor, lung tumor and diffuse large B-cell lymphoma from http://levis.tongji.edu.cn/gzli/data/mirror-kentridge.html.
20. en.wikipedia.org/wiki/Sensitivity_and_specificity.

11

A System for Melanoma Diagnosis Based on Data Mining

Jerzy W. Grzymala-Busse, Zdzislaw S. Hippe, and Lukasz Piatek

CONTENTS

11.1 Introduction

Melanoma is a dangerous skin cancer. It is less common than other skin cancers but much more dangerous if not detected early. Melanoma causes 75% of deaths of people affected by skin cancer. Annually, 160,000 new cases of melanoma are reported worldwide [1].

A common way to diagnose melanoma is through the *ABCD* rule. There exist two basic variations of the *ABCD* rule. In the former, every character, *A*, *B*, *C*, and *D*, stands for some symptom; the purpose of the rule is to raise awareness. The latter variation, which we will call the *ABCD* formula, is more precise; every character is associated with some numerical value, and the entire formula is used to compute the dermatoscopic score (TDS). Diagnosticians use some levels of TDS to diagnose melanoma. In both cases, *A* stands for asymmetry, *B* for border, *C* for color, and *D* for diversity (or differential structure). Note that there exist a few variations of the rule; for example, according to some, *D* stands for diameter (where lesions larger than 6 mm in diameter are suspicious). Also, there exists yet another rule, called *ABCDE*, where *E* means evolving (or changing) over time or lesion elevation.

In this chapter, we will adopt the *ABCD* formula as published in Nachbar et al. [2] and Stoltz et al. [3], where

- The variable *A* (asymmetry) has three different values: *symmetric* (counts as 0), *one-axial symmetry* (counts as 1), and *two-axial symmetry* (counts as 2).
- *B* (border) is a numerical attribute with values from 0 to 8. A lesion is divided into eight segments equal to each other, defined by four axes, all crossing in the lesion center. The border of each segment is evaluated; the sharp border contributes 1 to *border*; the gradual border contributes 0.

- *C* (color) has six possible values: black, blue, dark brown, light brown, red, and white.
- *D* (diversity) has five values: pigment dots, pigment globules, pigment network, structureless areas, and branched streaks. Examples of such structures may be found in the work of Stoltz et al. [4] and Wolff and Johnson [5].

Thus, the maximum value of *C* is 6, and the maximum value of *D* is 5. The TDS is computed by the following formula:

$$\text{TDS} = 1.3{*}\text{asymmetry} + 0.1{*}\text{border} + 0.5{*}\Sigma \text{ colors} + 0.5{*}\Sigma \text{ diversities},$$

where Σ colors represents the sum of all values of the six color attributes and Σ diversities represents the sum of all values of the five diversity attributes. For values of TDS smaller than 4.9, with lesions that do not contain blue color spots, the diagnosis is that the lesion is *benign nevus*; for the same TDS and lesions containing blue color spots, the diagnosis is *blue nevus*; for TDS between 4.9 and 5.5, the diagnosis is *suspicious melanoma*; and for TDS greater than 5.5, the lesion is *malignant melanoma*.

There exist other strategies, based on different ideas for diagnosis of melanoma than the *ABCD* formula [6,7].

11.2 Data Set

Many experiments were conducted on a data set collected at the Regional Dermatology Center in Rzeszow, Poland [8]. The data set has 410 cases, with 146 cases of *benign nevus*, 78 cases of *blue nevus*, 92 cases of *suspicious melanoma*, and 94 cases of *malignant melanoma*. In our data set, color and diversity were replaced by binary single-valued variables, so the original data set had 13 attributes. Obviously, TDS is a linear combination of the remaining 13 attributes. However, it was shown that the error rate as a result of tenfold cross-validation is smaller with TDS [9].

Our earlier results on optimization of the *ABCD* formula have been published [10–13].

Note that in some papers [14,15], claims were published that the *ABCD* formula does not improve diagnosis of melanoma. To validate these claims, we ran experiments on melanoma data with a removed attribute called TDS, which is a result of the *ABCD* formula.

11.3 Rule Induction and Validation

In our experiments, rules were induced by the Learning from Examples Module, version 2 (LEM2) algorithm. LEM2 is a part of the system Learning from Examples based on Rough Sets (LERS) [16]. Rough set theory was initiated in 1982 [17,18].

The most important performance criterion for methods of data mining is the total number of errors. For evaluation of an error number, we used tenfold cross-validation: All cases were randomly reordered, and then the set of all cases was divided into 10 mutually disjoint subsets of approximately equal size. For each subset, all remaining cases were used for training, that is, for rule induction, while the subset was used for testing. Thus, each case was used nine times for training and once for testing. Note that using different reorderings of cases causes slightly different error numbers. LERS may use *constant tenfold cross-validation* by using the same way of case reordering for all experiments. Also, LERS may perform tenfold cross-validation using different case reorderings for every experiment, called *variable tenfold cross-validation*.

11.4 Optimization of the *ABCD* Formula

Many experiments were conducted to find the optimal *ABCD* formula [9,10,12,13,19,20]. The main assumption was that the optimal *ABCD* formula is still a linear combination of 13 attributes:

$$\begin{aligned}\text{new\_TDS} = {} & c_1 * \text{asymmetry} + c_2 * \text{border} + c_3 * \text{color\_black} + c_4 * \text{color\_blue} + c_5 * \\ & \text{color\_dark\_brown} + c_6 * \text{color\_light\_brown} + c_7 * \text{color\_red} + c_8 * \\ & \text{color\_white} + c_9 * \text{diversity\_pigment\_dots} + c_{10} * \\ & \text{diversity\_pigment\_globules} + c_{11} * \text{diversity\_pigment\_network} + c_{12} * \\ & \text{diversity\_structureless\_areas} + c_{13} * \text{diversity\_branched\_streaks}.\end{aligned}$$

The objective was to find the optimal values of $c_1, c_2, \ldots, c_{13}$, where the optimization criterion was the smallest average error rate for five experiments of variable tenfold cross-validation for data with 13 old, unchanged attributes and with a new 14th attribute, new TDS, that replaced the original TDS attribute.

Basically, the selection of optimal values of $(c_1, c_2, \ldots, c_{13})$ was done using brute force, by searching through all possibilities, through more than 160,000 experiments, and evaluating every choice of the coefficients $c_1, c_2, \ldots, c_{13}$ by five-variable tenfold cross-validations. Additionally, a few discretization techniques were selected and combined with selection of $c_1, c_2, \ldots, c_{13}$. In phase 1, the total number of executed variable tenfold cross-validations was about 163,000. In phase 2 of our experiments, the total number of errors was computed using 30 experiments of variable tenfold cross-validation experiments. In the experiments reported in the work of Alvarez et al. [9], four used discretization techniques were based on polythetic divisive method of cluster analysis [21,22]. The best results were provided by the following formula:

$$\begin{aligned}\text{new\_TDS} = {} & 0.8 * \text{asymmetry} + 0.11 * \text{border} + 0.5 * \text{color\_black} + 0.8 * \text{color\_blue} + 0.5 * \\ & \text{color\_dark\_brown} + 0.6 * \text{color\_light\_brown} + 0.5 * \text{color\_red} + 0.5 * \\ & \text{color\_white} + 0.5 * \text{diversity\_pigment\_dots} + 0.6 * \\ & \text{diversity\_pigment\_globules} + 0.5 * \text{diversity\_pigment\_network} + 0.6 * \\ & \text{diversity\_structureless\_areas} + 0.6 * \text{diversity\_branched\_streaks}.\end{aligned}$$

TABLE 11.1

Comparison of Three Data Sets

Data Set With	Number of Errors	Standard Deviation
Original TDS	12.17	2.157
New TDS	13.97	1.520
No TDS	55.23	4.666

TABLE 11.2

Quality of Rule Sets, Evaluated by an Expert Diagnostician

Data Set With	Average Score per Rule
Original TDS	3.23
New TDS	2.72

Additionally, we excluded the 14th attribute, TDS, and computed an error rate again using 30 variable tenfold cross-validation experiments. Results are presented in Table 11.1. It is obvious that diagnosis of melanoma is significantly better with new TDS (level of significance 5%, two-tailed test) than with the original *ABCD* formula. Moreover, diagnosis of melanoma without taking TDS into account is much less accurate. Surprisingly, the discretization intervals for new TDS were 1..4.75, 4.8..5.45, 5.5..8.7, very close to the intervals for original TDS (1..4.85, 4.9..5.45, 5.5..8.7).

Finally, selected rule sets, induced from data sets obtained by addition of either original TDS or new TDS, were graded by an experienced melanoma diagnostician [9]. Every rule was graded on a scale from 0 to 5, with 0 meaning a useless rule (i.e., a rule that is misleading) and 5 meaning an excellent rule. Results are presented in Table 11.2. Note that rules with the original TDS received higher scores; most likely, the expert was used to diagnosis with the original TDS.

11.5 Internet Melanoma Diagnosing and Learning System

A system called the Internet Melanoma Diagnosing and Learning System (IMDLS) was developed by the Department of Expert Systems and Artificial Intelligence, University of Information Technology and Management, in Rzeszow, Poland [23–32]. In its current form, IMDLS supports diagnosis of melanoma in two languages: English and Polish. Five different classifiers are included in IMDLS:

- TDS computed using the original *ABCD* formula based on TDS (Section 11.1)
- New TDS computed using the optimized *ABCD* formula (Section 11.4)
- Decision tree created especially for melanoma diagnosis [27]
- A classifier based on a linear learning machine with genetic searching for the most relevant attributes [27,28]
- Belief network [27,29,30]

The five outcomes, produced by five classifiers of the IMDLS, are combined using a voting mechanism to output a unique result of diagnosis. The IMDLS is more accurate than all five individual classifiers [25].

11.6 Synthetic Skin Lesions

A system generating synthetic skin lesions was presented in the work of Chmielewski and Grzymala-Busse [33], Friedman et al. [34], and Mroczek et al. [35]. The system transforms a set of specifications (such as the following—create an image of the skin lesion that is symmetric, has 0 segments of sharp border, is partially light brown and partially dark brown, and contains two structures: pigmented globules and pigmented dots) into a synthetic skin lesion image. Such a system may be used for teaching in programs in which recognition of melanoma is important, such as public health, cosmetology, and so forth.

Examples of synthetic skin lesions are presented in Figures 11.1 through 11.4. Tables 11.3 through 11.6 present respective attribute values.

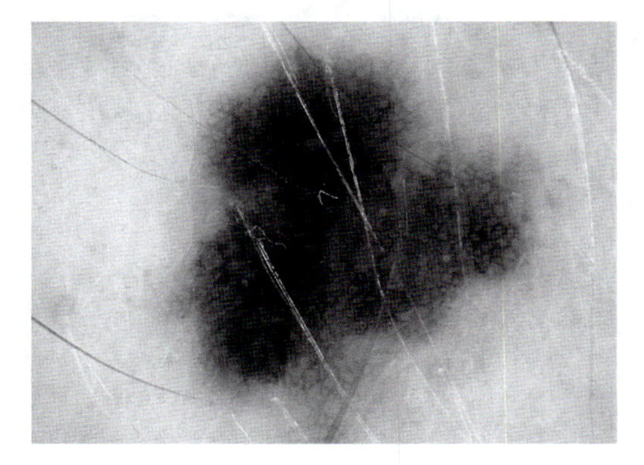

FIGURE 11.1
(See color insert.) First lesion.

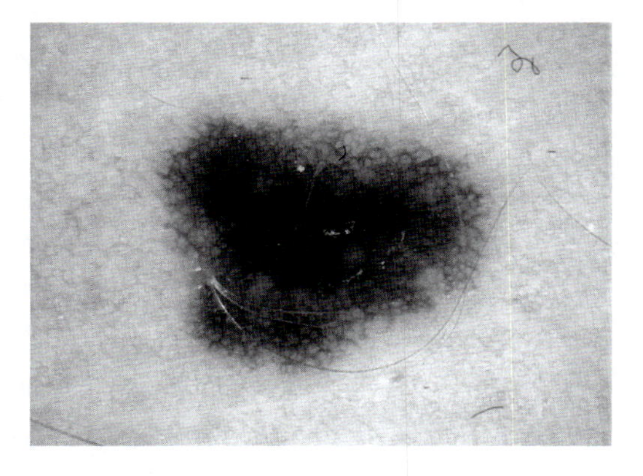

FIGURE 11.2
(See color insert.) Second lesion.

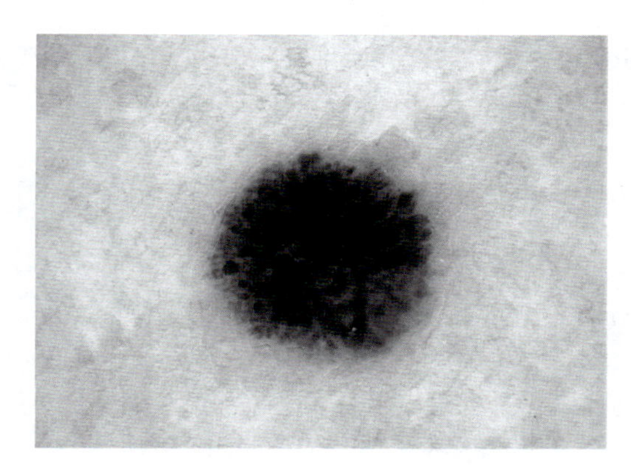

FIGURE 11.3
(See color insert.) Third lesion.

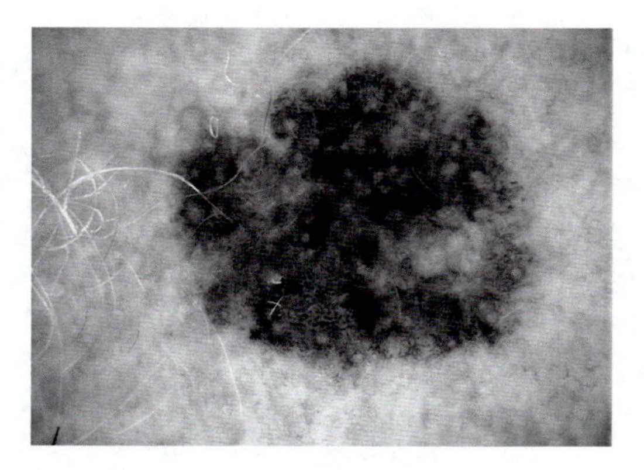

FIGURE 11.4
(See color insert.) Fourth lesion.

TABLE 11.3
Lesion 1

Attribute	Value
Asymmetry	Symmetric
Border	0
Color	Dark brown, light brown
Diversity	Pigment globules, pigment network

TABLE 11.4
Lesion 2

Attribute	Value
Asymmetry	Symmetric
Border	0
Color	Light brown
Diversity	Pigment globules, pigment network

TABLE 11.5
Lesion 3

Attribute	Value
Asymmetry	One-axial symmetry
Border	2
Color	Black, dark brown, light brown
Diversity	Pigment dots, pigment globules, pigment network, structureless areas

TABLE 11.6
Lesion 4

Attribute	Value
Asymmetry	Symmetry
Border	5
Color	Black, dark brown, light brown
Diversity	Pigment dots, pigment globules, pigment network, structureless areas

11.7 Conclusions

We presented selected topics related to diagnosis of melanoma based on data mining. Firstly, the original *ABCD* formula and our attempts to optimize the *ABCD* formula are discussed. Results of many experiments on our basic melanoma data set were conducted. The standard statistical test about the difference between two means shows that the new, optimized TDS is significantly better than the original TDS (5% level of significance, two-tailed test). Additionally, we presented the IMDLS, available on the Internet, for diagnosis of melanoma based on five different classifiers. The IMDLS is more accurate than any of these five classifiers. This is not a surprise since ensemble learning usually provides better results. Finally, four examples of synthetic skin lesions are presented. These pictures were automatically generated by a special system designed for that purpose.

References

1. Melanoma, Wikipedia, the free encyclopedia. http://en.wikipedia.org/wiki/Melanoma diagnosis.
2. Nachbar, F., Stoltz, W., Merkle, T., Cognetta, A. B., Vogt, T., Landthaler, M., Bilek, P., Braun-Falco, O., and Plewig, G. (1994) The ABCD rule of dermatoscopy: high prospective value in the diagnosis of doubtful melanocytic skin lesions. *J. Am. Acad. Dermatol.* 30, 551–559.
3. Stoltz, W., Riemann, A., Cognetta, A. B., Pillet, L., Abmayr, W. (1994) ABCD rule of dermatoscopy: a new practical method for recognition of malignant melanoma. *Eur. J. Dermatol.* 4, 521–527.
4. Stolz, W., Braun-Falco, O., Bilek, P., Landthaler, A. B., and Cogneta, A. B. (1993) *Color Atlas of Dermatology*, Blackwell Science, Cambridge, MA.
5. Wolff, K. and Johnson, R. (2009) *Fitzpatrick's Color Atlas and Synopsis of Clinical Dermatology*, Sixth Edition. McGraw-Hill, New York.

6. Argenziano, G., Fabbrocini, G., Carli, P., De Giorgi, V., Sammarco, E., and Delfino, M. (1998) Epiluminescence microscopy for the diagnosis of doubtful melanocytic skin lesions. Comparison of the ABCD rule of dermatoscopy and a new 7-point checklist based on pattern analysis. *Arch. Dermatol.* 134, 1563–1570.
7. Menzies, S. W. (1997) Surface microscopy of pigmented skin tumors. *Aust. J. Dermatol.* 38, 40–43.
8. Hippe, Z. S. (1999) Computer database NEVI on endangerment by melanoma. *Task Quarterly* 4, 483–488.
9. Alvarez, A., Bajcar, S., Brown, F. M., Grzymala-Busse, J. W., and Hippe, Z. S. (2003) Optimization of the ABCD formula used for melanoma diagnosis. In *Proceedings of the IIPWM'2003, International Conference on Intelligent Information Processing and WEB Mining Systems*, Zakopane, Poland, Springer-Verlag, 233–240.
10. Bajcar, S., Grzymala-Busse, J. W., and Hippe, Z. S. (2002) A comparison of six discretization algorithms used for prediction of melanoma. In *Proceedings of the Eleventh International Symposium on Intelligent Information Systems*, IIS'2002, Sopot, Poland, Physica-Verlag, 3–12.
11. Grzymala-Busse, J. P., Grzymala-Busse, J. W., and Hippe, Z. S. (2001) Melanoma prediction using data mining system LERS. In *Proceeding of the 25th Anniversary Annual International Computer Software and Applications Conference COMPSAC 2001*, Chicago, October 8–12, 2001, IEEE Computer Society, 615–620.
12. Grzymala-Busse, J. W. and Hippe, Z. S. (2002) Postprocessing of rule sets induced from a melanoma data set. In *Proceedings of the COMPSAC 2002, 26th Annual International Conference on Computer Software and Applications*, Oxford, England, August 26–29, 2002, *IEEE Computer Society*, 1146–1151.
13. Grzymala-Busse, J. W. and Hippe, Z. S. (2002) A search for the best data mining method to predict melanoma. In *Proceedings of the RSCTC 2002, Third International Conference on Rough Sets and Current Trends in Computing*, Malvern, PA, Springer-Verlag, 538–545.
14. Lorentzen, H., Weismann, K., Secher, L., Peterson, C. S., and Larsen, F. G. (1999) The dermatoscopic ABCD rule does not improve diagnostic accuracy of malignant melanoma. *Acta Derm. Venereol.* 79, 469–472.
15. Pizzichetta, M. A., Piccolo, D., Argenziano, G., Pagnanelli, G., Burgdorf, T., Lombardi, D., Trevisan, G., Veronesi, A., and Soyer, H. P. (2001) The ABCD rule of dermatoscopy does not apply to small melanocytic skin lesions. *Arch. Dermatol.* 137, 1376–1378.
16. Grzymala-Busse, J. W. (1997) A new version of the rule induction system LERS. *Fundam. Inform.* 31, 27–39.
17. Pawlak, Z. (1982) Rough sets. *Int. J. Comput. Inform. Sci.* 11, 341–356.
18. Pawlak, Z. (1991) *Rough Sets: Theoretical Aspects of Reasoning about Data*. Kluwer Academic Publishers, Dordrecht.
19. Andrews, R., Bajcar, S., Grzymala-Busse, J. W., Hippe, Z. S., and Whiteley, C. (2004) Optimization of the ABCD formula for melanoma diagnosis using C4.5, a data mining system. In *Proceedings of the RSCTC'2004, the Fourth International Conference on Rough Sets and Current Trends in Computing*, Uppsala, Sweden, June 15, 2004. *Lecture Notes in Artificial Intelligence* 3066, Springer-Verlag, 630–636.
20. Bajcar, S., Grzymala-Busse, J. W., Grzymala-Busse, W. P., and Hippe, Z. S., (2003) Diagnosis of melanoma based on data mining and ABCD formulas. In *Proceedings of the HIS 2003, Third International Conference on Hybrid Intelligent Systems*, Melbourne, Australia, *Design and Application of Hybrid Intelligent Systems*, ed. by A. Abraham, M. Koeppen, and K. Franke, IO Press, 614–622.
21. Everitt, B. (1980) *Cluster Analysis*, Second Edition. Heinemann Educational Books, London.
22. Peterson, N. (1993) Discretization using divisive cluster analysis and selected postprocessing techniques. Department of Computer Science, University of Kansas, Internal Report.
23. Cudek, P., Grzymala-Busse, J. W., and Hippe Z. S. (2009) Multistrategic classification system of melanocytic skin lesions: architecture and first results. In *Computer Recognition Systems 3. In Advances in Intelligent and Soft Computing*, vol. 57, Springer-Verlag, 381–387.

24. Cudek, P., Grzymala-Busse, J. W., and Hippe Z. S. (2011) Asymmetry of digital images describing melanocytic skin lesions. In *Proceedings of the 4th International Conference on Computer Recognition Systems Wroclaw*, Poland, May 2325, 2011, Springer Verlag, Berlin, Heidelberg, *Advances in Intelligent and Soft Computing*, 95, 605–611.
25. Hippe, Z. S., Grzymala-Busse, J. W., Blado, P., Knap, M., Mroczek, T., Paja, W., and Wrzesien, M. (2005) Classification of medical images in the domain of melanoid skin lesions. In *Proceedings of the CORES' 05, 4th International Conference on Computer Recognition Systems*, Springer-Verlag, Rydzyna, Poland, 519–526.
26. Melanoma.PL. http://ksesi.wsiz.rzeszow.pl.
27. Grzymala-Busse, J. W., Hippe, Z. S., Knap, M., and Mroczek, T. (2004) A new algorithm for generation decision trees. *Task Quarterly* 8, 243–247.
28. Hippe, Z. S., Wrzesien, M. (2002) Some problems of uncertainty of data after the transfer from multi-category to dichotomous problem space. In *Methods of Artificial Intelligence*, ed. by Burczynski, T., Cholewa, W., and Moczulski, W., Gliwice, Poland, Silesian University of Technology Press, 185–189.
29. Grzymala-Busse, J. W., Hippe, Z. S., Knap, M., and Mroczek, T. (2007) Deriving belief networks and belief rules from data: a progress report. *Trans. Rough Sets, Lecture Notes in Computer Science Journal Subline*, Springer-Verlag, 7, 53–69.
30. Hippe, Z. S. and Mroczek, T. (2003) Melanoma classification and prediction using belief networks. In *Computer Recognition Systems*, ed. by Puchala, E. and Wozniak, M., Wroclaw, Poland, Wroclaw University of Technology Press, 337–342.
31. Hippe, Z. S., Grzymala-Busse, J. W., and Piatek L. (2006) Randomized dynamic generation of selected melanocytic skin lesion features. In *Proceedings of the IIS'2006 International Conference on Intelligent Information Systems, New Trends in Intelligent Information Processing and WEB Mining*, Ustron, Poland, Springer-Verlag, 21–29.
32. Grzymala-Busse, J. W., Hippe Z. S., and Piatek, L. (2012) Synthesis of shapes in the domain of melanocytic skin lesions. In *Proceedings of the International Conference on Applied and Theoretical Information Systems Research*, Taipei, Taiwan, 1–10.
33. Chmielewski, M. R. and Grzymala-Busse, J. W. (1996) Global discretization of continuous attributes as preprocessing for machine learning. *Int. J. Approx. Reason.* 15, 319–331.
34. Friedman, R. J., Rigel, D. S., and Kopf, A. W. (1985) Early detection of malignant melanoma: the role of physician examination and self-examination of the skin. *CA Cancer J. Clin.* 35, 130–151.
35. Mroczek, T., Grzymala-Busse, J. W., and Hippe, Z. S. (2004) Rules from belief networks: a rough set approach. In *Proceedings of the RSCTC'2004, the Fourth International Conference on Rough Sets and Current Trends in Computing*, Uppsala, Sweden, June 15, 2004. *Lecture Notes in Artificial Intelligence 3066*, Springer-Verlag, 483–487.

12

Implementation and Optimization of a Method for Retinal Layer Extraction and Reconstruction in Optical Coherence Tomography Images

Marcos Ortega Hortas, Ana González López, Manuel Gonzalez Penedo, and Pablo Charlón Cardeñoso

CONTENTS

12.1 Introduction

Nowadays, most areas in medicine make use of specialized computer systems for disease diagnosis assistance, providing clinicians with a tool for a more efficient processing of information, making it more reliable and reproducible. Ophthalmology is no exception as many computerized tools have been developed to diagnose, control, and follow retinal pathologies such as glaucoma, age-related macular degeneration, or neovascularization. These tools rely on image acquisition techniques that provide images with adequate quality and information. Among these techniques, it is important to highlight optical coherence tomography (OCT), a relatively recent modality that enables the acquisition of retinal images to a level not achieved before. OCT allows the acquisition of images with differentiation of the layers of the retina. Layer distribution defines a number of surfaces in the retina, in particular, six surfaces of interest for several pathologies: *Surface 1* corresponds to the inner limiting membrane (ILM). *Surface 2* separates the nerve fiber layer (NFL) from the ganglion cell layer (GCL). *Surface 3* corresponds to the separation between inner plexiform layer (IPL) and the inner nuclear layer (INL). *Surface 4* separates the outer plexiform layer (OPL) from the outer nuclear layer (ONL). *Surface 5* is the union of the inner segments (IS) and outer segments (OS). *Surface 6* separates the OS from the retinal pigment epithelium (RPE).

Given that a layer is delimited internally by a surface and externally by another one, and a surface separates two layers, we can make no distinction between layer or surface detection, as they are equivalent.

As we mentioned, OCT allows acquiring images where layers can be visually distinguished. This technique was introduced by Fujimoto [1] in 1991, and it has become a relevant method for clinical practice in ophthalmology. For example, images from the interior of the retina are used more and more in recent years thanks to this technique. It is particularly used in the study of retinal pathologies and the macular vitreous–retinal interface, although it has been used in other fields such as intraocular tumor detection. Tomographic images facilitate the diagnosis of diseases difficult to detect with typical ophthalmoscope techniques but, also, enables the follow-up process of a disease and the response to a particular treatment.

Manual detection of retina layers in OCT images is a time-consuming task enabling the appearance of subjective errors. Therefore, automatic detection is an interesting task in terms of saving time and money as well as offering more reliable results.

Although each layer in the retina can be affected by pathology, most of them tend to appear in surfaces 1, 5, and 6.

There exist a variety of methods for this task, most of them focusing on ILM and RPE detection and ignoring the rest of the layers. In this group, we can find the method described in the work of Fabritius et al. [2] introducing two different algorithms for ILM and RPE detection, both of them based on intensity changes on the OCT signal. Hiroshi Ishikawa et al. [3] introduce a method for detection of five of the six retinal surfaces but omitting one of the most important, surface 5, which separates the IS from the OS. This approach searches edges using an adaptive threshold technique. Another approach is introduced in the work of Mircea Mujat et al. [4] where NFL width is measured by means of noise suppression and deformable splines. This spline determines the vitreous–NFL and NFL–GCL edges by exploration of reflectivity changes. An interesting method is presented by Mona Haeker et al. [5] where a solution for detection of all layers is introduced. This method uses macular OCT images in performing a series of six tomographies centered in the fovea with different angles of orientation. The best performance is achieved by a minimum of six series per eye and averaging each series, allowing the construction of a composed 3-D image per orientation. In each 3-D image, six surfaces can be searched for by performing a minimum cut search in an associated graph representing connections between points in the image. This approach has the advantage of combining several tomographies from the same area, increasing the quality of the results, but time and memory performance are very low because of the extensive computations and structures needed. Azadeh Yazdanpanah et al. [6] introduce a method based on active contours, achieving detection of retinal layers. This method is an adaptation of the general Chan–Vese algorithm for the particular case of the intraretinal region. Obtained results are good, but these were tested in synthetic tests using rodent eyes, so its performance has not been yet tested in humans. Another approach is the one introduced by Delia Cabrera Fernandez et al. [7] presenting satisfactory results for layer detection based on texture analysis, using a structural tensor combined with a complex diffusion filter and local information of the retina structure. However, it requires a very high computation capability and memory requirements to be used.

In this chapter, starting from graph-based approaches, we introduce a particular implementation of a technique for layer detection in OCT as well as its 3-D reconstruction and optimization. Experiments show good results in both quality of detection and performance improvement. Although each layer in the retina can be affected by pathology, most of them tend to appear in surfaces 1, 5, and 6. Thus, in this work, we present particular adaptations of the method for these surfaces, although the methodology is valid for extension to the remaining ones.

This chapter is structured as follows: The next section discusses basic methodology, with different modalities, for OCT layer detection. Section 12.3 offers some test results with this implementation. Section 12.4 introduces a series of optimization proposals in terms of computational and memory usage performance. Finally, Section 12.5 offers some conclusions on the work.

12.2 Retinal Layer Detection in OCT Images

In this section, we present the proposed approach for implementation of automatic layer detection in OCT retinal images. Based on what was mentioned in the introduction, we will transform this problem of layer detection and search for the minimum closed set in a 3-D graph built by local and edge information and some uniformity constraints. This transformation was previously proposed by Li et al. [8] and Wu and Chen [9], demonstrating its good functioning in this kind of problems.

To achieve the goal, the methodology is divided into three stages: stage I: preprocessing (image acquisition, noise filtering, segmentation); stage II: detection (2-D model vs. 3-D model, 3-D graph construction, minimum closed set search, reconstruction and storage of layer); stage III: representation (layer extraction, highlighting, 3-D modeling).

First, it will be necessary to acquire OCT images from the patient retina and preprocess them for a greater system performance. Next is layer detection by the minimum closed set search in a 3-D graph. Finally, detected layers will be extracted, and a representation of them will be obtained. These three stages are explained in more detail in the following sections.

12.2.1 Preprocessing

Given that OCT devices show more information than the simple image, a series of techniques must be applied to remove unnecessary information. First, we select the region of interest (ROI) of the acquired image by the OCT since the original image consists of two different ones: the specific OCT image and a representation of the position of that OCT acquisition relative to the whole retina. After the ROI has been automatically selected, a noise removal process must be done to alleviate the noisy original acquisition. For this task, we use a generic nonlinear diffusion filter. This filter, in contrast to a linear diffusion filter, preserves the original location of edges, even enhancing them, which is a very interesting property for our task. Figure 12.1 shows the results obtained for the preprocessing stage in an original OCT image.

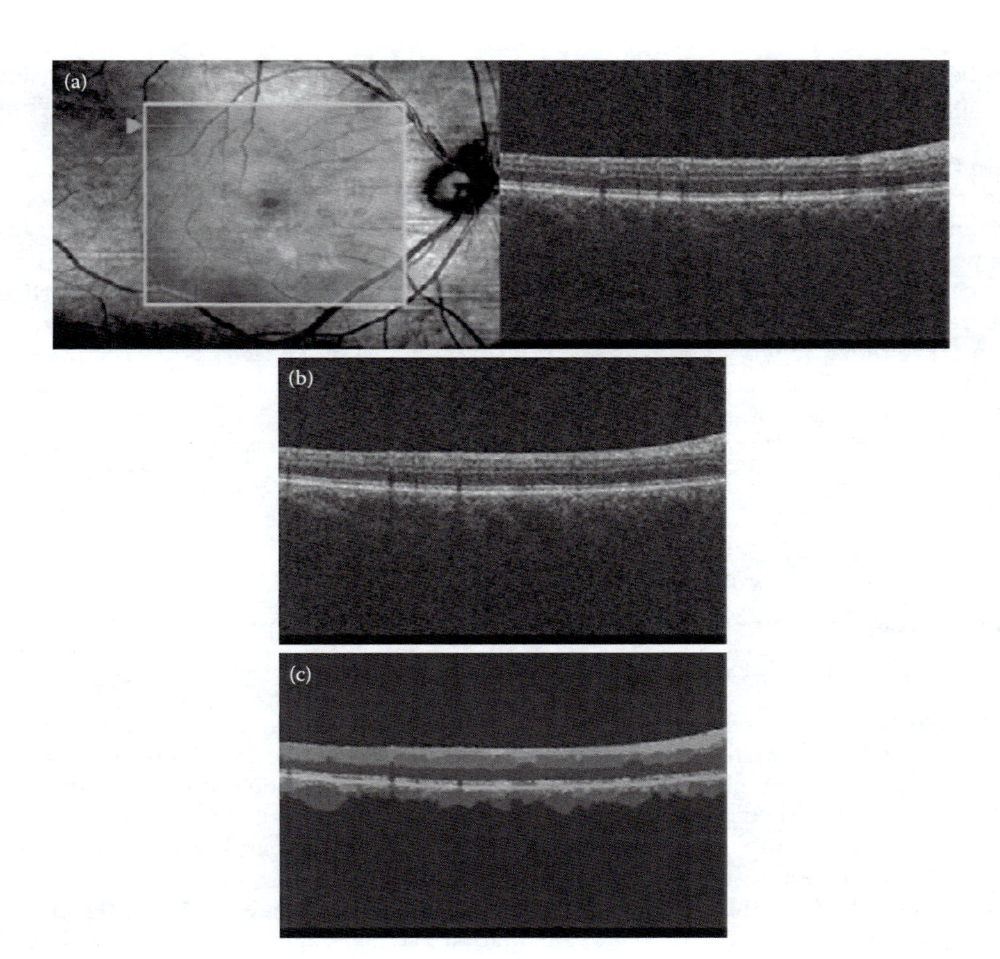

FIGURE 12.1
(See color insert.) Preprocessing of an OCT image. (a) Original acquired image. (b) ROI extracted. (c) Image after the application of the nonlinear diffusion filter.

12.2.2 Layer Detection

As explained earlier, this approach is based on the transformation of the layer detection problem into a search for the minimum closed set of an associated 3-D graph. This task is performed at this stage. It is important to note that the steps composing this stage must be performed for each of the layers to be detected and having all of them stored at the end of it. In this work, we focus on surfaces 1, 5, and 6.

Layer detection can be performed by two different approaches: using a 2-D model or a 3-D model. This offers versatility in terms of adaptation to particular scenarios. In the 2-D model, each tomography is processed individually, so after the layer detection in each image, a reconstruction is done at the end. In the 3-D model, all tomographies, or a group of them, are treated jointly, forming a global 3-D image, which is used for obtaining the detected layer as a whole.

There are certain situations where choosing between both models can be critical for good layer detection. An example of this would be when discontinuities appear in the surfaces due to the presence of vessels. If a pure 2-D model is used, discontinuity in a

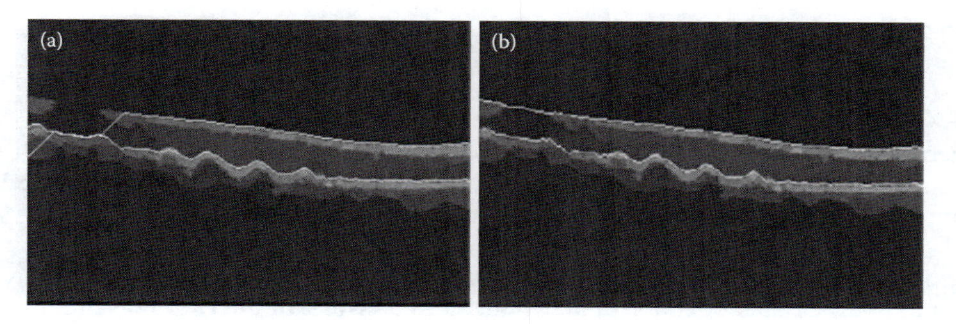

FIGURE 12.2
Detection of surfaces 1 and 5 presenting discontinuities in the ILM in a (a) 2-D model and (b) 3-D model.

particular image could not be resolved by using information of neighbor samples. Figure 12.2 shows an example of detection of surfaces 1 and 5 in a retina with a discontinuity in the ILM under both models. A more detailed study of differences between both models and their results will be discussed in a later section.

12.2.2.1 3-D Graph Construction

Many segmentation algorithms have made use of graphs as part of their implementation, some of them based on minimum spanning trees, as in the work of Xu et al. [10] and Felzenszwalb and Huttenlocher. [11]; others are based on the shortest paths, such as those introduced by Nyul et al. [12] and Milan Sonka et al. [13]; finally, others are based on minimum closed sets such as the ones in the work of Wu and Leahy [14] and Shi and Malik [15]. Among all of them, the latter ones can be considered the most powerful for segmentation of images. Considering this, for our particular domain of layer detection, they should result in a powerful tool.

The basis of this approach is the formation of a graph $G = (V, E)$, where V is the vertex set and E is the edge set. Each vertex $v \in V$ corresponds to an image pixel (or voxel in 3-D), and edge $<vi, vj> \in E$ connects vertices vi and vj based on some neighboring rule. Each edge also has an associated cost or weight representing somehow the degree to which the pixel pertains to the object of interest.

A surface S in an image I can be defined by a function $N: (x,y) \rightarrow N(x,y)$, where $x = \{0, \ldots, X - 1\}$, $y = \{0, \ldots, Y - 1\}$, and $N(x,y) = \{0, \ldots, Z - 1\}$, where X, Y, and Z are maximum sizes allowed for I. Then, any surface S of I intersects with a unique voxel of each column, being formed by exactly X × Y voxels. This model is known as a multicolumn model.

A surface is feasible if it satisfies a series of uniformity constraints, defined through two parameters, Δx and Δy, guaranteeing connectivity of the surface in 3-D. The first of them, Δx, preserves uniformity of the surface in x direction, while with Δy, some uniformity is obtained in adjacent tomographies. More precisely, if $I(x,y,z)$ and $I(x + 1,y,z')$ are two voxels from the same surface, then $|z - z'| \leq \Delta x$; in the same manner, $|z - z'| \leq \Delta y$ for $I(x,y,z)$ and $I(x,y + 1,z')$. By means of these constraints, we can control the rigidness of the surface (i.e., rigidness is inversely proportional to Δx and Δy).

The cost $c(x,y,z)$ of a voxel $I(x,y,z)$, is calculated applying to the image a combination of cost functions, each one of them focusing on a particular feature of the image. These cost functions will be discussed later in more detail. Generally, cost c is a real number inversely proportional to the probability that the voxel pertains to the surface looked for.

As mentioned earlier, the constraint of uniformity Δy deals with the uniformity between adjacent tomographies. In the case of a 2-D model, this parameter has no use as each tomography is processed independently of the rest, so the detected surface could be nonuniform.

12.2.2.1.1 *Basic Cost Functions*

Detection of good cost functions is an important part in the process as they will guide the search for the optimal surface. Four basic cost functions have been implemented here, grouped into edge information-based or local-information based.

12.2.2.1.1.1 Cost Function Based on Edge Information *Edge cost function (fEdge)*: A particular variation of the Sobel operator that also includes a Gaussian filter for noise removal and no-maxima suppression for edge refining. This function has a parameter that is the σ in the Gaussian function.

12.2.2.1.1.2 Cost Functions Based on Local Information *A cost function (fA)*: Sum of the intensities of pixels from a region above a possible voxel of the surface to be detected. This is useful for detection of those surfaces with darker regions above.

$$f_A = \sum_{(x,y,z-ws)}^{(x,y,z)} I(x,y,z) \tag{12.1}$$

This cost function possesses a parameter for adapting to a particular surface: region size where the sum is applied. This parameter is called window size. The greater the window, the more information will be included for the cost computation.

B cost function (fB): Negated sum of pixel intensities in a region below a possible voxel of the surface to be detected in order to detect those surfaces with whiter areas below. Similarly to *fA*, this function is also controlled by a window size.

$$f_B = \sum_{(x,y,z)}^{(x,y,z+ws)} -I(x,y,z) \tag{12.2}$$

C cost function (fC): Cumulated sum of pixel intensities above a possible voxel of the surface to be detected beginning from the upper part of the image and cumulating downwards.

$$f_C = \sum_{0}^{(x,y,z)} I(x,y,z) \tag{12.3}$$

Basic cost functions presented here are used to configure particular cost functions used in each surface detection process. These particular functions are established as a linear combination of the basic functions in order to configure the specific properties of each surface (Table 12.1).

Once each basic cost function is calculated, it is necessary to perform a value normalization to ensure a proper weighting of them. This is achieved by scaling and translation of the values. In the case of *fB*, as it is a negated sum, the normalized interval will be the same as in the other terms but with negative values.

TABLE 12.1

Basic Cost Functions Used (Marked as Yes) for Each Surface Detection

		Basic Cost Functions			
		Edge	A	B	C
Surfaces	1	Yes			Yes
	5	Yes	Yes	Yes	
	6	Yes	Yes	Yes	

12.2.2.2 Minimum Closed Set Search

For any feasible surface N of an image I, the set of vertices in N or under it, called $C = \{V(x,y,z) \mid z \geq N(x,y)\}$, form a closed set in G. This means that if a voxel $V(x,y,z)$ belongs to C, then all of the vertices under it in (x,y) will also belong to C.

This fact allows us to conclude that, given the cost assignation during graph construction, the costs of N and C are equal. In fact, any feasible surface N of I corresponds to a unique closed set C in G with the same cost. This is a key property to transforming the problem from an optimal surface search to a minimum closed set search in G.

A search for the minimum closed set C starts by making some modifications in the graph $G = (V,E)$, transforming it into a new graph G_{st}. First, we define sets $V-$ and $V+$ as those sets of vertices from G whose costs are negative and positive, respectively. The set of vertices of the new graph G_{st} will be the same of the original G, adding two new vertices: a source vertex (s) and a sink one (t); additionally, the edge set of G_{st} will be the same as that of G, adding a new set $E_{st} = \{\langle s, v\rangle \mid v \in V+\} \bigcup \{\langle v, t\rangle \mid v \in V-\}$:

$$G_{st} = (V \bigcup \{s, t\}\ E \bigcup [(E_{st})]) \tag{12.4}$$

G_{st} cost values will be defined by

$$c(S,T) = \begin{cases} \infty & (u,v) \in G \\ w(\nu) & u = s, \nu \in V^{+} \\ -w(u) & \nu \in V^{-}, \nu = t \end{cases} \tag{12.5}$$

Once the new graph is built, the search for the minimum closed set is performed by searching the minimum $s - t$ cut, splitting G_{st} into two closed sets, *S (source set)* and *T (sink set).* The cost for that cut is

$$c(S,T) = w(V^{+}) - \sum_{\nu \in S} w(\nu) \tag{12.6}$$

Thus, cost for cut $c(S,T)$ in G_{st} and the cost of the corresponding closed set in G differ by a constant, making the minimization task for cut cost (S,T) equivalent to a maximization of the cost of the source set.

There are plenty of algorithms for calculation of the minimum closed set in a graph G, such as those used by Picard [16] and Hochbaum [17]. Among the proposals presented in those works, we use a variety of the Ford–Fulkerson algorithm as it reduces resource usage. This algorithm is based on the search for paths along the graph where flow can be increased until a maximum flow value is obtained.

12.2.2.3 Reconstruction and Layer Storage

At this point, graph G_{st} is divided into two closed sets, the source set and the sink set. Those vertices belonging to the source set correspond to elements from the image above the detected surface; likewise, vertices in the sink set correspond to elements from the image below the surface. Keeping this in mind, we can consider as elements from the detected surface those lower vertices from the source set and upper vertices from the sink set, that is, those elements closest to the cut considering an element for each column (x,y). Here we assume that those elements belonging to the surface are those of the sink set, although it could be considered otherwise.

Once the surface is reconstructed with these considerations, the next step is to store it in a practical way to be processed in the following stages. There are many possibilities, but given that, as mentioned earlier, any surface intersects with one and only one voxel per column, the easiest way is storing the z coordinate for each one of those columns, considerably reducing the usage of memory and making posterior access easier.

12.2.3 Representation

The last stage focuses on the visual representation of each surface through models that provide the expert with tools for clinical diagnosis. The initial task consists of extracting detected layers from the rest of the information in order to work with them independently for the remaining steps. Next, proper representation is performed through the techniques of highlighting and modeling. Highlighting displays a superimposition of the detected layer on the original image in a way that facilitates the visualization for the clinical expert. It usually implies using colors in the original grayscale image. This visualization also allows the assessment on the detection process to correct possible errors and perform a finer adjustment. Figure 12.3 shows an example of the highlighting of the ILM.

In the second technique, 3-D modeling, a tridimensional representation of the detected layers is performed. For this purpose, each layer is constructed in a 3-D fashion by joining individually detected layers in each image. One of the most important advantages of this technique versus highlighting is that it enables detection of anomalies derived from pathology through the surface that are almost impossible to detect by exploring layers in

FIGURE 12.3
ILM highlighting.

FIGURE 12.4
ILM 3-D modeling.

each image individually. Additionally, being a 3-D model, it offers the possibility to visualize the surface model from different points of view, giving the expert a more powerful tool. An example of this visualization can be viewed in Figure 12.4, where a 3-D model of the ILM is shown.

12.3 Experiments and Results

In this section we present some results obtained through experimentation on the implemented system, using a representative set of OCT images for that purpose. These tests were performed using both 2-D and 3-D models, a previously discussed.

The system will be tested with real OCT images, assessing its performance in terms of quality and computation and memory usage. For this purpose, it is necessary to establish a study on the optimal configuration of system parameters.

12.3.1 Data Set

In order to test and validate the implemented methodology, we collected a set of images obtained by means of a Carl Zeiss Cirrus HD-OCT 400 able to acquire up to 128 tomographies by scanning the interest area. The data set consisted of a total of 512 OCT images obtained from evaluation of both eyes of two different patients. These images were previously adjusted according to the preprocessing stage earlier discussed. The image format is Portable GreyMap format (PGM), with a size of 300 × 200 pixels.

We performed initial tests using different parameter configurations. The optimal configuration is shown in Table 12.2.

TABLE 12.2
Parameter Values for Optimal Configuration of the System[a]

Configuration Parameters			
Parameter	**Value for Each Surface**		
Uniformity constraint Δx	1	1	1
Uniformity constraint Δy	1	1	1
Standard deviation σ	1	2	0.6
Window size ws	–	1	1

[a] Each surface has a particular configuration.

12.3.2 Results in the 2-D Model

For the OCT images of each patient, a 2-D model was applied, obtaining highlighting and 3-D modeling results.

Figure 12.5 shows an example of the obtained result for the three surfaces using highlighting. Figure 12.6 illustrates 3-D modeling for each surface from the same image as the previous figure.

As was commented on earlier, the 2-D model is not able to offer optimal results, presenting situations where these could be unacceptable in terms of quality.

One of these situations occurs when surface to be detected presents a notable discontinuity derived from the presence of vessels in the OCT area of acquisition. This issue arises due to the individual processing of each tomography, denying the possibility of establishing uniformity constraints between adjacent lines of the surface. This could result in irregular surfaces at the end of the process. In Figure 12.7, an OCT image can be visualized where a discontinuity was present along the layers, causing irregularities in their detection.

Another difficult situation, derived from the previous one, is the one present in surface 6 detection. This surface presents a particularly high number of discontinuities while being very close to surface 5, resulting in situations where the system is unable to distinguish both of them. Again, this situation comes from the individual processing of images in the 2-D model.

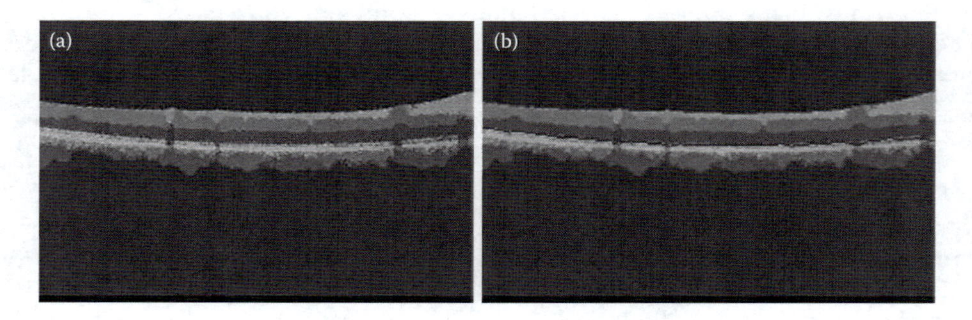

FIGURE 12.5
(See color insert.) (a) OCT image number 5 from the right eye of the first patient. (b) Three-layer highlighting.

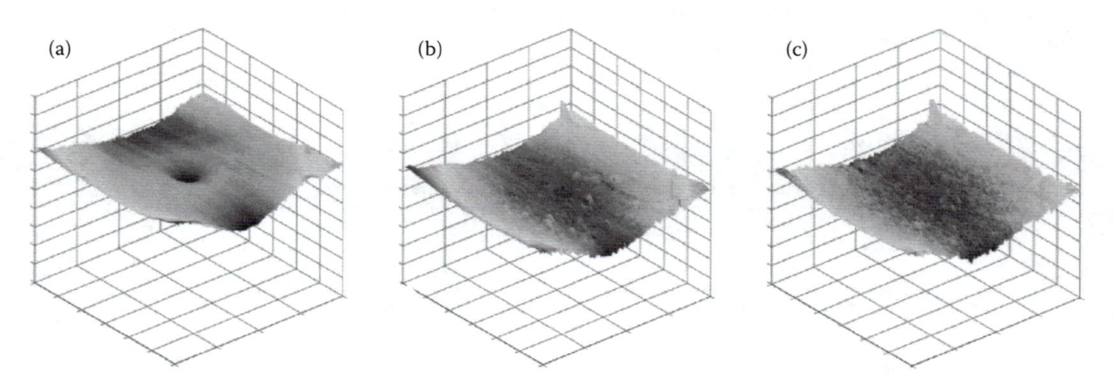

FIGURE 12.6
(See color insert.) 3-D modeling for each surface. (a) Surface 1, (b) surface 5, (c) surface 6.

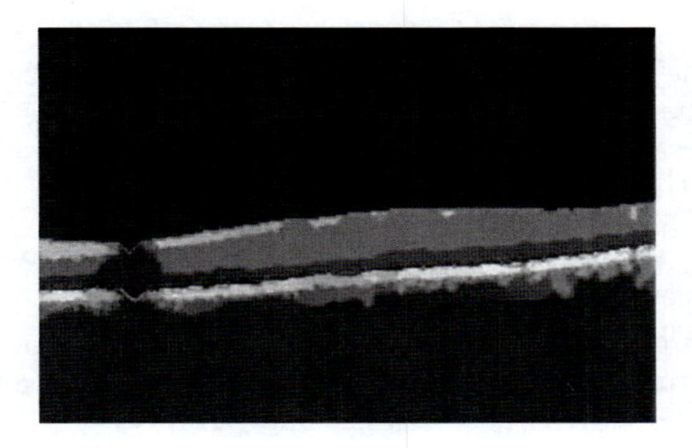

FIGURE 12.7
OCT image presenting a discontinuity due to the presence of a vessel.

12.3.3 Results in the 3-D Model

Given that this model works with a unique associated graph and that the computational complexity could potentially increase very quickly with the number of images, unlike the 2-D model where it is linear, the problem of computation time and memory usage is crucial in this model. We have performed an initial study about system requirements in terms of memory usage. Computational time will be studied in more detail in Section 12.4.

Afterwards, in the same manner as in the 2-D model, we show some samples during the representation stage using both highlighting and 3-D modeling.

In Figure 12.8, a graph can be observed representing memory consumption of this model as a function of the simultaneously processed images. The dotted line represents, as a reference, the amount of memory available in the computer where these calculations were performed. With a typical desktop PC, memory availability could be not enough or, at least, provoke performance degradation due to excessive use of swapping memory. To

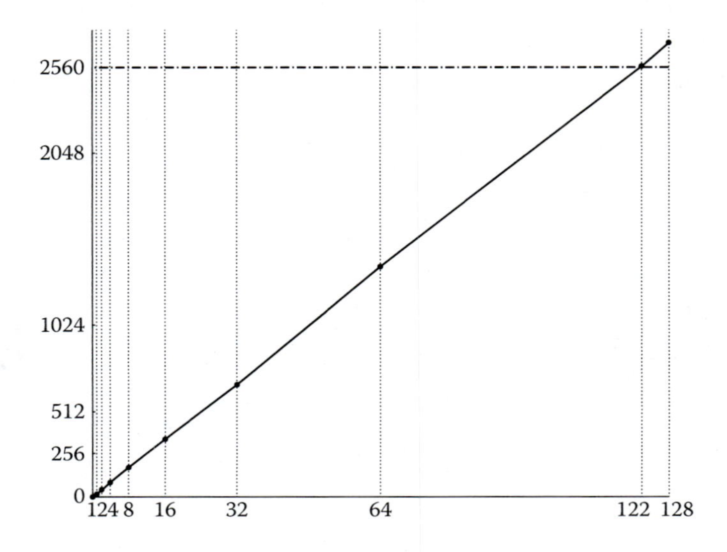

FIGURE 12.8
Memory usage (in MB) as a function of the number of images processed.

ease this problem, we propose a solution following the "divide-and-conquer" paradigm. With that in mind, the total number of images is divided into smaller groups; the system is run for each one of these groups, and the resulting image is reconstructed to the end of the process. The number of elements in each group was called group size (gs). Computation time grows exponentially with the group size, so it was established that sizes of 8 or more are to be avoided as the whole processing could take days on a desktop PC. Of course, the quality of results must be assessed too in order to pick a correct value of gs. It was established by visual inspection of the data set samples that with a gs of 4, results were already satisfactory in terms of correction of deviations from the 2-D model while keeping the process away from dramatically increasing computation times. Examples of the 3-D model results using gs = 4 are shown in Figures 12.9 and 12.10.

This model increases the quality of results compared to the 2-D model, although as a trade-off, it presents a limitation to take into account: computational performance. This raw performance could be a great limitation in setting up this process in a real environment for daily use. The quality of results allows the system to deal with the performance deviations from the 2-D model such as the irregularities in the layers. Figure 12.11 shows a comparison between the 2-D model (left) and 3-D model (right). Figure 12.12 shows a

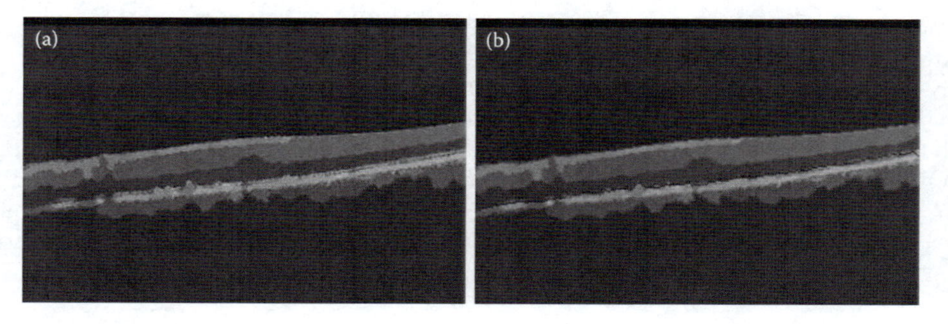

FIGURE 12.9
(a) OCT image number 36 from first patient's left eye. (b) Three-surface highlighting.

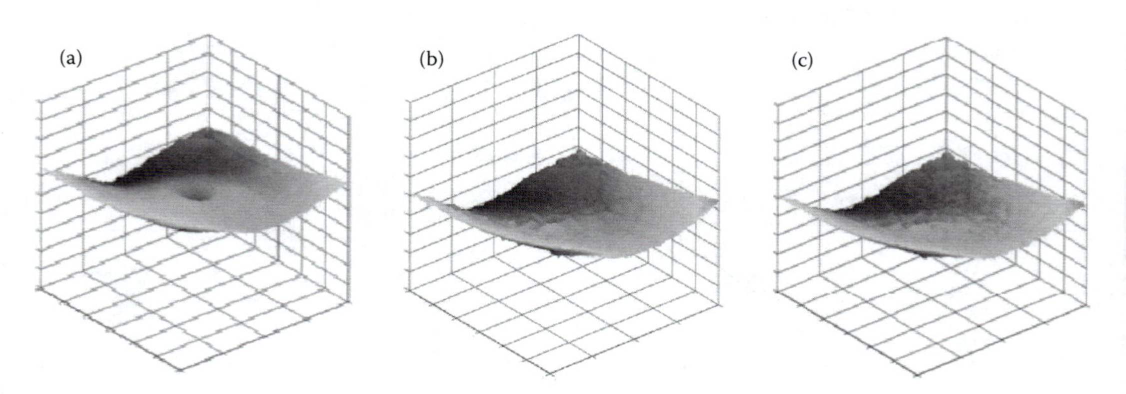

FIGURE 12.10
Individual 3-D modeling. (a) Surface 1, (b) surface 5, (c) surface 6.

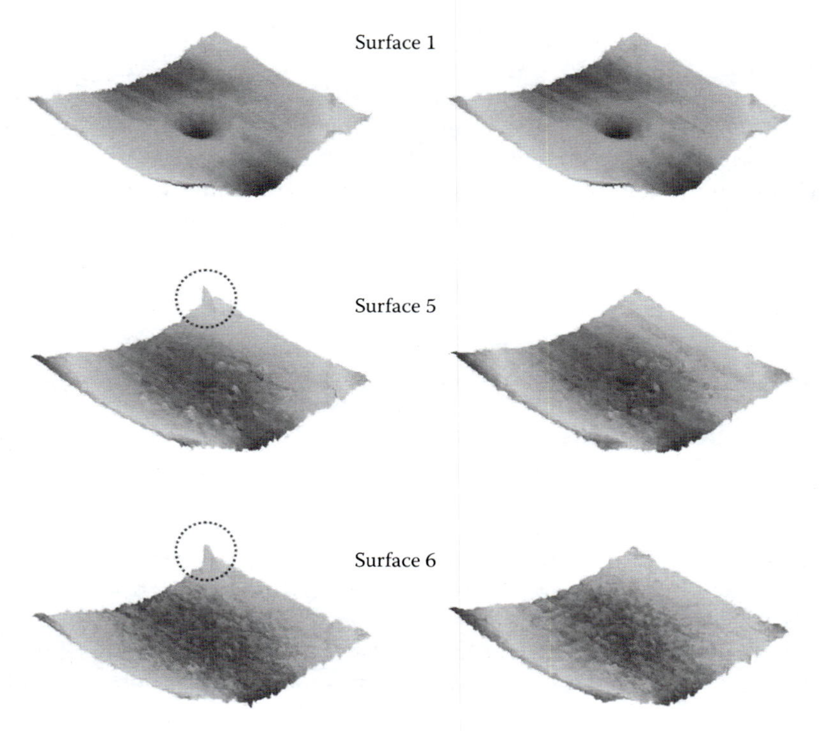

FIGURE 12.11
(See color insert.) 3-D modeling comparison for 2-D and 3-D processes for first patient's right eye.

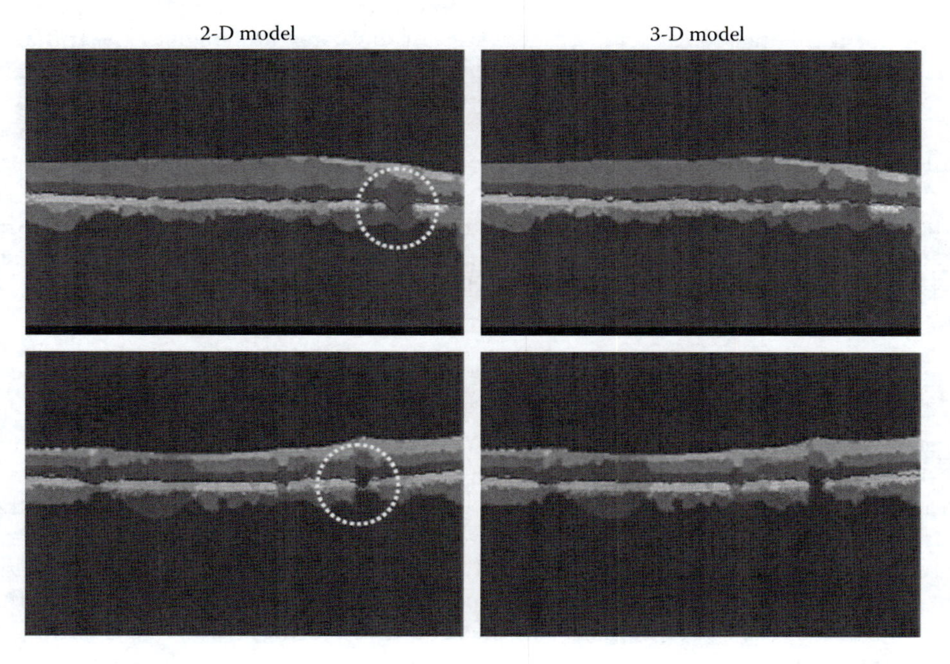

FIGURE 12.12
(See color insert.) Two samples of highlighting using 2-D model (left) and 3-D model (right). It can be observed how the discontinuities present in the 2-D model are avoided in the 3-D model.

comparison using highlighting. It can be observed that discontinuity problems are avoided in the 3-D model.

Up to this point, a series of results have been shown, illustrating how the 3-D model obtains better results than the 2-D model. Most of the structures in the OCT image layers derived from no pathology, such as intraretinal shadow or surface discontinuities, could be avoided. As mentioned earlier, the 3-D model presents limitations in terms of computational performance. In the next section, we introduce two optimizations that try to reduce the computation time and memory usage of the 3-D model.

12.4 Computational Optimizations to the 3-D Model

12.4.1 First Optimization

The retina is internally composed of 10 layers limited by the ILM in its internal part and the RPE in its external part. In those layers is where valuable information resides for this work, which is why any other structure outside those boundaries is to be automatically discarded, such as the vitreous humor or choroids. However, OCT devices are able to acquire more structures than those in the layers, so removing all the parts in the image of no interest to us could alleviate the computation needs of the system without altering its functioning.

Keeping this in mind, we propose a first optimization based on the removal of the structures outside the layers in the retina in OCT images, vitreous humor, and choroids, achieving smaller images and saving computation time and memory usage in the system.

To perform this task, we include a new step in the detection stage of the initially proposed methodology. This step is repeated for each image group entered in the system.

The procedure begins by an approximation of the ILM in each group image, by using the basic cost function *fedge* and removing all detected edges but the first in each column, which should correspond to the ILM. Next, RPE location is approximated by modifying *fedge* to search for bright–dark transition edges. Last, the upper ILM point and lower RPE point are localized, and the rectangle defined by them is selected as the ROI, plus a padding to ensure correct functioning of the surface 1 cost function, as information around the own surface is needed for a correct adjustment. Empirically, it was determined to set this padding to 20 pixels. Figure 12.13 shows an example of the results obtained with this approach.

12.4.2 Second Optimization

We previously introduced the uniformity constraints, Δx and Δy, that allow a natural adjustment of the surface shape. The second optimization proposes the elimination of the information outside the interval $[-(n \times \Delta y), (n \times \Delta y)]$ for each surface detected in the first group image. This enables a reduction in the image size by removing unnecessary data to be processed. Group size is kept to four, although this can be generalized to any size. The process begins with the goal surface detected, as explained in Section 12.2. Once the surface is located, its most internal and external points are located, and, like the first optimization, all points outside the rectangle area defined plus a padding are removed. Padding size comes determined, in this case by $n \times \Delta y$ plus a fixed term that, for the resolution of our images, was set to 20 pixels, similarly to previous optimization.

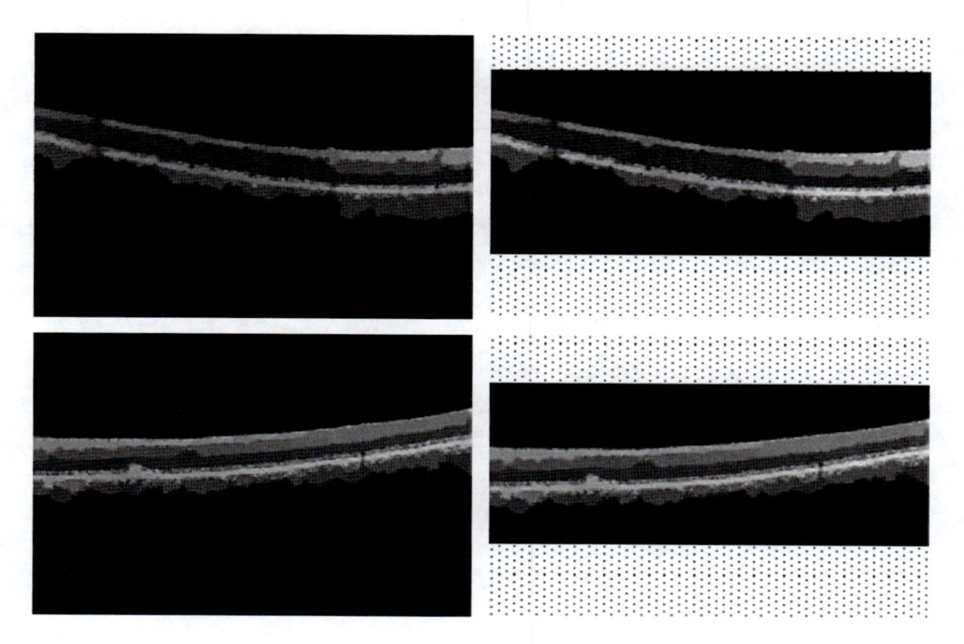

FIGURE 12.13
Result after applying first optimization in two preprocessed images. Left: Original images. Right: Images after optimization.

Only the removal of points is applied to each group image while the first stage of surface location is calculated in the first one. For each surface of interest, the ROI will be different, so they are computed and applied separately for each one of them. Once the ROIs are obtained, surfaces are detected again in the first image to avoid possible errors derived from discontinuities or other alterations in the surfaces. In Figure 12.14, the result of applying this optimization in an OCT image can be observed for our three surfaces of interest.

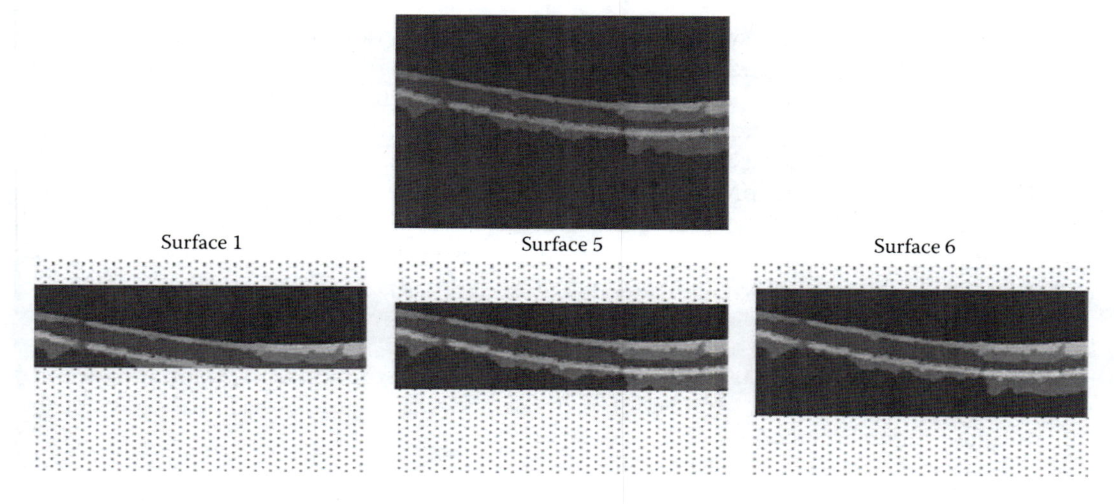

FIGURE 12.14
Result of applying second optimization to a preprocessed OCT image for each surface.

TABLE 12.3

Comparison of Computation Times with the Basic Proposed Implementation, Using First Optimization and Using Both of Them

Computation Time		
No Optimization	**+1st Optimization**	**+2nd Optimization**
25:08:04	17:32:56 (–30.18%)	11:19:44 (–53.71%)

TABLE 12.4

Comparison of Memory Usage in the Basic Proposed Implementation, Using First Optimization and Using Both of Them

Memory Usage		
No Optimization	**+1st Optimization**	**+2nd Optimization**
85.8 MB	48.53 MB (–43.4%)	34.05 MB (–60.31%)

12.4.3 Performance Analysis

In this section, a study of the performance using these optimizations is presented in terms of computation time and memory usage. As can be observed in Tables 12.3 and 12.4, each optimization significantly increases the performance of the proposed implementation. For reference, these computations were made on a basic desktop PC with an Intel Core 2.4 GHz with 2 MB of RAM.

12.5 Conclusions

In this work, we present a whole methodology for extraction of the main layers of the retina and its surface 3-D reconstruction. This methodology is based on previous works using a search of minimal graph cuts for similar tasks. We include here all the stages in the process of layer detection and reconstruction from the original acquisition.

Two paradigms or models are implemented and studied, 2-D and 3-D, where the 3-D model proves to overcome problems in the image due to alterations in the layers such as discontinuities caused by the presence of vessels. As a drawback, the 3-D model is far more inefficient, so we propose two optimizations as a starting point to refine the needed amount of calculations for the methodology.

As future lines of work, more complex optimizations seem to be needed in order to obtain a very fast method while keeping the current quality of the results. For instance, a multiscale approach can be used to obtain a coarser segmentation of layers in low resolution and, after that, to perform a refining of the layer at higher resolutions, focusing on nonregular interest regions.

References

1. W. Drexler and J. G. Fujimoto. State-of-the-art retinal optical coherence tomography. *Prog. Retin Eye Res.* vol. 27, no. 1, January 2008, 45–88.
2. T. Fabritius, S. Makita, M. Miura, R. Myllyla and Y. Yasuno. Automated segmentation of the macula by optical coherence tomography. *Opt. Express* vol. 17, no. 18, 2009, 15659–15669.
3. H. Ishikawa, D. M. Stein, G. Wollstein, S. Beaton, J. G. Fujimoto and J. S. Schuman. Macular segmentation with optical coherence tomography. *Invest. Ophthalmol. Vis. Sci.* vol. 46, no. 6, June 2005, 2012–2017.
4. M. Mujat, R. C. Chan, B. Cense, B. H. Park, C. Joo, T. Akkin, T. C. Chen and J. F. de Boer. Retinal nerve fiber layer thickness map determined from optical coherence tomography images. *Opt. Express* vol.13, no. 23, November 14, 2005, 9480–9491.
5. M. Haeker, M. Sonka, R. Kardon, V. A. Shah, X. Wu and M. D. Abramoff. Automated segmentation of intraretinal layers from macular optical coherence tomography images. *Proc. SPIE* 6512, *Medical Imaging 2007: Image Processing*, San Diego, CA, February 17, 2007, 651214.
6. A. Yazdanpanah, G. Hamarneh, B. Smith and M. Sarunic. Intra-retinal layer segmentation in optical coherence tomography using an active contour approach. In *Proceedings of the MICCAI '09*, London, September 2009, 20–24.
7. D. C. Fernandez, H. M. Salinas and C. A. Puliafito. Automated detection of retinal layer structures on optical coherence tomography images. *Opt. Express* vol. 13, no. 25, 2005, 10200–10216.
8. K. Li, X. Wu, D. Z. Chen and M. Sonka. Optimal surface segmentation in volumetric images – a graph-theoretic approach. *IEEE Trans. Pattern Anal. Mach. Intell.* vol. 28, no. 1, 2006 January, 119–134.
9. X. Wu and D. Z. Chen. Optimal net surface problems with applications. In *Proceedings of the 29th International Colloquium on Automata, Languages and Programming* (ICALP '02), Málaga, Spain, 2002, 1029–1042.
10. Y. Xu, V. Olman and E. Uberbacher. A segmentation algorithm for noisy images. In *Proceedings of the 1996 IEEE International Joint Symposia on Intelligence and Systems* (IJSIS '96). IEEE Computer Society, Washington, DC, 1996, 220–226.
11. P. F. Felzenszwalb and D. P. Huttenlocher. Efficient graph-based image segmentation. *Int. J. Comput. Vision,* vol. 59, no. 2, September 2004, 167–181.
12. L. G. Nyul, A. X. Falcao and J. K. Udupa. Fuzzy-connected 3D image segmentation at interactive speeds. *Graph. Models* vol. 64, no. 5, September 2002, 259–281.
13. M. Sonka, M. D. Winniford and S. M. Collins. Robust simultaneous detection of coronary borders in complex images. *IEEE Trans. Med. Imaging* vol. 14, no. 1, 1995, 151–161.
14. Z. Wu and R. Leahy. An optimal graph theoretic approach to data clustering: theory and its application to image segmentation. An optimal graph theoretic approach to data clustering: Theory and its application to image segmentation. *IEEE Trans. Pattern Anal. Mach. Intell.* vol. 15, no. 11, 1993, 1101–1113.
15. J. Shi and J. Malik. Normalized Cuts and Image Segmentation. *IEEE Trans. Pattern Anal. Mach. Intell.* 22, 8.
16. J. Picard. Maximal closure of a graph and applications to combinatorial problems. *Manage. Sci.* vol. 22, no. 11, July 1976, 1268–1272.
17. D. Hochbaum. A new-old algorithm for minimum-cut and maximum-flow in closure graphs. *Networks* vol. 37, no. 4, 2001, 171–193.

13

Deep Learning for the Semiautomated Analysis of Pap Smears

Kriti Chakdar and Brian Potetz

CONTENTS

13.1 Introduction

The precision in distinguishing a cancerous structure from a benign structure has the potential to immediately improve health outcomes in one of our most pressing diseases. It will also contribute to one of artificial intelligence's (AI's) biggest remaining frontiers—the automated extraction of complex discriminative features.

In 2012, the National Cancer Institute estimated that about 577,190 Americans were expected to die of cancer, more than 1500 people a day. Cancer is the second most common cause of death in the United States, accounting for nearly one of every four deaths. Cancer diagnosis has a very important role in the early detection and treatment of cancer. Automating the cancer diagnosis process can play a very significant role in reducing the number of falsely identified or unidentified cases. In this section, different machine learning approaches for cancer detection are demonstrated.

The process of cancer diagnosis is semiautomatic and is prone to human error. It is also time consuming. The Papanicolaou (Pap) smear—the most common test to identify precancerous or cancerous processes in the cervix of the human body—generates large digital images. Pathologists have a challenging task of manually monitoring the digital images. It has been found that the test may fail because of "inadequate samples, insufficient time devoted to screening, or human fatigue" during examination of the images or biopsy report (Koss 1989).

Talking about figures in a large hospital, a pathologist typically handles 100 grading cases per day, each consisting of about 2000 image frames. As the numbers show, it is a very tedious and time-consuming task. A computer system that performs automatic grading can assist pathologists by providing second opinions, reducing their workload, and alerting them to cases that require closer attention, allowing them to focus on diagnosis and prognosis.

Some more interesting facts regarding the Pap smear test are as follows:

- Over 500,000 women will develop cervical cancer worldwide, with an annual death rate of close to 300,000/year.
- 55 million Pap tests are performed on a yearly basis in the United Stated (source: NCI, www.cancer.gov).
- 3 million abnormal Pap tests are reported on a yearly basis in the United States.

Unfortunately, Pap smear preparations produce thick slides with many layers of cells. This complicates diagnosis, and because there are many focal planes, digitization of slides is impractical.

Dr. Tawfik, a pathologist from the University of Kansas Medical Center (KUMC) is the inventor of a novel pathology tissue slicer. Briefly, the technology is either a manual or an electrical/ultrasonic device used to cut surgical tissue specimens into standardized, even slices ready for preparation of tissue blocks. The device is composed of a manual or an electric/ultrasonic cutting knife with multiple replaceable disposable blades, a specialized cutting board, and a multi-pin specimen holder device.

This device produces cell blocks from Pap smears, which allow samples to be condensed and sliced thinly. Thin slices simplify diagnosis and permit accurate digitization. Digital slide images make it possible to consult with expert pathologists remotely and can be better used in education, either in published material or online. Archiving of digital images is substantially easier than glass slides. Finally, digital slide images permit computer-assisted diagnosis. Automated computer vision techniques are able to prescreen digital images to flag suspicious regions and to discard areas of images that contain clearly healthy tissue. Automated prescreening alleviates the fatigue faced by professional pathologists, who must often spend hours peering through microscopes. In this study, the digital slides are read, and different classifiers are used to gain one optimal performance on a cancer data set.

13.1.1 Focus Area

Within a Pap smear preparation, architectural features are not available. Also, because cells in the Cell Blocks are scattered spatially, the variance of morphological features is weaker than for histological preparations, where the variance of cell size and shape can be measured within a small region of interest. For these preparations, cytological features are much more important. Thus, it is essential to develop effective textural features that can fully capture the properties of cells that correlate most strongly with cancer diagnosis. While modern textural features have proven highly useful, it remains that these approaches are limited to very basic texture features that are chosen by hand. A more effective strategy would be to apply machine learning techniques to search for highly discriminative visual features that are maximally predictive of cancer and precancer grading. Until recently, such an approach had been difficult due to the vast size of the set of possible visual features. However, modern advances in computer vision and machine learning have produced new technologies that allow complex, nonlinear image features to be learned from sample images using greedy, layer-by-layer training strategies.

13.2 Background and Related Work

Automated cancer diagnosis, both at the cellular and tissue levels, is based on (1) extracting information from the histopathological images of stained biopsies and (2) examining this information by using either statistical analysis or machine learning algorithms. There are three main computational processes in automated cancer diagnosis: *preprocessing*, *feature extraction*, and *diagnosis*. Preprocessing is used to enhance the image quality. It tries to reduce or fully eliminate the background noise from the image. This enhancement helps in identifying the focal areas lying on the image. The preprocessing helps and also decides the subsequent processes of feature extraction and diagnosis. The second process of feature extraction is aimed at the cellular or tissue level (Doyle et al. 2008).

A group of diagnosis studies have applied machine learning algorithms to learn (from data) how to distinguish the different classes from each other. Neural networks, *k*-nearest neighborhood, logistic regression, fuzzy systems, linear discriminant analysis, and decision trees are among those. Putting aside the machine learning methods implemented before, in this study, the deep learning method is the main focus.

Deep learning is a new area of machine learning research, and it depends on an efficient, layer-by-layer procedure for learning. Using the deep learning method, an autoencoder is trained with a large data set consisting of 576 features, for dimensionality reduction or feature extraction, and then depending on the extracted features, the whole data set is classified into two groups—malignant and benign.

This study uses different classification methods, and then, their performances are compared to diagnose the cells into malignant and healthy classes. The comparative study of the performance of various classifiers can possibly help in reducing errors in the examination of digital slides generated after the Pap smear test.

13.2.1 Statistical Methods

The automated diagnosis uses the feature set to distinguish the malignant cell structures from other cells. In short, the process categorizes cells into healthy, benign, and cancerous. It also assigns the grading to classify the level of malignancy of the tissue or cell. A section of the research is conducted on statistical tests on the features. This method finds the existence of substantial differences in the value of at least one feature of interest in the different classes. It calls for extra precaution in case of "histopathological" images. The precaution is needed because the data set comprises various tissue images procured from one patient. These data sets are not independent and could lead to wrong results.

13.2.2 Machine Learning Methods

A major contribution in the area of machine learning methods using neural networks was suggested by Esgiar et al. (2002) and others for *k*-nearest neighborhood. Research by Demir et al. (2004), Gunduz et al. (2004), Zhou et al. (2002), and others were based on neural networks. Fuzzy systems were suggested by Blekas et al. (1998); logistic regression was studied extensively by Woolberg et al. (1995) and others. Decision trees were recommended by Wiltgen et al. (2003), and linear discriminant analysis was devised by Smolle (2000) and others. These studies utilize data from machine learning algorithms to distinguish categories or classes.

The above studies utilized various techniques for evaluation. Smolle and Anderson et al. used no separate evaluation set. Schnorrenberg et al. (1996), Spyridonos et al. (2001), and others found the leave-one-out approach effective. A considerable number of studies carried out by Blekas et al. (1998), Wiltgen et al. (2003), and others utilized separate training and test sets. Zhou et al. (2002) and others implemented *k*-fold cross-validation.

The classification system that has been divided into two stages posed a problem of using a limited set of data in both the stages. In the first stage, a classifier is trained to learn the parameters of the system. In the second stage, the system is tested to evaluate the classifier's success rate. The use of same set of data in both the stages led to the problem of memorization prone to errors. To solve this crisis, it has been recommended to use separate data sets for the two stages. For this purpose, the data set is divided into two groups.

13.3 Processing and Data Analysis

13.3.1 The Data Set

The data set was made of 10 full-size images from healthy patients (typically 40,000 × 40,000 pixels each) and 21 samples of regions of low-grade squamous intraepithelial lesion (LGSIL). The 21 LGSIL samples were hand-annotated regions within 11 images from 11 different subjects, outlined by pathologist Dr. Ossama Tawfik. All slides were prepared using a cell block preparation and stained with hematoxylin and eosin (H&E).

13.3.2 Identification and Location of Nuclei

The hematoxylin stain applied to each slide dyes cell nuclei blue, and eosin stains cytoplasm red. The first step of processing is to locate cell nuclei, starting by identifying image regions that are sufficiently stained blue by hematoxylin.

Traditionally, linear filters are used for this task. For a given H&E-stained image where each pixel is represented by red, green, and blue intensity values, a representative blue color is selected, given by w_r, w_g, w_b. Then, for each pixel in the stained image, we compute

$$h = rw_r + gw_g + bw_b + w_0 \tag{13.1}$$

where r, g, and b represent the color of the current pixel, and w_0 is some constant. A pixel is considered to be within a cell nucleus if h lies above some threshold.

Linear filters are very fast. However, we found that a linear filter was not sufficient to accurately separate nuclei colors from nonnuclei colors. We selected a subset of healthy and LGSIL sample images and hand-labeled over 1000 cells, specifying for each pixel in the image which pixels were within the nucleus boundary and which lay outside the cell nuclei. In Figure 13.1, we show the distribution of colors observed within cell nuclei. The distribution shows a curved shape; a linear boundary in RGB color space would be unable to divide nuclei from nonnuclei pixels with high accuracy.

Because of the nonlinearity boundaries of the colors of cell nuclei, we investigated quadratic color filters, of the form

$$h = rw_r + gw_g + bw_b + r^2w_{rr} + g^2w_{gg} + b^2w_{bb} + rgw_{rg} + rbw_{rb} + gbw_{gb} + w_0 \tag{13.2}$$

for some filter values w. Using the hand-labeled images, we trained a quadratic filter of the above form. We used Fisher linear discriminant analysis to train the filter, where 10 features were supplied for each pixel (one constant feature, three linear features, and six quadratic features, as given in Equation 13.2). Fisher discriminant analysis provided a quadratic curve

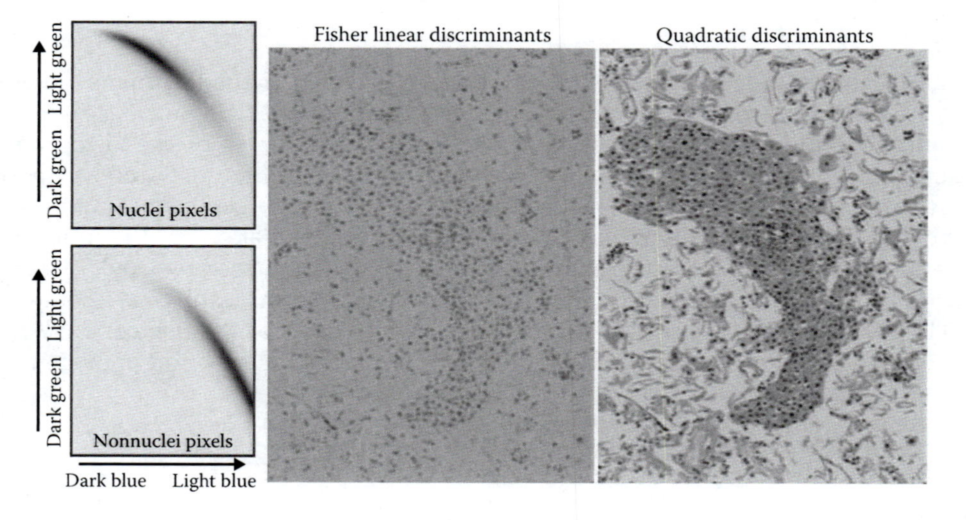

FIGURE 13.1
At the top left, we show the distribution of pixel colors observed within hand-labeled cell nuclei. This distribution sweeps out a curve in color space and cannot be easily separated from nonnuclei colors (shown bottom left) using a linear decision boundary. At the right, we show the results of filtering a sample LGSIL sample using linear filters (center) versus quadratic filters (right).

in RGB space that closely follows the contours seen in Figure 13.1. In Figure 13.1, we compare the results of linear and quadratic filters (each trained using Fisher discriminant analysis). Cell nuclei are more discernable after processing with the quadratic filter.

After the quadratic filter is applied, segmentation algorithms are applied to isolate individual cell nuclei, even in the case where neighboring nuclei overlap (Wahlby et al. 2004). We used the IdentifyPrimaryObjects module of CellProfiler to perform this operation.

After processing, the algorithm identified a total of 409,580 healthy cells and 2350 LGSIL cells.

13.3.3 Feature Extraction

Once cell nuclei are identified, we extract 54 quantitative features for each cell. These 54 features are designed to be discriminative toward LGSIL. Loosely, features can be divided into morphological, textural, and spatial categories. We describe each feature below.

13.3.3.1 Morphological Features

Morphological features describe physical properties of cells or the statistics of those properties over groups of cells. In this study, the morphological features are extracted using CellProfiler. In the data set there are five morphological features: Area, Perimeter, Area/Perimeter, Extent, and FormFactor of nuclei.

13.3.3.2 Nucleoli Features

Given the nuclei's major and minor axis and orientation, seven features of nucleoli were computed. Contrast, correlation, energy, and homogeneity were the basic ones, and depending on the circular window and patch number of gray pixels (between 0.1 and 0.9), the number of black pixels (less than or equal to 0.1) and nbhalf (less than or equal to 0.5) were computed.

13.3.3.3 Textural Features

Textural features describe low-level visual properties, which, in the case of histologic diagnosis applications, are designed to capture visual qualities including increases in nucleus size and the prominence of nucleoli. These features are measured by the histograms and co-occurrence matrices of pixel colors. Such features capture the first- and second-order statistics of cellular appearance. Textural properties have an advantage in that they cannot be disturbed by errors that may occur during cell segmentation or other intermediate processing steps. Similar feature extraction strategies are described by Doyle et al. (2007). Provided with the location of the cells in a sample, 12 textural features are computed in MATLAB®. Texture features are extracted separately from each "color band." With 25-, 50-, and 75-pixel widths at each radius, contrast, correlation, energy, and homogeneity are computed, and this makes the 12 textural features.

13.3.3.4 Color Features

During computing the color features, three different windows with radii of 5, 10, and 25 were taken, and six features were computed for each window. The first three are mean patch, and next three are variance patch.

Other than these, there are five features that are the mean of selected morphological features and five features that are the variance of morphological features.

So, as a whole, in the data set for each cell, there were a total of 54 features. Among these, 2 are nonclassification features like cell location, 5 are morphological features from CellProfiler, 12 are texture features computed in MATLAB, 18 are color features, 7 are nucleolus feature, 5 are features that are the mean of morphological feature, and 5 are features that are the variance of morphological features.

13.3.4 Outline of Data Set Processing

In general, any classification system consists of two stages:

1. Training the classifier to learn the system parameters
2. Testing the system to evaluate the success of the classifier

More data being used in training leads to better system designs, whereas more data being used in testing leads to more reliable evaluation of the system. Since the amount of available data was limited, and it is very important to test the network with extra data, the task to divide the whole data set into a training data set and a testing data set was also very important. To train the network, different sets of training data with different ratios were used, and then their performances were tested on the test set.

In statistics, machine learning, neural networks, and other related areas, one of the most discussed errors is accurate estimation of generalization while working on a finite data set. There are several kinds of estimation techniques proposed and examined in literature. Cross-validation is a strong technique that is used to examine the reliability of the results. But there are different kinds of cross-validation techniques that are being compared with each other to verify which is better for selecting a classifier. Leave-out-one cross-validation and *k*-fold cross-validation are prime among those.

It is important to evaluate the performance of a classifier accurately to predict the future assessment, and more than that, to choose a classifier for a given kind of data set among several classifiers. Choosing a classifier or combining classifiers to get better performance is one big task.

Performance was evaluated using a leave-out-two procedure: The network was trained on all but one healthy image and all but one LGSIL image and then tested on the remaining healthy image and LGSIL image.

13.3.5 Classification Methods Used

In this study, the whole data set is trained and tested on three kinds of classifiers:

1. Support vector machine (SVM)
2. Deep belief network (DBN) along with backpropagation
3. SVM on raw features appended with high-weight DBN features

In the first approach, the extracted features are used for training an SVM with Gaussian kernel function, and then the test data set was used to see the performance. For this method, the whole data set was divided into several training and testing data sets to measure the performance of the system.

In the second approach, the data set is modified a little bit (which will be explained in detail in the next section). The whole data set was divided into two parts: training and testing.

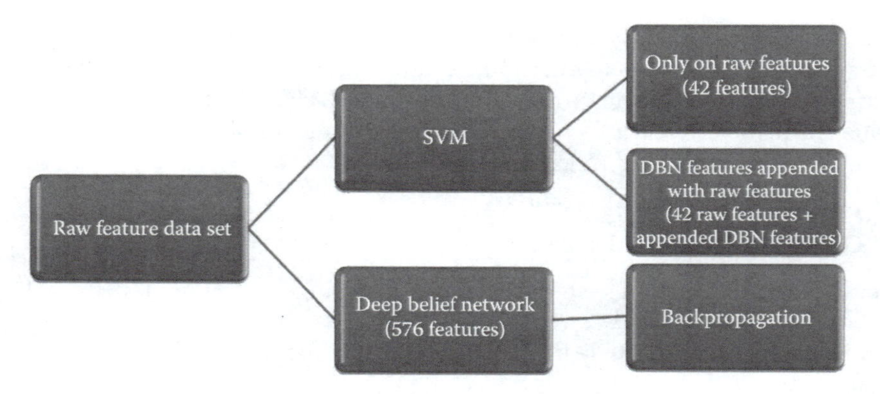

FIGURE 13.2
Flowchart of the method.

With the help of the training data set, the learning process for the DBN was achieved. Using three layers and after extracting features, two output nodes were assigned, and according to the targets, a backpropagation algorithm was used to fine-tune the weights.

For the third approach, high weight DBN features are appended with the raw features, and then an SVM with Gaussian kernel function is used to see the performance. All these methods are discussed in details in the next sections (Figure 13.2).

13.4 Applying SVM

13.4.1 Learning and Classification

SVMs are a popular method to distinguish between two classes (here, healthy vs. LGSIL cells) given numeric features and labeled training data. In this section, we train SVMs on a portion of our data set and test its performance on the remaining data. We use kernelized soft-margin SVMs, with a Gaussian kernel, which allows the boundary between the two classes to be nonlinear. We chose a value of $\sigma = 0.005$ for the Gaussian kernel, selected via cross-validation. SVMs were implemented using the SVM^{light} library (Joachims 1999, 2008).

Performance was evaluated using a leave-out-two procedure: The software was trained on all but one healthy image and all but one LGSIL image and then tested on the remaining healthy image and LGSIL image. This procedure was repeated 84 times (4 times for each LGSIL image), each time selecting a different pair of images for testing. All performance values were averaged over these 84 trials. The output of provides a numerical value for each cell: Higher values indicate LGSIL, while lower values indicate healthy cells. With a bias of −0.92, the software correctly labeled 97.5% of healthy cells and 92.4% of LGSIL cells. Other choices of bias produce different levels of trade-off between these two values. For example, a bias of −0.82 results in correct identification of 98.0% of healthy cells and 90.8% of LGSIL cells. Performance is shown under different bias levels in Figure 13.3.

13.4.2 Results

As a result, 438,194 cells within the healthy images and 2350 cells within the LGSIL images are located (Table 13.1). Typical results are shown in Figures 13.4 and 13.5.

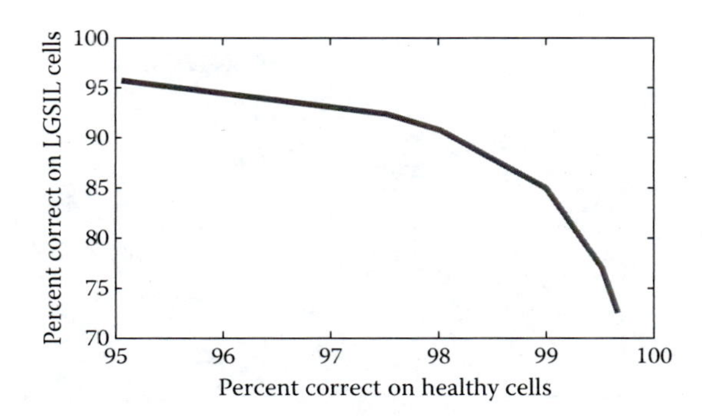

FIGURE 13.3
Performance of the algorithm, under different levels of bias. With a bias of –0.92, the software correctly labeled 97.5% of healthy cells and 92.4% of LGSIL cells.

TABLE 13.1
Results of Applying SVM

LGSIL Image	Healthy Image	Identified as LGSIL (True Positive)	Actually LGSIL (True Positive + False Negative)	Identified as Healthy (True Negative)	Actually Healthy (True Negative + False Positive)
LGSIL-HGSIL1_M35	K-144	12	15	23,098	23,098
LGSIL3_M13	K-144	138	158	23,098	23,098
LGSIL_M6	K-144	10	21		
LGSIL1_M12	K-20	155	355	75,149	75,216
LGSIL_K63	K-20	6	8	74,988	75,216
LGSIL1_M13	K-22	301	310	18,050	18,051
LGSIL_M12	K-22	130	181	18,051	18,051
LGSIL_M25	K-27	1	1	47,889	47,909
LGSIL_M13	K-27	53	55	47,891	47,909
LGSIL1_M26	K-31	110	128	40,442	40,442
LGSIL_M25	K-31	2	12	40,442	40,442
LGSIL1_M37	K-33-002	0	12	32,229	32,257
LGSIL_M26	K-33-002	126	153	3227	32,257
LGSIL2_M12	K-36	18	42	68,490	68,490
LGSIL_M27	K-36	10	40	68,489	68,490
LGSIL2_M13	K-56	540	557	43,329	44,694
LGSIL_M36	K-56	36	42	43,466	44,694
LGSIL2_M36	P-20	17	18	57,624	57,644
LGSIL_M37	P-20	37	131	57,625	57,644
LGSIL2_M37	P-5	11	87	29,162	29,162
LGSIL_M4	P-5	13	24	29,162	29,162

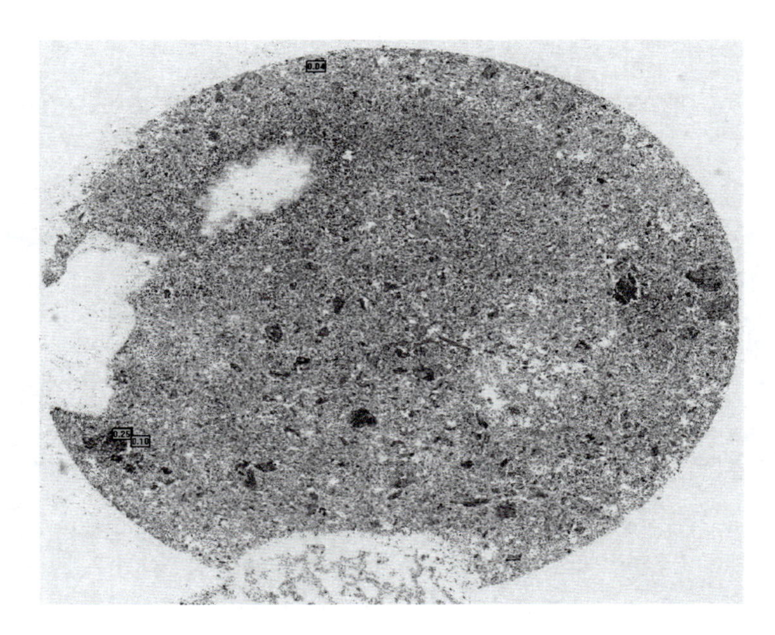

FIGURE 13.4
(See color insert.) Typical results for samples from a healthy patient. Four mildly suspicious cells (SVM response > 0) were discovered by the program. Only suspicious cells are shown.

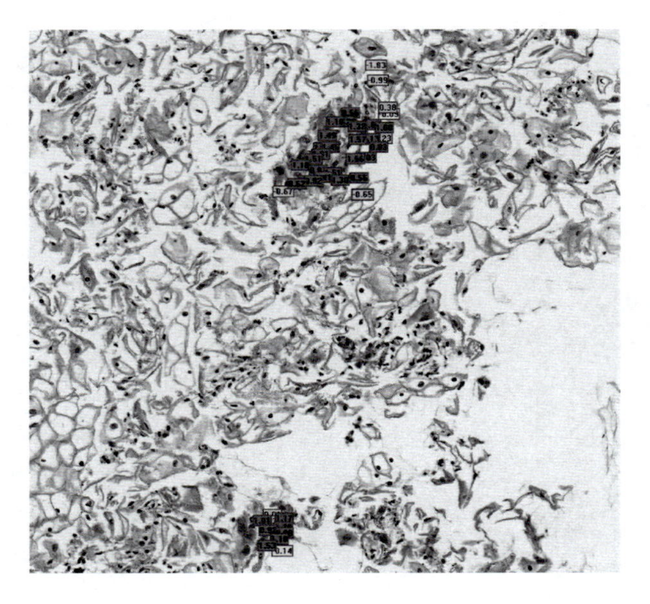

FIGURE 13.5
(See color insert.) Results for a sample LGSIL image. Two regions were graded LGSIL by Ossama Tawfik. Several strongly suspicious cells were discovered by the software in both regions. Only cells from within the known unhealthy regions are labeled.

13.5 Applying DBNs

13.5.1 Introduction and Motivation

DBNs are a kind of new innovation in the field of machine learning. They are capable of extracting distinctive features from a data set, and no label is needed for extracting features. So far, DBNs have not been applied to cytological cancer diagnosis. But they have shown promising results on similar methods. DBNs have been used for generating and recognizing images (Hinton et al. 2006; Ranzato et al. 2007; Bengio et al. 2006), video sequences (Sutskever and Hinton 2007), and motion-capture data (Taylor et al. 2007).

In most of the other classification systems, labeled data are used for training, and the task of finding labeled data is itself pretty tough. If meaningful features can be extracted without labeled data, it can be of great significance, even in the course of cancer diagnosis.

13.5.2 Data Set Modified

The data set used for SVM analysis had 54 features. For the analysis with DBNs 576 more features were added, totaling 630 features. In the successful applications of DBNs, in the visible layer, more raw features are fed, and from that input, with layer-by-layer processing, meaningful features are extracted.

For the same reason, to make the input layer of the DBN more complete and similar to previously used data sets, more features are added. But as the cancer cell samples are pretty huge and rich, adding the whole image pixel is not a very good idea. So to have a fewer dimensions, a different approach is taken. These features are mean color values within different regions centered at the nucleus center. Each feature is a section of an annulus that spans 45° (so, 8 sections), and annuli range from 2 pixels to 50 in steps of 2 (24 annuli). Since there are 3 colors, that is 576 features.

13.5.3 Feature Extraction Using DBN

While modeling a multi network, there are basically two questions that come up.

1. How many hidden layers are there?
2. How many neurons are there in each layer?

According to Hinton (2007), generative models with only one hidden layer are much too simple for modeling the high-dimensional and richly structured sensory data that arrive at the cortex, but they have been pressed into service because, until recently, it was too difficult to perform inference in the more complicated, multilayer, nonlinear models that are clearly required. There have been many attempts to develop multilayer, nonlinear models (Lewicki and Seinowski 1997; Hoyer and Hyya 2002; Portilla et al. 2003). In Bayes nets (also called belief nets), which have been studied intensively in AI and statistics, the hidden variables typically have discrete values. Exact inference is possible if every variable only has a few parents. This can occur in Bayes nets that are used to formalize expert knowledge in limited domains, but for more densely connected Bayes nets, exact inference is generally intractable. Extensive experiments by Yoshua Bengio's group suggest that several hidden layers are better than one. Results are fairly robust against changes in the size of a layer, but it is said that the top layer should be big.

Hinton et al. (2006) introduced a greedy, layer-by-layer unsupervised learning algorithm that consists of learning a stack of restricted Boltzmann machines (RBMs) one layer at a time. After the stack of RBMs has been learned, the whole stack can be viewed as a single probabilistic model, called a "deep belief network." In this study, depending on the raw feature set we planned to use, the DBN was structured with four layers.

Next, deciding on the number of neurons in each layer is a very important part of deciding on the overall network architecture. Using too few neurons in the hidden layers will result in something called underfitting. Underfitting occurs when there are too few neurons in the hidden layers to adequately detect the signals in a complicated data set.

Using too many nodes in the hidden layers can result in several problems. First, too many nodes in the hidden layers may result in overfitting. Overfitting occurs when the network has so much information processing capacity that the limited amount of information contained in the training set is not enough to train all of the nodes in the hidden layers. A second problem can occur even when the training data are sufficient. An inordinately large number of nodes in the hidden layers can increase the time it takes to train the network. The amount of training time can increase to the point that it is impossible to adequately train the network. Obviously, some compromise must be reached between too many and too few nodes in the hidden layers.

In the data set used for DBN structure (Figure 13.6), the input data are highly enriched with image information along with some extracted feature set, totaling 576 features. For this reason, we expect more information processing in the first layer, so we used 1000 nodes as the first layer. Then after extracting that information and using those data as the input for the next layer, 400 nodes were used as next hidden node layer. And in the next layer, 200 nodes were used. The idea was to extract some meaningful features that can be distinct enough to classify into two categories. After training the DBN with three layers, the last layer is the output layer, which has two nodes, one signifying an LGSIL cell and the other a healthy cell.

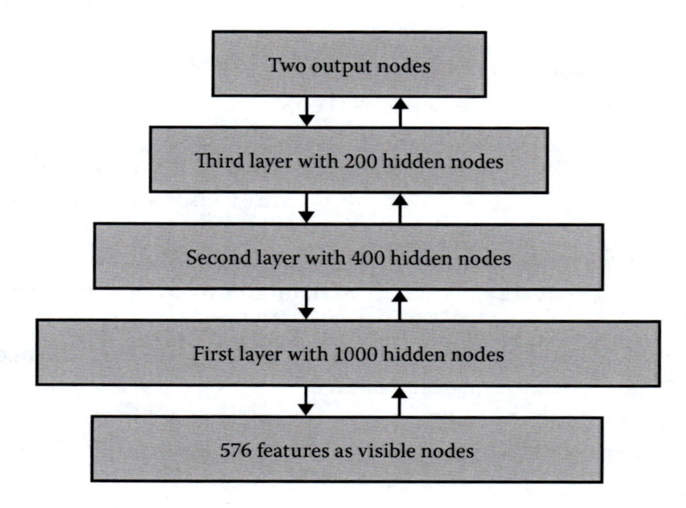

FIGURE 13.6
Structure of the deep belief network used in this study.

13.5.4 Partitions

The data set we have has more "healthy cell" samples than "LGSIL" cells. K-22 and LGSIL-M12 are left for testing, and the rest of all the cells are used for training. At this point, the data set consists of 409,580 healthy cell instances and 2,350 LGSIL cell instances. As DBNs perform better on a balanced data set, testing on the LGSIL cells was repeated 200 times to make the DBN data set balanced. So after the processing, the data set consisted of 409,580 healthy and 470,000 LGSIL instances, totaling 879,580 cells for training. And during testing, K-22 and LGSIL-M12 consisted of 18,051 healthy and 181 LGSIL cells, so overall, 18,232 cells.

In this study, only one partition is used. The backpropagation for 145 epochs took more than 1 week. So, it is tested on only one data set.

13.5.5 Normalization

Normalization is a very important step for any algorithm to perform well. This normalization is required because DBNs treat node values as activation probabilities. As the feature values range widely among different features, the data set was normalized to have feature values between 0 and 1. Simply, normalization is done so that the values are uniformly distributed within distinct range.

If A is a matrix, then the normalization techniques used in this study are as follows:

$$A = \frac{(A - \min(A(:)))}{(\max(A(:)) - \min(A(:)))}$$

13.5.6 Training the DBN

The data set used in this study has 630 raw features extracted from the digital slides, and it is hard to say which features have important information in predicting whether cells are cancerous or healthy. So the first objective is to reduce the dimensionality of the data set and take it down to fewer features that can be used easily for classifying purposes. So, removing the redundant features and making count of significant features are the main objective of training the DBN.

While feeding the raw features to the visible nodes of the DBNs, cell location features and mean and variance of the morphological features were not used. The objective was to extract meaningful features that play an important role in predicting whether a cell is cancerous or healthy, in which cell locations are not supposed to play any role. So, only 576 features were fed into the first layer or visible unit of the network. The DBN, as mentioned earlier, uses three layers with 1000 nodes, 400 nodes, and 200 nodes consecutively.

A DBN can be viewed as a composition of simple learning modules, each of which is a restricted type of Boltzmann machine that contains a layer of *visible units* that represent the data and a layer of hidden units that learn to represent features that capture higher-order correlations in the data (Figure 13.7).

So, in the first step, an RBM is constructed with 576 raw features (v) in the visible unit and 1000 hidden nodes (h). After training the first layer and extracting features, the second layer is to be trained.

So now, another 400 features are stacked on top of the RBM to make another RBM. So this, time the 1000 features (h1) that are extracted from the first layer of the DBN are used

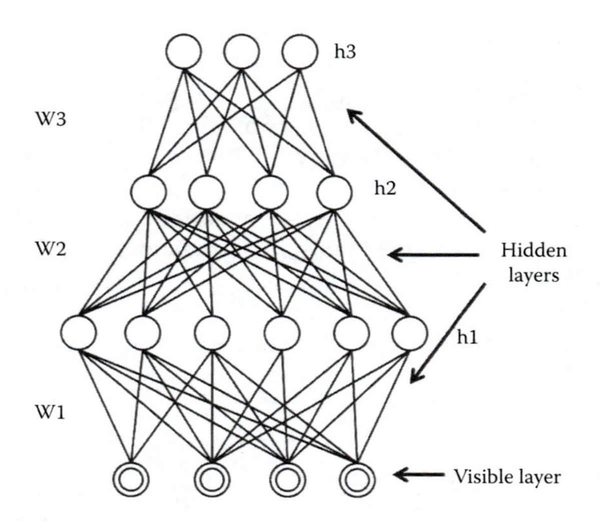

FIGURE 13.7
Hidden layers and visible layer (Hinton et al. 2006).

as the visible nodes for the RBM, and the second layer of the DBN, which is 400 hidden nodes (h2), is used as the hidden nodes for the RBM. And in the same way as the first layer, this second layer is trained.

Now another RBM is stacked on the existing network. So, 400 extracted features (h2) are used as the visible unit for the next layer of RBM, and this time, 200 hidden nodes are used.

While running the code, the whole data set is divided into batches, and in each batch, 100 cell instances are there. The whole data set is randomized so that in each batch, there is a random number of healthy and LGSIL cells.

This layer-by-layer learning procedure is called greedy layer-wise learning.

- It starts with the lowest level and stacks upward.
- It trains each layer of autoencoder on the intermediate code (features) from the layer below.
- The top layer can have a different output (e.g., softmax nonlinearity) to provide an output for classification.

13.5.7 Backpropagation

Backpropagation has many problems, like the gradient progressively becoming more dilute, it gets stuck in the local minima. But when used with greedy pretraining, it works better.

The backpropagation starts once sensible features are detected, which makes the initial gradients more sensible, and backpropagation only needs to perform a local search from a sensible starting point.

The backpropagation algorithm (Figure 13.8) is a supervised learning method for multi-layer feed-forward networks from the field of artificial neural networks and. more broadly, computational intelligence. The name refers to the backward propagation of error during the training of the network.

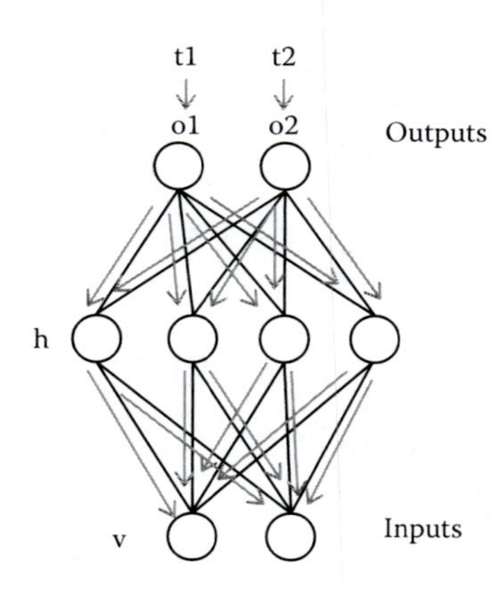

FIGURE 13.8
Backpropagation of error.

After extracting 200 features from the 536 raw features, two output nodes are added, one for LGSIL and one for healthy cells. The initializing weights for the output node layer are again randomly assigned. As we have labelled data, depending on the targets the error is backpropagated to fine-tune the weights.

During the training, the error calculated is backpropagated through the whole network. First, total error at the top is calculated, and then contribution to error at each step going forward is calculated.

In this study, the propagation was done for 145 epochs. For the first 10 epochs, the output node layer weights are updated, holding the other weights fixed, and then in rest of the 135 epochs, weights of all the layers are updated.

13.5.8 Combining SVM with DBN Features

Training of DBNs is unsupervised. So, the features extracted based on the first three layers (without backpropagation) extract some meaningful features, which may be useful for cancer diagnosis or may be totally irrelevant.

In this approach, from the extracted DBN feature nodes, nodes containing the highest information were selected and appended with the raw features and used for SVM classification. The intention is to observe how the data set will behave when unknown but important features are appended with some known features. The extracted features may play a significant role, or maybe they are not distinctive enough to predict a cell healthy or LGSIL. But the next question was how many features to select from the 200 features. The number of raw features was 52, so it does not make sense to append a high number of features with them. So, depending on the range of the weights in the connection, a threshold was decided, and the nodes whose connections had a value above that threshold were selected.

13.5.9 Selecting DBN Features

After DBNs extract important information in 200 feature nodes, two output nodes are used to classify the sample as healthy or LGSIL. With backpropagation of error, the whole network's weights are fine-tuned after that. In the last layer, high weight in the connection

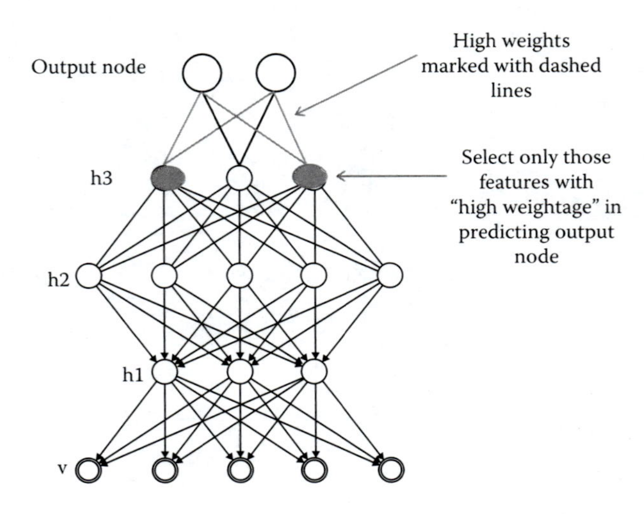

FIGURE 13.9
Selection features from DBN.

denotes that features have important information for classifying cells. So depending on that, those few features were selected from 200 features that had high weight in the last layer (200 nodes and 2 output node layers) to classify cells as LGSIL or healthy (Figure 13.9). We took the absolute value of the weights. If the absolute value was greater than 100, then 18 features were selected, and when we set the threshold at 120, 5 features were selected.

13.5.10 Results of DBN with Backpropagation on Set 1

The data set was divided into two parts: training and testing. One image of healthy cells (K22) and one image of LGSIL cells (LGSIL-M12) were left for testing, and of all the rest were taken to use for training. In the test data set, there are 181 LGSIL cells and 18,051 healthy cells.

After training the DBN on the 29 training images, we first tested classification performance based on the DBN output alone. The DBN correctly classified 159 out of 181 LGSIL cells and 17,733 out of 18,051 healthy cells, for an overall accuracy of 98.14%. On this same test set partitioning, SVM achieved an accuracy of 99.72%, correctly classifying 130 out of 181 LGSIL cells and 18,051 out of 18,051 healthy cells. The bias of each classifier can be adjusted to produce fewer false positives or fewer false negatives. However, SVM continued to outperform DBN features alone.

We next appended five DBN features (selected by the procedure described in a previous section) to the initial features used by SVM in Section 13.4. For this training/testing partitioning, the addition of DBN features allowed SVM to achieve 100% accuracy (Table 13.2).

In Table 13.2, FN, FP, TN, and TP refer to false negatives, false positives, true negatives, and true positives, respectively. Accuracy is the percentage of correctly classified cells: (TP + TN)/(TP + TN + FP + FN). Recall (also known as sensitivity) is the accuracy among LGSIL (positive) cells: TP/(TP + FN). Specificity is the accuracy among truly healthy (negative) cells: TN/(TN + FP). Precision is the accuracy among cells labeled as LGSIL: TP/(TP + FP). F-measure, or F1 score, is the harmonic mean of precision and recall: 2 × recall × precision/(precision + recall).

TABLE 13.2
Results of Set 1

	FN	FP	TN	TP	Accuracy	Recall	Specificity	Precision	F-Measure
SVM	44	0	18051	137	99.76%	75.69%	100.00%	100.00%	0.86
DBN	22	318	17733	159	98.14%	87.85%	98.24%	33.33%	0.48
SVM + 5DBN	0	0	18051	181	100.00%	100.00%	100.00%	100.00%	1.00

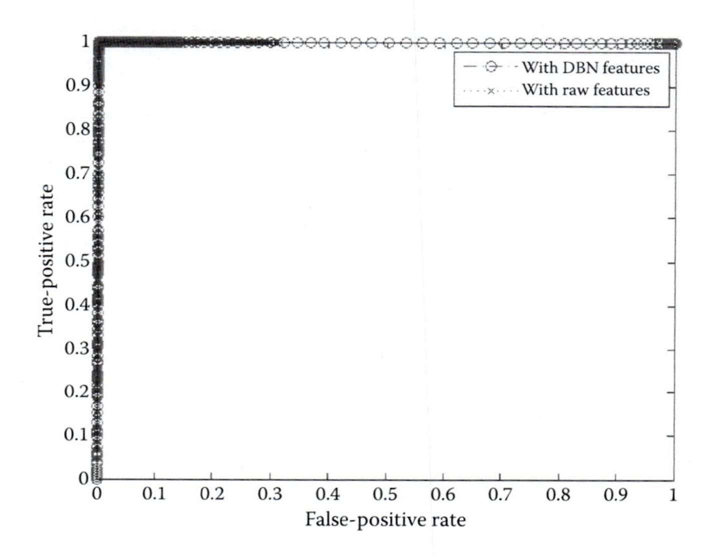

FIGURE 13.10
ROC curve for set 1.

For semiautomated cancer diagnosis, it is important to limit the rate of false negatives. The highest priority is to limit the number of LGSIL cells that are not identified by the system. Reducing the number of healthy cells incorrectly labeled as LGSIL is of secondary importance. SVM classifiers attempt to achieve the highest overall accuracy, which, given the large number of healthy cells in the training data, often results in classifiers that prioritize the false-negative rate over the false-positive rate, contrary to desired behavior.

The problem of prioritizing the false-negative rate is known as "cost-aware classification" and has many solutions. One simple approach is to include a bias on the SVM output as done in Figure 13.7, so that cells are only labeled as healthy if the SVM output is beyond some elevated threshold. In Figure 13.10, we plot the false-positive rate and true-positive rate of SVM, both with and without the five appended DBN features, as bias is adjusted. This is known as the receiver operating characteristic curve (ROC curve). The addition of DBN features allows SVMs to achieve a higher true-positive rate without increasing the false-positive rate.

13.5.11 Results of DBN with Backpropagation on Set 2

For the next data set, the whole data set was again divided into two parts: training and testing. One image of healthy cells (P-20) and one image of LGSIL cells (LGSIL-M37) were left for testing, and of all the rest were taken to use for training. In the test data set, there are 131 LGSIL cells and 54,507 healthy cells.

TABLE 13.3
Results of Set 2

	FN	FP	TN	TP	Accuracy	Recall	Specificity	Precision	F-Measure
SVM	94	0	54507	37	99.83%	28.24%	100.00%	100.00%	0.44
DBN	32	4071	50436	99	92.49%	75.57%	92.53%	2.37%	0.05
SVM + 5DBN	51	3	54504	80	99.90%	61.07%	99.99%	96.39%	0.75

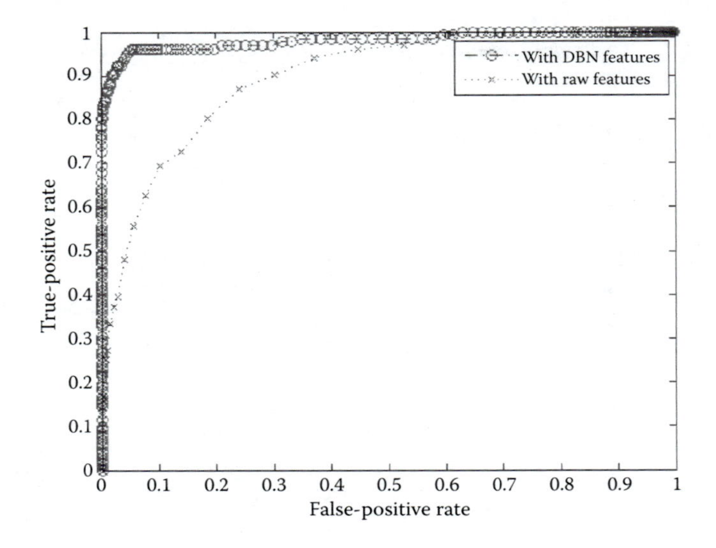

FIGURE 13.11
ROC curve for set 2.

Again, we trained the DBN on the 29 training images and examine classification performance based on the DBN output alone. The DBN correctly classified 99 out of 131 LGSIL cells and 50,436 out of 54,507 healthy cells, for an overall accuracy of 92.49%. On this same test set partitioning, SVM achieved an accuracy of 99.83%, correctly classifying 37 out of 131 LGSIL cells and 54,507 out of 54,507 healthy cells.

We next appended five DBN features to the initial features used by SVM in Section 13.4. The addition of five DBN features allowed SVM to achieve a performance of 99.90%, correctly classifying 80 out of 131 LGSIL cells and 54,504 out of 54,507 healthy cells (Table 13.3).

Again, we adjust the bias of the SVM to plot the ROC curve (Figure 13.11). The results show that when five DBN features are appended with the raw features, the number of misclassification of LGSIL decreases.

13.6 Conclusion and Discussions

The success of cancer treatment, to a large extent, depends on correct diagnosis. Automated cancer diagnosis has the capability to play a significant role in accurate diagnosis. In this study, we tried to analyze the data with some new prominent approaches and compare the results with some established classification methods.

For cancer diagnosis, the raw data set is high-dimensional, and the structure is too complicated to be represented by a simple model. Another challenge is the lack of labeled data for this kind of analysis. To represent the complicated structure is itself a big task. DBNs in simple terms stacked RBMs. And RBM provides a simple way to learn one layer of features without any supervision. And many layers of representation create a good generative model that can be fine-tuned.

13.6.1 Future Directions

In this study, the DBN approach is tried on only two sets because of the high computational time. If it can be tried on more than two training/testing splits, then it will be easy to conclude which DBN features are best. It will also help in selecting the normalization method that will work best.

Selecting DBN features seems like an important aspect of improvement, and there may be better ways of doing this. For example, maybe the variance of the features should have been considered in addition to the weight. Maybe features that were maximally discriminative could have been found using other means, like traditional feature selection techniques.

13.6.2 Recent Improvements in Deep Learning

DBN techniques themselves are continually improving—so there are some methodologies that can be applied for analyzing the data set of this study.

In their study, Swersky et al. (2010) provided a direct comparison of the stochastic maximum likelihood algorithm and contrastive divergence for training RBMs using the MNIST dataset is subset of NIST (National Institute of Standards and Technology) Dataset. They demonstrated that stochastic maximum likelihood is superior when using the RBM as a classifier and that the algorithm can be greatly improved using the technique of iterate averaging from the field of stochastic approximation. They further showed that training with optimal parameters for classification does not necessarily lead to optimal results when RBMs are stacked to form a DBN.

The study by Plahl et al. (2012) suggested the Sparse Encoding Symmetric Machine (SESM) as an alternative method for pretraining the DBN instead of the RBM. The RBM objective function cannot be maximized directly. Therefore, it is not clear what function to monitor when deciding to stop the training, leading to a challenge in managing the computational costs. In this paper, they explore SESM to pretrain DBNs and apply this to speech recognition. Their results indicate that pretrained DBNs using SESM and RBMs achieve comparable performance and outperform randomly initialized DBNs with SESM, providing a much easier stopping criterion relative to RBM. While the RBM relies on Contrastive Divergence (CD), SESM directly optimizes the objective function. Moreover, SESM provides easy criteria to stop the training, leading to less iteration. This technique maybe used with our data set to see if it improves.

References

Beck, A. H., Sangoi, A. R., Leug, S., Marinelli, R. J., Vijver, M., West, R. B., Matt, R., and Neilsen, T. O. (2011, December 15). Systematic analysis of breast cancer morphology uncovers stromal features associated with survival. *Science Transnational Medicine*, 3(108), 1–11.

Bengio, Y., Lamblin, P., Popovici, P., and Larochelle, H. (2006). *Greedy layer-wise training of deep networks*. Informally published manuscript, Universite de Montreal, Universite de Montreal, Montreal, Canada.

Blekas, K., Stafylopatis, A., Kontoravdis, D., Likas, A., and Karakitsos, P. (1998). Cytological diagnosis based on fuzzy neural networks. *Journal Intelligent Systems*, 8, 55–79.

Carpenter, A., Jones, T., Lamprecht, M., Clarke, C., Kang, I., Friman, O., Guertin, D., Chang, J., Lindquist, R., Moffat, J., Golland, P., and Sabatini. D. (2006). CellProfiler: image analysis software for identifying and quantifying cell phenotypes. *Genome Biology*, 7(10).

Cawley, C. C. and Talbot, N. L. (2004). Fast exact leave-one-out cross-validation of sparse least-squares support vector machines. *Neural Networks*, 17(10), 1467–1475.

Demir, C. and Yener, Y. (2009). Automated cancer diagnosis based on histopathological images: a systematic survey. Technical report, Rensselaer Polytechnic Institute, 5.

Demir, C., Gultekin, S. H., and Yener, B. (2004). Learning the topological properties of brain tumors, Rensselaer Polytechnic Institute Technical Report TR-04-14, Troy.

Doyle, S., Agner, S., and Madabhushi, A. (2008, May). Automated grading of breast cancer histopathology using spectral clustering with textural and architectural image features. Paper presented at *IEEE International Symposium on Biomedical Imaging, 5th IEEE International Symposium on Biomedical Imaging: From Nano to Macro*, Paris, France.

Doyle, S., Hwang, M., Shah, K., Madabhushi, A., Tomaszeweski, J., and Feldman, M. (2007). Automated grading of prostate cancer using architectural and textural image features. In *International Symposium on Biomedical Imaging*, pp. 1284–1287, Washington DC.

Esgiar, A. N., Naguib, R. N. G., Sharif, B. S., Bennett, M. K., and Murray, A. (2002). Fractal analysis in the detection of colonic cancer images. *IEEE Transactions on Information Technol*ogy in *Biomedicine*, 6, 54–58.

Gunduz, C., Yener, B., and Gultekin, S. H. (2004). The cell graphs of cancer. *Bioinformatics*, 20, i145–i151.

Hinton, G. E. (2007). Learning multiple layers of representation. *Science Direct*, 11(10), 428–433.

Hinton, G. E. and Salakhutdinov, R. R. (July 28, 2006). Reducing the dimensionality of data with neural networks. *Science Transnational Medicine*, 313, 504–507.

Hinton, G. E., Osindero, S., and Teh, Y. W. (2006). A fast learning algorithm for deep belief nets. *Neural Computation*, 18, 1527–1554.

Hoyer, P. O. and Hyya, R. A. (2002). A multi-layer sparse coding network learns contour coding from natural images. *Vision Research*, 1593–1605.

Joachims, T. (1999). *Making Large-Scale Support Vector Machine Learning Practical*, pp. 169–184. MIT Press, Cambridge, MA.

Joachims, T. (2008). Support vector machine. Informally published manuscript, Cornell University, Department of Computer Science. Retrieved from http://svmlight.joachims.org/.

Karklin, Y. and Lewicki, M. S. (2003). Learning higher-order structures in natural images. *Network*, 14, 483–499.

Koss, L. G. (1989). The Papanicolaou test for cervical cancer detection. *Journal of the American Medical Association*, 261(5), 737–743.

Lewicki, M. S. and Seinowski, T. J. (1997). Bayesian unsupervised learning of higher order structure. In M. Mozer (Ed.), *Advances in Neural Information Processing Systems* (Vol. 9, pp. 529–535). MIT Press, Cambridge, MA.

Malpica, N., Solrzano, C., Vaquero, J., Santos, A., Vallcorba, I., Garca-Sagredo, J., Malpica, N., de Solórzano, C. O., Vaquero, J. J., Santos, A., Vallcorba, I., García-Sagredo, J. M., and del Pozo, F. (1997). Applying watershed algorithms to the segmentation of clustered nuclei. *Cytometry*, 28(4), 289–297.

Plahl, C., Sainath, T. N., Ramabhadran, B., and Nahamoo, D. (2012). Improved pre-training of deep belief networks using sparse encoding symmetric machines. *ICASSP*, 978(1), 4165–4168.

Portilla, J., Strela, V., Wainwright, J., and Simoncelli, E. P. (2003). Image denoising using scale mixtures of Gaussians in the wavelet domain. *IEEE Transactions of Image on Processing*, 12(11), 1138–1150.

Roux, N. L. and Bengio, Y. (2008). Representational power of restricted Boltzmann machines and deep belief networks. *Neural Computation*, 20(6), 1631–1649.

Salakhutdinov, R. R. and Hinton, G. E. (2007). Semantic hashing. In *Proceedings of the SIGIR Workshop on Information Retrieval and Applications of Graphical Models*, Amsterdam.

Schnorrenberg, F., Pattichis, C. S., Schizas, C. N., Kyriacou, K., and Vassiliou, M. (1996). Computer-aided classification of breast cancer nuclei. *Technology and Health Care*, 4 147–161.

Schwartz, O., Senowski, T. J., and Dayan, P. (2006). Soft mixer assignment in a hierarchical generative model of natural scene statistics. *Neural Computation*, 18(11), 2680–2718.

Smolle, J. (2000). Computer recognition of skin structures using discriminant and cluster analysis. *Skin Research and Technology*, 6, 58–63.

Spyridonos, P., Ravazoula, P., Cavouras, D., Berberidis K., and Nikiforidis G. (2001). Computer-based grading of haematoxylin-eosin stained tissue sections of urinary bladder carcinomas. *Medical Informatics and the Internet in Medicine*, 26, 179–90.

Stanikov, A., Tsamardinos, L., Dosbayev, Y., and Aliferris, C. F. (2005). Gems: a system for automated cancer diagnosis and biomarker discovery from microarray gene expression data. *International Journal of Medical Informatics*, 74, 491–503.

Swersky, K., Chen, B., and Freitas, N. (2010). A tutorial on stochastic approximation algorithms for training restricted Boltzmann machines and deep belief nets. Informally published manuscript, University of British Columbia, Department of Computer Science, University of British Columbia, Canada.

Taylor, G. W., Hinton, and G. E., and Roweis, S. (2007). Modelling human motion using binary latent variables. In *Advances in Neural Information Systems* (Vol. 19), MIT Press, Cambridge, MA.

Wahlby, C., Sintorn, I. M., Erlandsson, F., Borgefors, G., and Bengtsson, E. (2004). Combining intensity, edge and shape information for 2D and 3D segmentation of cell nuclei in tissue sections. *Journal of Microscopy*, 215(1), 67–76.

Wolberg, W. H., Street, W. N., Heisey, D. M., and Mangasarian, O. L. (1995). Computer-derived nuclear features distinguish malignant from benign breast cytology. *Human Pathology*, 26, 792–796.

Wiltgen, M., Gerger, A., and Smolle, J. (2003). Tissue counter analysis of benign common nevi and malignant melanoma. *International Journal of Medical Informatics*, 69, 17–28.

Zhou, Z. H., Jiang, Y., Yang, Y. B., and Chen, S. F. (2002). Lung cancer cell identification based on artificial neural network ensembles. *Artificial Intelligence in Medicine*, 24, 25–36.

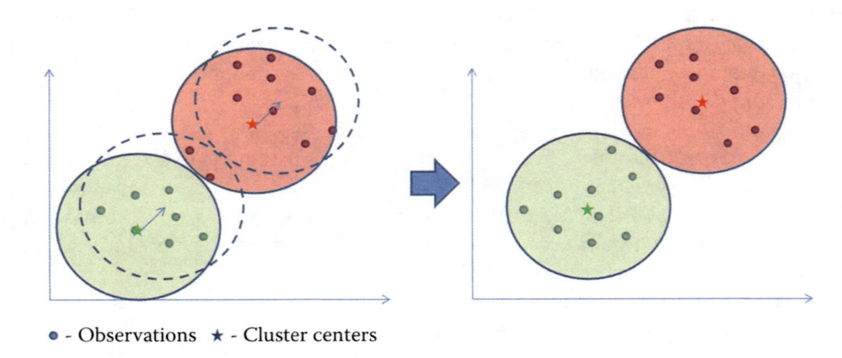

FIGURE 4.2
Changes in clusters' center points between iterations.

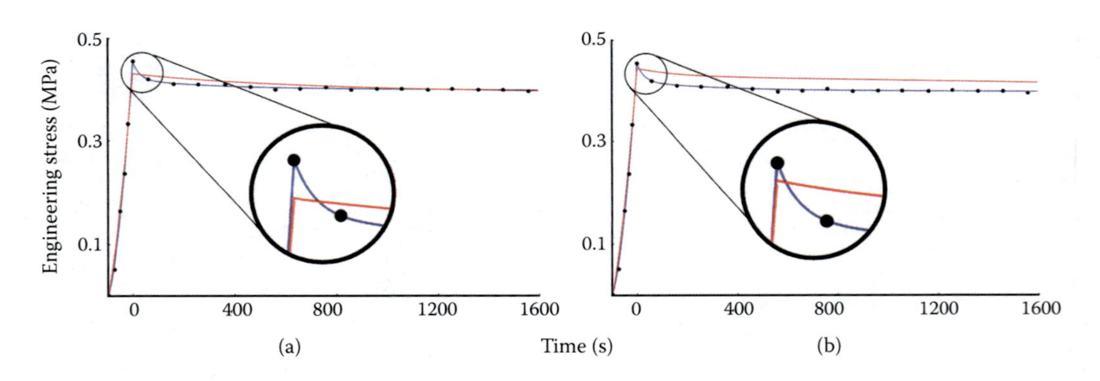

FIGURE 6.6
Engineering stress versus time for the *ACL*. The experimental data are depicted by the dots [Pioletti et al. 1998, 2000]. The fitted models at 2% strain with the focus on change point (peak stress point) (a) and relaxation portion (b) are indicated by the lines. (From Pioletti, D. P. et al., *Journal of Biomechanics*, vol. 31, 1998, pp. 753–757; Pioletti, D. P. and L. R. Rakotomanana, *Journal of Biomechanics*, vol. 33, 2000, pp. 1729–1732.)

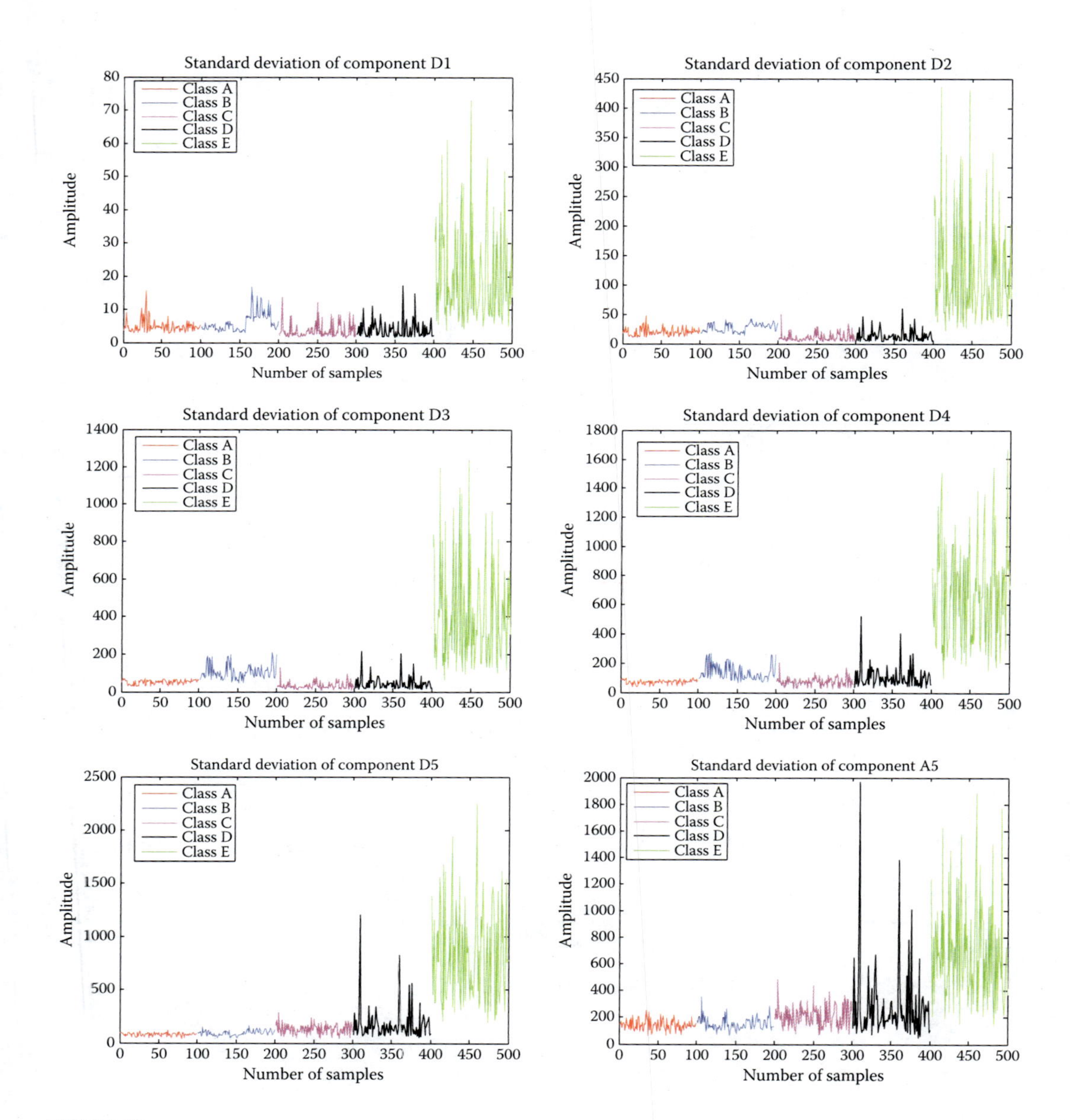

FIGURE 7.3
Distribution of the standard deviation over the wavelet coefficients for Db4 within the data set.

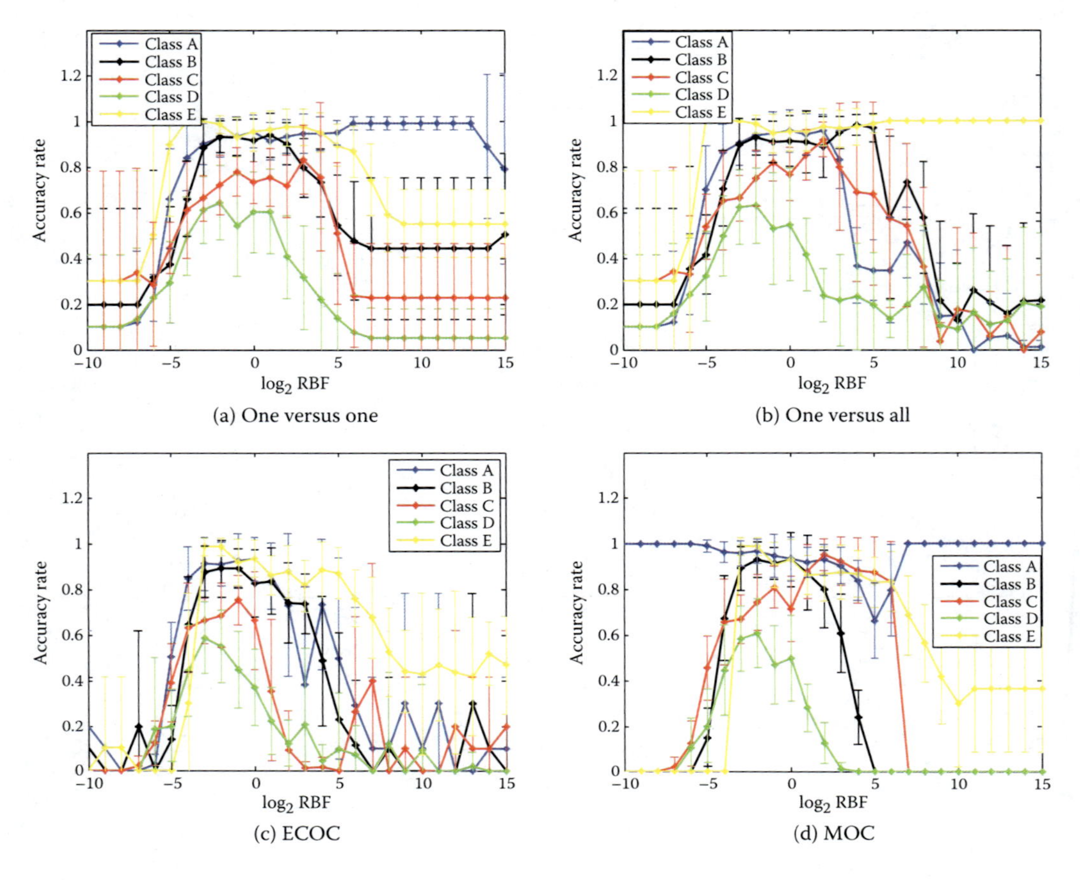

(a) One versus one

(b) One versus all

(c) ECOC

(d) MOC

FIGURE 7.4
Accuracy rate by class for different strategies of multiclass classification using the RBF kernel and feature vectors extracted through the standard deviation of the wavelet coefficients.

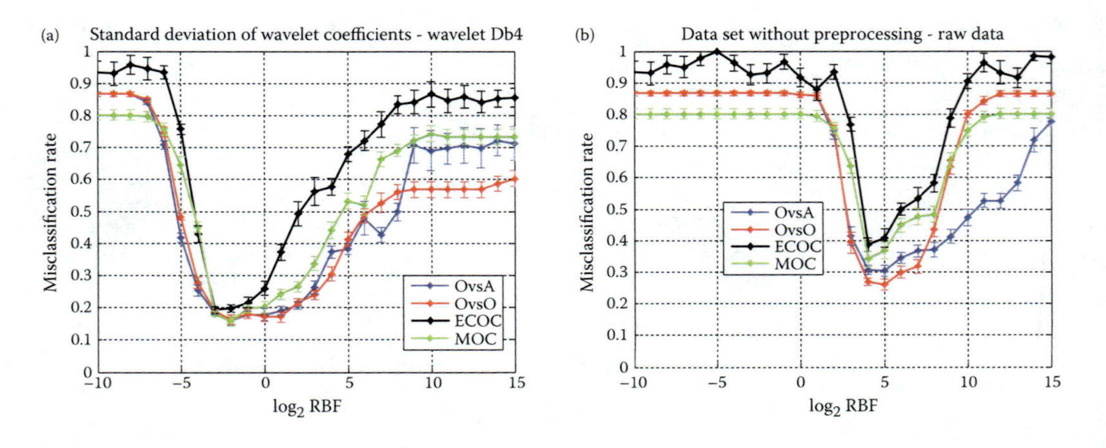

FIGURE 7.5
Error rate for different strategies for multiclass classification using the RBF kernel and feature vectors extracted through the standard deviation of the wavelet coefficients.

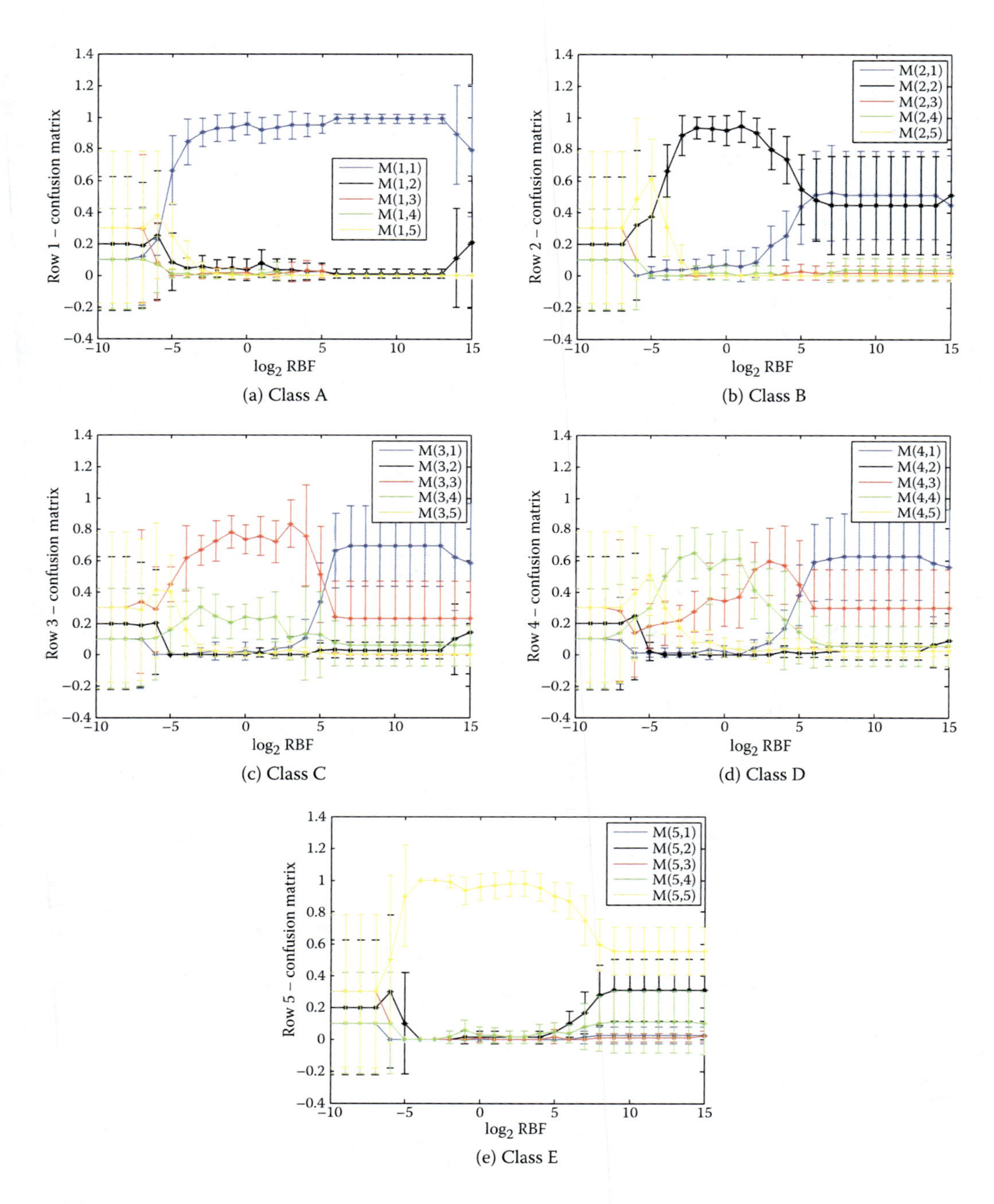

FIGURE 7.6
Effect of varying the kernel parameter in the classification of each class using standard deviation of wavelet coefficients and one-versus-one strategy.

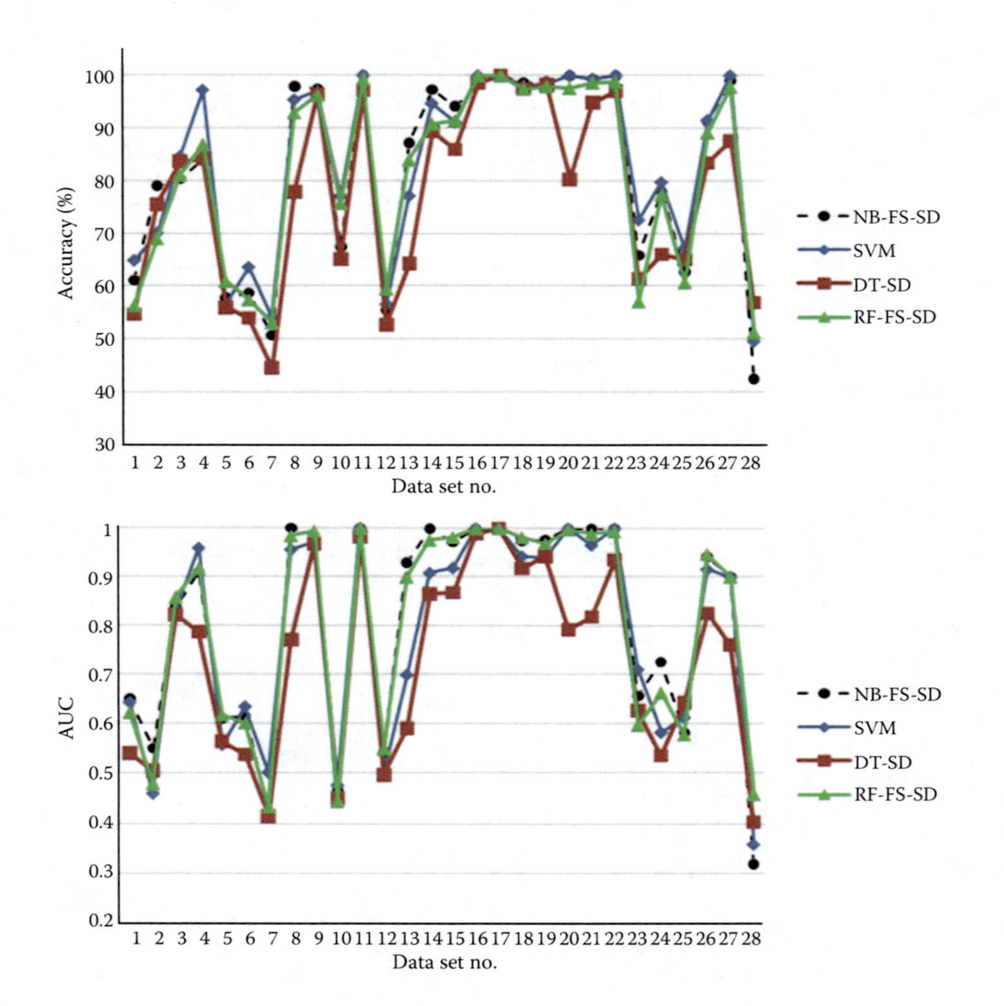

FIGURE 8.8
Accuracies and AUCs of NB-FS-SD and the other classifiers.

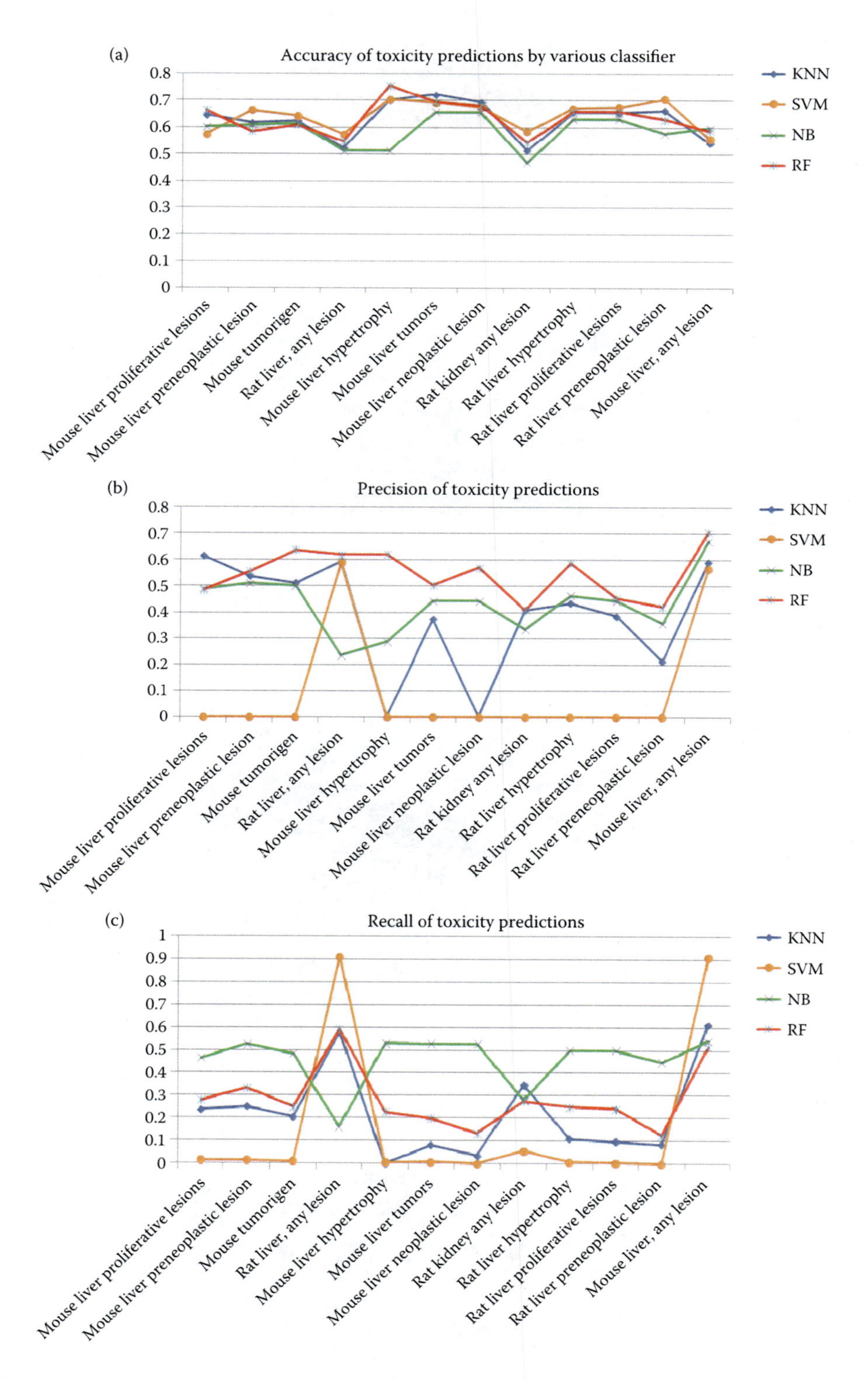

FIGURE 9.1
The (a) accuracy, (b) precision, and (c) recall values of the toxicity predictions made by different algorithms.

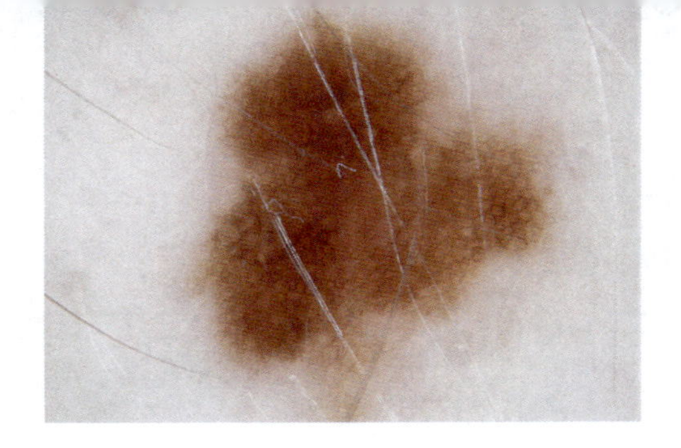

FIGURE 11.1
First lesion.

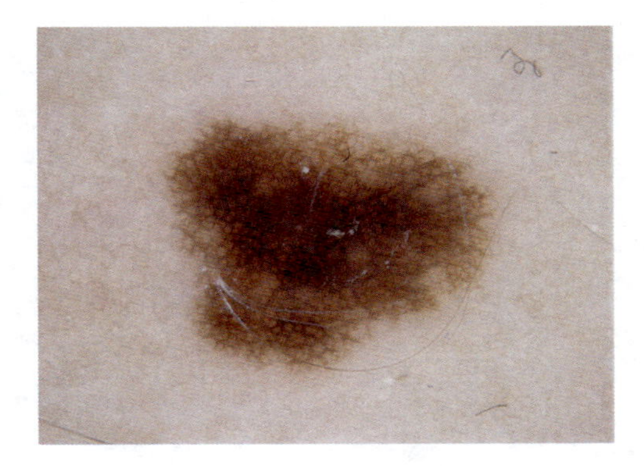

FIGURE 11.2
Second lesion.

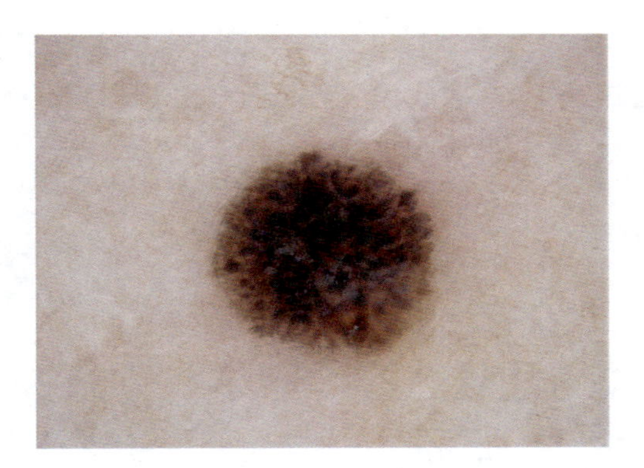

FIGURE 11.3
Third lesion.

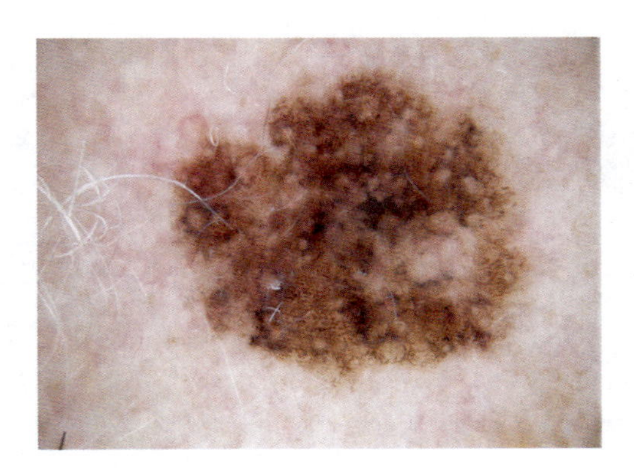

FIGURE 11.4
Fourth lesion.

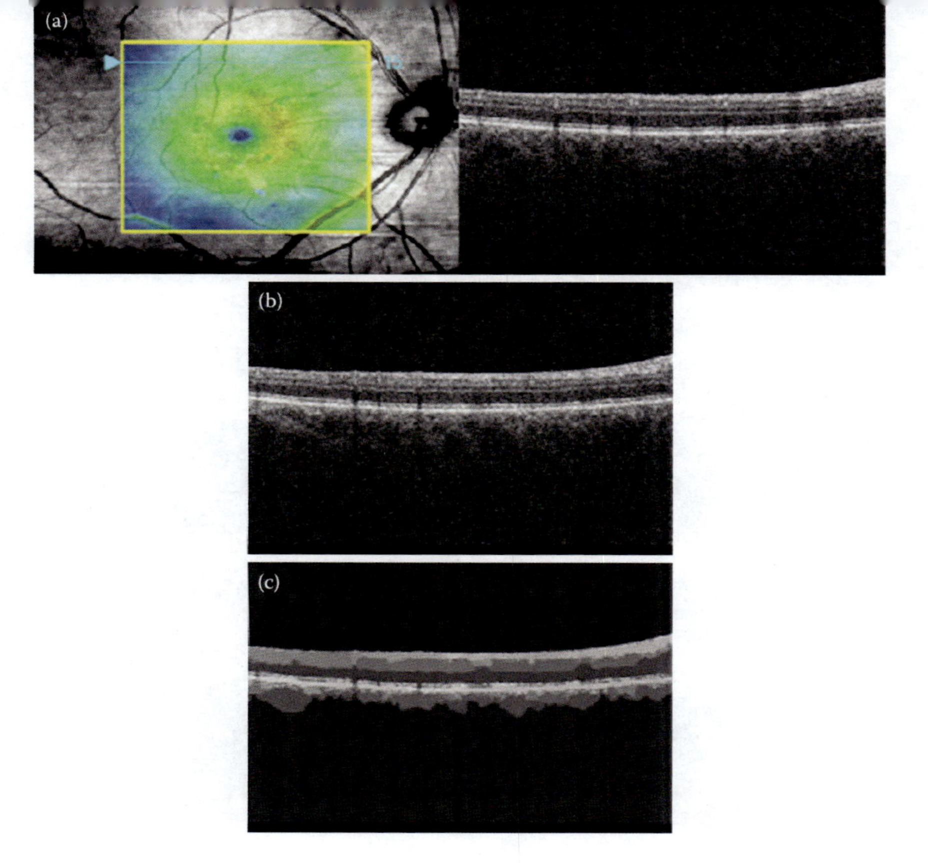

FIGURE 12.1
Preprocessing of an OCT image. (a) Original acquired image. (b) ROI extracted. (c) Image after the application of the nonlinear diffusion filter.

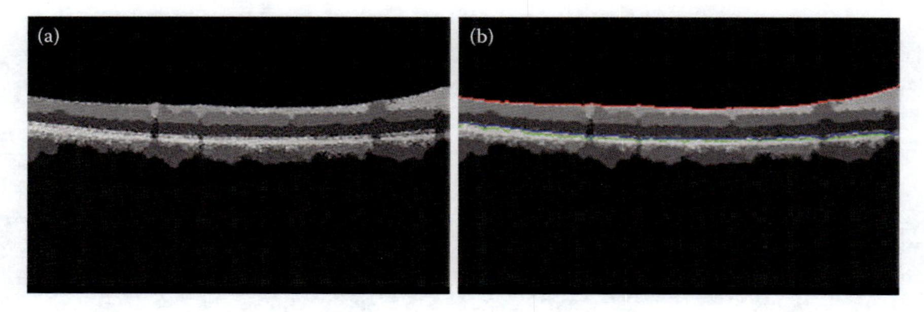

FIGURE 12.5
(See color insert.) (a) OCT image number 5 from the right eye of the first patient. (b) Three-layer highlighting.

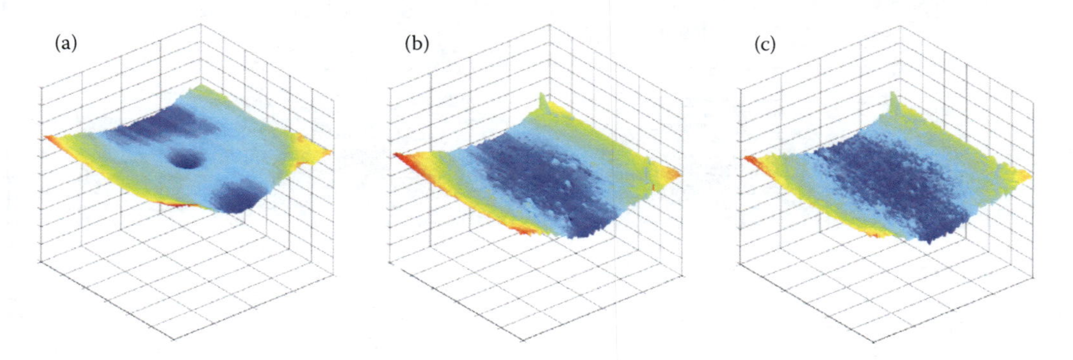

FIGURE 12.6
(See color insert.) 3-D modeling for each surface. (a) Surface 1, (b) surface 5, (c) surface 6.

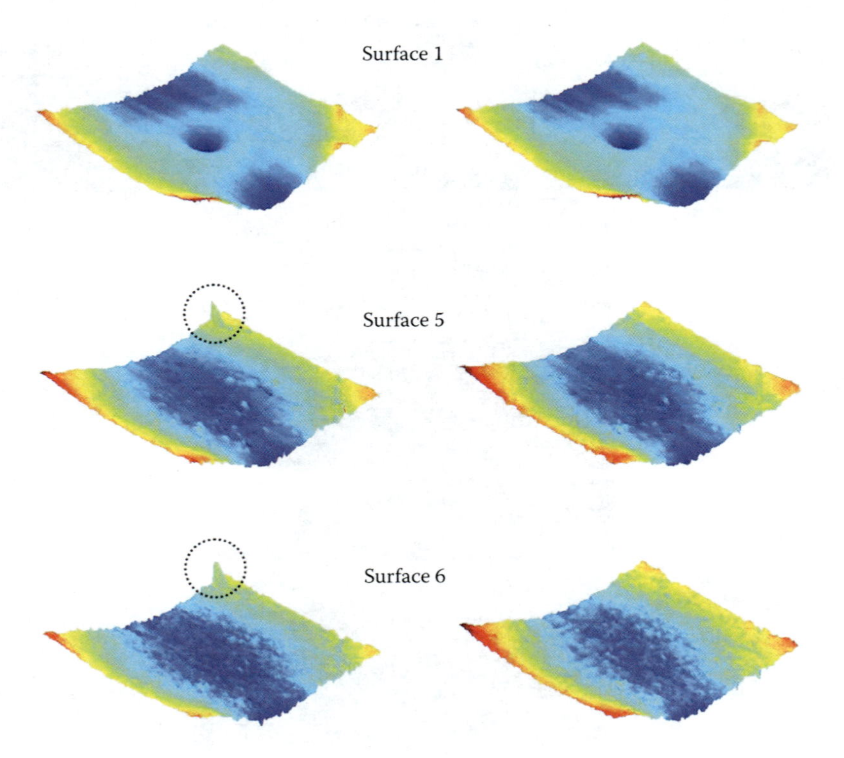

FIGURE 12.11
(See color insert.) 3-D modeling comparison for 2-D and 3-D processes for first patient's right eye.

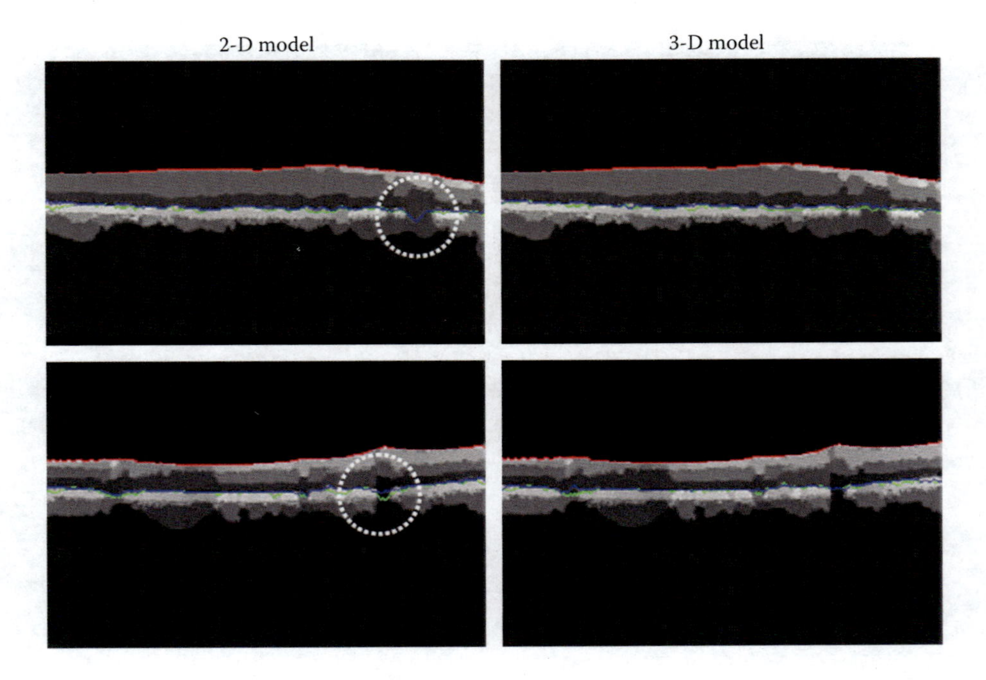

FIGURE 12.12
(See color insert.) Two samples of highlighting using 2-D model (left) and 3-D model (right). It can be observed how the discontinuities present in the 2-D model are avoided in the 3-D model.

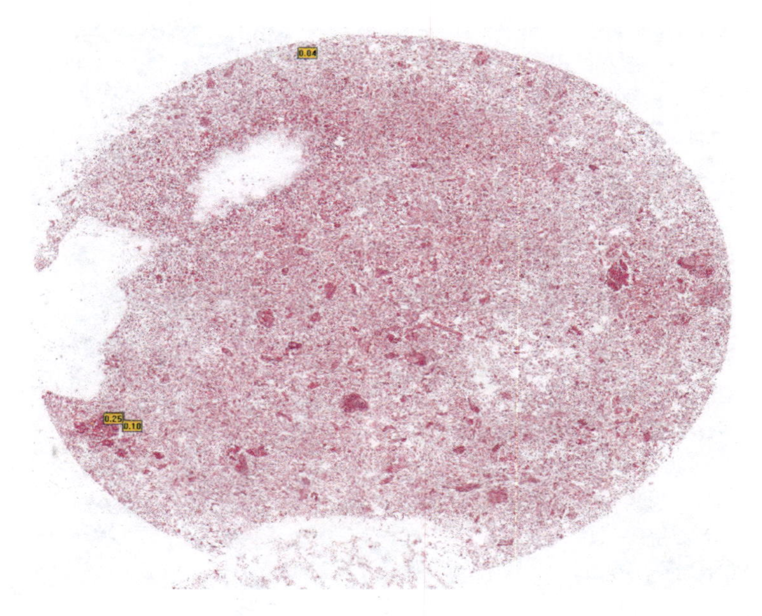

FIGURE 13.4
Typical results for samples from a healthy patient. Four mildly suspicious cells (SVM response > 0) were discovered by the program. Only suspicious cells are shown.

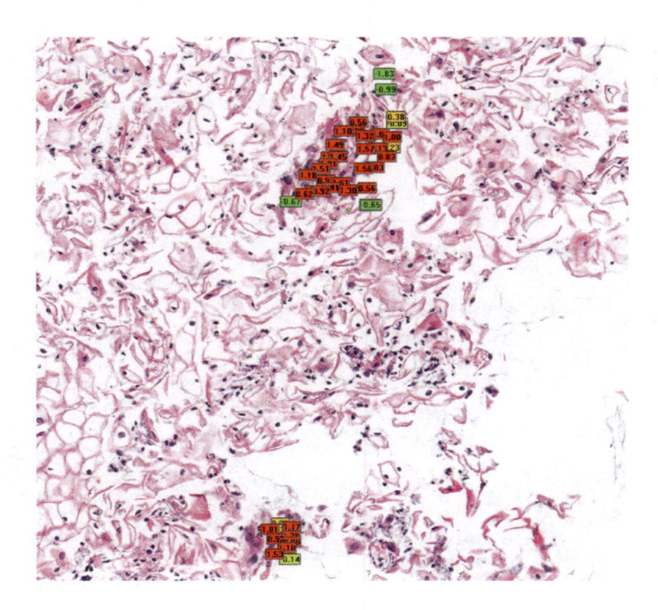

FIGURE 13.5
Results for a sample LGSIL image. Two regions were graded LGSIL by Ossama Tawfik. Several strongly suspicious cells were discovered by the software in both regions. Only cells from within the known unhealthy regions are labeled.

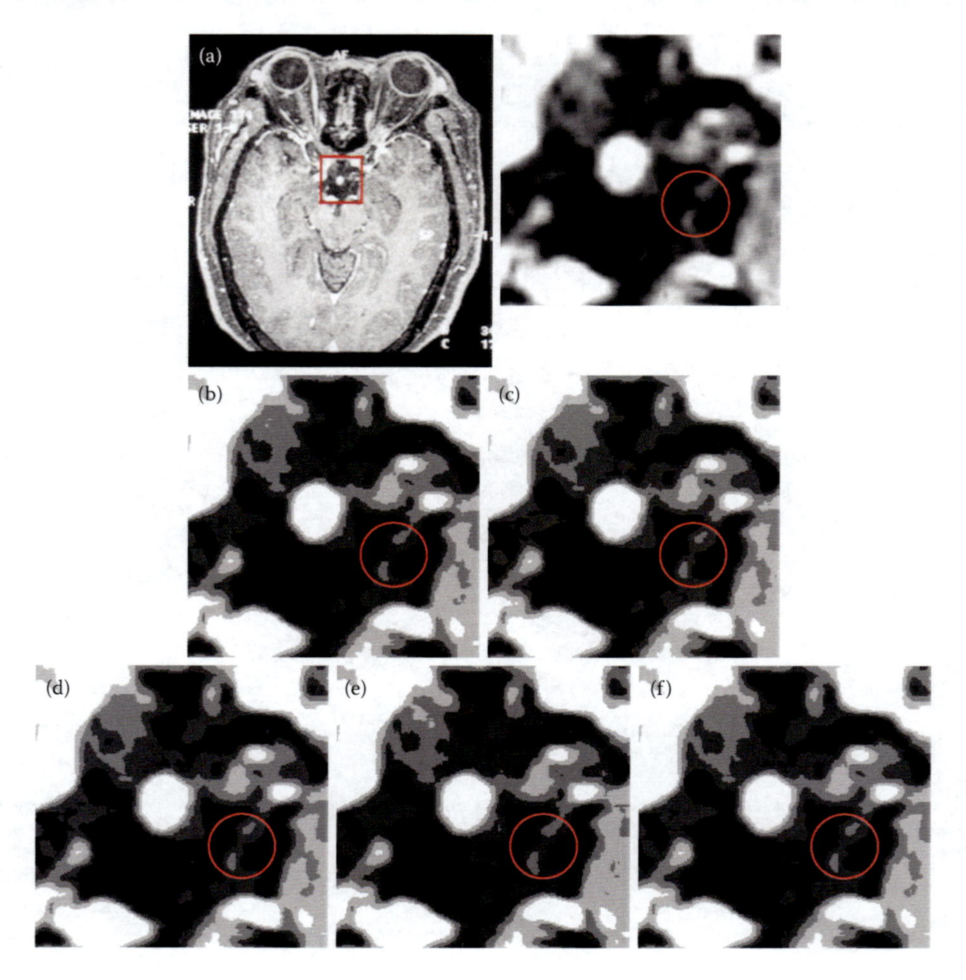

FIGURE 14.1
Original MR image and its window selection. (a) Segmentation result of AFCM (m = 2, S = 1, FN = 0, FP = 0). (b) Segmentation result of a (m = 2, S = 0.6250, FN = 0.3750, FP = 0), (c) Segmentation result of FSLVQ (S = 0.6667, FN = 0.3333, FP = 0). (d) Segmentation result of PICS (γ = 0.003, S = 0.7467, FN = 0, FP = 0.0477). (e) Segmentation result of PICS (γ = 0.0005, S = 0.7560, FN = 0.2440, FP = 0.0326).

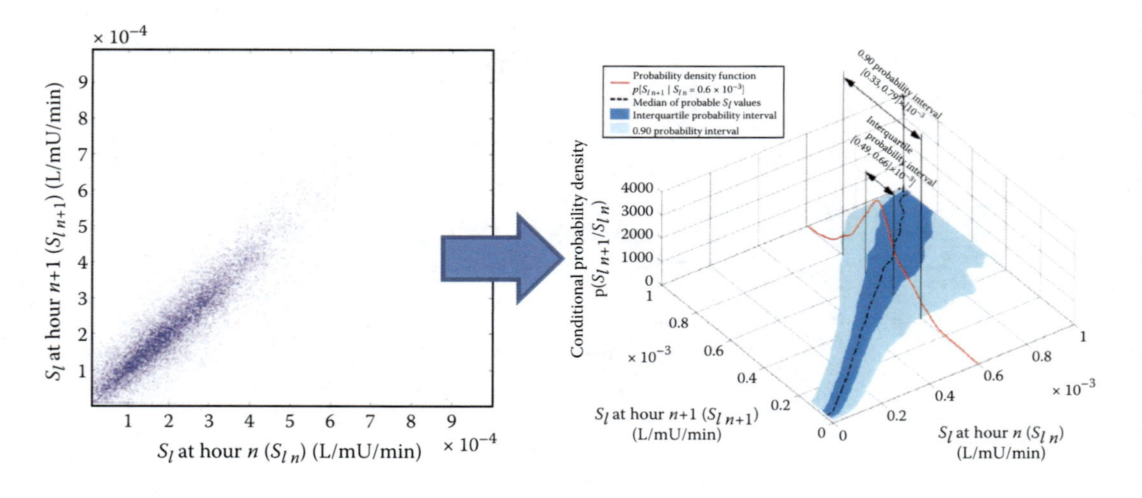

FIGURE 15.1
Stochastic model of S_I variability in the future for 1 h forward [31] and resulting risk distribution (right) is shown schematically for the interquartile and 90% confidence interval bands. The right plot shows both a patient-specific value of S_I and resulting distribution as well as these confidence ranges based on data from Chase et al. [5].

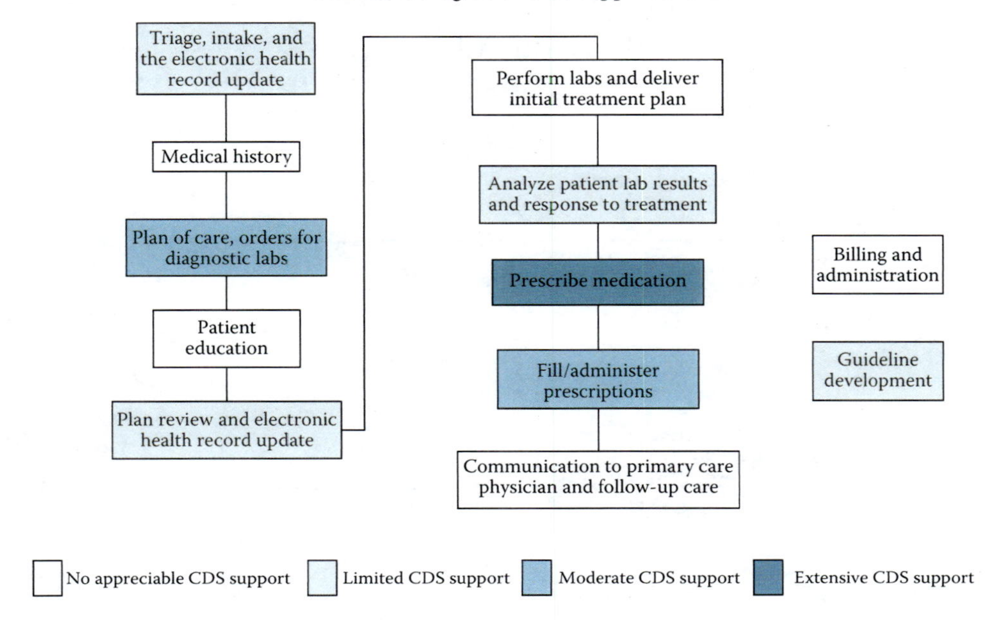

FIGURE 16.2
Availability of clinical decision support systems by work flow steps.

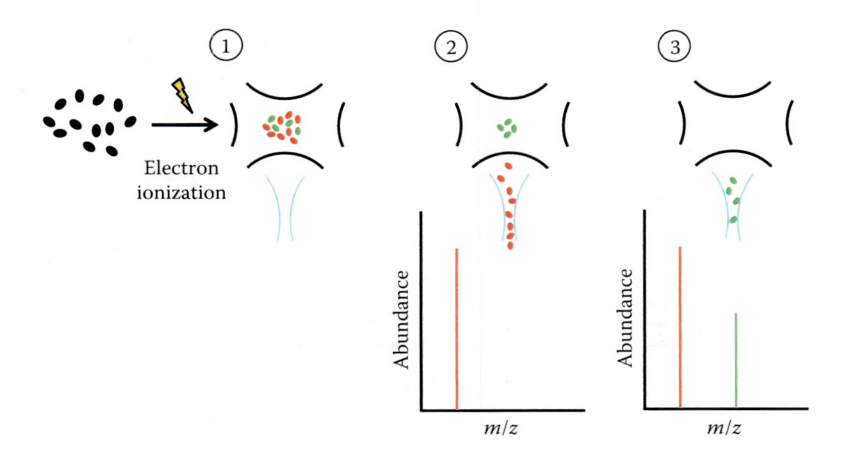

FIGURE 18.3
A schematic representation of quadrupole ion trap mass spectrometer operation. (1) A gas-phase chemical compound is introduced into the trap and is ionized via collisions with energetic electrons. The neutral molecules gain charge and produce structure-specific fragments. The ions are trapped in the quadrupole electric field. (2) The ions with different m/z are sequentially ejected from the trap and registered by a detector. (3) All ions are ejected out of the trap, and the detected ion current as a function of m/z comprises the mass spectrum.

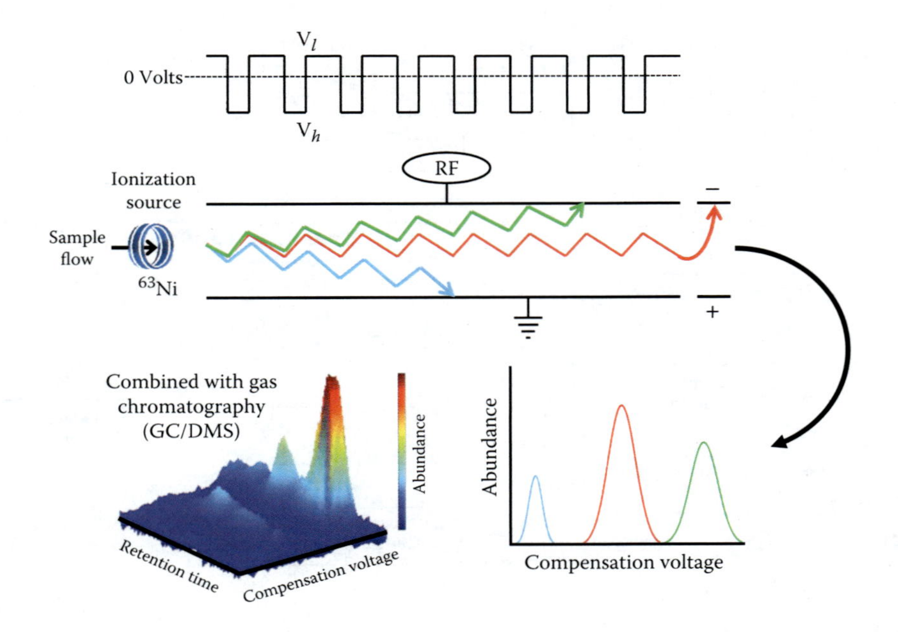

FIGURE 18.4
A schematic description of operation of high-field asymmetric-waveform ion mobility spectrometer (FAIMS), also known as the differential mobility spectrometer (DMS).

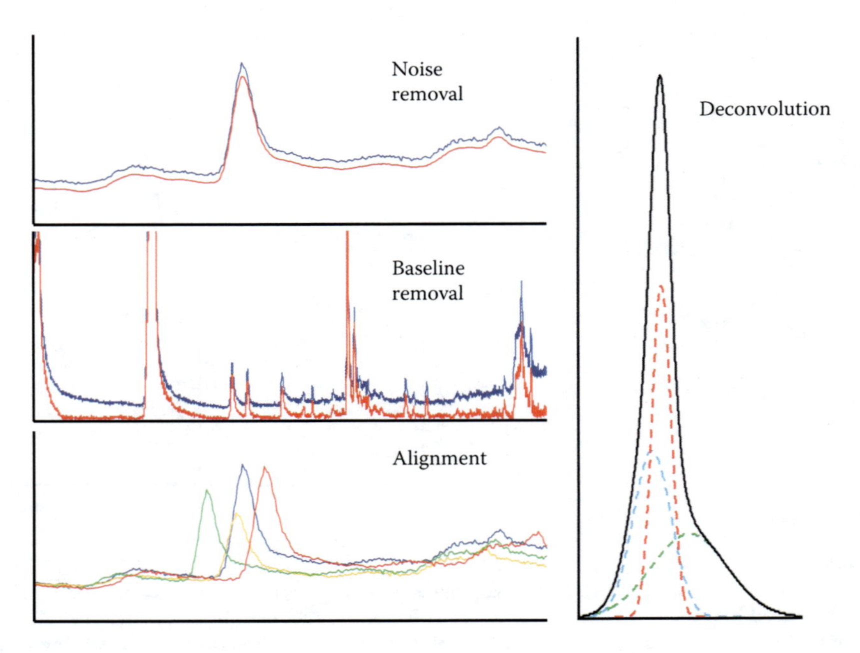

FIGURE 18.5
A diagram of some of the problems that data may have before preprocessing.

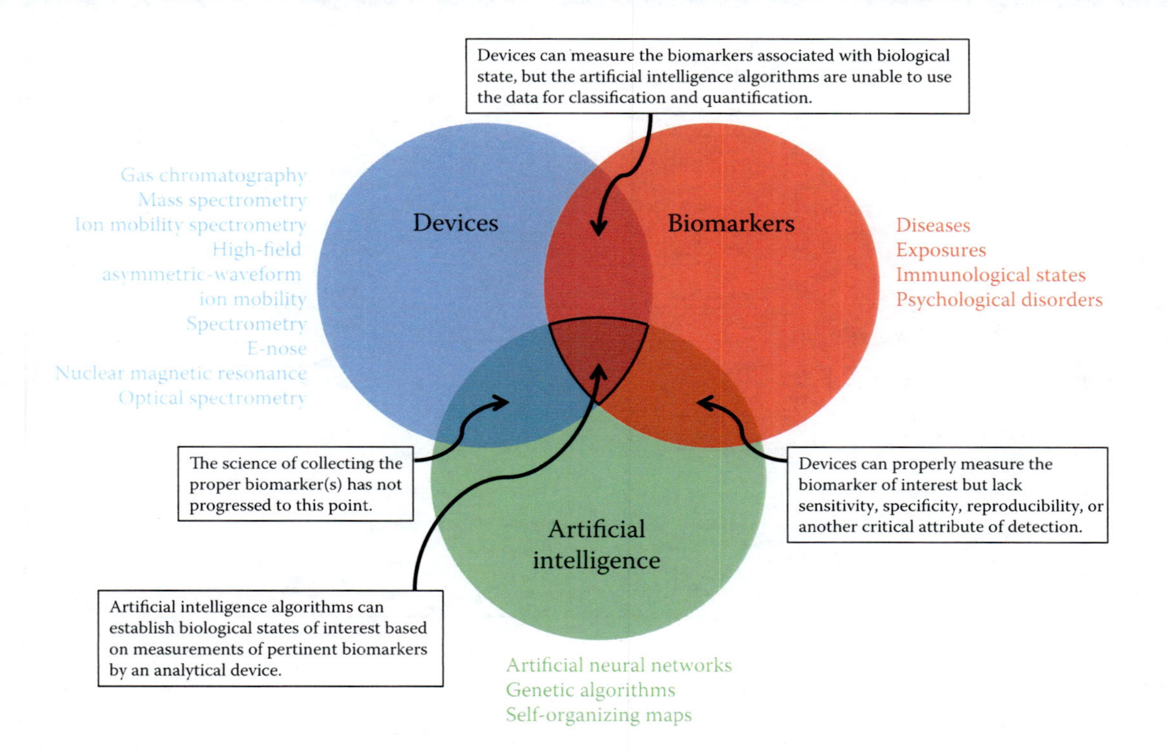

FIGURE 18.6
A conceptual representation of the interplay between the biological problem, the analytical chemistry instrumentation used to measure metabolic biomarkers of the state, and the data analysis approaches needed to interpret the result. The "Biomarkers" circle represents the body of biomarkers that is indicative of a certain biological state such as infection, malignancy, exposure, or another disorder. The "Devices" circle represents the multitude of analytical methods that are available to detect and measure the biomarkers of interest. The instrumentation options most adequate for the metabolite biomarkers' measurements may differ at the initial research/exploration stage (the biomarkers of interest are being established) versus the targeted analysis stage (the biomarkers of interest are known and documented). The developments in instrumentation, such as incremental improvements in performance, or introduction of conceptually new analysis approaches extends the circle. The "Artificial Intelligence" circle represents the variety of nonlinear data analysis approaches that allow discriminating of the biological states based on the output of the analytical instrumentation employed in the study. Incremental gains in knowledge of mechanistic details of a particular biological process would allow replacing nonlinear AI methods such as ANN with linear approaches. The necessity of a concerted approach required to solve the complex biological/medical challenges is represented by the intersection of all three circles.

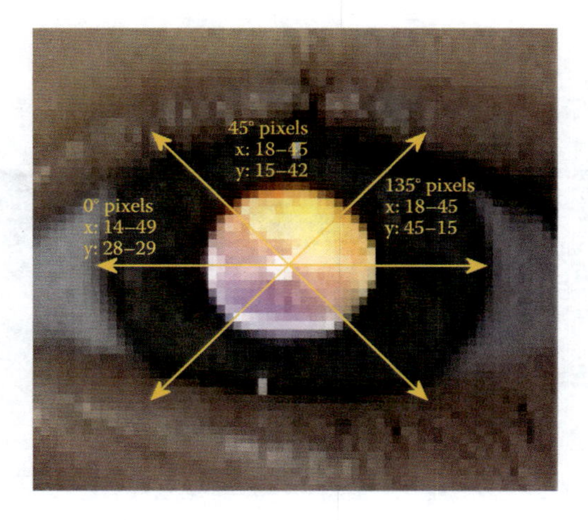

FIGURE 19.2
Illustration of iris and pupil color slope.

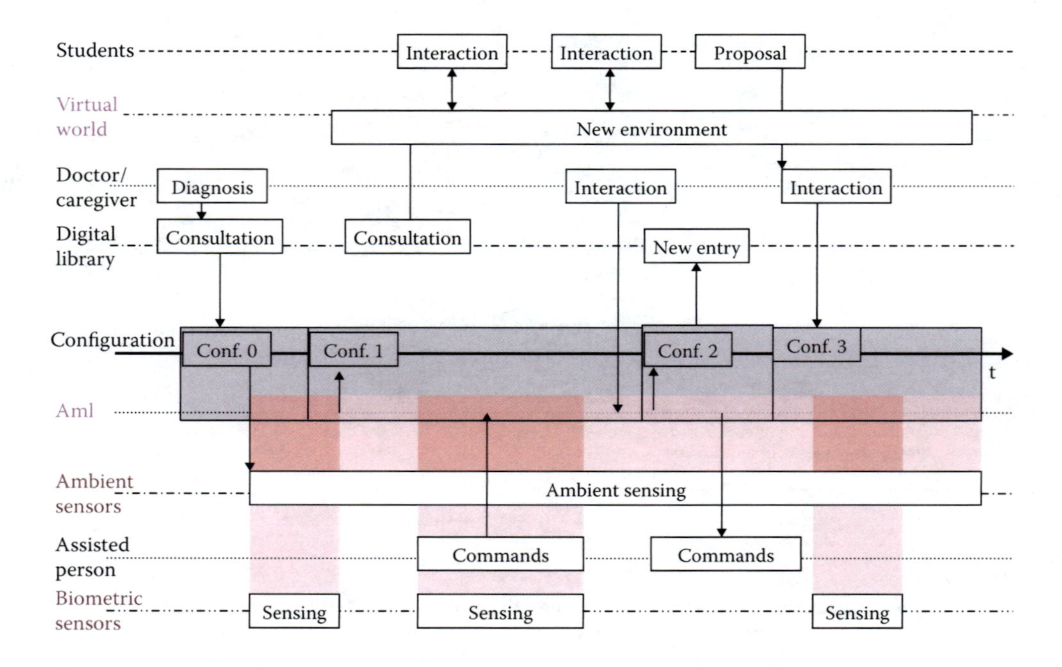

FIGURE 20.2
Full cycle of a smart assistive home.

FIGURE 21.2
Examples of dynamic lighting installed in Dutch nursing homes: Amadea by Derungs, Biosun by Van Doorn, and Strato by Philips.

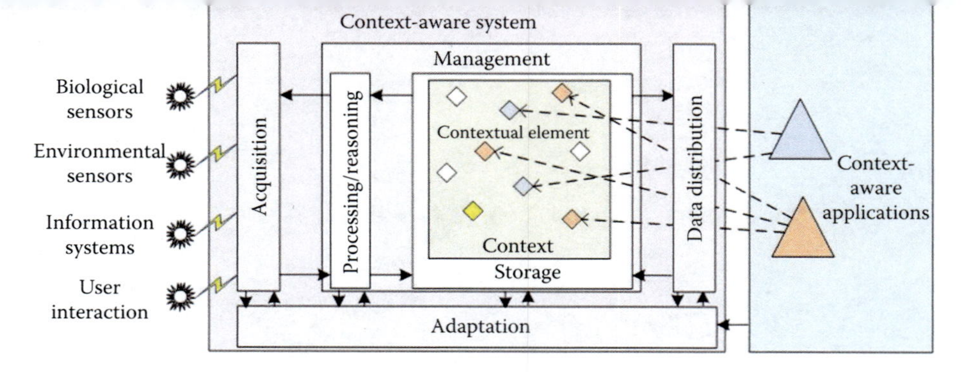

FIGURE 22.1
Architectural representation of context-aware systems.

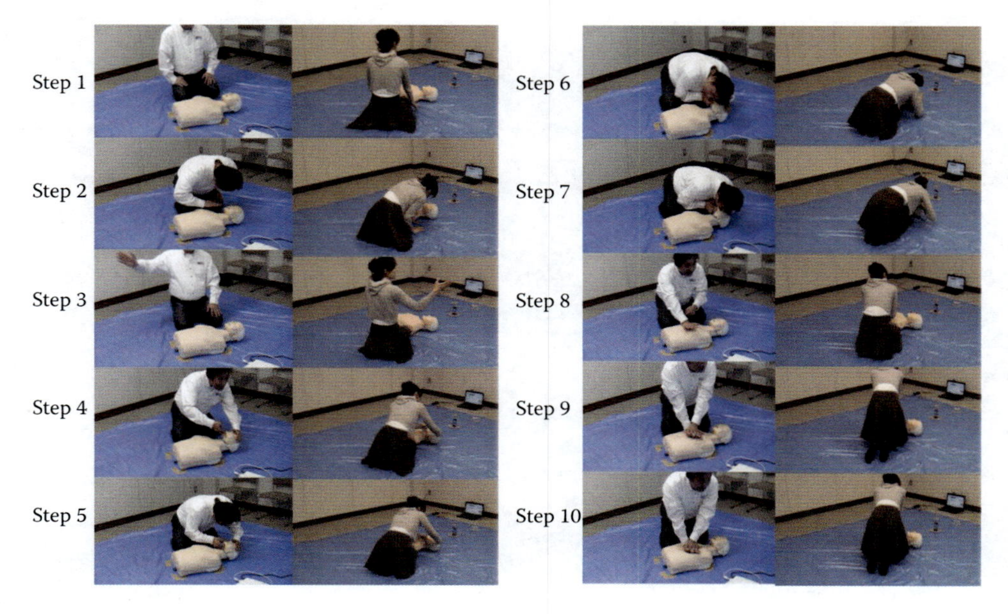

FIGURE 26.6
Setup of experiment using a smartphone.

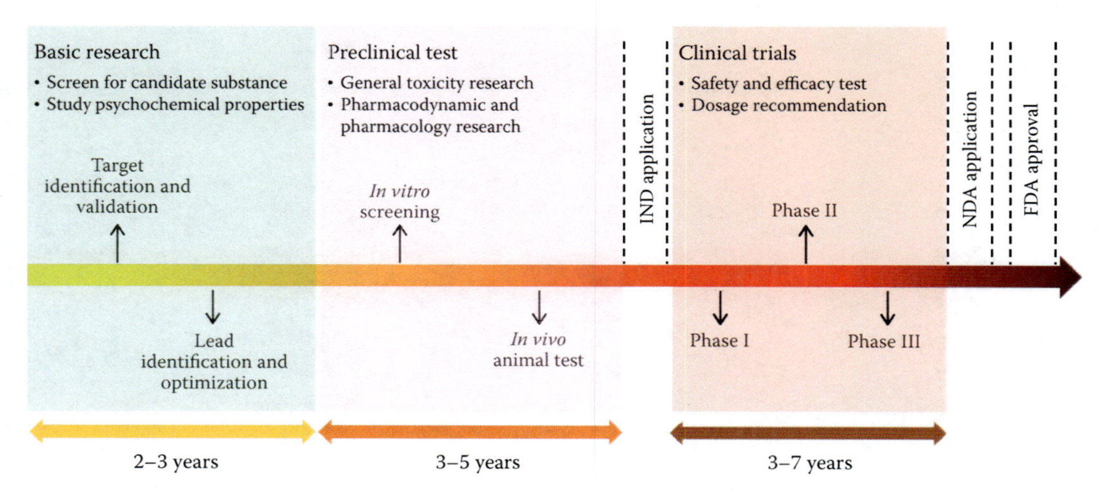

FIGURE 27.1
Overview of the drug development process. Once a new chemical compound passes preclinical test, pharmaceutical company files an IND with the US Food and Drug Administration (FDA) to obtain approval for testing the new drug in humans. After the drug passes all phases of clinical trials, a pharmaceutical company formally submits a proposal to the FDA to approve the new drug for sale and marketing. IND, investigational new drug; NDA = new drug application.

14

A Penalized Fuzzy Clustering Algorithm with its Application in Magnetic Resonance Image Segmentation

Wen-Liang Hung and Miin-Shen Yang

CONTENTS

14.1 Introduction

Magnetic resonance imaging (MRI) segmentation provides important information for detecting a variety of tumors, lesions, and abnormalities in clinical diagnosis. The segmentation can be described as the definition of clusters whose points are associated to similar sets of intensity values in the different images. An efficient analysis of dual-echo medical imaging volumes can be derived from a set of different diagnostic volumes carrying complementary information provided by medical imaging technology. The extraction of such volumes from imaging data is said to be segmentation, and it is usually performed, in the image space, defining sets of vowels with similar features within a whole dual-echo volume.

In general, medical images are obtained using different acquisition methods, including x-ray computer tomography (CT), single-photon emission tomography (SPET), positron emission tomography (PET), ultrasound (US), MRI and magnetic resonance angiographies (MRAs), and so forth. MRI systems are important in medical image analysis. MRI has the multidimensional nature of data provided from either one of two different pulse sequences. MRI segmentation is an important step in any image analysis system. Various segmentation methods for MRI have been used to differentiate abnormal and normal tissues [1–7].

Cluster analysis is a method of grouping data with similar characteristics into larger units of analysis. Image segmentation is a way to partition image pixels into similar regions. Thus, a clustering algorithm would naturally be applied in image segmentation. Since Zadeh [8] first articulated fuzzy set theory, which gave rise to the concept of partial membership, based on a membership function, fuzziness has received increasing attention. Fuzzy clustering, which produces overlapping cluster partitions, has been widely studied and applied in various areas [9]. In fuzzy clustering, the fuzzy c-means (FCM) clustering algorithm is the best-known and most powerful method used in cluster analysis [10].

As described by Yang et al. [5], most medical images always present overlapping grayscale intensities for different tissues. MRI medical imaging uncertainty is widely presented in data because of the noise and blur in acquisition and the partial volume effects originating from the low resolution of the sensors. In particular, borders between tissues are not clearly defined, and memberships in the boundary regions are intrinsically fuzzy. Therefore, fuzzy clustering methods such as the FCM are suitable for MRI segmentation [4,5,11,12]. In this chapter, we derive a penalized type of the FCM algorithm and then apply it to MRI segmentation in ophthalmology.

The idea of penalization is important in statistical learning. For example, ridge regression shrinks the regression coefficients by imposing a penalty on their size; the support vector machines (SVMs) can be regarded as a penalization method. Based on the penalty idea, we proposed the so-called penalized intercluster separation (PICS) by adding a penalty term to the ICS clustering algorithm [13]. Numerical comparisons are made with several fuzzy clusterings according to criteria of accuracy and computational efficiency. The remainder of the chapter is organized as follows. In Section 14.2, some penalized types of the FCM algorithm are reviewed. We then proposed the PICS clustering algorithm. In Section 14.3, the theoretical analysis on these penalized fuzzy clustering algorithms is investigated. We then give the parameter selection method for these algorithms. Section 14.4 presents the numerical comparisons of the proposed PICS with some other fuzzy clustering algorithms. An application to the ophthalmological MRI segmentation is presented in Section 14.5. Conclusions are made in Section 14.6.

14.2 Penalized Fuzzy Clustering Algorithms

Let $X = \{x_1,\ldots,x_n\} \subset R^s$ be a data set, and let c be a positive integer greater than one. A partition of X into c clusters is represented by mutually disjoint sets $X_1,\ldots,X_c$ such that $X_1 \cup \ldots \cup X_c = X$ or equivalently by the indicator functions $u_1,\ldots,u_c$ such that $u_i(x) = 1$ if x is in X_i and $u_i(x) = 0$ if x is not in X_i for all $i = 1,\ldots,c$. This is known as clustering X into c clusters $X_1,\ldots,X_c$ by a hard c-partition $\{u_1,\ldots,u_c\}$. A fuzzy extension allows $u_i(x)$ to take on values in the interval [0, 1] such that $\sum_{i=1}^{c} u_i(x) = 1$ for all x in X. In this case, $\{u_1,\ldots,u_c\}$ is called a fuzzy c-partition of X ([?]). Thus, the FCM objective function J_{FCM} is defined as [10]

$$J_{\text{FCM}}(u,a) = \sum_{j=1}^{n}\sum_{i=1}^{c} u_{ij}^{m}\left\|x_j - a_i\right\|^2$$

where $u = \{u_1,\ldots,u_c\}$ is a fuzzy c-partition with $u_{ij} = u_i(x_j)$ being the membership of the data point x_j in cluster i; $a = \{a_1,\ldots,a_c\}$ is the cluster centers; and the notation $\left\|x_j - a_i\right\|$ denotes the Euclidean distance between the data point x_j and the cluster center a_i. The parameter m is a weighting exponent on each membership and determines the amount of fuzziness of the resulting classification. Thus, the FCM clustering algorithm is an iteration through the necessary conditions for minimizing J_{FCM} with the following update equations:

$$a_i = \frac{\sum_{j=1}^{n} u_{ij}^m x_j}{\sum_{j=1}^{n} u_{ij}^m}, \text{ and } u_{ij} = \frac{\left\| x_j - a_i \right\|^{-2/(m-1)}}{\sum_{k=1}^{c} \left\| x_j - a_k \right\|^{-2/(m-1)}}.$$

The FCM algorithm is a well-known and powerful method in clustering analysis. One of important parameters in the FCM is the weighting exponent m. When m is close to 1, the FCM approaches the hard c-means algorithm. When m approaches infinity, the only solution of the FCM will be the mass center of the data set. Therefore, the weighting exponent m plays an important role in the FCM algorithm. Yu et al. [14] provided the theoretical analysis to selecting the weighting exponent m in the FCM.

The mixture maximum likelihood approach to clustering is a remarkable model-based clustering method. Scott and Symons [15] proposed the so-called classification maximum likelihood (CML) procedure, named first in Bryant and Williamson [16], that many of the commonly used clustering procedures correspond to applications of the maximum likelihood approach for normal groups with various restrictions on the covariance matrices and with the indicator classification variables of group membership associated with the data treated as unknown parameters. Yang [17] made the fuzzy extension of the CML procedure in conjunction with fuzzy c-partitions and called it a class of fuzzy CML procedures. On the other hand, the idea of penalization is important in statistical learning. For example, ridge regression shrinks the regression coefficients by imposing a penalty on their size; the SVMs can be regarded as a penalization method. Combining the CML procedure and penalty idea, Yang [17] added a penalty term to the FCM objective function J_{FCM} and then extended the FCM to the so-called penalized FCM (PFCM). Thus, the PFCM objective function is defined as follows:

$$\begin{aligned} J_{\text{PFCM}}(u,a) &= \sum_{j=1}^{n} \sum_{i=1}^{c} u_{ij}^m \left\| x_j - a_i \right\|^2 - w \sum_{j=1}^{n} \sum_{i=1}^{c} u_{ij}^m \ln \alpha_i \\ &= J_{\text{PFCM}}(u,a) - w \sum_{j=1}^{n} \sum_{i=1}^{c} u_{ij}^m \ln \alpha_i \end{aligned}$$

where $w \geq 0$, $\forall i$, $\alpha_i \geq 0$ and $\sum_{i=1}^{c} \alpha_i = 1$. The necessary conditions for a minimum of $J_{\text{PFCM}}(u, a)$ are

$$a_i = \frac{\sum_{j=1}^{n} u_{ij}^m x_j}{\sum_{j=1}^{n} u_{ij}^m}, \alpha_i = \frac{\sum_{j=1}^{n} u_{ij}^m}{\sum_{i=1}^{c} \sum_{j=1}^{n} u_{ij}^m}, \text{ and } u_{ij} = \frac{\left(\left\| x_j - a_i \right\|^2 - w \ln \alpha_i \right)^{-1/(m-1)}}{\sum_{k=1}^{c} \left(\left\| x_j - a_k \right\|^2 - w \ln \alpha_k \right)^{-1/(m-1)}}$$

Based on the numerical results of Yang and Su [18], the PFCM is more accurate than FCM. Furthermore, the PFCM has been applied in various areas (cf. [11,19,20]). The FCM algorithm is the best-known and most widely used technique in fuzzy quantization. Özdemir and Akarun [21] proposed the partition index maximization (PIM) algorithm, which minimizes the FCM objective function and simultaneously maximizes the partition coefficient. Thus, the PIM objective function becomes

$$J_{\mathrm{PIM}}(u,a)=\sum_{j=1}^{n}\sum_{i=1}^{c}u_{ij}^{m}\left\|x_j-a_i\right\|^2-\beta\sum_{j=1}^{n}\sum_{i=1}^{c}u_{ij}^{m}$$

where $\beta \geq 0$. The update equations for a minimizer of $J_{\mathrm{PIM}}(u, a)$ are

$$a_i=\frac{\sum_{j=1}^{n}u_{ij}^{m}x_j}{\sum_{j=1}^{n}u_{ij}^{m}},\ u_{ij}=\begin{cases}\dfrac{\left(\left\|x_j-a_i\right\|^2-\beta\right)^{-1/(m-1)}}{\sum_{k=1}^{c}\left(\left\|x_j-a_k\right\|^2-\beta\right)^{-1/(m-1)}}, & if\ \left\|x_j-a_i\right\|^2>\beta,\\ 1 & if\ \left\|x_j-a_i\right\|^2\leq\beta,\end{cases}$$

The PIM objective function can be thought of as a generalization of FCM objective function by adding to the penalty term $\left(-\beta\sum_{j=1}^{n}\sum_{i=1}^{c}u_{ij}^{m}\right)$. To implement the PIM algorithm, we need to determine the value of β that controls the amount of penalization. They suggested that $\beta=\delta\min_{i\neq j}\left\{\left\|a_i-a_j\right\|^2\right\}$, where $0 \leq \delta < 0.5$. In J_{PFCM}, we choose $\alpha_i = 1/c$, $\forall_i$. Then

$$\begin{aligned}J_{\mathrm{PFCM}}(u,a)&=J_{\mathrm{FCM}}(u,a)+w\ln c\sum_{j=1}^{n}\sum_{i=1}^{c}u_{ij}^{m}\\&=J_{\mathrm{FCM}}(u,a)-\beta\sum_{j=1}^{n}\sum_{i=1}^{c}u_{ij}^{m}\end{aligned}$$

where $\beta = -w \ln c \leq 0$. It means that the PIM algorithm is still meaningful when $\beta \leq 0$, and it is a special case of PFCM proposed by Yang [17]. By minimizing the FCM objective function and simultaneously maximizing the intercluster separation (ICS) measure, Özdemir and Akarun [13] proposed the ICS clustering algorithm with the objective function

$$J_{\mathrm{ICS}}(u,a)=\frac{1}{n}\sum_{j=1}^{n}\sum_{i=1}^{c}\left(\mu_{ij}^{m}\left\|x_j-a_i\right\|^2-\frac{\gamma}{c}\sum_{t=1}^{c}\left\|a_i-a_t\right\|^2\right)\tag{14.1}$$

where the parameter $\gamma \geq 0$. Yu and Yang [22] pointed out that the original update equations for the ICS algorithm were partially incorrect. The update equations for the ICS algorithm originally were corrected as follows [22].

$$a_i=\frac{\frac{1}{n}\sum_{j=1}^{n}\mu_{ij}^{m}x_{ij}-\frac{2\gamma}{c}\sum_{t=1}^{c}a_t}{\frac{1}{n}\sum_{j=1}^{n}\mu_{ij}^{m}-2\gamma},\ \mu_{ij}=\frac{\left\|x_j-a_i\right\|^{-2/(m-1)}}{\sum_{k=1}^{c}\left\|x_j-a_k\right\|^{-2/(m-1)}}\tag{14.2}$$

It was shown by Yang [17] and Yang and Su [18] that the PFCM algorithm is more accurate than the FCM method. It means that the penalty term can improve the performance of FCM. To improve the performance of ICS, we consider adding a penalty term to ICS. Because PFCM is a model-based approach and performs better than PIM (Section 14.4), we add the penalty term $\left(-w\sum_{j=1}^{n}\sum_{i=1}^{c} u_{ij}^{m} \ln \alpha_i\right)$ to ICS and call it the penalized ICS (PICS). Then the corresponding objective function is given by

$$J_{\text{PICS}}(u,a) = J_{\text{ICS}}(u,a) - w\sum_{j=1}^{n}\sum_{i=1}^{c} u_{ij}^{m} \ln \alpha_i$$

The update equations for a minimizer of $J_{\text{PICS}}(u, a)$ are

$$\alpha_i = \frac{\sum_{j=1}^{n} u =_{ij}^{m}}{\sum_{k=1}^{c}\sum_{j=1}^{n} u_{kj}^{m}}; i = 1,2,\ldots,c,$$

$$a_i = \frac{\frac{1}{n}\sum_{j=1}^{n} u_{ij}^{m} x_j - \frac{2\gamma}{c}\sum_{t=1}^{c} a_t}{\frac{1}{n}\sum_{j=1}^{n} u_{ij}^{m} - 2\gamma}; \ i = 1,2,\ldots,c,$$

$$u_{ij} = \frac{\left(\left\|x_j - a_i\right\|^2 - w \ln \alpha_i\right)^{-1/(m-1)}}{\sum_{k=1}^{c}\left(\left\|x_j - a_i\right\|^2 - w \ln \alpha_k\right)^{-1/(m-1)}}; i = 1,2,\ldots,c; j = 1,2,\ldots,n$$

14.3 Theoretical Analysis on Penalized Fuzzy Clustering Algorithms

The FCM algorithm is one of the most frequently used clustering algorithms. One of the most important parameters in the FCM is the weighting exponent m. Yu et al. [14] established the theoretical analysis for choosing m in FCM. Some empirical results for penalized fuzzy clustering algorithms exist in the literature. For example, Yang and Su [18] pointed out that $\omega \le 2.5$ when using PFCM. Özdemir and Akarun [21] indicated that β should be less that halfway between any two clusters when using PIM. In ICS, Özdemir and Akarun [13] experimented on 16-level quantization with $\gamma = 0.0005$ and 8-level quantization with $\gamma = 0.003$ with good results for most images. But they did not give a selection rule for these appropriate parameters. Next we consider a theoretical method for proper parameter selection, especially for the weighting exponent m, in these penalized fuzzy clustering algorithms.

First, we review the results of Yu et al. [14] for choosing the proper weighting exponent m in FCM. Let $X = \{x_1,\ldots,x_n\} \subset R^s$ be a data set. For a given $2 \le c < n$, $a = \{a_1,\cdots,a_c\}$ denotes the cluster centers, where $a_i \in R^S$. Let $u = [u_{ij}]_{c\times n} \in M_{fcn}$ be a partition matrix where

$$M_{fcn} = \left\{ u = [u_{ij}]_{c \times n} \middle| \forall i, \forall j, u_{ij} \geq 0, \sum_{i=1}^{c} u_{ij} = 1, 0 < \sum_{j=1}^{n} u_{ij} < n \right\}$$

Let G be a function with

$$G: u = [u_{ij}]_{c \times n} \in M_{fcn} \mapsto a = (a_1, \ldots, a_c)^T \in R^{cs}$$

where $a_i = \sum_{j=1}^{n} u_{ij}^m x_j \Big/ \sum_{j=1}^{n} u_{ij}^m$. Let F be a function with

$$\text{F}: a = (a_1, \ldots, a_c)^T \in R^{cs} \mapsto u = [u_{ij}]_{c \times n} \in M_{fcn}$$

where $u_{ij} = \left\| x_j - a_i \right\|^{-2/(m-1)} \Big/ \sum_{k=1}^{c} \left\| x_j - a_k \right\|^{-2/(m-1)}, i = 1, \ldots, c; j = 1, \ldots, n$, that is, $\text{F}(a) = u$. We then have a mapping T_m with $T_m(u, a) = (\hat{u}, \hat{a})$, where $\hat{u} = F(a)$, $\hat{a} = G(\hat{u})$. Thus, if $(u^{(\ell)}, a^{(\ell)}) = T_m(u^{(\ell-1)}, a^{(\ell-1)})$, $\forall \ell \geq$, then $(u^{(\ell)}, a^{(\ell)})$, $\ell = 1, 2, \ldots$ are called iteration sequences of the FCM algorithm, where $(u^{(\ell)}, a^{(\ell)})$ is an element of $M_{fcn} \times R^{cs}$. Let Ω^{FCM} denote the solution set of the FCM objective function $J_{\text{FCM}}(u, a)$ with

$$\Omega^{\text{FCM}} = \left\{ \begin{array}{l} (u^*, a^*) \in M_{fcn} \times R^{CS} \Big| \ J^{\text{FCM}}(u^*, a^*) \leq J^{\text{FCM}}(u, a^*), \forall u \in M_{fcn}, u \neq u^*, \\ \qquad\qquad J^{\text{FCM}}(u^*, a^*) < J^{\text{FCM}}(u^*, a), \forall a \in R^{cs}, a \neq a^* \end{array} \right\}$$

It is known that the point $(u^*, a^*) \in \Omega^{\text{FCM}}$ if and only if (u^*, a^*) is an FCM fixed point.

In other words, $(u^*, a^*) \in \Omega^{\text{FCM}}$ if and only if $u^* = F(a^*)$ and $a^* = G(u^*)$. Selim and Ismail [23] then gave the Hessian matrix H_u^{FCM} of $\psi_m(u) = \min_{a \in Rcs} J^{\text{FCM}}(u, a)$.

Clearly, $J^{\text{FCM}}(u, G(u)) = \psi_m(u)$ and $\psi_m(u)$ is convex at u if and only if H_u^{FCM} is positive semi-definite. Selim and Ismail [23] showed that this is a sufficient condition for a local minimum of $J^{\text{FCM}}(u, a)$. Based on the Hessian matrix H_u^{FCM}, Yu et al. [14] showed the following theorem for the FCM weighting exponent m.

Theorem 1

Let $U^* = [1/c]_{c \times n}$ and $\lambda_{\max}(F_{U^*})$ be the maximum eigenvalue of the matrix F_{U^*}. If $\lambda_{\max}(F_{U^*}) < 0.5$ and $m > 1/(1 - 2\lambda_{\max}(F_{U^*}))$, then $(U^*, \bar{x})$ is a strict local minimum of $J^{\text{FCM}}(u, a)$, where

$$F_{U^*} = \left[f_{kr}^{U^*} \right]_{n \times n}, f_{kr}^{U^*} = \frac{1}{n} \left(\frac{x_k - \bar{x}}{\left\| x_k - \bar{x} \right\|} \right)^T \left(\frac{x_r - \bar{x}}{\left\| x_r - \bar{x} \right\|} \right).$$

In Theorem 1, we can rewrite

$$F_{U^*} = \frac{1}{n} H^T \times H = C_0$$

where

$$H = \left[\frac{x_1 - \bar{x}}{\|x_1 - \bar{x}\|}, \ldots, \frac{x_n - \bar{x}}{\|x_n - \bar{x}\|} \right]$$

and

$$C_0 = \sum_{j=1}^{n} \frac{(x_j - \bar{x})(x_j - \bar{x})^T}{n\left(\|x_j - \bar{x}\|^2\right)}.$$

Hence, $\lambda_{\max}(F_{U^*}) = \lambda_{\max}(C_0)$. However, it is easy to compute $\lambda_{\max}(C_0)$ when $s < n$. Thus, if $\lambda_{\max}(C_0) > 0.5$, then the condition with $m \le 1/(1 - 2\lambda_{\max}(C_0))$ can prevent $\bar{x}$ from being a stable FCM fixed point. To obtain a simple and efficient Hessian matrix, Wei and Mendel [24] considered a function of cluster centers a_i and proposed $J^{\mathrm{FCM}}(F(a), a) = \min_{u \in fcn} J^{\mathrm{FCM}}(u, a)$. Using simple computation, we had that

$$J^{\mathrm{FCM}}(F(a), a) = \sum_{j=1}^{n} \left(\sum_{i=1}^{c} \|x_j - a_i\|^{\frac{-2}{m-1}} \right)^{1-m}.$$

Then, they considered the Hessian matrix H_a^{FCM} of $J^{\mathrm{FCM}}(F(a), a)$ based on the following theorem. ■

Theorem 2

(1) $$\frac{\partial J^{\mathrm{FCM}}(F(a), a)}{\partial a_i} = -2 \sum_{j=1}^{n} u_{ij}^m (x_j - a_i),$$

(2) If $i \neq \ell$, $$\frac{\partial^2 J^{\mathrm{FCM}}(F(a), a)}{\partial_{a_i} \partial_{a_\ell}} = \sum_{j=1}^{n} \frac{4m u_{ij}^m u_{\ell j}^m (x_j - a_i)(x_j - a_\ell)^T}{(m-1)\|x_j - a_i\|^2},$$

(3) $$\frac{\partial^2 J^{\mathrm{FCM}}(F(a), a)}{\partial_{a_i} \partial_{a_i}} = \sum_{j=1}^{n} \frac{4m\left(u_{ij}^{m+1} - u_{\ell j}^m\right)(x_j - a_i)(x_j - a_i)^T}{(m-1)\|x_j - a_i\|^2} + 2\left(\sum_{j=1}^{n} u_{ij}^m\right) \times I_{s \times s}$$

where $I_{s \times s}$ is a $s \times s$ identity matrix. Based on the Hessian matrix H_a^{FCM}, Wei and Mendel [24] proposed an efficient local optimality test for the FCM fixed points.

Based on the discussion of Section 14.2, we consider $\beta \neq 0$ in PIM algorithm. Then, according to the analysis of Wei and Mendel [24], the reduced objective function of PIM with respect to the cluster centers $\{a_1, \ldots, a_c\}$ is calculated as follows:

$$J^{\mathrm{PIM}}(F(a), a) = \sum_{j=1}^{n} \left(\sum_{i=1}^{c} \left(\|x_j - a_i\|^2 + \beta \right)^{\frac{-1}{m-1}} \right)^{1-m}.$$

Let Ω^{PIM} denote the solution set of the PIM objective function $J^{PIM}(u, a)$ with

$\Omega^{PIM} = \{u^*, a^* \in M_{fcn} \times R^{cs} \mid J^{PIM}(u^*, a^*) \le J^{PIM}(u, a^*), \forall u \in M_{fcn}, u \ne u^*, J^{PIM}(u^*, a^*) < J^{PIM}(u^*, a), \forall a \in R^{cs}, a \ne a^*\}$.

However, the point in Ω^{PIM} may be a saddle point of $J^{PIM}(u, a)$. To get the Hessian matrix of $J^{PIM}(F(a), a)$ is enough to determine whether or not a point in Ω^{PIM} is a saddle point. We set

$$t_j(a) = \sum_{j=1}^{c} \left(\|x_j - a_j\|^2 + \beta \right)^{\frac{-1}{m-1}}, \ u_{ij} = \frac{\left(\|x_j - a_i\|^2 + \beta \right)^{\frac{-1}{m-1}}}{t_j(a)}, \text{ and } h_{ij} = u_{ij}^m (x_j - a_i), \forall i, j \qquad ■$$

Thus, similar to Theorem 2, we can derive the following theorem for PIM.

Theorem 3

(1) $$\frac{\partial J^{PIM}(F(a), a)}{\partial a_i} = -2 \sum_{j=1}^{n} u_{ij}^m (x_j - a_i),$$

(2) If $i \ne \ell$, $$\frac{\partial^2 J^{PIM}(F(a), a)}{\partial_{a_i} \partial_{a_\ell}} = \frac{4m}{m-1} \sum_{j=1}^{n} (t_j(a))^{m-1} h_{ij} (h_{\ell j})^T,$$

(3) $$\frac{\partial^2 J^{PIM}(F(a), a)}{\partial_{a_i} \partial_{a_i}} = \frac{4m}{m-1} \sum_{j=1}^{n} (t_j(a))^{m-1} h_{ij} (h_{\ell j})^T + 2 \left(\sum_{j=1}^{n} u_{ij}^m \right) \times I_{s \times s}.$$
$$- \frac{4m}{m-1} \sum_{j=1}^{n} \left(\|x_j - a_i\|^2 + \beta \right)^{-1} h_{ij} (x_j - a_i)^T$$

Therefore, the Hessian matrix H_a^{PIM} of $J^{PIM}(F(a), a)$ can be decomposed to $H_a^{PIM} = A + B$ where $A = \text{diag}(A_1, \ldots, A_c)$, $B = [B_{i\ell}]_{c \times c}$, and

$$B_{i\ell} = \frac{4m}{m-1} \sum_{j=1}^{n} t_j(a)^{m-1} h_{ij} (h_{\ell j})^T,$$

$$A_i = 2 \left(\sum_{j=1}^{n} u_{ij}^m \right) \times I_{s \times s} - \frac{4m}{m-1} \sum_{j=1}^{n} \left(\|x_j - a_i\|^2 + \beta \right)^{-1} h_{ij} (x_j - a_i)^T.$$

Thus, we can decide if a point $(u, a) \in \Omega^{PIM}$ is a strictly local minimum of $J^{PIM}(u, a)$ using the Hessian matrix H_a^{PIM}. First, we define a set Ω of stationary points for $J^{PIM}(F(a), a)$ as

$$\Omega = \left\{ a = (a_1, \cdots, a_c) \middle| \forall i, \ \frac{\partial J^{PIM}(F(a), a)}{\partial a_i} = 0 \right\}.$$

Based on Theorem 3, we consider the second term of the Taylor series expansion for $J^{PIM}(F(a), a)$ about the point $a = (a_1,\ldots,a_c) \in \Omega$. Then, we obtain the following equation.

$$\phi_a^T \left[\frac{\partial^2 J^{PIM}(F(a),a)}{\partial_{a_i} \partial_{a_\ell}} \right] \phi_a = \phi_a^T B \phi_a + \sum_{i=1}^{c} \phi_{a_i}^T A_i \phi_{a_i}$$

$$= \frac{4m}{m-1} \sum_{j=1}^{n} (t_j(a))^{m-1} \left(\sum_{i=1}^{c} u_{ij}^m \left(\phi_{a_i}^T (x_j - a_i) \right) \right)^2 + 2 \sum_{i=1}^{c} \phi_{a_i}^T \left(\sum_{j=1}^{n} u_{ij}^m \right) \phi_{a_i}$$

$$- \frac{4m}{m-1} \sum_{i=1}^{c} \phi_{a_i}^T \left(\sum_{j=1}^{n} u_{ij}^m \frac{(x_j - a_i)(x_j - a_i)^T}{\left\| x_j - a_i \right\|^2 + \beta} \right) \phi_{a_i}$$

Adopting the same technique as in Yu et al. [14], we set

$$C_\beta = \sum_{j=1}^{n} \frac{(x_j - \bar{x})(x_j - \bar{x})^T}{n \left(\left\| x_j - \bar{x} \right\|^2 + \beta \right)}.$$

Then, we obtain the following results. The point $\bar{x}$ is a stable fixed point if $m > 1/(1 - 2\lambda_{\max}(C_\beta))$ and $\lambda_{\max}(C_\beta) < 0.5$. If $m < 1/(1 - 2\lambda_{\max}(C_\beta))$ or $\lambda_{\max}(C_\beta) \geq 0.5$, then $\bar{x}$ is not a stable PIM fixed point. Moreover, let $\lambda_1, \lambda_2, \ldots, \lambda_s$ be s eigenvalues of the matrix C_β. Note that

$$\sum_{i=1}^{s} \lambda_i = \sum_{j=1}^{n} \frac{\left\| x_j - \bar{x} \right\|^2}{n \left(\left\| x_j - \bar{x} \right\|^2 + \beta \right)}.$$

Set

$$F_\beta = \sum_{j=1}^{n} \frac{\left\| x_j - \bar{x} \right\|^2}{n \left(\left\| x_j - \bar{x} \right\|^2 + \beta \right)}.$$

Then

$$\frac{1}{s} F_\beta \leq \lambda_{\max}(C_\beta) \leq F_\beta$$

if

$$\beta > -\min_{1 \leq j \leq n} \left\{ \left\| x_j - \bar{x} \right\|^2 \right\}.$$

If $1 \leq m \leq s/(s - 2F_\beta)$ where $\beta > -\min_{1 \leq j \leq n} \left\{ \left\| x_j - \bar{x} \right\|^2 \right\}$, then $\bar{x}$ will not be a stable fixed point. It is obvious that F_β is a monotonically decreasing function of β if $\beta > -\min_{1 \leq j \leq n} \left\{ \left\| x_j - \bar{x} \right\|^2 \right\}$.

If β is large enough, then $\bar{x}$ is a stable fixed point for almost any $m > 1$. Similarly, if $\beta < 0$ and $|\beta|$ is large enough, then $\lambda_{\max}(C_\beta)$ is negative. In this case, $\bar{x}$ is also a stable fixed point. Hence, the selection of β is important for the PIM algorithm. Özdemir and Akarun [21] suggested that β should be a fraction of the distance between the closed two quantization centers, that is, $\beta = \delta \min_{i \neq j}\left\{\left\|a_i - a_j\right\|^2\right\}$, where $0 \le \delta < 0.5$. Since β is always nonnegative in PFCM, the Hessian matrix in PFCM is the same as that in PIM at the point $\bar{x}$. Thus, the above analysis for PFCM is also valid. A similar discussion can also be found in Yu and Yang [25].

In the ICS algorithm, the determination of the parameter is an essential issue. Next, we give a theoretical explanation for this. We know that if

$$1 - 2\gamma(c+1)c^{m-1} - \frac{2m}{m-1}\lambda_{\max}(C_0) > 0,$$

then the point $\bar{x}$ is a stable fixed point. If γ or c is large, then not only is $1 - 2\gamma(c+1)c^{m-1} - \frac{2m}{m-1}\lambda_{\max}(C_0)$ negative, but also, the Hessian matrix of the ICS algorithm is not positive semidefinite. If the Hessian matrix of ICS is not semidefinite, then the ICS algorithm is meaningless because it cannot produce any local minimum. More details can be founded in Yu and Yang [22]. Thus, when c increases, γ should decrease dynamically and become very small positive. Özdemir and Akarun [13] experimented on 16-level quantization with $\gamma = 0.0005$ and 8-level quantization with $\gamma = 0.003$ with good results for most images. This is completely consistent with our theoretical analysis. However, if γ is negative, when c increases, the point $\bar{x}$ is always a stable fixed point in ICS. This is not expected and should be avoided in clustering. Therefore, it is reasonable that $\gamma > 0$.

Combining the theoretical results of PIM, PFCM, and ICS, we have that $\bar{x}$ is a stable PICS fixed point if

$$1 - 2\gamma(c+1)c^{m-1} - \frac{2m}{m-1}\lambda_{\max}(C_\beta) > 0 \text{ and } \lambda_{\max}(C_\beta) < 0.5.$$

■

14.4 Numerical Comparisons

In this section, we make a comparison of five different algorithms: FCM, PFCM, PIM, ICS, and PICS, according to the bivariate normal mixtures of two classes under the accuracy and computational efficiency criteria. The accuracy of an algorithm is measured by the mean-squared error (MSE), which is the average sum of squared error between the true parameter and its estimate in N repeated trials. The computational efficiency of an algorithm is measured by the average number of iterations (NI) in N repeated trials.

Let $N_2(a, \Sigma)$ represent the bivariate normal with mean vector a and covariance matrix Σ. As the separation between subpopulations is determined by varying the parameters of subpopulations, without loss of generality, we give that one subpopulation bivariate normal is mean vector $\mathbf{a}_1 = 0$ and identity covariance matrix **I**, and the other is mean vector $\mathbf{a}_2$ and identity covariance matrix **I**. That is, we consider the random sample of data drawn from $\alpha_1 N_2(0, \mathbf{I}) + \alpha_1 N_2(\mathbf{a}_2, \mathbf{I})$ with $\alpha_2 = 1 - \alpha_1$. We also design various bivariate normal mixture distributions, shown in Table 14.1. In tests A1 and A2, we consider a well-known clustering problem [26] where there is an inordinate difference in the number of members in each cluster. But test A3 has almost equal size in each cluster, and test A4 has well-separated clusters.

In each test, we consider the sample size $n = 100$, $\in = 0.0001$, and $N = 500$. The MSE is calculated by

$$\frac{1}{2N}\sum_{k=1}^{N}\sum_{i=1}^{2}\left\|\hat{a}_i^{(k)} - a_i\right\|^2$$

where $\hat{a}_i^{(k)}$ is the estimated mean vector for the kth trial and a_i is the true mean vector. First, we consider which value of m is suitable when implementing FCM. According to the theoretical analysis in Section 14.3, we must compute $\lambda_{\max}(C_0)$. Because MSE is evaluated by all trails, we also compute $\lambda_{\max}^{(k)}(C_0)$ for the kth trial. Then, we use the mean and standard deviation of $\lambda_{\max}^{(1)}(C_0),\cdots,\lambda_{\max}^{500}(C_0)$, to choose m. That is,

$$\bar{\lambda}_{\max} = \sum_{k=1}^{N}\lambda_{\max}^{k}(C_0),\ \mathrm{SD} = \frac{1}{499}\sum_{k=1}^{N}\left(\lambda_{\max}^{(k)}(C_0) - \bar{\lambda}_{\max}\right)^2.$$

We work out the following: (1) $\lambda_{\max} = 0.55$, SD = 0.02 in test A1; (2) $\lambda_{\max} = 0.55$, SD = 0.03 in test A2; (3) $\lambda_{\max} = 0.55$, SD = 0.03 in test A3; and (4) $\lambda_{\max} = 0.70$, SD = 0.03 in test A4. These results illustrate that the values of $\lambda_{\max}(C_0)$ are stable in each trail. Since these mean values are greater than 0.5, there is no theoretically invalid weighting exponent of the FCM for tests A1 through A4. In this case, how to select m depends on the user. Because most researchers have proposed $m = 2$, we also choose $m = 2$ in this section. Next, we choose $w = 0.5, 1$, and 2 in PFCM; $\delta = 0.01, 0.449$ in PIM; and $\gamma = 0.003, 0.0005$ in ICS based on the results

TABLE 14.1

Various Bivariate Normal Mixture Distributions for Numerical Tests

Test	Mixture Model
A1	$0.1N_{2(0,\mathrm{I})} + 0.9N_2\left(\begin{pmatrix}1\\0\end{pmatrix}, I\right)$
A2	$0.1N_{2(0,\mathrm{I})} + 0.9N_2\left(\begin{pmatrix}1\\0\end{pmatrix}, I\right)$
A3	$0.1N_{2(0,\mathrm{I})} + 0.9N_2\left(\begin{pmatrix}1\\0\end{pmatrix}, I\right)$
A4	$0.1N_{2(0,\mathrm{I})} + 0.9N_2\left(\begin{pmatrix}1\\0\end{pmatrix}, I\right)$

TABLE 14.2

Accuracy and Computational Efficiency for Different Clustering Algorithms

			PFCM			PIM		ICS		PICS ($\gamma = 0.003$)			PICS ($\gamma = 0.0005$)		
TesT		**FCM**	$w = 0.5$	$w = 1.0$	$w = 2.0$	$\delta = 0.01$	$\delta = 0.449$	$\gamma = 0.003$	$\gamma = 0.0005$	$w = 0.5$	$w = 1.0$	$w = 2.0$	$w = 0.5$	$w = 1.0$	$w = 2.0$
A1	MSE	0.3557	0.3154	0.2576	0.3902	0.3719	0.3793	0.3560	0.3948	0.3241	0.2455[a]	0.3457	0.2986	0.2647	0.3832
	NI	84.5	96.8	109.4	35.6	85.3	82.3	84.6	118.4	76.3	113.8	35.4	83.3	109.8	36.8
A2	MSE	0.1763	0.1346	0.0987	0.2615	0.1919	0.1883	0.1918	0.1885	0.1369	0.0999	0.2547	0.1407	0.0850[a]	0.2441
	NI	49.2	46.1	64.5	50.1	50.3	44.4	64.9	64.1	51.4	74.6	53.7	44.3	67.5	48.6
A3	MSE	0.1424	0.0958	0.0720	0.1893	0.1388	0.1603	0.1362	0.1400	0.1009	0.0677	0.1869	0.1023	0.0628[a]	0.1706
	NI	39.9	37.1	65.9	65.6	39.3	37.6	39.7	36.3	37.2	56.1	70.9	42.5	51.3	60.6
A4	MSE	0.0142	0.0128	0.0135	0.0118	0.0162	0.0132	0.0150	0.0156	0.0126	0.021	0.0126	0.0128	0.0125	0.0112[a]
	NI	8.2	9.1	10.4	16.2	8.5	8.6	8.5	8.2	9.1	10.6	16.1	8.9	10.3	16.3

[a] Represents the smallest value.

of Section 14.3. Furthermore, we compute the mean and standard value of $\lambda_{max}(C_\beta)$ in PIM when $\delta = 0.01$. We obtain the following: (1) $\bar{\lambda}_{max} = 0.48$, SD = 0.02 in test A1; (2) $\bar{\lambda}_{max} = 0.49$, SD = 0.02 in test A2; (3) $\bar{\lambda}_{max} = 0.49$, SD = 0.03 in test A3; and (4) $\bar{\lambda}_{max} = 0.67$, SD = 0.03 in test A4. These results illustrate that the values of $\lambda_{max}(C_\beta)$ are stable in each trail. According the analysis of Section 14.3, we find that $m = 2$ is suitable for each test.

The numerical results are shown in Table 14.2. Comparing PFCM with FCM, we see that PFCM with $w = 1.0$ can lead to an MSE reduction of 27.6% in test A1, 44.0% in test A2, 49.4% in test A3, and 4.9% in test A4. But PIM cannot reduce MSE. These results illustrate that the penalty term $\left(-w\sum_{j=1}^{n}\sum_{i=1}^{c}u_{ij}^{m}\ln\alpha_i\right)$ to the FCM objective function can improve the accuracy of FCM, but PIM fails. This is why we add the penalty term $\left(-w\sum_{j=1}^{n}\sum_{i=1}^{c}u_{ij}^{m}\ln\alpha_i\right)$ to the ICS objective function. Comparing PICS with ICS, we see that (1) PICS ($\gamma = 0.003$) with $w = 1.0$ can lead to an MSE reduction of 31% in test A1, 47.9% in test A2, 50.3% in test A3, and 19.3% in test A4; and (2) PICS ($\gamma = 0.0005$) with $w = 1.0$ can also lead to an MSE reduction of 33.0% in test A1, 54.9% in test A2, 55.1% in test A3, and 19.9% in test A4.

Based on the above results, we find that the reduction percentage of PICS is greater than PFCM. It illustrates that the effect of the penalty term $\left(-w\sum_{j=1}^{n}\sum_{i=1}^{c}u_{ij}^{m}\ln\alpha_i\right)$ on ICS algorithm is significant.

Moreover, we also find that (1) PICS ($\gamma = 0.003$) with $w = 1.0$ has the smallest MSE in test A1; (2) PICS ($\gamma = 0.0005$) with $w = 1.0$ has the smallest MSE in tests A2 and A3; (3) PFCM and PICS ($\gamma = 0.003$) with $w = 1.0$ have good accuracy in tests A2 and A3; and (4) as we expected, five different algorithms have good accuracy in test A4.

From the above discussion, we conclude that PICS ($\gamma = 0.003, 0.0005$) with $w = 1.0$ has good accuracy in most of our numerical tests.

14.5 Application to Ophthalmological MRI Segmentation

In this section, we use PICS ($\gamma = 0.003, 0.0005$ and $w = 1.0$) in a real case study of MRI segmentation to differentiate between normal and abnormal tissues in ophthalmology. Segmentation of the medical images obtained from MRI is a primary step in most applications of computer vision to medical image analysis. In this section, we apply PICS ($\gamma = 0.003, 0.0005$ and $w = 1.0$) in a real case study of MRI segmentation to differentiate between normal and abnormal tissues in ophthalmology. This case is from a 55-year-old woman who developed complete left-side oculomotor palsy immediately after a motor vehicle accident. Her brain MRI with MRA, skull routine, orbital CT, and cerebral angiography did not reveal a brainstem lesion, skull fracture, or vascular anomaly.

The MRI data set was analyzed using fuzzy-soft Kohonen's competitive learning (FSKCL) and fuzzy-soft learning vector quantization (FSLVQ) algorithms proposed by Lin et al. [6] and Yang et al. [7], respectively. According to the hypothesis from the neurologic specialist and ophthalmologist, the pathogenesis is the nerve root avulsion and damage at the exit site from the midbrain. To detect a small lesion on the oculomotor nerve, we used window segmentation (Figure 14.1) to enhance the nerve root under the recommendations of neurospecialists. The window selection picture shown in the right

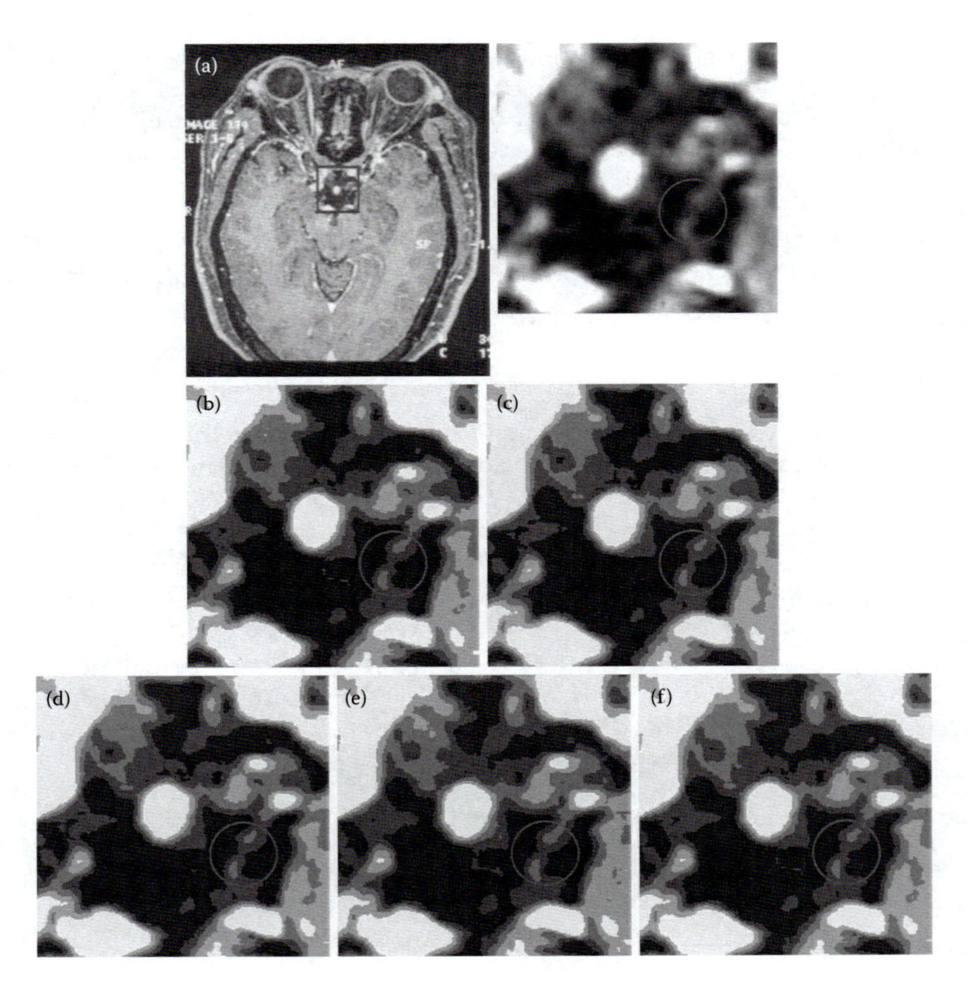

FIGURE 14.1
(See color insert.) (a) Original MR image and its window selection. (b) Segmentation result of AFCM (m = 2, S = 1, FN = 0, FP = 0). (c) Segmentation result of a (m = 2, S = 0.6250, FN = 0.3750, FP = 0), (d) Segmentation result of FSLVQ (S = 0.6667, FN = 0.3333, FP = 0). (e) Segmentation result of PICS (γ = 0.003, S = 0.7467, FN = 0, FP = 0.0477). (f) Segmentation result of PICS (γ = 0.0005, S = 0.7560, FN = 0.2440, FP = 0.0326).

panel of Figure 14.1 was processed at 203 × 209 pixels. The two dimension MRA picture was obtained using 3-dimensional fast spoiled gradient recalled acquisition in steady state (3-D FSPGR) to reduce the image noise and obtain more segmentation among the damaged tissue. The picture was grouped into five tissue classes: edema, gray matter, nerve tissue, white matter, and cerebrospinal fluid. We used these segmentation algorithms in this case. With internal ophthalmoplegia immediately after a closed head injury, the patient had complete oculomotor palsy. The brain MRI showed no abnormalities in the brainstem. Oculomotor palsy from head trauma is uncommon. A rotator force sufficient to cause nerve root avulsion is frequent in traumatic subarachnoid hemorrhage or skull fracture. Internal ophthalmoplegia has a poorer recovery prognosis than external ophthalmoplegia.

Few recover completely [27]. However, not all neurologic deficits show corresponding radiographic evidence. The radiopaque signal on the oculomotor nerve fibers through the midbrain is obscured due to very small lesion size or low radioactivity of radio image reaction. For all these considerations, these algorithms should be helpful in enhancing oculomotor nerve root avulsion at the exit site from the midbrain. We detected a shady signal structural abnormality on the oculomotor nerve at the exit site from the midbrain that actually gives support to the neurologic specialist's diagnosis. The red circles in Figure 14.1a through f, the fuzzy shadows on the nerve roots are suspected oculomotor nerve abnormalities.

To evaluate the detection of abnormal tissue, it is necessary to make a quantitative comparison of the image, segmented by each algorithm, with a reference image. Because the segmentation results with Alternative Fuzzy C-mean (AFCM) ($m = 2$, [5]) can successfully differentiate the tumor from the normal tissues, Figure 14.1a is considered as a reference image. The comparison score S [28,29] for each algorithm is defined as

$$S = \frac{|A \cap A_{ref}|}{|A \cup A_{ref}|}$$

where A represents the set of pixels belonging to the tumor tissue found by the algorithm and A_{ref} represents the set of pixels belonging to the tumor tissue in the reference segmented image. Moreover, adopting the similar idea of false negative and false positive from Fernández-García et al. [30], we may also define the following two error types based on A and A_{ref}:

$$\text{false negative (FN)} = \frac{|A_{ref} \cap A_c|}{|A_{ref}|}, \text{ false positive (FP)} = \frac{|A_{ref}^c \cap A|}{|A_{ref}^c|},$$

where $A_{ref} \cap A^c$ represents the set of pixels in A_{ref} that has not been detected to be tumor tissue by the algorithm; $A_{ref}^c \cap A^c$ represents the set of pixels in A_{ref}^c that has been detected to be tumor tissue by the algorithm; and A^c and A_{ref}^c represent the complements of A and A_{ref}, respectively. From Table 14.3, showing the values of S, FN, and FP corresponding to Figure 14.1b through e, we can see that PICS performs well.

TABLE 14.3

Quantitative Comparison of the AFCM, FSKCL, FSLVQ, and PICS ($\gamma = 0.003, 0.0005$) Algorithms

Criterion	AFCM	FSKCL	FSLVQ	PICS ($\gamma = 0.003$)	PICS ($\gamma = 0.0005$)
S	1	0.6250	0.6667	0.7467	0.7560
FN	0	0.3750	0.3333	0	0.2440
FP	0	0	0	0.0477	0

14.6 Conclusion

MRI and MRA segmentations are also helpful in the diagnosis of brain trauma, a signature of traumatic brain injury. In this chapter, we discuss a patient with oculomotor palsy immediately after a motor vehicle accident; her brain MRI with MRA, skull routine, orbital CT, and cerebral angiography did not reveal brainstem lesions, skull fracture, or vascular anomalies. For early treatment with radiotherapy and surgery, the clustering algorithm is a method that makes many groups of similar individuals from a data set. The purpose of image segmentation is to divide the image space into separate regions with similar intensity values. Clustering segmentation can then be successfully used in discriminating different tissues from the medical images.

Furthermore, we added a penalty term to the ICS algorithm [13] and then extended the ICS to the so-called PICS. Numerical comparisons are made for several fuzzy clusterings according to criteria of accuracy and computational efficiency. The results show that the PICS is impressive. Finally, the PICS algorithms are applied in the segmentation of the MRI of an ophthalmic patient. In these MRI segmentation results, we find that PICS provides useful information as an aid to diagnosis in ophthalmology. Overall, we recommend that those concerned with applications in cluster analysis try the proposed PICS algorithm.

References

1. L.P. Clarke, R.P. Velthuizen, M.A. Camacho, J.J. Heine, M. Vaidyanathan, L.O. Hall, R.W. Thatcher and M.L. Silbiger, MRI segmentation: methods and applications, *Magnetic Resonance Imaging*, 13, 343–368, 1995.
2. R.C. Gonzalez and R.E. Woods, *Digital Image Processing*, Massachusetts: Addison-Wesley, 1992.
3. D.L. Pham and J.L. Prince, Adaptive fuzzy segmentation of magnetic resonance images, *IEEE Transactions on Medical Imaging*, 18, 737–752, 1999.
4. W.E. Phillips, R.P. Velthuizen, S. Phuphanich, L.O. Hall, L.P. Clarke and M.L. Silbiger, Application of fuzzy *c*-means segmentation technique for differentiation in MR images of a hemorrhagic glioblastoma multiforme, *Magnetic Resonance Imaging*, 13, 277–290, 1995.
5. M.S. Yang, Y.J. Hu, K.C.R. Lin and C.C.L. Lin, Segmentation techniques for tissue differentiation in MRI of ophthalmology using fuzzy clustering algorithms, *Magnetic Resonance Imaging*, 20, 173–179, 2002.
6. K.C.R. Lin, M.S. Yang, H.C. Liu, J.F. Lirng and P.N. Wang, Generalized Kohonen's competitive learning algorithms for ophthalmological MR image segmentation, *Magnetic Resonance Imaging*, 21, 863–879, 2003.
7. M.S. Yang, K.C.R. Lin, H.C. Liu and J.F. Lirng, Magnetic resonance imaging segmentation techniques using batch-type learning vector quantization algorithms, *Magnetic Resonance Imaging*, 25, 265–277, 2007.
8. L.A. Zadeh, Fuzzy sets. *Information and Control*, 8, 338–353, 1965.
9. J.C. Bezdek, J.M. Keller, R. Krishnapuram and N.R. Pal, *Fuzzy Models and Algorithms for Pattern Recognition and Image Processing*, Boston: Kluwer Academic Publishers, 1999.
10. J.C. Bezdek, *Pattern Recognition With Fuzzy Objective Function Algorithms*, New York: Plenum Press, 1981.
11. J.S. Lin, K.S. Cheng and C.W. Mao, Segmentation of multispectral magnetic resonance image using penalized fuzzy competitive learning network, *Computers and Biomedical Research*, 29, 314–326, 1996.

12. J. Suckling, T. Sigmundsson, K. Greenwood and E.T. Bullmore, A modified fuzzy clustering algorithm for operator independent brain tissue classification of dual echo MR images, *Magnetic Resonance Imaging*, 17, 1065–1076, 1999.
13. D. Özdemir and L. Akarun, Fuzzy algorithm for combined quantization and dithering, *IEEE Transactions on Image Processing*, 10, 923–931, 2001.
14. J. Yu, Q.S. Cheng and H.K. Huang, Analysis of the weighting exponent in the FCM, *IEEE Transactions on Systems, Man, and Cybernetics*, 34, 634–639, 2004.
15. A.J. Scott and M.J. Symons, Clustering methods based on likelihood ratio criteria, *Biometrics*, 27, 387–397, 1971.
16. P.G. Bryant and J.A. Williamson, Maximum likelihood and classification: a comparison of three approaches, in: W. Gaul and M. Schader, Eds., *Classification as a Tool of Research*, North Holland, Amsterdam: Elsevier Science Ltd., pp. 35–45, 1986.
17. M.S. Yang, On a class of fuzzy classification maximum likelihood procedures, *Fuzzy Sets and Systems*, 57, 365–375, 1993.
18. M.S. Yang and C.F. Su, On parameter estimation for normal mixtures based on fuzzy clustering algorithms, *Fuzzy Sets and Systems*, 68, 13–28, 1994.
19. S.H. Liu and J.S. Lin, Vector quantization in DCT domain using fuzzy possibilistic *c*-means based on penalized and compensated constraints, *Pattern Recognition*, 35, 2201–2211, 2002.
20. M.S. Yang and N.Y. Yu, Estimation of parameters in latent class models using fuzzy clustering algorithms, *European Journal of Operational Research*, 160, 515–531, 2005.
21. D. Özdemir and L. Akarun, A fuzzy algorithm for color quantization of images, *Pattern Recognition*, 35, 1785–1791, 2002.
22. J. Yu and M.S. Yang, A note on the ICS algorithm with correction and theoretical analysis, *IEEE Transactions on Image Processing*, 14, 973–978, 2005.
23. S.Z. Selim and M.A. Ismail, On the local optimality of the Fuzzy ISODATA clustering algorithm, *IEEE Transactions on Pattern Analysis and Machine Intelligence*, 8, 284–288, 1986.
24. W. Wei and J.M. Mendel, Optimality tests for the fuzzy *c*-means algorithms. *Pattern Recognition*, 27, 1567–1573, 1994.
25. J. Yu and M.S. Yang, Optimality test for generalized FCM and its application to parameter selection, *IEEE Transactions on Fuzzy Systems*, 13, 164–176, 2005.
26. R.O. Duda and P.E. Hart, *Pattern Classification and Scene Analysis*, New York: Wiley, 1973.
27. P. Muthu and P. Pritty, Mild head injury with isolated third nerve palsy. *Emergency Medical Journal*, 18, 310–311, 2001.
28. F. Masulli and A. Schenone, A fuzzy clustering based segmentation system as support to diagnosis in medical imaging. *Artificial Intelligence in Medicine*, 16, 129–147, 1999.
29. D.Q. Zhang and S.C. Chen, A novel kernelized fuzzy *c*-means algorithm with application in medical image segmentation. *Artificial Intelligence in Medicine*, 32, 37–50, 2004.
30. N.L. Ferńandez-Garcia, R. Medina-Carnicer, A. Carmona-Poyato, F.J. Madrid-Cuevas and M. Prieto-Villegas, Characterization of empirical discrepancy evaluation measures. *Pattern Recognition Letters* 25, 35–47, 2004.

15

Uncertainty, Safety, and Performance: A Generalizable Approach to Risk-Based (Therapeutic) Decision Making

J. Geoffrey Chase, Balazs Benyo, Thomas Desaive, Liam Fisk, Jennifer L. Dickson, Sophie Penning, Matthew K. Signal, Attila Illyes, Noeimi Szabo-Nemedi, and Geoffrey M. Shaw

CONTENTS

15.1 Introduction

15.1.1 Overall Problem Definition

From a perspective of applying deterministic, physiological models to clinical scenarios, the application of artificial intelligence (AI) in medicine is currently focused on decision support (DS) tools that provide guidance and "human-in-the-loop" feedback for diagnosis and control of therapeutic delivery. There are two key differentiators when differentiating emerging DS applications: (1) the model and computational methods used and (2) the decision-making framework used with these models. It is on this second aspect that this chapter focuses.

Clinically, safety and performance are the two primary goals of any DS system, in that order. In this, they match the fundamental "first, do no harm" philosophy in medicine. However, such decisions are complicated by the main element that differentiates clinical AI applications and decision making from traditional engineering applications, specifically, the significant patient variability in condition and disease state that increases the risk of any therapeutic decision, resulting in harmful treatment choices, reducing both safety and performance. The problem is further complicated by the fact that clinical therapy is often targeted to a band, where too much or too little of a dose or a measured outcome results in a worsened condition or harm.

All of these issues interact with the main DS or control elements of measurement and/or actuator error, and measurement interval. In particular, while most medical "actuators" can deliver therapy at very high resolution, measurement errors can be large, and the interval between measurements in some cases can be far longer than the dynamics that one wishes to control, increasing the potential for variability to have an impact. Hence, there is significant risk in achieving a target range due to uncertainty in both the patient (the "plant") and the measurement (the "sensor").

More specifically, what is really required is maximization of time in a band, while minimizing (to zero preferably) dangerous, out-of-band events. This approach maximizes performance and safety and must be accomplished in the absence of reliable continuous measurements. Hence, the main goal is decision making under uncertainty to minimize risk.

While there is significant theoretical and engineering-based literature in this area, there is little work in applying it to medicine. In particular, there is often no definition or quantification of patient variability that can be made patient specific or state specific. More specifically, interpatient variability (differences between patients and their response to care) and intrapatient variability (evolution of a patient over time) are often conflated and mixed in medical literature and analyses. Hence, there is no existing framework for using such variability definitions in clinical use to find best solutions between sometimes-conflicting goals. This article presents a risk-based approach to managing patient-specific variability for safety and performance in the context of a specific medical application but as a template for a broader class of problems.

15.1.2 Clinical Problem Definition

High blood sugar or hyperglycemia is prevalent in intensive care unit (ICU) patients [1]. Hyperglycemia averaging 125–145 mg/dL (7.0–8.0 mmol/L) or higher can reduce the odds of surviving by a factor of up to 4× [2]. Thus, glycemic control (GC) to lower levels closer to normal can reduce mortality [3–5], organ failure [6,7], and cost [8,9]. However, some studies have failed to repeat these results [10–12], and many others have seen no clear result [13], while all but one of these studies [5] reported increased risk of hypoglycemia and increased glycemic variability, which are associated with increased risk of mortality [14–18].

GC is a typical clinical drug dosing problem where the goal is to maintain blood glucose (BG) in a tight band between 80 and 145 mg/dL (4.0–8.0 mmol/L) with minimal variability to reduce harmful effects. Hypoglycemia, or low BG, as well as hyperglycemia have harmful effects, which can be traced to cumulative exposure to chronic high levels or variability [19]. The failure to repeat successful results or avoid hypoglycemia in many studies can be attributed to significant hour-to-hour variability in patient state [20,21], where measurements are typically two to five hourly in many protocols, combined with not accounting for nutritional inputs in balance with nutrition [21]. Hence, it is an area where model-based GC can offer significant opportunity to provide good control [22–24] with the caveat that it must be able to account for patient variability under uncertainty.

15.1.3 Problem Definition

The problem goals are formally defined as follows:

- **Performance**: Maximize BG levels time in a band of 80–145 mg/dL (4.4–8.0 mmol/L)
- **Safety**: Minimize moderate hypoglycemia [%BG < 72 mg/dL (4.0 mmol/L)] and severe hypoglycemia [number of patients with BG < 40 mg/dL (2.2 mmol/L)]

Within achieving these goals, clinically, it is desirable to also meet goals related to

- **Patient-centered quality of care**: Maximize nutritional intake of the patient (grams of carbohydrate per day)
- **Clinical burden**: Minimize the number of measurements per day

These goals must be achieved within the uncertainty defined by interpatient variability and differences in response to therapy, and intrapatient variability and evolution of their response to therapy. In addition, relatively infrequent measurement, as compared to the system dynamics, with associated measurement error, adds further uncertainty.

It is important to note that all four goals can be in conflict and are not necessarily independent. For example, maximizing nutrition requires more insulin to maintain time in desired glycemic bands, which in turn increases the risk of hypoglycemia (reduced safety) resulting from intrapatient variability in response to therapy. Solving this problem requires a quantifiable means of managing both interpatient and intrapatient variability, upon which, with a relevant physiological model, one can search in a heuristic fashion for an optimum solution within a series of feasible treatments. Thus, this problem and all similar dosing problems are well suited to AI methods.

This chapter will present a GC application and decision-making framework to manage uncertainty directly with quantified risk. It is a generalizable approach that could serve as one template for such AI problems in medicine.

15.2 Method

15.2.1 Main Elements of Model-Based GC and Decision Making

The main elements of model-based control are (1) a clinically validated model and (2) a model or method of quantifying future variability (or risk) in a patient-specific and patient-state-specific fashion. These elements can be combined to create a decision-making framework to manage uncertainty.

15.2.2 A Clinically Validated Model

The model used in this work is the ICING model defined as follows [25]:

$$\dot{G} = -p_G G(t) - S_I G(t) \frac{Q(t)}{1+\alpha_G Q(t)} + \frac{P(t) + EGP_b - CNS + PN(t)}{V_g} \tag{15.1}$$

$$\dot{I} = -\frac{n_L I(t)}{1+\alpha_I I(t)} - n_K I(t) - (I(t)-Q(t))n_I + \frac{u_{ex}(t)}{V_I} + (1-x_L)\frac{u_{en}(G)}{V_I} \quad (15.2)$$

$$\dot{Q} = (I(t)-Q(t))n_I - n_C \frac{Q(t)}{1+\alpha_G Q(t)} \quad (15.3)$$

$$\dot{P}_1 = -d_1 P_1 + P(t) \quad (15.4)$$

$$\dot{P}_2 = -\min(d_2 P_2, P_{max}) + d_1 P_1 \quad (15.5)$$

$$u_{en} = \max\left(16.67, \frac{14G(t)}{1+0.0147G(t)} - 41\right) \quad (15.6)$$

where $G(t)$ [mmol/L] is the total plasma glucose [the model is run in mmol/L but data are reported here in mg/dL for ease of use (mmol/L is the European standard, and mg/dL is more commonly used in the United States)]; $I(t)$ [mU/L] is the plasma insulin; and interstitial insulin is represented by $Q(t)$ [mU/L]. Exogenous insulin input is represented by $u_{ex}(t)$ [mU/min], and endogenous insulin production is estimated with u_{en} [mU/min], modeled as a function of plasma glucose concentration determined from critical care patients with a minimum pancreatic output of 1 U/h. First-pass hepatic insulin clearance is represented by x_L, and n_I [1/min] accounts for the rate of transport between plasma and interstitial insulin compartments. Patient endogenous glucose clearance and insulin sensitivity are p_G [1/min] and S_I [L/(mU min)], respectively. The parameter V_I [L] is the insulin distribution volume, and n_K [1/min] and n_L [1/min] represent the clearance of insulin from plasma via renal and hepatic routes, respectively. Basal endogenous glucose production (unsuppressed by glucose and insulin concentration) is denoted by EGP_b [mmol/min], and V_G [L] represents the glucose distribution volume. CNS [mmol/min] represents non–insulin-mediated glucose uptake by the central nervous system. Michaelis–Menten kinetics are used to model saturation, with α_I [L/mU] used for the saturation of plasma insulin clearance by the liver and α_G [L/mU] for the saturation of insulin-dependent glucose clearance and receptor-bound insulin clearance from interstitium. P_1 [mmol] represents the glucose in the stomach, and P_2 [mmol] represents glucose in the gut. The rate of transfer between the stomach and gut is represented by d_1 [1/min], and the rate of transfer from the gut to the bloodstream is d_2 [1/min]. Enteral glucose input is denoted $P(t)$ [mmol/min], and P_{max} represents the maximum disposal rate from the gut.

Model values are found in the work of Fisk et al. [23], and the model has been clinically validated using independent matched patient cohort data [26]. Patient data from the work of Chase et al. [5] are also freely available [27]. The key parameter identified to make the model patient specific is insulin sensitivity, S_I, which is found using an integral-based method and clinical data [28]. Hence, the model can be made patient specific to a specific patient state (S_I) at any measurement.

15.2.3 Quantifying Future Variability in S_I

For a given patient-specific state, S_I, an insulin dose or insulin and nutrition dose can be calculated to achieve a specific desired BG level, assuming there is no change in S_I.

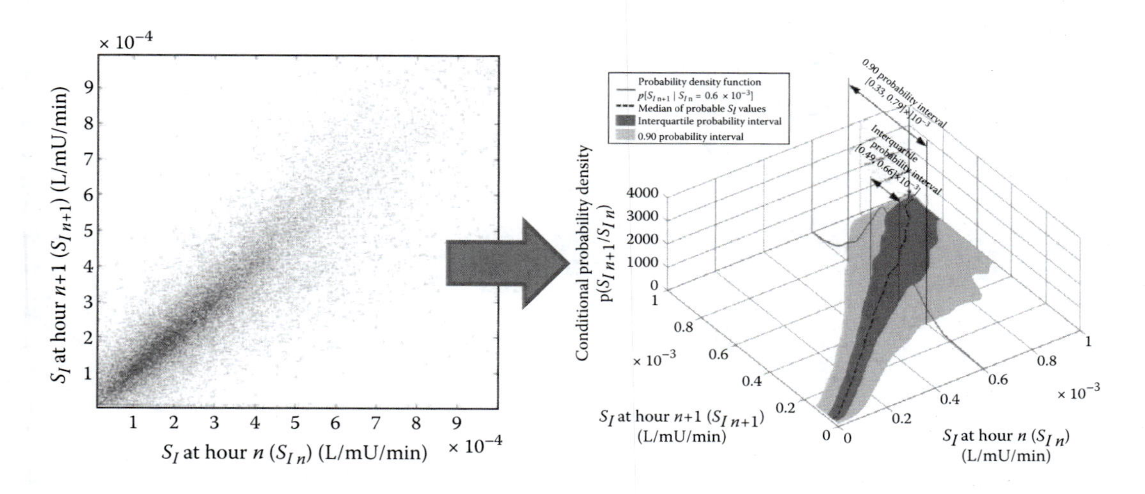

FIGURE 15.1
(See color insert.) Stochastic model of S_I variability in the future for 1 h forward [31] and resulting risk distribution (right) is shown schematically for the interquartile and 90% confidence interval bands. The right plot shows both a patient-specific value of S_I and resulting distribution as well as these confidence ranges based on data from Chase et al. [5].

However, S_I can change hourly or more rapidly both acutely (very fast) or chronically (slow evolution) due to changes in patient condition and other drug therapy [20,29,30]. However, given significant clinical data, a cohort-wide stochastic model of variation over one or several hours can be created, an example of which is shown in Figure 15.1 [31,32]. This stochastic model of S_I shows how patient-specific state at time "n" can be assessed as a range of possibilities at time "$n + 1$" or more hours forward, creating a probabilistic assessment of the risk of change in the patient's metabolic condition.

15.2.4 Decision Making under Uncertainty—Stochastic Targeting

Stochastic targeting (STAR) is a form of decision making under uncertainty designed to quantify and manage risk in determining the proper intervention of insulin and/or nutrition for a given 1–3 h interval (forward). STAR aims to maximize performance (time in glycemic bands) with a minimum, clinically specified risk of mild hypoglycemia (5% for BG < 80 mg/dL, ≈1% for BG < 72 mg/dL), as defined earlier in the problem definition.

When a BG measurement is taken, integral-based fitting is used to identify the current insulin sensitivity parameter from the model, and the stochastic model is used to generate a range of insulin sensitivity values to use for prediction over the next 1 to 3 h. The predicted insulin sensitivity ranges are used in conjunction with the model to generate BG outcomes for potential insulin/nutrition combinations. Optimal treatments and outcomes balance clinical workload, nutrition rates, and BG variability. Specifically, insulin cannot be administered if it poses a significant risk of hypoglycemia.

This balance of objectives, as defined in the problem definition, requires a heuristic method of sorting between feasible solutions for an optimal intervention. Thus, starting from a target goal nutrition rate and stepping down, the 5th percentile (Figure 15.2, points A, B, and C) and 95th percentile BG outcomes (Figure 15.2, points D, E, and F) are simulated for every allowable insulin rate (0–6 U/h in 0.5 U/h increments). Importance is placed on maximizing nutrition rates [33,34] but not at the risk of exacerbating hyperglycemia.

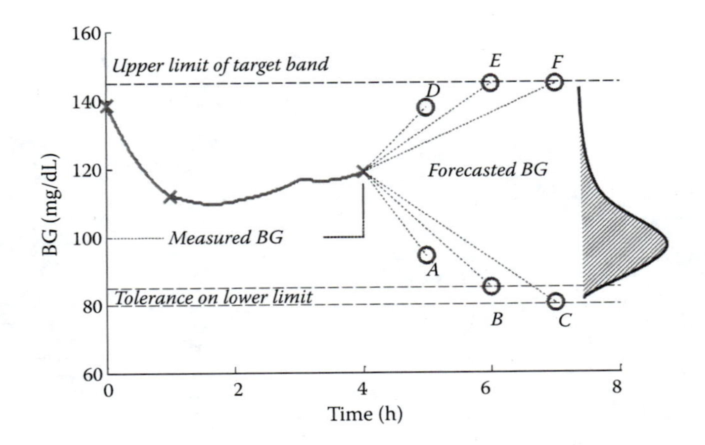

FIGURE 15.2
Controller forecast schematic for BG a target range of 80–145 mg/dL. A BG measurement has been taken at 10 h, and forecasts of BG have been generated (points A through F). The depicted distribution indicates the skewed nature of BG forecasts within the 5th–95th percentiles.

For each insulin rate at a specific nutrition rate, if the one hourly 5th percentile BG outcome is below the target range, the combination is discarded, and the algorithm moves on. Otherwise, if the prediction is in tolerance (Figure 15.3, panel A), the combination is saved (but does not overwrite if a previous combination is already saved). Finally, if the prediction is not in tolerance but is closer to the lower bound than any previous combinations, it is saved as a possible treatment. If no insulin/nutrition combination corresponds to a 5th percentile BG outcome above the lower bound of the target range for 1 h, the controller defaults to maximum nutrition and zero insulin.

A similar procedure occurs at 2 and 3 h, with admissibility determined by the heuristic rules described in Figure 15.3. If a treatment is admissible under Figure 15.3, condition A, the treatment is saved, whereas if it meets condition B, it is saved as a possible treatment. Possible treatments are not overridden but are instead saved/cleared after all insulin rates are tested for a given nutrition rate. If no allowable treatments exist, these longer time periods cannot be offered. If an allowable treatment is found, the selected nutrition rate is used as a lower limit for shorter treatment intervals. This approach ensures treatment consistency across all intervention and measurement intervals to maximize transparency and clinical acceptance, and thus compliance [35]. Specifically, it ensures an intuitive combination of treatment options, where longer measurement intervals generally yield wider stochastic forecasting bounds and thus more conservative (lower insulin) treatment choices.

The 5th percentile target is prioritized for control (Figure 15.3) due to the skewed nature of the BG outcome distribution, as depicted in Figure 15.2, which ensures that BG outcomes best overlap with the lower (80–125 mg/dL) desired portion of the 80–145 mg/dL range. This 80–125 mg/dL range is associated with better outcomes [4,15] and is also associated with reduced rate and severity of organ failure [6]. Tightness of the GC provided by STAR is further ensured by the treatment of the 95th percentile forecast outcomes. Monitoring the likelihood of BG rising above the target range allows for more explicit direct control

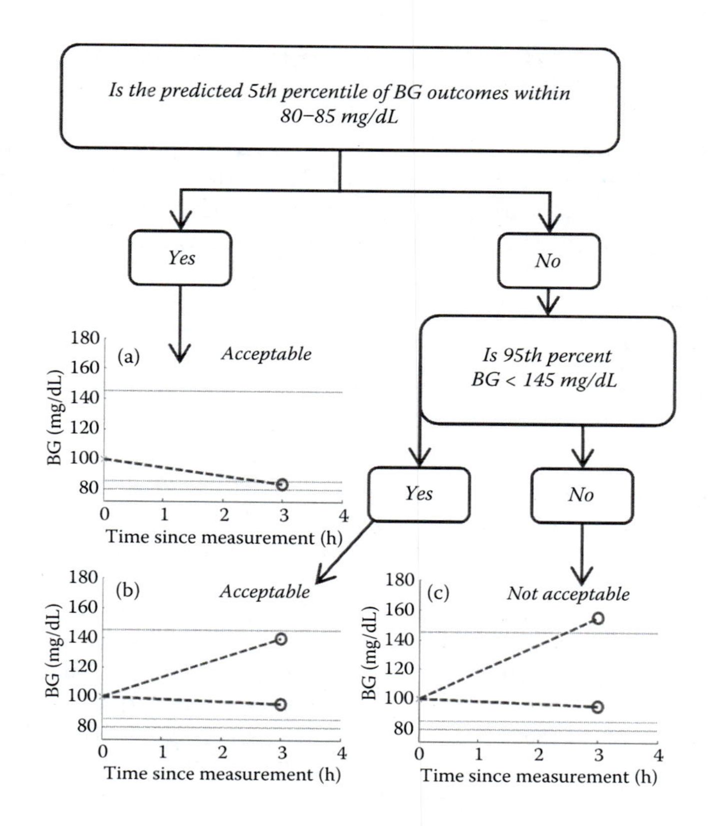

FIGURE 15.3
Heuristic selection logic for possible treatments at a three-hourly treatment interval. The lower blue line depicts the 5th percentile of BG outcomes, and the upper depicts the 95th percentile of BG outcomes. Target glycemic range (including tolerance on the lower bound) is shown by the horizontal black lines.

over intrapatient variability and therefore directly limits the risk and occurrence of mild hyperglycemia, as well.

Overall, it is decision making or dose selection under uncertainty accomplished by direct management of risk. It mixes a deterministic physiological model that characterizes interpatient variability by identifying patient-specific insulin sensitivity with a stochastic model that quantifies intrapatient variability in outcome from an insulin intervention. Finally, a heuristic method is used, as shown in Figures 15.2 and 15.3, to select an optimum treatment that best meets all goals defined in the problem definition from a selection of all feasible possible treatments.

15.2.5 Analyses and Assessment

STAR is the successor to a system that titrated using an effective insulin sensitivity metric. Derived and designed using essentially the same model, the Specialized Relative Insulin Nutrition Titration (SPRINT) method [5] protocol essentially targeted the median (no change) insulin sensitivity value as it had no ability to quantify risk like STAR. It thus was not able to maximize nutritional intake as part of its goals and relied on restricting this intake with moderate insulin to minimize risk. Thus, a comparison of these two protocols' clinical results, where SPRINT was the only published protocol to reduce both mortality and hypoglycemia, will indicate the relative success of the overall approach to decision making presented.

The protocols are compared on the basis of performance (percentage of time in relevant glycemic bands) and safety (percentage of BG < 72 mg/dL or moderate hypoglycemia and number of patients with BG < 40 mg/dL or severe hypoglycemia). Additionally, given the different approach to nutrition forced upon these protocols, comparing nutrition intake is also of interest.

15.3 Results

Table 15.1 shows the clinical results for STAR and SPRINT. Note that there is an extra column for the *in silico* virtual trials used to design STAR, which will allow assessment of the design approach and model, as well. Figure 15.4 shows the GC results as well in a cumulative distribution format.

It is clear from Table 15.1 and Figure 15.4 that there is good correlation of design to clinical outcome. Equally, performance and safety are, for the patients to date, improved. In particular, the safety metrics and BG < 72 mg/dL are almost three times lower. The nutrition is higher as desired, showing how STAR is able to quantify risk and thus be more aggressive where it is safe to maximize nutritional input without hyperglycemia or hypoglycemia. In addition, the ability to manage risk meant far fewer measurements and thus far less clinical burden, despite the improved safety and performance. The overall results show that STAR is better able to quantify and manage risk using computational and AI-based decision making.

TABLE 15.1

Clinical and *In Silico* Results for STAR and Comparison to Clinical Results for SPRINT

Whole Cohort Statistics	STAR Pilot Trial	STAR (Predicted *In Silico*)	SPRINT Clinical Data
# patients	35	371	371
Total hours	3762 h	40,101 h	39,841 h
# BG measurements	1948	20,050	26,646
Measures per day	12.2	12.0	16.1
Median BG [IQR] (resampled)	6.13 [5.60–6.81]	6.15 [5.65–6.75]	5.60 [5.00–6.40]
% resampled BG within 80–125 mg/dL	**77.79**	**80.96**	**78.48**
% resampled BG within 80–145 mg/dL	89.42	91.61	85.95
% resampled BG within 145–180 mg/dL	6.14	4.69	4.45
% resampled BG > 180 mg/dL	2.99	1.69	2.00
% resampled BG < 80 mg/dL	**1.54**	**2.27**	**7.83**
% resampled BG < 72 mg/dL	**0.87**	**0.83**	**2.89**
% resampled BG < 40 mg/dL	0.000	0.020	0.040
# patients <40 mg/dL	0	4	8
Median insulin rate [IQR] (U/h):	2.5 [1.0–5.5]	2.5 [1.5–4.0]	3.0 [2.0–4.0]
Median glucose rate [IQR] (g/h):	4.8 [0.2–6.4]	5.0 [2.2–6.4]	4.1 [1.9–5.6]

Note: Interquartile range (IQR).

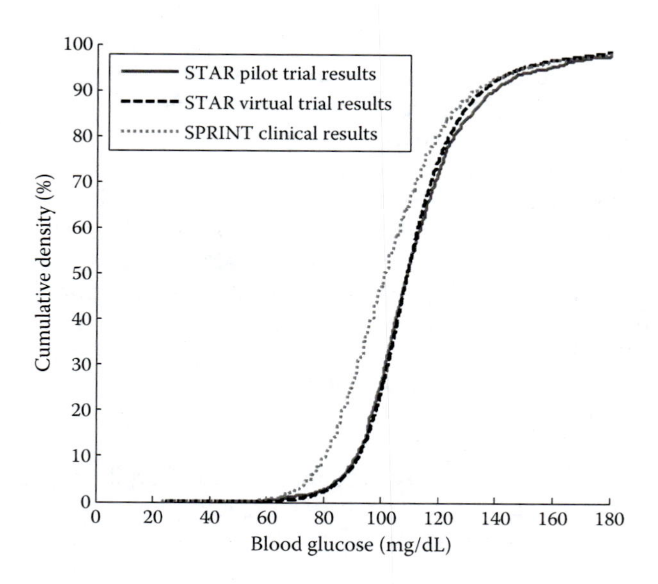

FIGURE 15.4
Cumulative distribution results for SPRINT clinical results compared to results of STAR clinical and *in silico* virtual trial design (done before implementation, so they are a prediction).

15.4 Discussion and Outcomes

Foremost, this work presents a control-focused definition of a typical and commonly occurring clinical control problem, particularly relevant to a wide range of drug dosing problems. Thus, the problem presented is relatively general or generalizable. The problem definition attempts to cast this clinical problem in a fashion that is easily interpreted and translated to engineering and AI analysis and design, particularly with respect to the mix of deterministic, stochastic, and heuristic elements. Hence, the goal was to place the problem in this context to gain a different perspective.

The main issue in such problems is typically not the dosing or optimization, nor even the model, but the ability to manage patient-specific variability in response to treatment, as well as in their patient-specific evolution (of this response) over time due to changes in patient condition. Thus, there is significant interpatient and intrapatient (between and within patients) variability to manage, and it defines the performance of any protocol or method applied to the problem [21]. As such, the main requirement for an effective decision-making framework is the ability to quantify these variabilities.

In this case, and typical to many medical scenarios, the overall approach is separated. Interpatient variability is treated as a patient-specific model problem. Hence, this approach identifies the insulin sensitivity metric (S_I) every time a measurement is taken, making the model patient specific for determining the next intervention dose. In this way, interpatient variability at this point in time is identified and managed directly.

Intrapatient evolution, over time, is more difficult. This chapter presents a method using stochastic models to forecast such variation and an associated dosing STAR decision-making method that uses this quantified risk of change to maximize safety and performance in the insulin and nutrition dose selected. The results show the impact where safety is significantly improved from a proven, highly effective protocol (SPRINT), while also maximizing nutrition further. These outcomes are a direct effect of the ability to manage and quantify risk and to use these values to be more aggressive where it is safe to do so, something that SPRINT was not able to do by design.

More specifically, the STAR approach is unique in the field. This uniqueness stems from the future forecasting enabled by the stochastic model. The ability to guarantee a level of risk is particularly relevant in medicine where outcomes are typically variable and expressed in similar odds ratio statistics. More importantly, it is a very general approach that could be adapted to a wide range of similar problems in medicine.

There are potential limitations to this approach. The stochastic model is cohort based and requires significant data to create. Without such data, a different approach or estimation must be used to create a similar stochastic model. However, for many common dosing problems, such data or estimations are either available in the literature or can be created from clinical testing.

In particular, when generalizing any probabilistic function describing different (future) patient states, one must also be concerned with ensuring that it is general and covers all possible patients. In this case, the stochastic model is made using data from 371 patients, totaling over 40,000 h [32]. In addition, it has been shown to match independent data from the Glucontrol study cohort in Europe [26,36]. However, in any new application, it is important to ensure such generality over all patient types to ensure that the stochastic model accurately captures all potential patient states, making possible the reliable estimation of intrapatient variability and associated risk that are critical to this approach.

Potentially more problematic is the cohort nature of this approach. A patient-specific approach using autoregressive modeling (AR) techniques could be used, such as in the eMPC approach [24]. However, AR methods require excessive measurement to "dial in" and become accurate, increasing the clinical burden. In addition, while such approaches offer better resolution, it is only useful until the patient state changes acutely, which requires further measurements before the AR model is again accurate. Since such changes may occur, on average, one to two times per day in the critical first 3–5 days of care, such methods may not be valid. Overall, the choice of approach should be based on the specific problem and patient variability (acute or chronic) that is likely be encountered in that specific problem.

Finally, the problem is similar to many drug dosing scenarios and several other areas of ICU medicine, such as the delivery of mechanical ventilation, circulatory management, or sedation delivery [37–41]. In every case, the requirements are

- A clinically validated deterministic model that can accurately predict outcomes and be used to identify critical patient-specific parameters to mitigate interpatient variability
- A stochastic model or similar means of predicting the range of (patient-specific) intrapatient variability in response to therapy as patient condition evolves over a relevant time frame
- A heuristic method of searching between all feasible solutions to find an optimum solution that best balances a range of potentially conflicting physiological, clinical, and patient-centered goals

The case study presented meets these requirements, and many other medical scenarios are quite similar and could be fit within this framework. Hence, the overall approach is generalizable to a range of similar clinical problems and offers a template for decision making in medicine under uncertainty.

References

1. McCowen, K.C., A. Malhotra, and B.R. Bistrian, Stress-induced hyperglycemia. *Crit Care Clin*, 2001. **17**(1): 107–124.
2. Krinsley, J.S., Association between hyperglycemia and increased hospital mortality in a heterogeneous population of critically ill patients. *Mayo Clin Proc*, 2003. **78**(12): 1471–1478.
3. Krinsley, J.S., Effect of an intensive glucose management protocol on the mortality of critically ill adult patients. *Mayo Clin Proc*, 2004. **79**(8): 992–1000.
4. Van den Berghe, G., P. Wouters, F. Weekers, C. Verwaest, F. Bruyninckx, M. Schetz, D. Vlasselaers, P. Ferdinande, P. Lauwers, and R. Bouillon, Intensive insulin therapy in the critically ill patients. *N Engl J Med*, 2001. **345**(19): 1359–1367.
5. Chase, J.G., G. Shaw, A. Le Compte, T. Lonergan, M. Willacy, X.W. Wong, J. Lin, T. Lotz, D. Lee, and C. Hann, Implementation and evaluation of the SPRINT protocol for tight glycaemic control in critically ill patients: a clinical practice change. *Crit Care*, 2008. **12**(2): R49.
6. Chase, J.G., C.G. Pretty, L. Pfeifer, G.M. Shaw, J.C. Preiser, A.J. Le Compte, J. Lin, D. Hewett, K.T. Moorhead, and T. Desaive, Organ failure and tight glycemic control in the SPRINT study. *Crit Care*, 2010. **14**(4): R154.
7. Van den Berghe, G., P.J. Wouters, R. Bouillon, F. Weekers, C. Verwaest, M. Schetz, D. Vlasselaers, P. Ferdinande, and P. Lauwers, Outcome benefit of intensive insulin therapy in the critically ill: Insulin dose versus glycemic control. *Crit Care Med*, 2003. **31**(2): 359–366.
8. Van den Berghe, G., P.J. Wouters, K. Kesteloot, and D.E. Hilleman, Analysis of healthcare resource utilization with intensive insulin therapy in critically ill patients. *Crit Care Med*, 2006. **34**(3): 612–616.
9. Krinsley, J.S., and R.L. Jones, Cost analysis of intensive glycemic control in critically ill adult patients. *Chest*, 2006. **129**(3): 644–650.
10. Finfer, S., D.R. Chittock, S.Y. Su, D. Blair, D. Foster, V. Dhingra, R. Bellomo, D. Cook, P. Dodek, W.R. Henderson, P.C. Hebert, S. Heritier, D.K. Heyland, C. McArthur, E. McDonald, I. Mitchell, J.A. Myburgh, R. Norton, J. Potter, B.G. Robinson, and J.J. Ronco, Intensive versus conventional glucose control in critically ill patients. *N Engl J Med*, 2009. **360**(13): 1283–1297.
11. Brunkhorst, F.M., C. Engel, F. Bloos, A. Meier-Hellmann, M. Ragaller, N. Weiler, O. Moerer, M. Gruendling, M. Oppert, S. Grond, D. Olthoff, U. Jaschinski, S. John, R. Rossaint, T. Welte, M. Schaefer, P. Kern, E. Kuhnt, M. Kiehntopf, C. Hartog, C. Natanson, M. Loeffler, and K. Reinhart, Intensive insulin therapy and pentastarch resuscitation in severe sepsis. *N Engl J Med*, 2008. **358**(2): 125–139.
12. Preiser, J.C., P. Devos, S. Ruiz-Santana, C. Melot, D. Annane, J. Groeneveld, G. Iapichino, X. Leverve, G. Nitenberg, P. Singer, J. Wernerman, M. Joannidis, A. Stecher, and R. Chiolero, A prospective randomised multi-centre controlled trial on tight glucose control by intensive insulin therapy in adult intensive care units: the Glucontrol study. *Intensive Care Med*, 2009. **35**(10): 1738–1748.
13. Griesdale, D.E., R.J. de Souza, R.M. van Dam, D.K. Heyland, D.J. Cook, A. Malhotra, R. Dhaliwal, W.R. Henderson, D.R. Chittock, S. Finfer, and D. Talmor, Intensive insulin therapy and mortality among critically ill patients: a meta-analysis including NICE-SUGAR study data. *CMAJ*, 2009. **180**(8): 821–827.
14. Bagshaw, S.M., R. Bellomo, M.J. Jacka, M. Egi, G.K. Hart, and C. George, The impact of early hypoglycemia and blood glucose variability on outcome in critical illness. *Crit Care*, 2009. **13**(3): R91.

15. Egi, M., R. Bellomo, E. Stachowski, C.J. French, and G. Hart, Variability of blood glucose concentration and short-term mortality in critically ill patients. *Anesthesiology*, 2006. **105**(2): 244–252.
16. Egi, M., R. Bellomo, E. Stachowski, C.J. French, G.K. Hart, G. Taori, C. Hegarty, and M. Bailey, Hypoglycemia and outcome in critically ill patients. *Mayo Clin Proc*, 2010. **85**(3): 217–224.
17. Finfer, S., B. Liu, D.R. Chittock, R. Norton, J.A. Myburgh, C. McArthur, I. Mitchell, D. Foster, V. Dhingra, W.R. Henderson, J.J. Ronco, R. Bellomo, D. Cook, E. McDonald, P. Dodek, P.C. Hebert, D.K. Heyland, and B.G. Robinson, Hypoglycemia and risk of death in critically ill patients. *N Engl J Med*, 2012. **367**(12): 1108–1118.
18. Krinsley, J.S., Glycemic variability: a strong independent predictor of mortality in critically ill patients. *Crit Care Med*, 2008. **36**(11): 3008–3013.
19. Signal, M., A. Le Compte, G.M. Shaw, and J.G. Chase, Glycemic levels in critically ill patients: are normoglycemia and low variability associated with improved outcomes? *J Diabetes Sci Technol*, 2012. **6**(5): 1030–1037.
20. Pretty, C.G., A.J. Le Compte, J.G. Chase, G.M. Shaw, J.C. Preiser, S. Penning, and T. Desaive, Variability of insulin sensitivity during the first 4 days of critical illness: implications for tight glycemic control. *Ann Intensive Care*, 2012. **2**(1): 17.
21. Chase, J.G., A.J. Le Compte, F. Suhaimi, G.M. Shaw, A. Lynn, J. Lin, C.G. Pretty, N. Razak, J.D. Parente, and C.E. Hann, Tight glycemic control in critical care-The leading role of insulin sensitivity and patient variability: A review and model-based analysis. *Comput Methods Programs Biomed*, 2011. **102**(2): 156–171.
22. Van Herpe, T., D. Mesotten, P.J. Wouters, J. Herbots, E. Voets, J. Buyens, B. De Moor, and G. Van den Berghe, LOGIC-insulin algorithm-guided versus nurse-directed blood glucose control during critical illness: the LOGIC-1 single-center randomized, controlled clinical trial. *Diabetes Care*, 2013. **36**(2): 189–194.
23. Fisk, L., A. Lecompte, S. Penning, T. Desaive, G. Shaw, and G. Chase, STAR development and protocol comparison. *IEEE Trans Biomed Eng*, 2012. **59**(12): 3357–3364.
24. Plank, J., J. Blaha, J. Cordingley, M.E. Wilinska, L.J. Chassin, C. Morgan, S. Squire, M. Haluzik, J. Kremen, S. Svacina, W. Toller, A. Plasnik, M. Ellmerer, R. Hovorka, and T.R. Pieber, Multicentric, randomized, controlled trial to evaluate blood glucose control by the model predictive control algorithm versus routine glucose management protocols in intensive care unit patients. *Diabetes Care*, 2006. **29**(2): 271–276.
25. Lin, J., N.N. Razak, C.G. Pretty, A. Le Compte, P. Docherty, J.D. Parente, G.M. Shaw, C.E. Hann, and J. G. Chase, A physiological Intensive Control Insulin-Nutrition-Glucose (ICING) model validated in critically ill patients. *Comput Methods Programs Biomed*, 2011. **102**(2): 192–205.
26. Chase, J.G., F. Suhaimi, S. Penning, J.C. Preiser, A.J. Le Compte, J. Lin, C.G. Pretty, G.M. Shaw, K.T. Moorhead, and T. Desaive, Validation of a model-based virtual trials method for tight glycemic control in intensive care. *Biomed Eng Online*, 2010. **9**: 84.
27. Chase, J., A. LeCompte, G. Shaw, A. Blakemore, J. Wong, J. Lin, and C. Hann, A benchmark data set for model-based glycemic control in critical care. *J Diabetes Sci Technol (JoDST)*, 2008. **24**(4): 584–594.
28. Hann, C.E., J.G. Chase, J. Lin, T. Lotz, C.V. Doran, and G.M. Shaw, Integral-based parameter identification for long-term dynamic verification of a glucose-insulin system model. *Comput Methods Programs Biomed*, 2005. **77**(3): 259–270.
29. Pretty, C., J.G. Chase, J. Lin, G.M. Shaw, A. Le Compte, N. Razak, and J.D. Parente, Impact of glucocorticoids on insulin resistance in the critically ill. *Comput Methods Programs Biomed*, 2011. **102**(2): 172–180.
30. Pretty, C., J.G. Chase, A. Le Compte, J. Lin, and G. Shaw, Impact of metoprolol on insulin sensitivity in the ICU. *Trauma*, 2011. **4**: 4.
31. Lin, J., D. Lee, J.G. Chase, G.M. Shaw, C.E. Hann, T. Lotz, and J. Wong, Stochastic modelling of insulin sensitivity variability in critical care. *Biomedical Signal Processing Control*, 2006. **1**(3): 229–242.

32. Lin, J., D. Lee, J.G. Chase, G.M. Shaw, A. Le Compte, T. Lotz, J. Wong, T. Lonergan, and C.E. Hann, Stochastic modelling of insulin sensitivity and adaptive glycemic control for critical care. *Comput Methods Programs Biomed*, 2008. **89**(2): 141–152.
33. Alberda, C., L. Gramlich, N. Jones, K. Jeejeebhoy, A.G. Day, R. Dhaliwal, and D.K. Heyland, The relationship between nutritional intake and clinical outcomes in critically ill patients: results of an international multicenter observational study. *Intensive Care Med*, 2009. **35**(10): 1728–1737.
34. Heyland, D.K., N.E. Cahill, R. Dhaliwal, M. Wang, A.G. Day, A. Alenzi, F. Aris, J. Muscedere, J.W. Drover, and S.A. McClave, Enhanced protein-energy provision via the enteral route in critically ill patients: a single center feasibility trial of the PEP uP protocol. *Crit Care*, 2010. **14**(2): R78.
35. Chase, J.G., S. Andreassen, K. Jensen, and G.M. Shaw, Impact of human factors on clinical protocol performance: a proposed assessment framework and case examples. *J Diabetes Sci Technol*, 2008. **2**(3): 409–416.
36. Suhaimi, F., A. Le Compte, J.C. Preiser, G.M. Shaw, P. Massion, R. Radermecker, C. Pretty, J. Lin, T. Desaive, and J.G. Chase, What makes Tight Glycemic Control (TGC) tight? The impact of variability and nutrition in 2 clinical studies. *J Diabetes Sci Technol*, 2010. **4**(2): 284–298.
37. Sundaresan, A. and J.G. Chase, Positive end expiratory pressure in patients with acute respiratory distress syndrome-The past, present and future. *Biomed Signal Process Control*, 2012. **7**(2): 93–103.
38. Revie, J.A., D.J. Stevenson, J.G. Chase, C.E. Hann, B.C. Lambermont, A. Ghuysen, P. Kolh, P. Morimont, G.M. Shaw, and T. Desaive, Clinical detection and monitoring of acute pulmonary embolism: proof of concept of a computer-based method. *Ann Intensive Care*, 2011. **1**(1): 33.
39. Chase, J.G., A.J. Le Compte, J.C. Preiser, G.M. Shaw, S. Penning, and T. Desaive, Physiological modeling, tight glycemic control, and the ICU clinician: what are models and how can they affect practice? *Ann Intensive Care*, 2011. **1**(1): 11.
40. Shaw, G., J. Chase, A. Rudge, C. Starfinger, Z. Lam, D. Lee, G. Wake, K. Greenfield, and R. Dove, Rethinking sedation and agitation management in critical illness. *Crit Care Resusc*, 2003. **5**: 198–206.
41. Shaw, G.M., J.G. Chase, C. Starfinger, B.W. Smith, C.E. Hann, T. Desaive, and A. Ghuysen, Modelling the cardiovascular system. *Crit Care Resusc*, 2007. **9**(3): 264–269.

16

Clinical Decision Support in Medicine: A Survey of Current State-of-the-Art Implementations, Best Practices, and Gaps

Sylvia Tidwell Scheuring, Wanda Larson, Jerome Scheuring, and Thomas Harlan

CONTENTS

16.1 Introduction

Clinical decision support systems (CDSs) are increasingly relied upon by medical practitioners to provide timely, accurate support to patient diagnosis and treatment [1]. Effective CDSs are considered essential to improving health care accessibility, transparency, efficiency, and efficacy of patient care [2–4]. However, there remain significant challenges to the acceptance and use of such systems in clinical settings. Chief among these challenges are ease-of-use issues and perceived threats to professional autonomy by clinicians [5–7]. We hypothesize that new perspectives on the role of CDSs within the clinical work flow can take advantage of developing technologies to reduce the extent of these challenges.

One such perspective leads to a paradigm shift away from thinking of the CDS as reference tool technology, isolated from human–human interaction, and toward an understanding of the CDS as a medium for cooperative communication between the multiple interprofessional teams involved in patient care. Patient care team members are inclusive of the patient and their point-of-care providers in diverse roles and settings, such as nurses, physicians, specialists, laboratory and other medical technologists, administrative and financial personnel, as well as the combined team of software developers, medical researchers, subject matter experts, and informatics professionals who build and maintain CDS systems. In this new paradigm, we essentially advocate that interprofessional teams adopt the CDS system as a surrogate team member or consultant available to *dialogue with* the clinical staff. In addition, we posit that CDS systems should be capable of learning through team interactions and able to liaison *between* the clinical staff and their supporting administrative and information technology (IT) staff.

Currently, many CDS systems serve a medication management function as a reference and safety alert tool. As such, the principal roles of these CDS systems are as *front-end reference tools* (e.g., drug-to-drug interaction checking, weight-based dose calculations) and *back-end tools* (e.g., prompts for data entry elements, end-user notifications, system efficacy reports) [8]. Current interactions between the electronic health (or medical) record (EHRs, also known as EMRs) and multiple other CDS systems supporting health care, include safe and effective medication management. These were perceived as "state-of-the-art" uses of CDS systems as of 2009 [1].

To explore current practices and clarify our proposed perspective shift, we have created a case-based narrative structure that follows a mock patient, Ruby Melrose, throughout

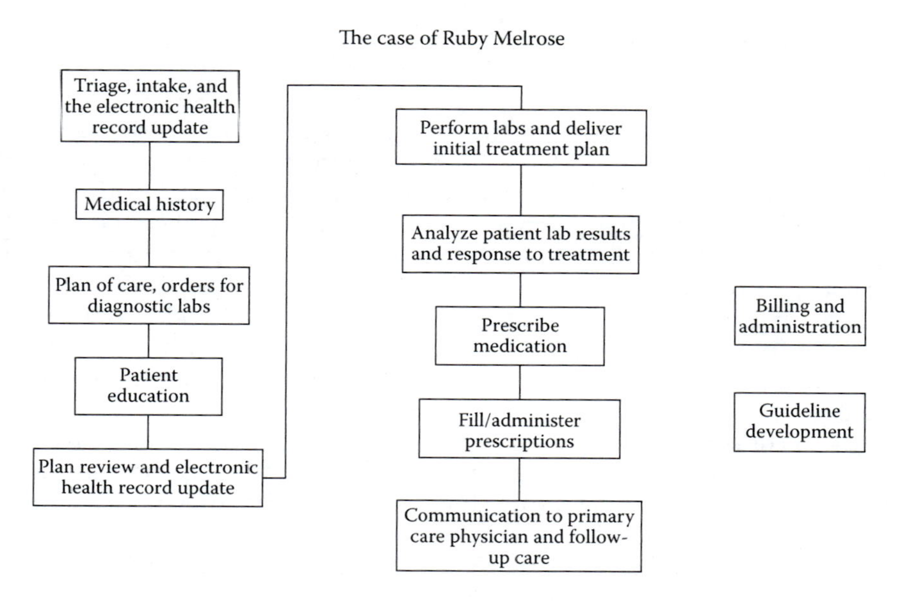

FIGURE 16.1
Work flow-associated structure of CDS paper review.

her care (Figure 16.1). Through her story, we will contextualize and categorize our review of selected meta-analyses and original research results related to CDS systems from 2003 to 2012. We will present a discussion of Ruby Melrose's case primarily as it relates to the current medication management process flow. We will highlight the extent to which cooperative communication with CDS systems has enhanced safety, eased end-user concerns, and supported utilization. This narrative structure will help us to identify gaps and work flow issues with current implementations.

In Section 16.4, we will highlight future opportunities to improve health care by optimizing the partnership between CDS system development teams and health care delivery teams. Today's dynamic online reference text view of CDS systems will be contrasted with a new paradigm that advocates adopting a holistic team-member perspective for CDS systems. We postulate that designers and end users of CDS systems may improve health care (and Mrs. Melrose's story) by thinking of these intelligent information systems in the context of the entire patient care process and by interacting with their CDS systems as surrogate team members within that process.

16.2 Case-Based Narrative Review of Selected Clinical Support Systems (2003–2012)

A case of congestive heart failure (CHF): Ruby Melrose is a 69-year-old female who presented to the emergency department (ED) with a complaint of shortness of breath that had worsened gradually over the past week, particularly when she attempted to recline in her bed. She had been sleeping propped up by several pillows.

16.2.1 Triage, Intake, and the EHR Update

16.2.1.1 The Case of Ruby Melrose

The triage nurse at the front desk in the ED asked Mrs. Melrose why she had come to the emergency room, and Mrs. Melrose explained her concerns regarding shortness of breath while reclining. The triage nurse identified that Mrs. Melrose was breathing faster than normal and had increased effort to breathe. She immediately weighed the patient, identified that she had "wet"-sounding lungs, obtained an initial set of vital signs, and took her to an ED patient care room.

When accessing medication information, ensuring accurate and current information helped to identify that Mrs. Melrose had been a prior patient in this hospital. The nurse verified that her medical history, recent labs, and prescribed medications were a part of her current EHR.

During their conversation, the ED triage nurse confirmed with Mrs. Melrose the list of allergies that had been entered by the nurse at Mrs. Melrose's primary care clinic. These were available because the primary care clinic that Mrs. Melrose uses is a part of the same health care system as the hospital. A validation of Mrs. Melrose's prior and new medications, her current weight, and allergies were integrated in her ongoing record (EHR).

The triage nurse hands off care of Mrs. Melrose to the ED nurse assigned to her room.

16.2.1.2 CDS Investigations, 2003–2012 (Triage, Intake, and the EHR Update)

Data-centered CDS systems require data on which to operate. While virtually every hospital in the United States runs on at least a computerized billing system with a wealth of patient data, implementations of full EHRs are less common.

HealthIT.gov provides a definition of basic and full implementations of electronic health care records [9]. For hospitals, basic implementations include patient demographics, physician notes, nursing assessments, problems lists, medication lists, discharge summaries, advanced directives, computer provider order entry (CPOE) for medications, and results management for lab and radiology reports. Full implementation includes all of the basic capabilities with the exception of advanced directives and adds CPOE for lab reports, radiology tests, consultation requests, nursing orders, and results management for radiology images, diagnostic images, and consult reports. EHR with full CDS provides support for guidelines, reminders, drug allergies, drug interactions, drug–lab interactions, and drug dosing. Basic and full implementations have a more limited definition for physicians than for hospitals.

EHRs are central to the operation of CDS systems [10]. Since the development of the first generation of EHRs in the late 1960s, adoption of EHRs has been slowly increasing. A survey by the National Coordinator for Health Information Technology found that as of 2011, only 34.8% of nonfederal acute care hospitals had adopted an EHR system. This was, however, nearly double the adoption rate in the survey conducted in 2009 (16.1%) [11]. A survey of 2758 responding physicians conducted by DesRoches et al. in 2008 [12] concluded that 83% of physicians in the United States had not yet adopted an EHR system, though many (16% of survey respondents) had acquired but not yet implemented an EHR system [12].

While adoption rates are expected to increase dramatically in clinics as the result of financial incentive revisions to Medicaid and Medicare reimbursements [American Recovery and Reinvestment Act (ARRA)/Health Information Technology for Economic and Clinical Health Act [HITECH] title IV)], two research studies published in 2012 of physicians in Texas (long-term care) and Maryland (physicians of inpatients) suggest that even when EHR is adopted by the hospital or clinic, not all aspects of the EHR systems are utilized equally by all participating physicians. Factors found to contribute to utilization differences of various EHRs and associated computerized order entry (CPOE) systems included employment relationship, gender, patient ratio, and size of the institution [13,14]. Our personal experience is that the best way to boost adoption rates is to mandate usage as a requirement to maintain privileges. When such requirements are enacted, we have seen utilization rates approach 100%.

The reasons for the relatively low rate of adoption by even large hospitals have been surveyed; leading reasons as reported by the Healthcare Information Systems Society include significant start-up costs, low perceived or recognized efficiency gains, and the perception that EHR systems are difficult to use [10]. It has been observed that despite clear system-wide gains in reliability, small clinics frequently see productivity decline as they adopt electronic record systems [5]. However, many in the health care industry continue to encourage adoption of electronic records as a means of promoting communication between clinics and facilities [15]. According to the survey by DesRoches et al. [12], many of the hospitals adopting EHR have chosen to do so as a result of funding changes from the United States government, in that continued federal financing from certain programs (e.g., particular Medicare reimbursement levels in the ARRA) is contingent on hospitals adopting EHR systems [12]. Based on these funding changes, we expect an exponential growth in the adoption of EHRs.

Unfortunately, systemic use of CDS systems is somewhat limited by the same sorts of data design issues that have plagued other fields. EHRs are not all created equally—the data field definitions and implementations differ across different networks. If a clinic and a hospital use a compatible EHR, and share a network or repository, then the EHR of the clinic and the hospital can effectively be treated as the same record. This cannot be assumed to be the case when the EHR provider for the hospital and the clinic differ or use different networks [15]. Given that EHR records for the same patient can be created at multiple separate facilities, the research seems oddly silent on the issue of how conflicts between EHR source records are handled or how systems will be merged in the future as facilities join larger networks or share data.

Worldwide, the advantages of using electronic records have been recognized in diverse environments. One program, among clinics in northwestern Kenya, adopted an EHR and CDS running over the cellular telephone network in the area. This system was considered effective and easy to use by doctors in a difficult clinical setting [10]. These doctors in Kenya saw value in the CDS system even though there were documented unresolved challenges in keeping paper and EHR in synch with one another.

Within clinical practices in the United States, EHR is coming to be seen as a medium of communication replacing handwritten or typed notes in physical folders, allowing physicians and other care providers easy access to explanatory notes or supplementary information for which they might otherwise have to request physical materials [1,3]. According to the 2010 study by Metzger et al. [16], one of the motivations for the adoption of EHR systems is increasing collaboration in patient care between primary care clinics, specialists, and hospitals. Sharing a common record format and repository allowed for significant gains in the timeliness and accuracy of patient information shared between these care providers [15]. One of the reasons for the relatively slow adoption of EHR systems in small (single-physician) practices is that individual physicians do not always perceive a need to communicate with other clinicians, in that their notes are often simply reminders of observations or conclusions that they have already made or drawn [17].

Even when full electronic documentation is captured in the EHR, machine readability of the documentation can vary across clinics and even across care providers. Ideally, a CDS would be able to use natural language processing to read free text entered by any care provider, but the reality is that such free text can often not even be interpreted by care providers within the same practice, and that structure text is more likely to be accurately interpreted by both CDS systems and other providers unfamiliar with the particular context [18,19]. As a result, free text is not reportable. Some progress has been made by some vendors in setting up SmartNotes, which are boilerplate physician notes where the provider picks the boilerplate and then clicks through the fields picking responses. The responses are then stored separately and tied to the template. These may be more reportable than free-text notes.

16.2.2 Medical History

16.2.2.1 The Case of Ruby Melrose

Back in the patient care room, the ED treatment nurse and ED physician continued to ask questions and examine Mrs. Melrose. The nurse and physician verified with her that she had a significant past medical history that included an enlarged heart and the chronic condition of CHF. They verified that she continues to remain on her medications, as prescribed by her primary care physician (PCP). They verified that her home medications included ones to control her heart rhythm and to keep her body's water balanced. The medication containers she had brought with her were consistent with the medications listed in her medical record, which was recently updated in Mrs. Melrose's EHR at her PCP's office. They verified substantive changes in her most current chest x-ray, completed a week ago in an outpatient clinic of her PCP. Through the EHR, they also accessed her latest lab work and the electrocardiogram (ECG) performed the prior week. They confirmed this information with Mrs. Melrose.

16.2.2.2 CDS Investigations, 2003–2012 (Medical History)

EHR systems can store all of the information about a patient that would normally be included in a patient's physical chart. An international overview by Eichelberg et al. in 2006 [20] noted that international standards for EHR systems specifically included diagnostic data such as lab results and imaging results. Not all EHR systems included the capability to display high-resolution medical images [1,20]. Picture archiving and communication system (PACS), while often separate from the EHR, do allow care providers to view results from workstations within a given facility. At this time, there is currently no medical imaging clearinghouse.

As we mentioned earlier, not all EHR records are alike, and at times, the content in them is dependent on where the records are created. Prescriptions written by a physician, nurse practitioners, or physician's assistant are typically a part of a patient's EHR; it is much less common for commercial pharmacy records regarding dispensing medication to be incorporated into the same record [21]. Dispensaries and hospital-associated pharmacies are typically tied in to the hospital's computerized provider order entry (CPOE) systems and report on medication dispensed for particular patients [5]. Pharmacy chains maintain comprehensive records within the chain, but few pharmacies are tied in to the same EHR providers used by clinics and hospitals, a possibly significant gap in information about patients [18]. There are regional prescription clearinghouses (operated by Allscripts) into which health systems can tie. Use of this system is a Meaningful Use metric for stage one of the HITECH, which was a part of the ARRA implementation.

Radiology reports are typically a component of a patient's EHR, and while provision for actual radiology data [e.g., the patient's x-rays, computerized axial tomography (CAT) scans, and MRI] are present in most EHR standards and implementations, actual use of these provisions is rare in a clinical setting [20]. In conversation with various specialists, we noted that the specialists did not consider it uncommon for outpatients to have to hand-carry printouts, films, and DVDs of these records to and from various specialists in a multiphysician/multiclinic process of diagnostics due to the variance in participation in EHR networks. Increasing subscription to larger systems such as Cerner and Epic that provide services to multiple clinics may result in a decrease of such issues for patients who seek advice from multiple clinical groups. The industry has agreed to use the same standards (ISO 13606-2), but no agreement was reached on a central cloud repository. As a result, EHR systems do not always support the same fields in the same way, so interoperability between systems is not necessarily transparent [22]. This interoperability issue could be a challenge to authors of CDS systems seeking to learn from larger data sets.

16.2.3 Plan of Care, Orders for Diagnostic Labs

16.2.3.1 The Case of Ruby Melrose

The immediate plan of care determined by the ED physician and ED nurse included treatments to support Mrs. Melrose's oxygenation; reassessment to determine if there were new changes in her heart's or lungs' condition (i.e., x-rays, ECG, blood and urine tests); and augmenting the medications she was already using with additional medications designed to reduce excessive retention of fluids from her body that were likely congesting her lungs due to her chronic heart failure condition.

16.2.3.2 CDS Investigations, 2003–2012 (Plan of Care, Orders for Diagnostic Labs)

Computerized provider order entry (CPOE) is one of the most recognized applications of medical IT and one of the primary areas of focus for decision support systems [1,15,23]. According to a 2010 survey by Metzger et al. [16], over 40% of large US hospitals have adopted CPOE systems. Adoption rates are generally lower for smaller hospitals, with less than 13% of hospitals with fewer than 200 beds having adopted CPOE. Adoption rates of CPOE by clinics, even those that have otherwise implemented EHRs, remains low, with less than 10% of clinics with five or fewer physicians on staff having adopted CPOE [5].

There are a number of reasons for the relatively low adoption rates of CDS systems in general and CPOE systems in particular. Leading issues include the costs involved, especially personnel costs, because adoption of such systems involves numbers of personnel over relatively long periods of time, and the perceived threat to a physician's autonomy in the decision-making process [5,7]. Costs, however, appear to be declining as more competitive vendors enter the field [1,23], and autonomy issues diminish as more physicians are exposed to the usefulness of these systems [6,24].

CPOE systems typically incorporate decision support in the ability to generate ancillary orders; for example an order to administer gentamicin results in an alert asking the physician if he/she also wants to order comparative blood levels 1 h before and 3 h after administration. If the physician agrees, the orders for the blood levels are automatically added to the EHR, thus improving compliance with established standards of care [25]. Especially in settings where cost controls are important, such CPOE systems also provide automated access to an array of frequently updated clinical guidelines with recommendations concerning the context of particular medications or procedures.

Providing this information during the process of order entry increases the probability that providers will comply with the most recent guidelines. Many, but not all, of these systems are "active systems," allowing for a two-way conversation with the provider and guideline developers where conflicts arise.

CDS systems facilitate a unique contemporaneous view of the information that goes into decisions that providers make in ordering specific procedures or medications. Providers' professional judgment occasionally conflicts with official or research-based guidelines concerning medication or procedures, and active CDS systems enable hospitals and clinics to see the patterns by which clinicians override or agree with guidelines [16,25,26]. Designers of some active systems often require the ordering provider to document the context at the time of the order (e.g., Does the patient meet the guidelines?) [23,25]. Other designers support requesting the input asynchronously (e.g., at shift change) in their systems to minimize the effect on real-time patient care needs yet still get the needed feedback when guidelines have been overridden [27].

Our opinion is that guidelines need to be proactive for these reasons and because guidelines can be affected by Food and Drug Administration (FDA) warnings or as the results of research trials that can be released at any time and may therefore be more current than guidelines. If the CDS systems are not kept current, they will not be trusted by care providers.

It seems unfortunate that, for the most part, CPOE systems are seen as "the computer"; except when engaged in the development process itself, few clinicians are aware of the existence, let alone presence, of the development team building the knowledge base and presentation systems that drive CPOE and related guidelines [7,28]. Some guideline developers are using such actively collected data to revise guidelines when they are first implemented [28]. The asynchronous system described in the previous paragraph automatically aggregates guideline overrides and compares overrides across physicians and cases to aggregate common features of the cases for the guideline developers. This intercase analysis feature proposes updated guidelines that are evaluated and modified as needed by guideline developers [27].

16.2.4 Patient Education

16.2.4.1 The Case of Ruby Melrose

Back in the room, the ED physician and ED nurse explained to Mrs. Melrose that weakened contractions of her heart, coupled with imbalances to her body's salts and fluids, had likely contributed to the increased backlog of fluids into her lungs—causing shortness of breath. They entered their findings and comments about teaching Mrs. Melrose into her EHR via the bedside computer.

16.2.4.2 CDS Investigations, 2003–2012 (Patient Education)

While many CDS systems provide guidance on patient education, it is rare that such systems follow up or record the education actually provided or what was learned by the patient. In the case of nonadherence to the recommendations by the patient, such records would be useful for determining why. This can be very important to hospitals as the insurance companies and Medicare start holding hospitals financially accountable for patient adherence to recommendations. Such accountability is already occurring in the cases of heart failure [29].

16.2.5 Plan Review and EHR Update

16.2.5.1 The Case of Ruby Melrose

The ED physician recorded Mrs. Melrose's symptoms and the ED team's proposed treatment plan via the bedside computer into Mrs. Melrose's EHR, and reviewed the resulting CDS system reminders about treatment protocols and orders. These reminders and alerts were triggered automatically from presumptive information being entered about heart failure symptoms. The physician reviewed and acknowledged the alerts; overrode or accepted the reminders offered by the CDS system, and submitted the reviewed orders into Mrs. Melrose's EHR.

16.2.5.2 CDS Investigations, 2003–2012 (Plan Review and EHR Update)

One of the key roles of CDS systems is their ability to alert clinical staff to potential issues with the treatment plan entered for the patient, for example, potential side effects of medications, possible interactions with other medications, and similar issues. Overuse of alerts in clinical settings can lead to "alert fatigue": when alerts are systematically ignored or overridden by the receiving staff without even being reviewed. If an important alert, meaning one that has clinical relevance to this patient, is overridden, this can result in suboptimal patient care.

CDS system designers are aware of these issues and are actively working with behavioral researchers and clinical staff to address them [25,26,30,31]. Just as in human–human dialogue, situationally varying the conversation between the CDS system and users can help to improve communication. Designers can decrease alert fatigue by contextualizing the alerts in a number of ways: reducing alert frequency for content within a physician's specialty, tailoring alerts to the current patient condition, emphasizing serious alerts and deemphasizing less serious alerts, providing alternatives rather than constraints, and providing certain alerts to other staff and within the work flow rather than at the time of order entry [16,31]. When CDS systems have timely, relevant, proactive, and accurate information, care providers are more likely to trust them and see using them as invaluable.

In addition to alerting the clinical staff to potential issues and concerns, CDS systems are effective at providing timely reminders for routine or expected activities, such as recommended times for blood tests (see, for example, the postadministration blood tests associated with gentamicin, above) [6,32]. These reminders can be thought of as facilitating indirect communication between the ordering physician and the nursing or direct patient care staff, though little of the literature takes this view directly.

In most cases, regardless of when such reminders are provided, designers currently generate them in response to entries through the CPOE. Thus, reminder technical models are generally not continuous time-based models; the effects of medicines are only evaluated when the orders are made and do not take into account changing patient conditions unless orders are regenerated.

16.2.6 Performing of Labs and Delivering the Initial Treatment Plan

16.2.6.1 The Case of Ruby Melrose

The respiratory therapist and the radiology technician had arrived at the bedside, triggered by the orders from Mrs. Melrose's EHR. The nurse, physician, respiratory therapist, radiology technician, and radiologist updated their actions and findings in the EHR. All orders based on the team's new plan of care were now in progress or had been marked as completed. The CDS system was now tracking the timing of orders, the completion of orders, and Mrs. Melrose's responses to treatments and noting them in her EHR.

16.2.6.2 CDS Investigations, 2003–2012 (Performing of Labs and Delivering the Initial Treatment Plan)

In many hospital settings, bedside systems are capable of monitoring the administration of medications, for example, through the use of bar-code scanners to record administration of medications or smart IVs for administration [33,34]. Recent research has suggested that various administration and treatment events could be automatically detected by monitoring short-term changes in blood pressure [35]. However, most systems that track actual patient care are only done for research purposes and do so by asking patient care staff to confirm that particular activities were, in fact, carried out [16,17]. Bedside patient monitoring systems are not integrated into CPOE or medication alert systems. Systems that track patient care compliance with orders do not often provide alerts or reminders for compliance exceptions. In those cases where they are noted, they appear on regular reports to hospital or clinic administrators and generally not to clinicians [1,23]. Due to the complexity of the data, such compliance reports are generally not designed into CDS systems; rather, they are custom-built for individuals or groups of institutions and are run in response to specific concerns.

A proactive system could identify when changing patient conditions suggest reconsidering orders or detect when entries regarding orders have been omitted based on the passage of time. A proactive system could, for example, alert patient care staff when orders suggest that a blood draw should be done after the administration of a particular medication, and no evidence has been provided of such a draw within the expected time. Future designers of CDS systems might want to investigate how a continuous or time-triggered event-learning model could be integrated with location-based services to increase the effectiveness of access to useful information during the patient care work flow. Such proactivity will require integration of multiple devices. For instance, bedside monitoring systems could automatically trigger CPOE or medication management systems when monitored vital signs go outside of specified ranges. There is little in the literature to suggest that such systems are currently in development, though some authors are considering how to gather data for such models in the home and in hospital intensive care units [33,35,36]. Proactive tracking systems have been used in other industries, such as those employed by UPS, FedEx, and manufacturing [37,38].

16.2.7 Analysis of Patient Lab Results and Response to Treatment

16.2.7.1 The Case of Ruby Melrose

Her ED team could see that Mrs. Melrose's treatments had eased her respiratory distress. Results from her new diagnostic studies had been entered into her EHR directly by the team members from the appropriate departments (radiology, respiratory, electrocardiology, and pathology). The ED physician and others on the ED health care team reviewed the results of these diagnostic studies and Mrs. Melrose's responses to treatment. The results indicated that there were minor adverse changes in her underlying conditions contributing to her increased heart failure.

16.2.7.2 CDS Investigations, 2003–2012 (Analysis of Patient Lab Results and Response to Treatment)

One author suggests that there may be a role for CDS systems in coordinating team members during this stage of patient care [32], but we could find no evidence that such coordination is actually currently being done using CDS systems.

While there are diagnostic systems that use artificial intelligence to help support differential diagnosis based on test results and patient histories, there does not appear to be an emphasis on integrating such capabilities into the EHR and CDS systems by the authors we reviewed. It would be interesting to investigate the role that such systems could play if interaction with all of the care providers could be incorporated into the diagnostic decision support process.

16.2.8 Prescription of Medication

16.2.8.1 The Case of Ruby Melrose

CDS systems used for Mrs. Melrose's medication management included CPOE, which were integrated with the pharmacy's computerized medication management, medication dispensing, and administration systems. When entering the medication orders into the CPOE, the ED physician was able to use alert triggers displayed about Mrs. Melrose to avoid the complete list of medication allergies, which had been confirmed by the triage nurse, and to use Mrs. Melrose's current weight to select medications, and, in this case, automatically calculate appropriate dosages.

16.2.8.2 CDS Investigations, 2003–2012 (Prescription of Medication)

One of the primary roles of CDS systems, and a primary motivation for their adoption, is the ability of the CDS system to determine quickly the potential for side effects and drug interactions that could adversely affect the present, individual patient, for example, administration of medications that are contraindicated for the elderly [16,39]. In the 2009 survey by Amarasingham et al. [40], this was the most frequently cited nonfinancial reason for the adoption of CPOE in hospital settings. A 2009 randomized controlled trial by Terrell et al. [41] found significantly fewer inappropriate prescriptions for elderly patients when physicians were guided by CDS systems.

One of the advantages of CDS systems is their ability to provide immediate feedback or contact with the clinician: If the physician is ordering a medication "off label" or with side effects or contraindications particular to this patient, the system may refuse to accept the order without further justification. In some cases, this may require that the order be seconded by one of the physician's administrative superiors. Such systems can reduce costs by limiting the use of costly medications in situations for which there is no supporting medical research [25].

16.2.9 Filling/Administration of Prescriptions

16.2.9.1 The Case of Ruby Melrose

After noting new medications ordered in the computer, the hospital's pharmacist was able to establish the patient's medication profile in the dispensing system. The pharmacist also completed a safety check of the medication orders, determining that the orders were appropriate and safe specifically for Mrs. Melrose. She uploaded an authorization to the medication storage and dispensing computer (e.g., Pyxis, Omnicell) in the ED's medication room. This allowed the nurse to withdraw only the ordered medications from the dispenser, enabling the medications, dosages, and alert flags about Mrs. Melrose to be displayed to her nurse.

For instance, when the nurse prepared to remove a medication from the dispenser, an alert flag reminded him to administer an ordered medication, furosemide, at a very slow intravenous rate. He had also acknowledged an alert to monitor the heart rate closely for the heart medication he was about to administer. He had acknowledged an alert to recheck the patient's lab work to make sure he could safely give the medication based on the results of Mrs. Melrose's blood test. He had also acknowledged the alert to double-check medications when the drawer opened in the dispenser with Mrs. Melrose's medication because one of her medications was flagged as a "look-alike/soundalike" medication, for which there was a higher risk that the medication dispenser could have been incorrectly stocked with the wrong "look-alike/soundalike" medication.

At the bedside, the nurse first used two identifiers to verify Mrs. Melrose's identity. Then he used a handheld electronic scanner to match Mrs. Melrose's armband to the medication that he was about to deliver, which ensured that the right medication was being administered to Mrs. Melrose. He also programmed the bedside medication administration pump (the smart IV) to deliver the cardiac infusion to her, which alerted the nurse that he was about to administer a safe dosage and that the rate of medication delivery was appropriate. He also acknowledged the safety flag on the medication pump, reminding him to keep the patient properly monitored during infusion.

The efforts of Mrs. Melrose and her medical team (nurses, physicians, laboratory technicians, and pharmacists) helped avoid unsafe orders or the chance of administering medications to her that were likely to cause an adverse reaction. The provider orders in the CDS accounted for allergies, appropriate weight, past medical history, current medications, and current laboratory findings before the patient medications could be released from the dispenser. The nurse interacted with several alerts and reminders before infusing the medications into Mrs. Melrose. The CDSs stored information was shared with pharmacy and administrative personnel, generating reports supporting safety and adherence to medication management protocols, timeliness and efficacy, as well as cumulative costs and billing-related data about the care that Mrs. Melrose received.

16.2.9.2 CDS Investigations, 2003–2012 (Filling/Administration of Prescriptions)

While monitoring compliance with provider orders is routine in filling prescriptions, for example, from a hospital pharmacy, relatively little is normally done to verify administration [26]. Hospitals have been relatively slow to adopt administration verification systems, and primary care clinics even slower. However, the perceived benefits of tracking administration of medications have been improving adoption rates [42]. Even point-of-service scanning does not actually guarantee that the medication was properly administered; there is still a human element involved, but the widespread acceptance of the practice strongly encourages care providers to be more attentive to the task at hand [34].

16.2.10 Communication to PCP and Follow-Up Care

16.2.10.1 The Case of Ruby Melrose

Upon discharge from the hospital, Mrs. Melrose and her PCP could access all of the up-to-date information about the diagnostics, treatments, responses to therapy, and new medications used to control her condition and improve her health status.

The discharge nurse was also able to use the computerized clinic scheduling system to schedule an appointment with Mrs. Melrose for reevaluation and follow-up with her PCP the day after discharge.

The community pharmacy, having received an electronic copy of her new prescriptions, had the medications ready for Mrs. Melrose to pick up on her way home.

16.2.10.2 CDS Investigations, 2003–2012 (Communication to PCP and Follow-Up Care)

Except in cases where the patient is subject to chronic conditions requiring constant in-home observation, it is extremely unusual for patients to have access to their own EHR [36,39]. Clinics operating in limited-resource settings have special needs with regard to follow-up with the patient or with the patient's PCP. The personal attention called for by follow-up reminders directed to the PCP can encourage patients to pay more attention to their own health and seek medical care more frequently [10]. Some CDS systems, particularly in environments that emphasize preventative care (e.g., Kaiser Permanente), do distribute expected follow-up information, but this appears to be the exception rather than the rule.

In a 2011 survey by Fuji and Christoffersen [21], only 12% of non-hospital-based pharmacies had an EHR, and only two-thirds of these reported using an EHR shared by other health providers. The survey also suggested that the number of pharmacies with EHR systems is expected to double in the near term [21].

With the exception of the Google Health experiment and Microsoft HealthVault, neither of which had much traction in the health care industry, consumers have not been encouraged to access and control their own health records [43]. Despite this seeming lack of interest in supporting direct access by patients to their EHR, for long-term care, there does appear to be interest in CDS systems operating in the patient's home. These systems typically interact directly with the patient, providing education and behavioral recommendations designed to improve health monitoring and the rate of adherence to medical recommendations. For example, some system designs in the United States and Taiwan send text message alerts or reminders to the patient's and family members' cell phones and collect responses either through text or videos. Some designs use links over the Web to collect data from electronic health monitors such as blood pressure, body temperature, heart rate, and ECGs [33,36,39] As the population ages, this is likely the largest area for potential growth in the use of CDS systems.

16.2.11 Billing and Administration

16.2.11.1 The Case of Ruby Melrose

All information about her hospitalization was immediately accessible by Mrs. Melrose, her PCP, and third-party payers.

16.2.11.2 CDS Investigations, 2003–2012 (Billing and Administration)

The first electronic records associated with patients were nearly universally associated with billing and insurance filling. It is safe to say that all EHR implementations support billing. There does not appear to be a focus on CDS systems designed to identify cost reduction as a priority other than through guideline development. Billing is, however, the stick that medical insurance companies, the Centers for Medicare and Medicaid Services (CMS), are using to drive adoption of full implementations of EHR and CDS systems. Without proper billing and cost control as drivers, it is unlikely that CDS systems would have the data required to operate. It is equally unlikely that with these drivers present, CDS systems will languish: Rather, these financial forces will drive CDS systems to be both more prevalent and effective in the future.

16.2.12 Guideline Development

16.2.12.1 The Case of Ruby Melrose

As Ruby Melrose is being prepared for discharge from the hospital, the hospital follows quality review and improvement guidelines set forth nationally aimed at avoiding readmission of heart failure patients within 30 days of discharge. If Mrs. Melrose requires readmission within 30 days, payment for her readmission could be challenged under CMS reimbursement guidelines and affect the hospital's credentialing. If she is readmitted, physicians, nurses, and quality improvement staff in the hospital will review Mrs. Melrose's case to determine if her readmission, due to another exacerbation of her heart failure within 30 days of her discharge, was anticipated or unanticipated. The hospital strives for a low readmission rate and coordinated her care at discharge to avoid readmission within 30 days.

16.2.12.2 CDS Investigations, 2003–2012 (Guideline Development)

Some developers have taken the communicative view and actively engage clinicians in guideline development. This has typically resulted in high performance, that is, more frequently accepted guidelines [28]. Clearly, the CDS system information must be maintained as new medications become available and as pharmaceutical researchers discover more potential interactions, side effects, and contraindications [15]. Maintenance and distribution of common guideline "knowledge bases" has become an important part of CDS systems research [8,33]. An issue in maintenance of guideline knowledge bases has been customization to a particular clinical setting; for example, many hospitals have developed their own guidelines for particular procedures or medications, which are not expected to be shared with other facilities [17]. A few systems address the need for a meta-CDS that assesses the changes needed to improve CDS systems themselves [27].

16.3 Summary and Conclusions

As can be seen from this analysis, and is shown in Figure 16.2, CDS systems have differentially penetrated the market in very few places in the work flow. The majority of uses apply to order entry and writing prescriptions, and very little attention has been focused on

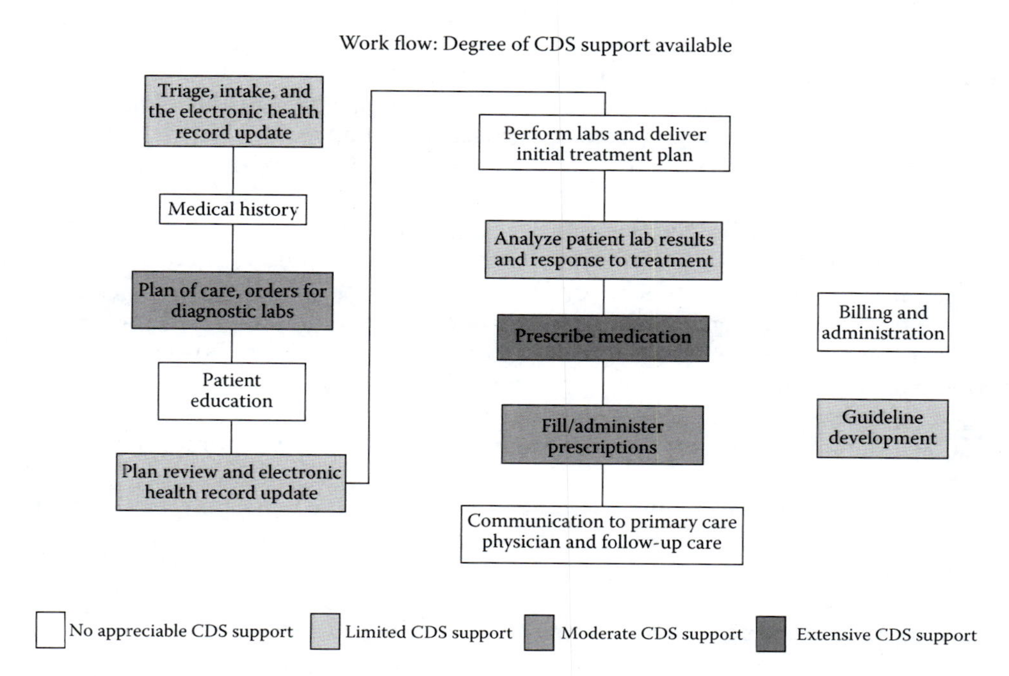

FIGURE 16.2
(See color insert.) Availability of clinical decision support systems by work flow steps.

other parts of the work flow. We hope that through the use of this narrative structure and the experiences of our fictitious Mrs. Melrose, the gaps in CDS usage can be more clearly seen, and that developers and users will both see advantages in filling them.

16.4 Future Directions

We suggest that looking at CDS and EHR systems as partners with other members of the health care team could help both users and developers of these systems. Because of the challenges with integrating old records, designing guidelines that make effective and contextual use of the information available in the EHR, and differences in work flow and knowledge levels across various facilities and care providers, it may be beneficial for us to "humanize" these systems.

A balance needs to be struck between recording structured and unstructured information. Increasing the communication between designers and users directly and through innovative designs, and a recognition of the limitation changes as CDS systems grow, is vital for their future development and utilization. Distinguishing between information that is human-readable-only and that which is also machine-readable is important for both designers and users.

For example, future systems could acknowledge that not all historical records will be translated into machine-readable form. Perhaps, developers need to accept this fact and allow users to scan and attach older patient records explicitly identified as human-readable-only records. The size of these records and utilization of them by the care providers could be treated by the system as evidence of the extent to which the system lacks knowledge of the complete patient history. When care providers realize that such records indicate conflict with system assumptions, they could provide machine-readable data on a case-by-case basis for future use. The same approach might apply to information in various nonintegrated systems. Perhaps linking through the human care providers to other data sources could be a first step toward total information access by CDS systems. To avoid the expectation of magic, or the assumption of system stupidity, it is vitally important for systems designers to explicitly communicate to users what is and is not machine-readable at any given point in the systems life cycle.

Increasing the asynchronous and ongoing nature of the communication between users and the recommendation systems also shows promise. Market saturation of existing handheld or wearable smart devices (e.g., smart phones, tablets, and Google Glasses) could enable a new generation of CDS systems that blend more graciously with the patient care work flow, enabling reminders that are prompt and contextually appropriate for all team members.

Such devices could enable new ways of thinking about the work flow. For example, the on-duty nurse could be reminded of a missing check of the patient's blood pressure within the required time window as she passes or enters the patient's room or, if she does not pass it, with enough time to get there from where she currently is. If she is unavailable, the designer could support patient care by allowing her to send the reminder to another nearby nurse.

The nearby nurse could use the picture or short video of the patient, taken during triage or during the last time someone checked on her, to help him identify the patient. Such identification could help with increased accuracy as well as decreased insurance fraud, since a visual record of who was treated would be available. The nurse helping out could also compare the photos or short video to see if the patient is looking better or worse since the last time she was checked on.

In addition to the professional health care team, patients and family members could be routinely invited into the data gathering process and receive timely reminders to increase adherence to recommendations for improved health. This is already done in some long-term patient care settings [36].

The resulting continuous flow of accurate and complete information on what actions actually occur, how the patient is doing, and the accuracy of current guideline recommendations could help us to distinguish out-of-bounds data that should be used by CDS designers and patient care teams to improve guidelines and reduce the cost of effective health care.

References

1. Berner, Eta S. Clinical decision support systems: State of the art. *AHRQ Publication* No. 09-0069-EF (2009).
2. Wachter, Robert M., Sumant Ranji, Christopher Moriates, Kaveh G. Shojania, and Russ Cucina, AHRQ patient safety network – glossary, accessed January 14, 2012, http://www.psnet.ahrq.gov/glossary.aspx?indexLetter=C.
3. Yong Pierre L. and Leigh Anne Olsen. The healthcare imperative: Lowering costs and improving outcomes: workshop series summary. National Academy Press, Washington, DC, 2010.

4. Berwick, Donald M., Thomas W. Nolan, and John Whittington. The triple aim: Care, health, and cost. *Health Affairs* 27, no. 3 (2008): 759–769.
5. Cresswell, K., M. Ali, A. Avery, N. Barber, T. Cornford, S. Crowe et al. *The Long and Winding Road: An Independent Evaluation of the Implementation and Adoption of the National Health Service Care Records Service (NHS CRS) in Secondary Care in England*. Birmingham: CFHEP (2011).
6. Moore, Mary and Kimberly A. Loper. An introduction to clinical decision support systems. *Journal of Electronic Resources in Medical Libraries* 8, no. 4 (2011): 348–366.
7. Walter, Zhiping and Melissa Succi Lopez. Physician acceptance of information technologies: role of perceived threat to professional autonomy. *Decision Support Systems* 46, no. 1 (2008): 206–215.
8. Wright, Adam, Dean F. Sittig, Joan S. Ash, Joshua Feblowitz, Seth Meltzer, Carmit McMullen, Ken Guappone, Jim Carpenter, Joshua Richardson, Linas Simonaitis, R. Scott Evans, W. Paul Nichol, and Blackford Middleton. Development and evaluation of a comprehensive clinical decision support taxonomy: Comparison of front-end tools in commercial and internally developed electronic health record systems. *Journal of the American Medical Informatics Association* 18, no. 3 (2011): 232–242.
9. Health IT Adoption Rates | Policy Researchers and Implementers | HealthIT.gov, Office of the National Coordinator for Health Information Technology, accessed November 9, 2012, http://www.healthit.gov/policy-researchers-implementers/health-it-adoption-rates.
10. Anokwa, Yaw, Nyoman Ribeka, Tapan Parikh, Gaetano Borriello, and Martin C. Were. Design of a phone-based clinical decision support system for resource-limited settings. In *Proceedings of the Fifth International Conference on Information and Communication Technologies and Development*, pp. 13–24. ACM, 2012.
11. National Coordinator for Health Information Technology, Electronic Health Record Adoption and Utilization: 2012 Highlights and Accomplishments, accessed November 9, 2012, http://www.healthit.gov/sites/default/files/highlights_accomplishments_ehr_adoptionsummer2012_2.pdf.
12. DesRoches, Catherine M., Eric G. Campbell, Sowmya R. Rao, Karen Donelan, Timothy G. Ferris, Ashish Jha, Rainu Kaushal, Douglas E. Levy, Sara Rosenbaum, Alexandra E. Shields, and David Blumenthal. Electronic health records in ambulatory care—a national survey of physicians. *New England Journal of Medicine* 359, no. 1 (2008): 50–60.
13. Hudson, John S., James A. Neff, Miguel A. Padilla, Qi Zhang, and Larry T. Mercer. Predictors of physician use of inpatient electronic health records. *American Journal of Managed Care* 18, no. 4 (2012): 201.
14. Wang, Tiankai and Sue Biedermann. Adoption and utilization of electronic health record systems by long-term care facilities in Texas. *Perspectives in Health Information Management/AHIMA, American Health Information Management Association* no. 9, Spring (2012): 1–4.
15. Ash, Joan S., Dean F. Sittig, Richard Dykstra, Adam Wright, Carmit McMullen, Joshua Richardson, and Blackford Middelton. Identifying best practices for clinical decision support and knowledge management in the field. *Studies in Health Technology and Informatics* 160, no. Pt 2 (2010): 806–810.
16. Metzger, Jane, Emily Welebob, David W. Bates, Stuart Lipsitz, and David C. Classen. Mixed results in the safety performance of computerized physician order entry. *Health Affairs* 29, no. 4 (2010): 655–663.
17. Bright, Tiffani J., Anthony Wong, Ravi Dhurjati, Erin Bristow, Lori Bastian, Remy R. Coeytaux, Gregory Samsa et al. Effect of clinical decision-support systems: A systematic review. *Annals of Internal Medicine* 157, no. 1 (2012): 29–43.
18. D'Avolio, Leonard W., Thien M. Nguyen, Sergey Goryachev, and Louis D. Fiore. Automated concept-level information extraction to reduce the need for custom software and rules development. *Journal of the American Medical Informatics Association* 18, no. 5 (2011): 607–613.
19. Demner-Fushman, Dina, Wendy W. Chapman, and Clement J. McDonald. What can natural language processing do for clinical decision support? *Journal of Biomedical Informatics* 42, no. 5 (2009): 760.

20. Eichelberg, Marco, Thomas Aden, Jörg Riesmeier, Asuman Dogac, and Gokce B. Laleci. Electronic health record standards—a brief overview. In *Information Processing in the Service of Mankind and Health: ITI 4th International Conference on Information and Communications Technology: Information Technology Institute, Giza, Egypt*, pp. 7–19, 2006.
21. Fuji, Kevin T., Kimberly A. Gait, Mark V. Siracuse, and J. Scott Christoffersen. Electronic health record adoption and use by Nebraska pharmacists. *Perspectives in Health Information Management*/AHIMA, American Health Information Management Association no. 8, Summer (2011): 1–11.
22. Sachdeva, Shelly and Subhash Bhalla. Semantic interoperability in standardized electronic health record databases. *Journal of Data and Information Quality (JDIQ)* 3, no. 1 (2012): 1.
23. Ash, Joan S., James L. McCormack, Dean F. Sittig, Adam Wright, Carmit McMullen, and David W. Bates. Standard practices for computerized clinical decision support in community hospitals: a national survey. *Journal of the American Medical Informatics Association* 19, no. 6 (2012): 980–987.
24. Kerollos, Joseph. The management and sustainability of organizational change in primary care adoption of electronic medical record systems. (2012).
25. Bates, David W., Gilad J. Kuperman, Samuel Wang, Tejal Gandhi, Anne Kittler, Lynn Volk, Cynthia Spurr, Ramin Khorasani, Milenko Tanasijevic, and Blackford Middleton. Ten commandments for effective clinical decision support: Making the practice of evidence-based medicine a reality. *Journal of the American Medical Informatics Association* 10, no. 6 (2003): 523–530.
26. Moxey, Annette, Jane Robertson, David Newby, Isla Hains, Margaret Williamson, and Sallie-Anne Pearson. Computerized clinical decision support for prescribing: Provision does not guarantee uptake. *Journal of the American Medical Informatics Association* 17, no. 1 (2010): 25–33.
27. Montani, Stefania. Case-based reasoning for managing noncompliance with clinical guidelines. *Computational Intelligence* 25, no. 3 (2009): 196–213.
28. Trafton, Jodie A., Susana B. Martins, Martha C. Michel, Dan Wang, Samson W. Tu, David J. Clark, Jan Elliott, Brigit Vucic, Steve Balt, Michael E. Clark, Charles D. Sintek, Jack Rosenberg, Denise Daniels, and Mary K. Goldstein. Designing an automated clinical decision support system to match clinical practice guidelines for opioid therapy for chronic pain. *Implementation Science* 5, no. 1 (2010): 1–11.
29. Readmissions Reduction Program, Centers for Medicaid and Medicare Services, accessed November 9, 2012, http://www.cms.gov/Medicare/Medicare-Fee-for-Service-Payment/AcuteInpatientPPS/Readmissions-Reduction-Program.html.
30. Wipfli, Rolf, and Christian Lovis. Alerts in clinical information systems: Building frameworks and prototypes. *Studies in Health Technology and Informatics* 155 (2010): 163.
31. Sijs, Ida Helene van der. *Drug Safety Alerting in Computerized Physician Order Entry: Unraveling and Counteracting Alert Fatigue*. Erasmus University Rotterdam, 2009.
32. Karsh, Ben-Tzion. Clinical practice improvement and redesign: How change in workflow can be supported by clinical decision support. *Agency for Healthcare Research and Quality, US Department of Health and Human Services. Publication* 09-0054 (2009).
33. Heldt, Thomas, Bill Long, George C. Verghese, Peter Szolovits, and Roger G. Mark. Integrating data, models, and reasoning in critical care. In *Engineering in Medicine and Biology Society, 2006. EMBS'06. 28th Annual International Conference of the IEEE*, pp. 350–353, IEEE, 2006.
34. Patterson, Emily S., Michelle L. Rogers, and Marta L. Render. Fifteen best practice recommendations for bar-code medication administration in the Veterans Health Administration. *Joint Commission Journal on Quality and Patient Safety* 30, no. 7 (2004): 355–365.
35. Aleks, Norm, Stuart Russell, Michael G. Madden, Diane Morabito, Kristan Staudenmayer, Mitchell Cohen, and Geoffrey Manley. Probabilistic detection of short events, with application to critical care monitoring, In *Proceedings of NIPS 2008: 22nd Annual Conference on Neural Information Processing Systems*, Vancouver, Canada, December 2008.
36. Chen, Yen-Lin, Hsin-Han Chiang, Chao-Wei Yu, Chuan-Yen Chiang, Chuan-Ming Liu, and Jenq-Haur Wang. An intelligent knowledge-based and customizable home care system framework with ubiquitous patient monitoring and alerting techniques. *Sensors* 12, no. 8 (2012): 11154–11186.

37. Rao, Bharat, Ziv Navoth, and Mel Horwitch. Building a world-class logistics, distribution and electronic commerce infrastructure. *Electronic Markets* 9, no. 3 (1999): 174–180.
38. Bhatnagar, Rohit, and S. Viswanathan. Re-engineering global supply chains: Alliances between manufacturing firms and global logistics services providers. *International Journal of Physical Distribution and Logistics Management* 30, no. 1 (2000): 13–34.
39. Gurwitz, Jerry H., Terry S. Field, Paula Rochon, James Judge, Leslie R. Harrold, Chaim M. Bell, and Monica Lee. Effect of computerized provider order entry with clinical decision support on adverse drug events in the long-term care setting. *Journal of the American Geriatrics Society* 56, no. 12 (2008): 2225–2233.
40. Amarasingham, Ruben, Laura Plantinga, Marie Diener-West, Darrell J. Gaskin, and Neil R. Powe. Clinical information technologies and inpatient outcomes: a multiple hospital study. *Archives of Internal Medicine* 169, no. 2 (2009): 108.
41. Terrell, Kevin M., Anthony J. Perkins, Paul R. Dexter, Siu L. Hui, Christopher M. Callahan, and Douglas K. Miller. Computerized decision support to reduce potentially inappropriate prescribing to older emergency department patients: a randomized, controlled trial. *Journal of the American Geriatrics Society* 57, no. 8 (2009): 1388–1394.
42. Crane, Jacquelyn and Frederick G. Crane. Preventing medication errors in hospitals through a systems approach and technological innovation: A prescription for 2010. *Hospital Topics* 84, no. 4 (2006): 3–8.
43. Kahn, James S., Veenu Aulakh, and Adam Bosworth. What it takes: Characteristics of the ideal personal health record. *Health Affairs* 28, no. 2 (2009): 369.

17

Fuzzy Naïve Bayesian Approach for Medical Decision Support

Kavishwar B. Wagholikar and Ashok W. Deshpande

CONTENTS

17.1 Introduction

In the medical domain, a vast amount of knowledge is required even to solve seemingly simple problems. Health care providers are required to remember and apply knowledge of a vast array of clinical disease presentations, diagnostic criteria, and drug therapies. However, the providers' mental abilities are limited [1], making it impossible for them to provide the best possible/optimal care. For instance, the information a general practitioner in the UK is required to know forms a 68 cm pile, and the practitioner is required to recall and apply this information within the few minutes of the patient consultation [2]. These limitations result in an estimated 30% of morbidity and mortality, which is avoidable [3].

Realization of this problem has led to attempts to computationally discover the ideal medical decision-making algorithm. The potential of computer-based tools to address the medical decision-making problems was realized half a century ago, and several algorithms have been developed to construct medical decision support systems (MDSs) for a variety of specialties. MDSs have been defined as "active knowledge systems which use two or more items of patient data to generate case-specific advice." Figure 17.1 shows the logical components of the diagnostic process, as outlined by Ledley and Lusted [4]. Accordingly,

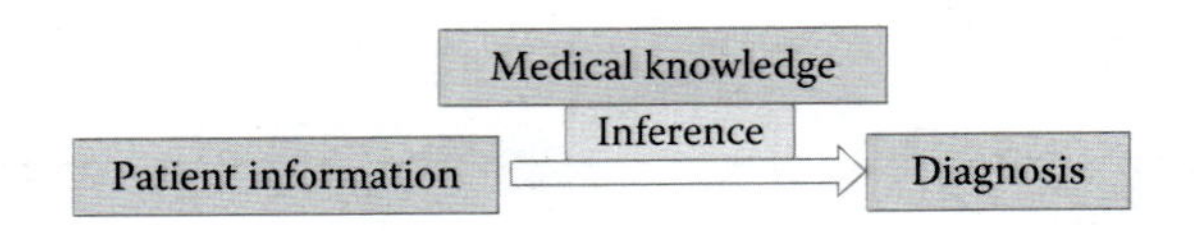

FIGURE 17.1
Logical components of the process of medical diagnosis.

MDSs typically have two components—medical knowledge base and an inference engine, to generate advice specific to querying patient cases. Medical knowledge is composed of symptom–disease, intersymptom, and interdisease relationships. Inference is the process of combining medical knowledge with patient data to decide on the diagnosis.

Such systems may be referred to as expert systems, in the sense that they have sophisticated inference abilities, though the original use of the term referred to the ability of systems to capture expert knowledge. The modern approach is to support the decision making, rather than to supplant it, which was the aim initially. The motivation now is to free the care providers from routine work and allow them to focus on important tasks.

17.2 Overview of Modeling Techniques

A wide variety of modeling techniques have been developed for medical decision making [5]. A brief overview is as follows.

The naïve Bayes (NB) model was the basis of the earliest MDS systems. This model makes assumptions of conditional independence between symptoms and mutual exclusivity of diseases. One of the earliest studies with this approach was by Homer Warner at the University of Utah [6] for diagnosing congenital heart disease. deDombal's application of this approach is especially notable [7]. DXplain is a current system based on this approach [8], and several modifications of NB have also been used [9]. Although NB systems were found to perform as well as, and at times better than, experts, NB did not receive wider application. The assumptions of NB were considered as oversimplistic, and the quantitative approach of NB was considered to be inappropriate representation of human reasoning, which seems to be qualitative.

Around that time, artificial intelligence (AI) methods emerged. These included rule-based and heuristic systems. MYCIN was a rule-based system for prescribing antimicrobial therapy that used the if–then form of rules with certainty factors (Figure 17.2). EXPERT [10] and Psyxpert [11] are some of the notable examples. Heuristics systems prominently used ad hoc scoring schemes for inference. For instance, INTERNIST and Quick Medical Reference (QMR) [12] were systems developed with a concerted effort, and they had a wide

IF the morphology of the organism is rod
AND the Gram stain of the organism if negative
AND the site of the culture is blood
THEN there is suggestive evidence (0.6) that the organism is Pseudomonas

FIGURE 17.2
Example of a rule in MYCIN.

domain of application. AI systems appeared to provide alternative solutions to bypass the disadvantages with the NB approach. For instance, the confidence factor [13] used in MYCIN and the QMR inference model used heuristic methods for inference [14]. The focus of AI methods was on representing the expert's knowledge. A major disadvantage of rule-based systems is that they are increasingly difficult to manage when the number of rules increases. However, the advantage is that the decisions computed by rule-based systems are easy to explain. Rule-based systems have gained popularity in the last few decades due to the development of clinical guidelines, which embed expert knowledge of good clinical practices [15].

The parameters for the AI systems are more likely to be erroneous as compared to the NB model. This is because the assessment of symptom–disease relationships for these systems required the care providers to provide judgments in terms of likelihood of a disease given a finding in a direction (from symptom to disease) that is considered cognitively challenging. This is in contrast to the Bayes rule, which involves the estimation of probabilities in terms of likelihood of a symptom given a disease, which is natural to humans.

These developments led researchers in the last two decades to revert to the principled probability-based approach for inference. But now, they began coupling it with expressive schemes for knowledge representation, which led to the development of Bayesian networks (BN). Other methods like support vector machines (SVMs), artificial neural networks (ANNs), and information theory were concurrently developed (Figure 17.3). However, despite the development of these new complex models [16], no significant improvement in accuracy has been achieved, and many comparative studies have evaluated NB as near optimal [17,18]. NB is still the most widely researched approach. Consequently it is considered by many researchers as the reference standard, and it used as a benchmark for comparing the performance of any new algorithm.

Fuzzy set theory (FST)-based methods were also applied to MDS beginning in the 1980s. They were developed in parallel to the AI methods, with which they shared many similarities. We describe the historical developments of FST applications later in the chapter.

Overall, a wide choice of modeling techniques is available for developers of MDS tools. However, there is a dearth of studies that compare the performance of different techniques [19]. Each method has its advantages and disadvantages. In the following section, we examine the advantages of FST.

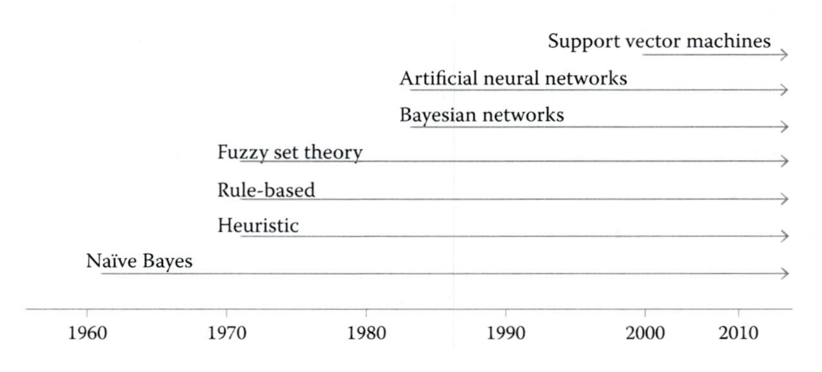

FIGURE 17.3
Approximate timeline of evolution of medical decision support models.

17.3 Why FST is Useful

The medical domain is characterized by use of linguistic terms that are inherently vague. Vagueness or ambiguity is sometimes described as "second-order uncertainty," where apart from the uncertainty of a particular outcome, there is uncertainty even about the definition of the outcome. Use of FST has been shown to provide improved characterization of vague medical concepts and patient information, which can help reduce the epistemic uncertainty in medical decision making.

17.4 Conventional Naïve Bayes

In this chapter, we describe a method to use FST to model medical knowledge as well as the inference process. Ledley and Lusted [20] described the application of the Bayes formula to estimate the probability of a diagnosis. This approach is based on the assumption that the findings/symptoms for a particular disease are not interdependent. For example, the probability that a patient with symptoms of cough and fever has a diagnosis of pneumonia is computed as

$$P(pnuemonia \mid cough \cap fever) = \frac{P(pnuemonia) \times P(cough \mid pnuemonia) \times P(fever \mid pnuemonia)}{P(cough \cap fever)}$$

where *P(cough|pnuemonia)* indicates the probability of cough given the diagnosis of pneumonia. The assumption here is that there is no dependence between fever and cough.

A more general form of the model is as follows. Given a set of symptoms *S* and diseases *D*, using a set of patient data, marginal probabilities of symptoms $P(s_i)$ and diseases $P(d_j)$ and conditional probabilities of symptoms on all diseases $P(s_i \mid d_j)$ are calculated by counting frequencies in the data. The posterior probability for each diagnosis for the patient is calculated as

$$P(d_j \mid S) = P(d_j) \prod_{s_i \in S} \frac{P(s_i \mid d_j)}{P(s_i)}$$

Since the denominator $\prod_{s_i \in S} P(s_i)$ is common in the computation of posterior probabilities for all diagnoses, it is dropped, and a diagnostic score is computed for each diagnosis as $P(d_j) \prod P(s_i \mid d_j)$.

Conditional probability of symptom s_i for disease d_j is

$$P(s_i \mid d_j) = \frac{f(s_i \cap d_j)}{f(d_j)}$$

where $f(d_j)$ is the number of patients in the data set with disease d_j and $f(s_i \cap d_j)$ is the frequency count of patients with both s_i and d_j. Symptoms that are not reported in a particular disease have zero conditional probability for the disease. While calculating the diagnostic scores, the zero probabilities wipe out the information from other symptoms, and hence, to avoid this problem, their zero conditional probabilities are corrected to $0.5/m$, where m is the total number of patients [21]. Differential diagnosis is output by ranking diagnoses in descending order of their corresponding computed diagnostic score and excluding the diagnoses below a cutoff rank.

17.5 Fuzzy Naïve Bayes

However, in real life, the information about the presence of a particular symptom in a patient is rarely certain. Such information can be said to be fuzzy, and the particular piece of symptom information can be referred to as fuzzy symptom description or simply fuzzy symptom value. In contrast, crisp symptom values are those that are certain. Each fuzzy symptom value (indicated by underscore) is defined as a fuzzy set on the set of classical/crisp symptom values. The membership value of crisp symptom s_i in fuzzy symptom $\underline{s_k}$ is obtained by interviewing physicians with the following question: "What is your degree of belief in $\underline{s_k}$ when a patient asserts s_i?" The membership value is represented as $\mu_{\underline{s_k}} s_i$.

For instance, fever can be described as absent (no), as present (yes), or in terms of its grades—low and high. When symptom descriptions are directly recorded from the patient, without an elaborate examination by the physician to establish the symptom, their values/grades are likely to be incorrect due to loose interpretations of the used vocabulary. The uncertainty resident in such information is vagueness or fuzziness [22], which is modeled by defining fuzzy sets for fuzzy descriptions of fever on the crisp descriptions of fever. For instance, $\mu_{\underline{low-fever}} high-fever$ is the degree of confidence that a patient has "low fever" when the patient reports "high fever." The fuzzy memberships values need to be normalized across all crisp symptoms that have a nonzero membership in the fuzzy symptom using the following operation: $\frac{\mu_{\underline{s_k}} s_i}{\sum \mu_{\underline{s_k}} s_j} = \mu'_{\underline{s_k}} s_i$ gives the orthogonalized membership function where s_j is a nonzero member of $\underline{s_k}$. This operation is referred to as orthogonalization, as the sum of the memberships of all crisp symptoms in the fuzzy symptom now equals 1.

The marginal probabilities of crisp symptoms $P(s_i)$ and diseases $P(d_j)$ and the conditional probabilities of crisp symptoms on all diseases $P(s_i \mid d_j)$ are calculated as in the conventional NB method. The training data used for this purpose are required to have accurate symptom information. Conditional probabilities of fuzzy symptoms $\underline{s_k}$ for a particular diagnosis $P(d_j)$ is calculated as

$$P\left(\underline{s_k} \middle| d_j\right) = \sum_i P(s_i \mid d_j) \mu'_{\underline{s_k}} s_i$$

When the information given for a patient test case is fuzzy, $\underline{S} \equiv \{\underline{s_k}\}$, the posterior probability for each diagnosis for the case is calculated using

$$P(d_j \mid \underline{S}) = P(d_j) \prod_{\underline{s_k} \in \underline{S}} \frac{P(\underline{s_k} | d_j)}{P(\underline{s_k})}$$

Since denominator $\prod_{\underline{s_k} \in \underline{S}} P(\underline{s_k})$ is common for all diagnoses, it is dropped and a diagnostic score is computed for each diagnosis as $P(d_j) \prod_{\underline{s_k} \in \underline{S}} P(\underline{s_k} | d_j)$. As in the conventional NB, differential diagnosis is output by ranking diagnoses, in descending order of their computed diagnostic score and using a rank cutoff.

Fuzzy naïve Bayes (FNB) [22] is an extension of the fuzzy Bayesian approach [23,24]. A disadvantage of this method is the labor required to obtain the fuzzy membership values from the experts. An alternative approach to compute the membership functions is to use genetic algorithms to automatically discover the optimal membership functions from a data set [25].

FNB has been found to be optimal over NB for diagnosing patients with imprecise fuzzy information. The advantages of FNB are that (1) it can model the information that values of some attributes are semantically closer than values of other attributes, and (2) it offers a mechanism to temper exaggerations in patient information. Precise data is preferable for training the algorithm, but a large training set of fuzzy data may be used, as the noisy individual cases in the fuzzy data will cancel out each others mislabeled attributes.

17.6 Sources of Fuzziness in Patient Data

Fuzziness in patient data may be contributed by several factors, in different stages of translation of disease signals into data. Signals indicating disease originate from alteration in bodily processes at one or more levels (cellular, organelle, organ, organ system). These stimulate sensory receptors for modalities like pain, taste, pressure, and temperature. Receptor stimulation leads to perception, followed by encoding and transmission.

We examine the factors that could contribute to fuzziness in these stages:

- **Perception:** The receptor stimulation leads to the formation of a perception of the changed process. However, as the distribution and structure of the receptors vary from one individual to another, the same stimulus can lead to varied degrees of perceptions in different individuals. The variation in receptors for pain and heat [26] and taste and hearing is especially well known. Also, other neuroanatomical structures are known to have a role in individual variation of emotional response to adverse stimuli [27].
- **Encoding:** The expression and perhaps even the formation of the perception itself both depend on the vocabulary of the patient. The patient encodes the perception

in terms of the words known to him/her. For instance, an infant has very limited ability to express his/her problems. Moreover, the definition of the words varies from person to person. For instance, the epigastric region may be referred to by different people as stomach, abdomen, or chest.

- **Transmission:** Transmission of the encoded information to the physician is modulated by the personality factors of the patient. For instance, the pain represented by an apprehensive housewife's expression of "very severe pain in leg" is likely to be referred to as "some pain in leg" by a soldier. Many patients are known to give exaggerated descriptions of their symptoms [28].

17.7 Other Fuzzy Set Theoretic Approaches

The FNB approach attempts to model noise in the patient information but uses probability to model medical knowledge and uses Bayes rule for inference. FST-based methods can also be used to model medical knowledge [especially using fuzzy relations (FRs)] as well as the process of inference from knowledge (compositional rule). The earliest applicability of FST to medicine is ascribed to Zadeh [29]. Early applications were largely based on FRs. However, in the last decade, fuzzy rule-based systems have been popular, and several other FST-based approaches have been developed, as described below.

17.7.1 FRs

Sanchez modeled medical knowledge as FRs between symptoms and diseases [30] and used the compositional rule for inference [31]. Sanchez's model was extended to develop CADIAG systems [32]. The CADIAG knowledge base was constructed in the form of FRs acquired from physicians, from medical literature, and semiautomatically from patient records (Figure 17.4). CADIAG captured uncertainty in information about a patient's symptoms and medical knowledge as binary FRs, and it used a heuristic algorithm for diagnostic inference. These systems were evaluated to have high accuracy for rheumatic and pancreatic diseases. Ciabattoni et al. [33] have defined a formal logical calculus to perform a consistency check of the CADIAG knowledge base. FRs have also been applied for diagnosis of cardiopathy and arteritis. Recently, Wagholikar et al. [34] developed and evaluated an FR algorithm on a variety of medical data sets and found the algorithm to be particularly useful for noisy and incomplete data sets.

'Probability of finding fever in patient having malaria is 0.8' translates to 'Fuzzy occurrence relation of fever with malaria is 0.8'

'Probability of diagnosing malaria in patient who has fever is 0.2' translates to 'Fuzzy confirmatory relation of fever with malaria is 0.2'

FIGURE 17.4
Translation of probabilistic knowledge to fuzzy relations in CADIAG.

If temperature is 'high' and leukocyte count is 'high'
then confidence for viral infection is 0.25

FIGURE 17.5
Fuzzy rule that models linguistic input variables.

17.7.2 Fuzzy Rules

In these systems, the knowledge base is generally derived as rules by inducing decision trees from data (Figure 17.5). The crisp rules are then fuzzified by computing fuzzy sets. Such systems have been extensively applied for a variety of conditions [35,36].

17.7.3 Other Approaches

Neuro-fuzzy systems combine FST with the theory of ANN. The concept of a fuzzy neuron was proposed by Kuncheva and was extended by Barreto and de Azevedo [37]. The neuro-fuzzy approach has been applied for diagnosing erythemato-squamous diseases, heart disease, and hemorrhage. Milord is an approximate reasoning-based system for rheumatology (Renoir) and pneumonia (Pneumon-IA) [38]. Other FST applications include the fuzzy-genetic approach [39], fuzzy classification method [40], fuzzy cognitive maps for pathology [41], fuzzy similarity classifier [42], intutionistic fuzzy sets [43], and fuzzy influence diagrams [44]. Overall, FST offers a wide variety of formalisms for modeling medical decision making and is a fertile ground for MDS research.

17.8 Directions for Future Work

FST facilitates the representation of uncertainties in medical knowledge as well as patient data. Furthermore, it offers principled inference methods to model the medical reasoning process. The increasing availability of medical information in the electronic form offers great opportunity to investigate the utility of FST. However, there is a lack of research comparing FST with other techniques, which has been a major limiting factor in educating the larger informatics community of the advantages offered by FST. Such comparative studies are necessary.

FST provides the power to represent medical information at a higher level of granularity. However, as the models become increasingly complex, it is useful to combine them with evolutionary algorithms to discover optimal models. The availability of cheap computational power will facilitate such approaches.

FST can be especially useful in modeling linguistic terms that indicate temporal aspects of events in the medical data. Fuzzy ontologies represent a promising area of research for medical applications.

References

1. G. A. Miller, The magical number seven plus or minus two: some limits on our capacity for processing information, *Psychological Review*, vol. 63, pp. 81–97, Mar 1956.

2. A. Hibble, D. Kanka, D. Pencheon, and F. Pooles, Guidelines in general practice: the new Tower of Babel?, *British Medical Journal*, vol. 317, pp. 862–3, Sep 26 1998.
3. M. L. Graber, N. Franklin, and R. Gordon, Diagnostic error in internal medicine, *Archives of Internal Medicine*, vol. 165, pp. 1493–9, Jul 11 2005.
4. R. S. Ledley and L. B. Lusted, Probability, logic and medical diagnosis, *Science*, vol. 130, pp. 892–930, Oct 9 1959.
5. K. B. Wagholikar, V. Sundararajan, and A. W. Deshpande, Modeling paradigms for medical diagnostic decision support: a survey and future directions, *Journal of Medical Systems*, vol. 36, pp. 3029–49, Oct 2012.
6. H. R. Warner, A. F. Toronto, L. G. Veasey, and R. Stephenson, A mathematical approach to medical diagnosis. Application to congenital heart disease, *JAMA: The Journal of the American Medical Association*, vol. 177, pp. 177–83, Jul 22 1961.
7. F. T. deDombal, Computer-aided diagnosis and decision-making in the acute abdomen, *Journal of the Royal College of Physicians of London*, vol. 9, pp. 211–18, Apr 1975.
8. G. O. Barnett, E. P. Hoffer, M. J. Feldman, R. J. Kim, and K. T. Famiglietti, Senior member presentation proposal. DXplain—20 years later—what have we learned?, In *AMIA... Annual Symposium Proceedings/AMIA Symposium. AMIA Symposium*, Washington, DC, pp. 1201–2, 2008.
9. M. Ben-Bassat, R. W. Carlson, V. K. Puri, M. D. Davenport, J. A. Schriver, M. Latif, R. Smith, L. D. Portigal, E. H. Lipnick, and M. H. Weil, Pattern-based interactive diagnosis of multiple disorders: the MEDAS system, *IEEE Transactions on Pattern Analysis and Machine Intelligence*, vol. 2, pp. 148–60, Feb 1980.
10. R. N. Goldberg and S. M. Weiss, An experimental transformation of a large expert knowledge base, *Journal of Medical Systems*, vol. 6, pp. 41–52, Feb 1982.
11. M. A. Overby, Psyxpert: an expert system prototype for aiding psychiatrists in the diagnosis of psychotic disorders, *Computers in Biology and Medicine*, vol. 17, pp. 383–93, 1987.
12. R. A. Miller, M. A. McNeil, S. M. Challinor, F. E. Masarie, Jr., and J. D. Myers, The INTERNIST-1/QUICK MEDICAL REFERENCE project—status report, *The Western Journal of Medicine*, vol. 145, pp. 816–22, Dec 1986.
13. E. H. Shortliffe, S. G. Axline, B. G. Buchanan, T. C. Merigan, and S. N. Cohen, An artificial intelligence program to advise physicians regarding antimicrobial therapy, *Computers and Biomedical Research*, vol. 6, pp. 544–60, 1973.
14. B. N. Grosof, Evidential confirmation as transformed probability, in *First International Workshop on Uncertainty in Artificial Intelligence*, Amsterdam, Netherlands, 1986.
15. K. B. Wagholikar, K. L. MacLaughlin, M. R. Henry, R. A. Greenes, R. A. Hankey, H. Liu, and R. Chaudhry, Clinical decision support with automated text processing for cervical cancer screening, *Journal of the American Medical Informatics Association: JAMIA*, vol. 19, pp. 833–9, Sep–Oct 2012.
16. G. Schwarzer, W. Vach, and M. Schumacher, On the misuses of artificial neural networks for prognostic and diagnostic classification in oncology, *Statistics in Medicine*, vol. 19, pp. 541–61, Feb 29, 2000.
17. J. L. Liu, J. C. Wyatt, J. J. Deeks, S. Clamp, J. Keen, P. Verde, C. Ohmann, J. Wellwood, M. Dawes, and D. G. Altman, Systematic reviews of clinical decision tools for acute abdominal pain, *Health Technology Assessment*, vol. 10, pp. 1–167, iii–iv, Nov 2006.
18. B. S. Todd and R. Stamper, The relative accuracy of a variety of medical diagnostic programs, *Methods of Information in Medicine*, vol. 33, pp. 402–16, Oct 1994.
19. E. S. Berner, G. D. Webster, A. A. Shugerman, J. R. Jackson, J. Algina, A. L. Baker, E. V. Ball, C. G. Cobbs, V. W. Dennis, E. P. Frenkel, L. D. Hudson, E. L. Mancall, C. E. Rackley, and O. D. Taunton, Performance of four computer-based diagnostic systems, *The New England Journal of Medicine*, vol. 330, pp. 1792–6, Jun 23, 1994.
20. R. S. Ledley and L. B. Lusted, Reasoning foundations of medical diagnosis; symbolic logic, probability, and value theory aid our understanding of how physicians reason, *Science*, vol. 130, pp. 9–21, Jul 3, 1959.

21. R. Kohavi, B. Becker, and D. Sommerfield, Improving simple Bayes, presented at the *European Conference on Machine Learning*, Prague, Czech Republic, 1997.
22. K. B. Wagholikar, S. Vijayraghavan, and A. W. Deshpande, Fuzzy Naive Bayesian model for medical diagnostic decision support, *IEEE Engineering in Medicine and Biology Society*, vol. 2009, pp. 3409–12, 2009.
23. T. Okuda, H. Tanaka, and K. Asai, A formulation of fuzzy decision problems with fuzzy information using probability measures of fuzzy events, *Information and Control*, vol. 38, pp. 135–47, 1978.
24. G. Klir and B. Yuan, *Fuzzy Sets and Fuzzy Logic: Theory and Applications*, Upper Saddle, NJ: Prentice Hall, 1995.
25. K. B. Wagholikar, V. Sundararajan, and A. W. Deshpande, GA-Fuzzy Naive Bayes method for medical decision support, presented at the *World Conference on Soft Computing*, San Francisco CA, 2011.
26. H. Kim, D. P. Mittal, M. J. Iadarola, and R. A. Dionne, Genetic predictors for acute experimental cold and heat pain sensitivity in humans, *Journal of Medical Genetics*, vol. 43, p. e40, Aug 2006.
27. D. J. Scott, M. M. Heitzeg, R. A. Koeppe, C. S. Stohler, and J. K. Zubieta, Variations in the human pain stress experience mediated by ventral and dorsal basal ganglia dopamine activity, *The Journal of Neuroscience: The Official Journal of the Society for Neuroscience*, vol. 26, pp. 10789–95, Oct 18 2006.
28. L. J. Kirmayer, J. M. Robbins, and J. Paris, Somatoform disorders: personality and the social matrix of somatic distress, *Journal of Abnormal Psychology*, vol. 103, pp. 125–36, Feb 1994.
29. L. A. Zadeh, Biological applications of the theory of fuzzy sets and systems, in *Proceedings of the International Symposium on Biocybernetics of the Central Nervous System*. Boston: Little, Brown & Co., pp. 199–212, 1969.
30. E. Sanchez, Medical diagnosis and composite fuzzy relations, in *Mediinfo*, Amsterdam: North-Holland, 1979, pp. 437–44.
31. L. A. Zadeh, Ed., *Towards a Theory of Fuzzy Systems* (Aspects of Networks and Systems Theory), New York: Holt, Rinehart and Winston, 1971.
32. H. Leitich, H. P. Kiener, G. Kolarz, C. Schuh, W. Graninger, and K. P. Adlassnig, A prospective evaluation of the medical consultation system CADIAG-II/RHEUMA in a rheumatological outpatient clinic, *Methods of Information in Medicine*, vol. 40, pp. 213–20, Jul 2001.
33. A. Ciabattoni, T. Vetterlein, and K. P. Adlassnig, A formal logical framework for Cadiag-2, *Studies in Health Technology and Informatics*, vol. 150, pp. 648–52, 2009.
34. K. Wagholikar, S. Mangrulkar, A. Deshpande, and V. Sundararajan, Evaluation of fuzzy relation method for medical decision support, *Journal of Medical Systems*, vol. 36, pp. 233–9, Feb 2012.
35. T. M. Akbarzadeh and M. Moshtagh-Khorasani, A hierarchical fuzzy rule-based approach to aphasia diagnosis, *Journal of Biomedical Informatics*, vol. 40, pp. 465–75, Oct 2007.
36. M. G. Tsipouras, T. P. Exarchos, D. I. Fotiadis, A. P. Kotsia, K. V. Vakalis, K. K. Naka, and L. K. Michalis, Automated diagnosis of coronary artery disease based on data mining and fuzzy modeling, *IEEE Transactions on Information Technology in Biomedicine: A Publication of the IEEE Engineering in Medicine and Biology Society*, vol. 12, pp. 447–58, Jul 2008.
37. J. M. Barreto and F. M. de Azevedo, Connectionist expert systems as medical decision aid, *Artificial Intelligence in Medicine*, vol. 5, pp. 515–23, Dec 1993.
38. L. Godo, R. L. de Mantaras, J. Puyol-Gruart, and C. Sierra, Renoir, Pneumon-IA and Terap-IA: three medical applications based on fuzzy logic, *Artificial Intelligence in Medicine*, vol. 21, pp. 153–62, Jan–Mar 2001.
39. C. A. Pena-Reyes and M. Sipper, A fuzzy-genetic approach to breast cancer diagnosis, *Artificial Intelligence in Medicine*, vol. 17, pp. 131–55, Oct 1999.
40. R. Jain and A. Abraham, A comparative study of fuzzy classification methods on breast cancer data, *Australasian Physical and Engineering Sciences in Medicine/Supported by the Australasian College of Physical Scientists in Medicine and the Australasian Association of Physical Sciences in Medicine*, vol. 27, pp. 213–18, Dec 2004.

41. V. C. Georgopoulos and C. D. Stylios, Diagnosis support using fuzzy cognitive maps combined with genetic algorithms, In *Conference proceedings:... Annual International Conference of the IEEE Engineering in Medicine and Biology Society. IEEE Engineering in Medicine and Biology Society Conference*, vol. 2009, pp. 6226–9, 2009.
42. P. Luukka and T. Leppalampi, Similarity classifier with generalized mean applied to medical data, *Computers in Biology and Medicine*, vol. 36, pp. 1026–40, Sep 2006.
43. V. Khatibi and G. A. Montazer, Intuitionistic fuzzy set vs. fuzzy set application in medical pattern recognition, *Artificial Intelligence in Medicine*, vol. 47, pp. 43–52, Sep 2009.
44. H. Y. Kao, Diagnostic reasoning and medical decision-making with fuzzy influence diagrams, *Computer Methods and Programs in Biomedicine*, vol. 90, pp. 9–16, Apr 2008.

18

Approaches for Establishing Methodologies in Metabolomic Studies for Clinical Diagnostics

Daniel J. Peirano, Alexander A. Aksenov, Alberto Pasamontes, and Cristina E. Davis

CONTENTS

18.1 Introduction

Modern analytical chemistry instrument development and new ideas for methods of use for these machines have flourished over the last decades, producing many different sophisticated instrument platforms capable of providing diagnostic information for modern medicine. In particular, mass spectrometry (MS) instruments have the capability to monitor many different chemicals in human effluents, and this information can sometimes be correlated with a clinical disease or disorder diagnosis. The concepts of "metabolomics" and "proteomics" both use spectrometry-like instruments to characterize processes ongoing within the human body, with the final purpose of potentially providing a new clinical tool to aid patient care and disease diagnosis and management.

Typically, MS (and related) instruments produce complex data sets, providing a unique challenge for data analysis methods. While single-compound identification is sometimes appropriate in spectrometry data sets, often, more than one chemical is important to a clinical disease state. This is where artificial intelligence (AI) applications can provide tools needed for feature selection and classification. An emerging field of "chemometrics" attempts to understand this complex chemical information and has made heavy use of advanced algorithms, including many familiar to the AI community [1]. Here we review some of the basic information from the field and provide an update on some of the more recent significant advances.

Figure 18.1 shows the typical work flow in a metabolomics study, from the study design to the AI implementation of the data analysis and interpretation. First, we start by clearly outlining the clinical problem to be solved, such as diagnosis of a disease or specific disorder. This is important to determine which categories or types of biomarkers are likely to be present and the best instrumentation methods to utilize. For example, a study of lung cancer biomarkers in breath might focus on volatile organic compounds (VOCs) as the primary target of investigation, and therefore, researchers may select gas chromatography/mass spectrometry (GC/MS) as an analytical technique. The chemical information is essential to guide the choice of the device that will be used to analyze the samples. Once the human clinical samples are analyzed, preprocessing the data is needed prior to implementing any AI methods. Preprocessing combines a set of computational tools that reduces the complexity of the original data and improves data quality by applying noise and baseline removal and/or aligning the data files. In GC, for example, sample alignment is necessary due to temporal shifts of retention time (RT) values in the time series. Once that step is completed, deconvolution is necessary to resolve peaks of co-eluting compounds, a common occurrence in complex clinical samples. Finally, AI methodologies such as artificial neural networks (ANNs), genetic algorithms (GAs), or self-organizing maps (SOMs) are applied in order to build a nonlinear prediction model of clinical outcomes. These

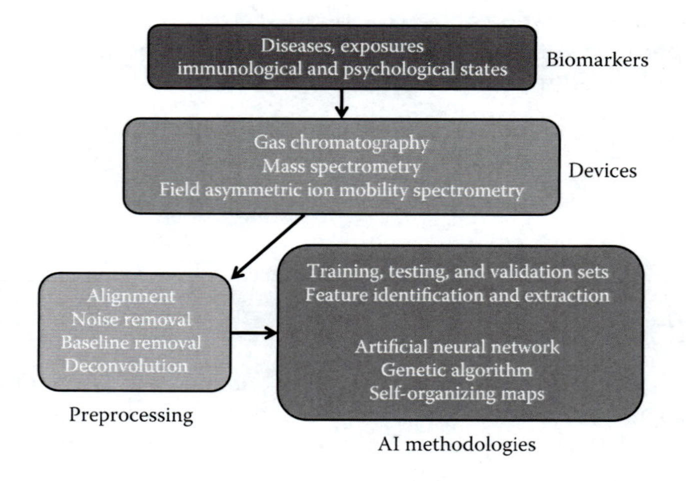

FIGURE 18.1
A typical work flow required to provide a clinical diagnosis from a metabolite monitoring perspective. The "Biomarkers" box lists examples of different biological states that may be manifested by changes in chemical composition of the system. The "Devices" box contains examples of analytical instrumentation that are commonly used in biomedical research. The "Preprocessing" box lists the preprocessing procedures that may be required before data analysis. The "AI methodologies" box lists a few examples of AI methods considered in the present chapter.

developed models allow us to discriminate samples into separate groups (i.e., healthy controls or diseased). Finally, the model is validated in order to confirm that it works and would function correctly as a medical diagnostic tool.

18.2 Medical Applications of Analytical Chemistry Sensors and Instruments

Metabolites are found in all biological systems and are thought to be the end-stage result of chemical processes going on within the organism. These small molecules can provide a tremendous amount of information related to both normal and disease processes. The study of these metabolites is generally termed "metabolomics" and frequently addresses both common and uncommon diseases [2–5]. Other applications are wide ranging and include examining human responses to drug therapies [6], discovering new cancer biomarkers [7], and identifying exposure to radiation [8] and environmental contaminants as well [9]. Given that there are thousands of small metabolites in biological samples, it is not trivial to determine which biomarkers are of interest and which are not. Thus, the last decade has seen an explosion of data analysis methods applied to this topic [10–13], and these methods will be explored later in this chapter after introducing the instrumentation platforms on which the clinical data were obtained.

18.3 Introduction to Analytical Instrumentation for Clinical Applications

There are many different sensors and instrumentation platforms available for use in medical diagnostics. Some of these first evolved as single instruments and subsequently were coupled together into so-called "hyphenated" platforms for advanced chemical discrimination and identification. These hyphenated techniques are powerful discovery tools for complex biological mixtures and are required to measure the many metabolites in human effluents. To fully understand how to mine these data sets for biomarker data using AI methods, it is necessary to appreciate how the instruments function.

18.3.1 Gas Chromatography (GC)

GC emerged as a new analytical tool in the 1950s, when Martin and James first introduced the method to the scientific field [14]. In September 1952, Martin spoke at the First International Congress on Analytical Chemistry on this topic, and a few weeks later, he and his colleague Synge were awarded the Nobel Prize in Chemistry for the first paper written on the topic [15]. The award stated that it was "for the invention of partition chromatography" [16].

GC is a straightforward method to understand. A gas stream (often referred to as a carrier gas) flows through a heated tube, usually fused silica, which is sub-millimeter-diameter size. This tube may range in size to as long as tens of meters and is usually wound into a compact "column," allowing it to be manipulated easily in the laboratory. An analyte of interest is injected into the stream and flows through the tube. If the tube and gas are at a temperature higher than the vaporization temperature of the analyte, then the analyte will diffuse through the carrier gas and will emerge from the far end of the tube based on the rate of speed of the carrier gas, unaffected by the makeup of the injected analyte [17].

Inside the GC column is a nonvolatile chemical coating. This coating is typically referred to as a "stationary phase" and has unique chemical properties of its own. It interacts with certain analytes, slowing down their advancement along the column. If the temperature is low, the analytes may be retained by the nonvolatile chemical coating. The temperature is raised in a predetermined manner, and the analytes break off from the stationary phase. The rising temperature creates a stronger pull on the analytes to gas phase than the chemical coating to liquid phase. This temperature is consistent for each analyte but can vary greatly between different types of analytes [17]. If the GC column is kept at a constant temperature, the volatile compounds elute very quickly, while less volatile compounds elute later and appear as wider peaks in the signal due to greater diffusion. Analyte mixtures can be run at different gradients of increasing temperature to ensure proper identification, and this temporal resolution of compounds as they elute off the column is one of the physicochemical properties that provides partial compound identification [18].

If a sample has multiple compounds flowing through the tube, they will interact with the chemical coating at low temperatures. As the temperature rises, separate chemicals desorb (change phase) at different temperatures and, therefore, different times. The carrier gas will sweep the released analytes to the end of the tube, where a detector can identify if the gas mixture changes from the base carrier gas, indicating that an analyte is flowing through. If a mixture change is identified, the time is recorded, known as the retention time (RT), which can be compared against other RTs under similar analytical conditions (specifically, the chosen type of nonvolatile chemical coating and detector sensitivity) to identify the makeup of the initial sample [17]. This concept is highlighted in Figure 18.2.

RT as a solitary variable can be effective at discerning analytes, especially if certain sets of chemicals are expected. However, as the range of analytes to be accurately identified increases, the possibility of misidentification through overlapping RTs also increases. A second parameter is necessary to differentiate between chemicals in a large possible set or to determine if the analyte is an unknown biomarker, of which there is no previous identification.

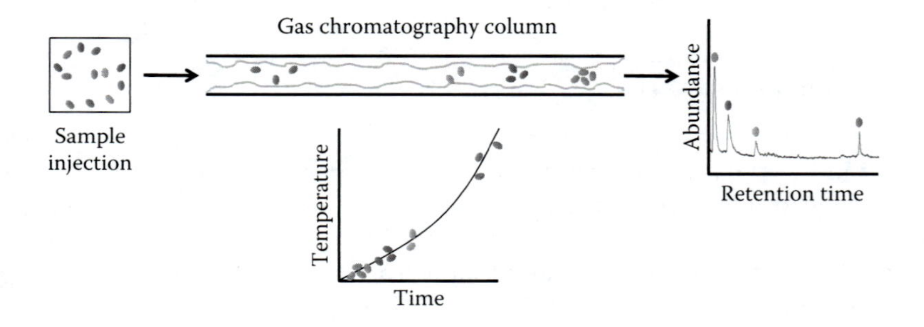

FIGURE 18.2
A schematic representation of gas chromatography (GC) separation. A sample containing mixture of compounds is introduced into a GC column. The flow of analytes through the column with the carrier gas is slowed down by interaction of molecules with the stationary phase. Increasing the temperature accelerates analyte elution. The compounds elute in the order of their boiling point and affinity to the stationary phase.

18.3.2 Mass Spectrometry (MS)

The first mass spectrometer that was developed in 1919 by Francis W. Aston (Cambridge, UK), under the tutelage of Joseph J. Thomson [19], was able to differentiate between the two isotopes of neon [20,21]. Work continued in the identification of singular isotopes, until commercial instruments were developed and sold in the 1940s. Early applications were for petroleum analysis, namely, assisting in refinement [22], but have since branched out to include many fields.

In 1954, John H. Beynon identified that MS had achieved a precision high enough to identify unknown substances by matching mass spectra to those of known substances and estimating the amount and type of similarities and divergences within the mass spectra measurement [23]. At this point, organic chemists began testing samples from humans, starting with fatty acids from human hair [24]. Expanding further on his previous observations, Beynon identified how combining GC [or liquid chromatography (LC)] with MS could be even more effective at separation and identification of compounds within an unknown mixture [22,25].

There are many types of MS devices, but they all share four common elements: an inlet, an ionization source, a method to analyze or separate the ions according to their mass-to-charge (m/z) ratio, and an ion detector [26]. The different types of mass spectrometers each have their own strengths and weaknesses. For example, the most common MS instruments are quadrupole ion traps (Paul traps) [27]. These types of mass spectrometers operate by trapping ions in oscillating electric field inside a vacuum system. The ions can be manipulated in the trap to achieve selective ion storage or fragmentation and then selectively ejected towards an ion detector (usually an electron multiplier) for ion detection, as demonstrated in Figure 18.3.

These instruments are comparatively inexpensive and easy to operate but do not provide very high resolution. For example, resolution may be insufficient to discern the isotopes of doubly charged peptides. A better resolution (defined as the m/z value divided by the peak width at half height) of ~10,000 can be achieved using time-of-flight (TOF) MS [28].

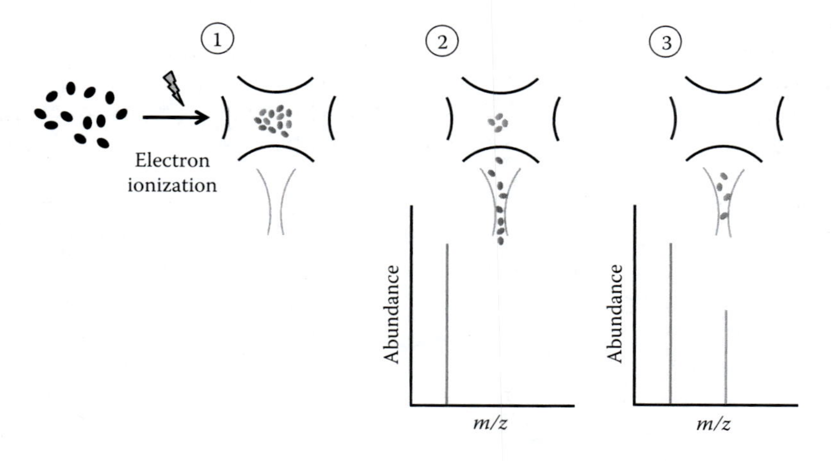

FIGURE 18.3
(See color insert.) A schematic representation of quadrupole ion trap mass spectrometer operation. (1) A gas-phase chemical compound is introduced into the trap and is ionized via collisions with energetic electrons. The neutral molecules gain charge and produce structure-specific fragments. The ions are trapped in the quadrupole electric field. (2) The ions with different m/z are sequentially ejected from the trap and registered by a detector. (3) All ions are ejected out of the trap, and the detected ion current as a function of m/z comprises the mass spectrum.

In TOF instruments, the ions are launched from an ion source towards a detector through an evacuated flight tube by a pusher pulse. Larger ions travel slower than smaller ones, which will result in ions arriving at the detector at different times, so their m/z can then be inferred from the flight time. TOF instruments, however, provide lower mass accuracy and require frequent calibration.

The ultimate resolution is provided by the "Fourier-transform ion cyclotron resonance" MS (FTICR-MS or FTMS), also called the Penning trap [29]. In FTICR, the ions are confined by a high magnetic field of several Tesla (typically 1–10 T) of a superconducting magnet. The ions oscillate with frequencies that are inversely proportional to their m/z value, so the m/z can be inferred by measuring an alternating current induced in the metal detector plates that comprise the trap. This time-varying current constitutes a frequency spectrum of the ion motion and is converted via Fourier transformation from time domain into m/z domain to obtain a mass spectrum. The FTICR has unrivalled high resolution, from 100,000 to 1,000,000, and great mass accuracy (a few parts per million). However, such instruments are extremely expensive and carry very high operation costs. Other types of mass spectrometers such as magnetic sector [30], Orbitrap [31], the triple quad [32], and others are also currently in use and have certain advantages and disadvantages.

More recently, combination instruments such as a combination of quadruple ion trap and TOF mass analyzer (qTOF) may bring combined advantages and produce data sets appropriate for AI applications. For qTOF, these advantages are the ability to manipulate ions inside the trap and the high resolution and speed of TOF mass analyzers. Furthermore, combinations of certain types of mass spectrometers with different ionization sources may be beneficial to achieve specific goals. For example, application of ionization sources such as matrix-assisted laser desorption/ionization (MALDI) and electrospray ionization (ESI) is common in studies of biomolecules, especially proteins, where the softest and least destructive ionization methods are required to preserve the structure of a biological molecule while transferring it from solution (ESI) or solid support (MALDI) into the gas phase. Electron ionization (EI) can be used to fragment the molecules of interest in order to ionize them while producing structure-specific fragmentation products, which can be later used for structural identification. Other considerations may also come into play. For example, the MALDI ion source is most commonly used in conjunction with a TOF instrument as both operate in pulse mode. There are ion sources that were specifically developed for certain applications. For example, the direct analysis in real time (DART) [33] ion source is an atmospheric pressure ion source that can instantaneously ionize gases, liquids, and solids under ambient conditions and does not require sample preparation, so solid and liquid materials can be analyzed by MS in their native state. This allows sampling in an exposed environment such as untreated surfaces rather than inside of a clean machine. This method found applications within forensics and other forms of environmental analysis. Another type of ion source, based on the proton transfer process from hydronium ions (H_3O^+) directly connected to a drift tube and an analyzing system such as a quadrupole mass analyzer, is proton-transfer-reaction mass spectrometry (PTR-MS) [34]. These instruments have a very short response time (~100 ms) and thus can be used for real-time monitoring of chemical composition of a (usually gaseous) sample. These instruments are commonly used for VOC monitoring in ambient air for medical research and diagnostic applications.

As mentioned before, combinations of separation and detection techniques produce "hybrid" or "hyphenated" techniques, such as GC/MS. Despite the term "hyphenated," a hyphen is usually used between an ionization source and an analysis device, while slashes are used between two chemical analysis devices [35]. Application of hyphenated techniques has revolutionized chemical analysis. In terms of utility and availability of chemical information, a hyphenated technique can greatly exceed the sum of its parts. For example, in GC/MS, the chemicals in a complex mixture are preseparated by GC, and each chemical can then be identified with MS. This is an important advantage, because GC alone does not provide unambiguous chemical identification, while the mass spectrum of a complex mixture would be extremely difficult to interpret. Many hyphenated techniques, such as liquid chromatography/mass spectrometry (LC/MS), capillary electrophoresis/mass spectrometry (CE/MS), and ion mobility spectrometry/mass spectrometry (IMS/MS), are commonly used and are essential research tools for many applications. Other analytical methods beyond MS can be hyphenated as well, such as GC-infrared spectroscopy, LC-NMR spectroscopy, and others. New hybrid techniques are being developed for various applications.

18.3.3 Differential Mobility Spectrometer

Originally published in Russia, I. A. Buryakov's 1993 paper identified high-field asymmetric-waveform ion mobility spectrometry (FAIMS) as "a new method of separation of multi-atomic ions by mobility at atmospheric pressure" in the first English publication of the technique [36]. His work derived from a little-explored patent developed in the former Union of Soviet Socialist Republics in 1982 by M. P. Gorshkov (Inventor's Certificate of USSR No. 966583, The Method of Impurity Analysis in Gases) [37]. Initially used in North America during the 90s as a method of explosives detection by Mine Safety Appliances (MSA), the project was not immediately successful but was further developed, and slight changes in device geometry produced better results [38]. At the same time, in the United States, researchers worked to develop a miniature device based on a parallel plate geometry termed differential mobility spectrometry (DMS). DMS and FAIMS have the same scientific theory behind their operation, with the FAIMS devices using a nonplanar geometry and DMS devices using flat surface electrodes to separate chemicals in the gas phase. DMS can be interfaced easily with miniature GC platforms and made field deployable, allowing for much wider applications [38]. It operates at atmospheric pressure and ambient temperature, making it a robust field tool for compound identification.

Differential mobility spectrometers derive from the family of IMS methods. DMS is able to sample a continuous flow of ions while providing very close to 100% [39] complementary/orthogonal information (meaning that the information is collected using a different scientific method that presents new ways of differentiation) to the mass of the ions [40]. The science behind DMS (and FAIMS) is based on the mobility of an ion (referred to as K) in an electric field (referred to as E). In a simplistic form, K is a function of E, and the function varies based on the ion. In other words, $K = f_i(E)$, where f_i is a function that varies based on the ion structure. f_i is actually a Taylor series with multiple values for α, ($K_0[1 + \alpha_1(E/N)^2 + \alpha_2(E/N)^4 \ldots]$), where K_0 is the mobility of the ion in a zero field, E is the electric field, and N is the gas number density.

The DMS consists of two parallel horizontal plates, with a carrier gas sweeping an ion between them. This gas applies a horizontal force to the ion and, without interference, would push all ions between the horizontal plates to the ion detector on the far end. To differentiate between the ions, a square waveform is introduced, with a high value (above +1000 V/cm or below −1000 V/cm) of E_h [also known as the dispersion voltage or (DV)] and a low value (greater than −1000 V/cm and less than 1000 V/cm) of E_l. The waveform is also defined so that the time for E_h (referred to as t_h) is multiplied by E_h to equal the product of E_l and t_l. In other words, the total amount of energy exerted by the high-field portion within the step function is equal to the total amount of energy exerted by the low-field portion, or $E_h t_h = -E_l t_l$. An ion in the high field will experience a mobility of $K_h = f_i(E_h)$ and therefore travel a distance horizontally (orthogonal to the horizontal force exerted by the motivator) of $d_h = K_h E_h t_h$. In order for the ion to not intersect with the wall, $d_h + d_l$ must equal 0, or $d_h = -d_l$. This means that $K_h E_h t_h = -K_l E_l t_l$, and since $E_h t_h = -E_l t_l$, K_h must equal K_l for the ion to successfully travel through the two horizontal plates and be detected on the far side.

However, if there is a difference between K_h and K_l for a specific ion, the ion will begin to drift towards a plate as the distance traveled due to $K_h E_h t_h$ is greater than $K_l E_l t_l$ (or vice versa), and the ion will fail to reach the sensor at the end of the parallel plates. In order to compensate for this drift away from a horizontal path, a constant DC voltage known as a compensation voltage (CV) is applied to one of the plates. The CV can be scanned through a range of voltages, and the successful detection of ions can be attributed to the corresponding CV. DMS separation used in conjunction with a GC separation defines two orthogonal parameters that can assist in identifying an analyte, namely, RT and CV, with the measured abundance from the sensor in the DMS device used as an input to calculate the concentration of the analyte. This concept is illustrated in Figure 18.4.

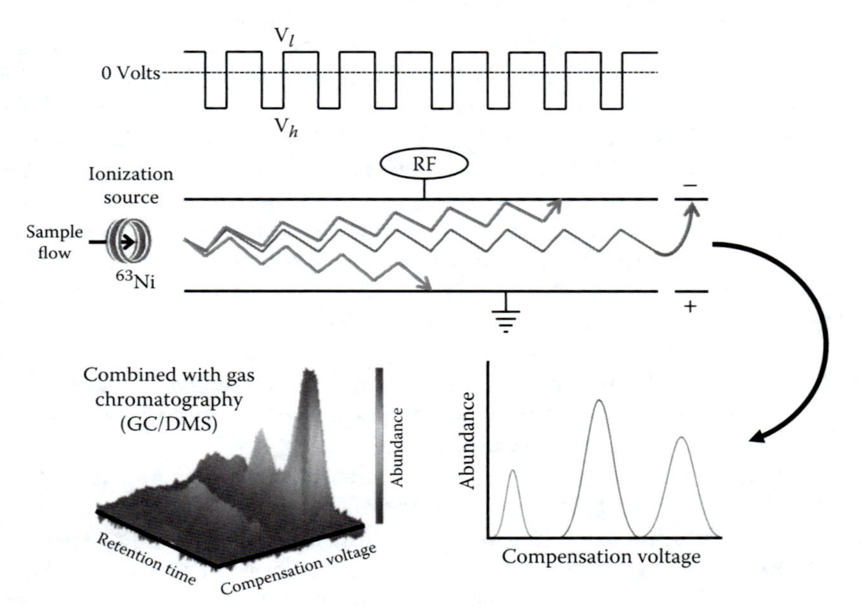

FIGURE 18.4
(See color insert.) A schematic description of operation of high-field asymmetric-waveform ion mobility spectrometer (FAIMS), also known as the differential mobility spectrometer (DMS). The analyte is introduced into the device by a carrier gas. The neutral particles are ionized (in the present example by β-electrons emitted by a ^{63}Ni ionization source). The ions are swept by carrier gas between the DMS electrodes, one of which is grounded, while the other has an asymmetric voltage waveform applied to it. The high and low parts of the waveform cycle are selected to create a magnitude of an electric field within the gap to be below (E_l) and above (E_h) 1000 V/cm. The ions with mobilities different under high and low field conditions will drift towards one of the electrodes and be annihilated. The trajectory of an ion can be corrected by a DC voltage offset [compensation voltage (CV)] applied to the RF electrode. When a range of CV values are scanned over time (ranging from microseconds to minutes), the ions that are able to traverse the device at each CV value will be recorded. The plot of ion current (abundance) versus CV defines the "CV spectrum" of the sample. If the CV is maintained at a constant value rather than scanning a CV range, only the ions that transmit at that value will be able to pass the device. This is referred to as ion filtering mode, while scanning a CV range is referred to as analytical mode. The GC/DMS plot in the lower left corner of the figure shows the separation of a complex sample in two dimensions, RT (GC) and CV (DMS).

The ability to deconvolute analytes that overlap in GC gives GC/DMS a very high specificity to accurately and reproducibly identify key biomarkers in many different types of mixtures. FAIMS (DMS) devices are able to achieve an extremely low limit of detection (LOD) by having a high signal-to-noise ratio and the ability to identify targeted biomarkers in two dimensions. In one report, carcinogens of bromate and chlorate were detected and quantified using FAIMS at 13 ng/L and chlorate at 3.0 ng/L levels [41]. To put this in perspective, since an Olympic size swimming pool holds 2.5 million liters, the device would have detected 1 gram of bromate (approximately a teaspoon) diluted into 30 pools and one gram of chlorate diluted into 133 pools. FAIMS devices have also identified specific analytes in mixtures in which they would otherwise be convoluted. In one report, FAIMS/MS was able to differentiate 0.1 μM perchlorate from a mixture containing 50 μM sulfate and 50 μM phosphate, a combination that would confound an MS-only device [42]. In another application, it enabled successful detection of pharmaceutical drugs in urine. A solid-phase microextraction (SPME)-ESI-FAIMS/MS device was able to detect amphetamine, methamphetamine, 3,4-methylenedioxyamphetamine [MDA, also known as tenamphetamine (INN)], 3,4-methylenedioxy-*N*-methylamphetamine (MDMA), and 3,4-methylenedioxy-*N*-ethylamphetamine (MDEA) in concentrations less than 7.5 ng/mL [43]. MDEA was accurately detected at a concentration of 200 pg/mL [43]. Morphine and codeine were detected in urine at 60 and 20 ng/mL, respectively [38,44]. In yet another report, FAIMS was applied during the selected reaction monitoring (SRM) transition, where an isobaric ion can be generated that can be indistinguishable from the desired drug by LC/MS/MS machines, which do not enable the proper identification and quantification without properly resolving the isobaric ion. FAIMS was able to differentiate between the desired drug and its identical molecular ion [45].

DMS has the ability to deconvolute similar analytes that are not differentiable by comparable technologies and works well in tandem, such as GC/DMS. This was made very clear when xylene isomers were selected to confound time of flight (TOF-IMS) yet presented distinct peaks separated by over 1 V on the CV axis in DMS [46]. The orthogonal separation capability when matched with GC, allowed for identification of homologous series of aldehydes, ethers, esters, and alcohols. All but one of the aldehydes, ethers, and esters had levels of detection less than 0.1 ng, and most of the alcohols had levels of detection less than 0.5 ng [47]. An additional point of interest for the DMS is the speed at which it can differentiate between analytes. When compared with GC, which can take upwards of 20 min depending on the oven, DMS can iterate through a target range of CV values within a second [46].

18.4 Applications of Advanced Data Analysis to Metabolite Data Sets

Given that analytical instruments can produce large and comprehensive data sets for chemicals in complex mixtures, it is not surprising that they have been invaluable tools in metabolomics-based research. But before AI algorithms can be applied to the data, it is necessary to preprocess the data in multiple steps. In Katajamaa's seminal paper "Data Processing for Mass Spectrometry-Based Metabolomics," the method of obtaining results from raw data is composed of two parts: data processing and data analysis [13]. Data processing (also known as preprocessing) centers on preparing the data to be analyzed, including translating file types, alignment, and normalization. Data analysis is frequently the statistical portion where normalized and aligned data are analyzed for patterns that can indicate a desired result [13].

Unfortunately, the data processing portion of analysis is often underutilized in research. Researchers are unfamiliar with the impact that attempting to identify patterns in misaligned data will have on results, often unwittingly settling for lower accuracy and reproducibility due to analyzing data that have not been appropriately preprocessed. We will review the major steps used in analyzing these types of complex data sets generated from the instruments described above.

18.4.1 Alignment

Often, signals from chemical analysis systems, like GC/MS, are temporally shifted, which results in variation of features within the signal over time. GC separates chemicals that elute at different temperatures. While this method can be considered reproducible, there is a variance that occurs between sampling iterations in relation to the time of measurement. Therefore, it is critical that signals between sequential instrument sample analyses are properly aligned prior to classification attempts on the data sets. The goal of this activity is illustrated in Figure 18.5.

A common method of alignment is correlated optimal warping (COW), which uses a Wallis filter as a mechanism to score alignment [48]. A "template sample" is chosen, which is used as the base sample to which all other samples are aligned. The template sample is broken up into many small segments based on the segment width, also referred to as window size. An example segment width would be 20 out of 2500 readings. Each of the other samples is then broken into segments that vary by a warping factor (a value of 7 would work with the earlier example) and range from the window width minus the warping factor to the window width plus the warping factor, or 13 to 27 points long in the current example. Each sample's segments are optimized, based on interpolations of varying length, to best match the segments in the template sample. The scoring system from the Wallis filter normalizes the template sample segment and each warped segment to an average of 0 with a standard deviation of 1 [49]. It then calculates the dot product of the two normalized segments. A high dot product indicates a high correlation. To find the optimal warping for each segment, dynamic programming is used from either the front of the signal or the end of the signal. This reduces the polynomial complexity from exponential to cubic but is dependent on correctly slicing the signal at the reference end to ensure that the signals are aligned at that point.

There are a few drawbacks to using this form of alignment. If the signals are not aligned at the reference end, the warping will be incorrect until the segments finally align. If segments can be removed, or if the warping factor is the same size as the segment width, then the extraneous segments from a misaligned reference point may be able to be removed but would require many more calculations. A more important drawback is that the knowledge of the spectra signals is not taken into account. COW works on a one-dimensional signal, which, in the case of GC/MS, would be the sum of the values of the mass spectrum peaks for each RT, known as the total ion current (TIC) chromatogram. The TIC reflects the abundances of the chemicals that elute from the GC column but ignores the m/z ratios of the spectra measured by the mass spectrometer; it is therefore less effective than if it incorporated this information.

Another large drawback of this method, as well as many other alignment methods, is the use of a single template sample to define the alignment of all other samples. In preprocessing for a binary experiment where there are two types of samples, for instance, sick or healthy, there may exist peaks in sick samples that do not exist in "healthy" samples and vice versa. If a healthy sample were chosen as the template sample, the peaks in sick

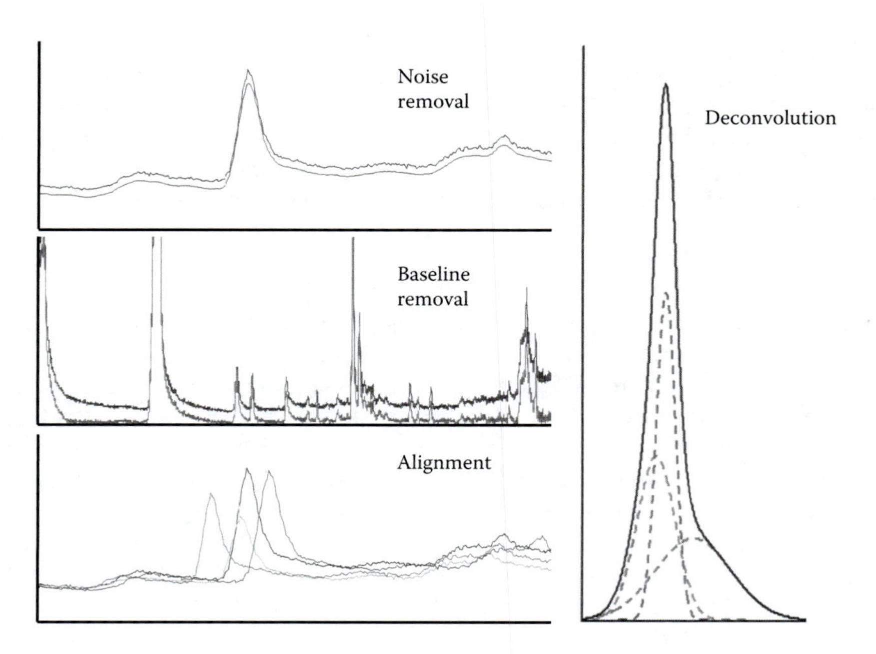

FIGURE 18.5
(See color insert.) A diagram of some of the problems that data may have before preprocessing. The noise removal figure shows the Savitzky–Golay method of smoothing (with an offset from the initial data for clarity), maintaining much of the initial data but losing minor details such as the sharp peaks on the top of the two larger peaks. Baseline removal demonstrates the top-hat algorithm, coming close to zero outside of the noise floor on many of the locations that lack activity. There are discrepancies, however, as the tail of the first peak is primarily flattened, while the tail of the second peak appears to have been amplified. The alignment plot shows three (artificially) warped images of the initial (blue) data. These warpings show dilations, constrictions, and shifts in the x-axis and y-axis, which comprise much of the misalignments that may appear in samples. There is also an obvious offset in the slicing of the data in the rear, with data missing from some of the samples, which will cause problems for certain methods of alignment. The deconvolution figure was created by summing three Gaussian distributions with different central locations into one peak. The combined peak appears primarily Gaussian due to the similar locations of the two largest peaks but possesses an unexpected tail from the location of the third component. If the m/z ratios that make up the three real components do not overlap, Stein's method of deconvolution (derived from Biller–Biemann) may be able to successfully deconvolute this data, though the first two peaks may be combined, as the sharpnesses and locations in the x-axis are so similar.

samples that do not exist in the healthy sample would align to noise or to an incorrect peak. During later analysis, these misaligned peaks may not indicate a strong signal of differentiation from the "healthy" peaks, and that information would be lost. A corrective approach may be to align a single sick sample to a single healthy sample and then proceed to use these two samples as template samples for two separate sets of COW alignments, but this would introduce strong biases in the data using *a priori* knowledge of the categorization and should be used with caution.

There are other methods of alignment that are commonly used. An example is dynamic time warping (DTW) [50], which relaxes the constraints, allowing for greater movement during alignment but at a possible cost of accuracy if peaks are found to align better with locations farther from the initial peak location. In comparison, parametric time warping (PTW) [51] is extremely fast and concentrates on a consistent shift for alignment, which may reflect certain offsets very well. However, this method may fail if the alignment shift is stochastic. There has also been research in landmark identification within the signal as a method of alignment in such algorithms as Interval Correlation Optimised Shifting (Icoshift) [52] and peak alignment using reduced set mapping (PARS) [53]. These require much less computational power than COW but suffer from some of the liabilities of DTW in that the best-aligned peak shapes may vary greatly from initial location and subsequently suffer from misalignment. A comprehensive testing of alignment, including visual inspection, is necessary before implementing each of these algorithms into a proposed data processing strategy.

18.4.2 Noise Removal

Given that the abundance of human metabolites is frequently at or near the noise floor of many modern analytical chemistry instruments due to the low concentration of these metabolites, it is often necessary to remove extraneous signal noise from the data sets prior to classification attempts. Without this removal, peaks indicating components for analysis may be indistinguishable from noise peaks, and analysis is likely to suffer. As with alignment, there is an art to smoothing stochastic noise without removing signal. The goal of this activity is illustrated in Figure 18.5.

The most primitive method of smoothing is a moving average window. In the simplest case, a window size of an odd integer is chosen. This "window" is moved through every point of the one-dimensional signal, finding the average of the values and assigning it to the central location. For example, if the window size is 7, then the "smoothed" value at point 15 would be the average of points 12–18. The larger the window, the smoother the resulting signal. However, the moving average method clearly shows the pitfalls of all smoothing techniques, namely, that *all* points become less informative than they were prior to the smoothing. Peaks and valleys are dampened by a moving average, losing the precious quantitative information they may provide. Slopes of peaks, which are necessary for component identification, will be subdued. Meanwhile, small peaks (in both amplitude and width) will be substantially flattened or removed altogether depending on the width of the window.

The desire for an ideal smoothing technique is to have no effect on the meaningful signals and to completely remove any stochastic or systematic noise. While this ideal smoothing technique would require preternatural knowledge of the data, there are techniques that can have less of an effect on the meaningful signals while still substantially removing the noise. One of these techniques was developed by Abraham Savitzky and Marcel J. E. Golay [54] and was cited by *Analytical Chemistry* as one of the top 10 seminal works in the history of the publication, with the accolade that "...the dawn of the computer-controlled analytical instrument can be traced to this article" [55]. The method derives from the moving average approach but modifies the weights assigned to points inside the window to strengthen the impact of values closer to the center of the window and, therefore, closer to the resulting location. Additionally, the method allows for the user to define an expected polynomial value of the true curve with which to define the possible movement that could occur inside the window [56].

For instance, the weighted window of Savitzky–Golay with a cubic (third degree) polynomial within a window of five values results in the coefficients (−3, 12, 17, 12, −3). The sum of these coefficients is 35, so after multiplying (x_{i-2}, x_{i-1}, x_i, x_{i+1}, x_{i+2}) by the corresponding coefficients, the sum of the values is divided by 35 to give the resulting value located at x_i. To compare, a moving average with a window of 5 would have the coefficients (7, 7, 7, 7, 7) in order to be divided by 35 and to find the average, so the central three points combined have an impact of just under two times the central three points of the moving average [(12 + 17 + 12)/(7 + 7 + 7)]. This allows for peaks and valleys to maintain most of their prominence during smoothing, without losing the initial maximum and minimum values, and ensures a similar derivative value to the true signal.

There are drawbacks to the Savitzky–Golay method. As can be seen in the example above, the edge values of the window are negative values. This allows for a large impact of the central values on the resulting value located in the center, while maintaining that a straight line will result in a straight line. However, if there is a point with "low signal" two points before a large peak (as will happen with GC data), the large peak will have a substantial negative impact on the location with low signal, sometimes creating a negative value. It is important to handle these occurrences in a consistent manner. One method may be to identify these locations and provide a different type of smoothing on those locations only. Another method may be to simply set these values to zero, or an average of the previous unaffected values, and expect them to be handled appropriately during baseline removal. It is important to note that while Savitzky–Golay is very powerful, it is not always an ideal filter.

There are other filters that can handle this drawback by attempting to recreate the true signal rather than remove the noise affecting it. These methods often are based on a fast Fourier transform (FFT) of the overall signal and a separation of low frequency signal from high frequency noise [57]. While this method can effectively maintain accurate peaks and valleys, it is up to the user to define the frequency threshold between noise and signal. This proves troublesome on GC/MS data as small peaks (in RT) may indicate a component if they appear in conjunction with other small peaks in other m/z ratios. By removing some or all of these signals in their respective m/z ratios, a component may end up misidentified or completely absent.

For applications to two-dimensional data such as GC/DMS, the best practice may be to iteratively smooth in only one direction. In other words, a method of smoothing may be to separate each CV into a one-dimensional signal with RT as the x-axis in preparation for the smoothing. After this method of smoothing is complete, the GC/DMS data should probably not be rotated around the z-axis and the same procedure executed in relation to the CV as the x-axis and abundance as the y-axis. This type of double processing often strengthens the weaknesses of a method, negatively impacting the ability to analyze the data. Choosing the single axis from which to remove the noise requires understanding the generation of the data. Removing noise in different axes will have different impacts, and it is hard to predict *a priori* which axis to choose. As in all methods of preprocessing, a sharp eye and a deep understanding of the meaning behind the data are necessary.

18.4.3 Baseline Removal

While noise is often stochastic, there can be a consistent offset from a machine between a signal portion that contains no information and true zero. This offset will often vary or drift between readings and may also vary during a reading. This may cause a constant slope that is easily identifiable by a user but can be computationally challenging to identify. Without baseline removal, peak perception and identification could still be possible, but quantification will be suspect, and classification based on this quantification will suffer. The goal of this activity is illustrated in Figure 18.5.

One method that is commonly used is asymmetric least squares [58]. This method iterates through multiple least-squares representation of raw data (which can be looked at as an extreme form of smoothing) but strengthens the impact of the lowest values on the result. This creates a polynomial function, which defines the baseline based on the lowest values of the signal in relation to this empirically determined baseline. It is a very elegant solution, but it does have limitations.

The first concern is that the lowest of the values defining the least-squares baseline will be below the baseline. This may cause problems during some of the subsequent analyses, but it also cannot exist in real life. By definition, there should never be a negative signal in many of the machines used for metabolomics-based diagnostics, and by creating this occurrence; the user has ensured that s/he has not found the proper baseline. One way to handle this situation would be offsetting the baseline to a point that all values are above it, but this offset will create many nonzero values for points in the signal that should be equal to zero. An incorrect approach would be to set all resulting negative values to zero, but this may unnecessarily remove small intensity peaks.

Another, and in some ways more aggressive, method uses a top-hat transformation [59]. A top-hat operator works by running an imaginary shape (or kernel) along the bottom of a signal. The shape can be anything (usually a square or a circle), but for this example, a square will be used. The kernel creates an *erosion* of the bottom of the signal, which is a second signal that maps the center of the shape as it follows the signal. If the kernel does not fit into a peak, then it maps the erosion of where it does fit. From this erosion, a *dilation* is mapped that recreates the path of the square as it follows the bottom of the signal. This combination of an erosion and dilation is called an *opening*. The size of the kernel is then increased, and the opening is remapped. As the kernel gets larger, it does not "fit" into some of the peaks as well as it initially did, and the opening becomes smaller. When the opening stays consistent over two size increases, the resulting map is considered the baseline.

Top-hat creates a well-contoured map of the smallest values in each portion of the signal while maintaining the relative intensity of small peaks. However, this method also has limitations. The baseline plot of top-hat is very jagged compared to least squares, in a manner that does not suggest a true baseline. This can affect large peaks and have a large impact on peak slopes in the lower portions of peaks. One method of handling this is to map the erosion and dilation of the top-hat signal itself using the same kernel size, creating a signal smaller than top-hat and a signal larger than top-hat [60]. The average of these two signals is then measured. The minimum value of the new signal and the original top-hat signal is the map of the final baseline. This allows for smoothing the jagged transitions that occur near large peaks while still maintaining a tight fit to the baseline of segments without peaks.

18.4.4 Deconvolution

Certain devices return signals that do not obviously differentiate between compounds. GC/MS is a good example of this type of data since compounds, such as isomers, will elute from the GC column at similar times yet be separate analytes that have distinguishable mass spectra. The overlapping signals do not have clearly separate peaks, and if the peaks indicate separate classifications, they will present conflicting information for categorization analysis. An additional concern with convoluted data is quantification values when multiple peaks are present. A single peak from a GC column should have a Gaussian appearance, but "shoulder humps" could indicate that multiple compounds are present within the signal, and these would need to be separated for accurate quantification. In order to address both of these concerns from convoluted data, the signal needs to be broken up into a series of single peaks.

A popular methodology for GC/MS deconvolution was created by the National Institute of Standards and Technology (NIST), under the supervision of Stephen Stein, and this method is available in a free software package called Automated Mass Spectral Deconvolution and Identification System (AMDIS) [61]. Stein based his method on the work of Biller and Biemann [62] and took into account the improvements made by Colby [63] and Dromey et al. [64]. This method assumes that peaks at different m/z ratios, and behaving in a similar manner at the same RT, would be indicative of a single component existing at that location. Isolated peaks, or peaks with widely varying behavior between m/z ratios, do not indicate the presence of a component centralized at that location. What is unexpected about Stein's method is that it works better without prior noise removal by calculating a noise factor (N_f), which it uses to differentiate between peaks that represent components and "junk" peaks. If the noise is artificially removed, Stein's method determines a much smaller noise factor and then retains many more artificial peaks that are not necessarily meaningful.

With the noise factor calculated, Stein's method separates GC/MS data into separate signals for each m/z ratio. Each of the peaks in these signals is then analyzed to identify if it differentiates itself from the surrounding values by a large enough multiple of the noise factor to be classified as a possible component location. The true centroid of the remaining peaks is found through parabolic interpolation, and a sharpness value for each peak is calculated, which represents the slope of each peak. The sum of the sharpnesses of all peaks that exist at each tenth of a second in RT is calculated into a large histogram, and the largest sums of these locations in relation to the nearby sharpness sums are identified as possible peaks.

At this point, a model is defined based on the largest sharpness value of each peak. Overlapping models are removed on an m/z basis to ensure that peak heights are properly defined, and flags are placed on peaks that may have questionable relevance as an indicator of a component. From each model, a constant and linear baseline is removed, and the remaining peak heights define the final spectra of the component.

Stein's method is both elegant and effective, providing a clear application and reasoning behind every step of the deconvolution and producing crisp spectra that can immediately be used for component identification within a database. It has the ability to differentiate between peaks less than one measurement apart, and it has been shown to break up one peak with a half-height width of five scans into four separate components.

The drawbacks are very small compared to the capabilities of the method, but they do exist and can be seen when used in conjunction with a library method. The multiple of the noise factor used to define the initial filter of the peaks may not be the best constant for every unique instrument and should be tested appropriately before subsequent steps of analysis. Additionally, Stein's method should take a lenient approach to defining a component, allowing for the user to visually inspect matches from a spectra database to indicate if a true match is found or if a meaningless peak was allowed through. This obviously requires user input for every analysis and is therefore untenable in complex samples such as biogenic samples, but it is the only way to ensure that all real components are identified.

18.5 Machine Learning Applications for Metabolomics on Hyphenated Analytical Chemistry Tools

18.5.1 Training/Testing/Validation Sets and Repetition

It is important to consider the path an unknown sample would take through a numerical analysis methodology to ensure reproducibility without biasing. If steps can be run on samples individually, then they can be reproduced on an unknown sample, but steps that require the knowledge of other samples would introduce a possible biasing towards the categorization of those other samples. The previous steps of preprocessing are run on each sample individually. Baseline removal, noise removal, and deconvolution iterate through every sample and do not use information from other samples to complete that particular phase. Alignment does require a template sample that all other samples are compared to, but if an unknown sample needed to be accurately processed, it could be aligned to the same template sample as the rest of the known samples and expect similar biasing, as described in Section 18.4.1.

The analysis steps, however, require the samples to be contrasted to each other in order to extract the features that differentiate between classifications and allow quantification. The strength of the methods is identifying the differences between many samples and the relation of those differences towards the goal of analysis, and if an unknown sample is used, it is unclear how capable the methodology would be able to accurately classify it.

The main problem with these nonlinear (AI) analysis methods is the potential for overfitting as we model a system. As an example, 200 features are extracted (a very large number) from 20 samples (a very small number), and the purpose of analysis is to discriminate between two groups. If an ANN is used with 200 input nodes, after a while, the network would converge with 100% accuracy on a Boolean output for each sample. If a 21st sample (unknown in the initial model) is then introduced, the ANN will most likely output a value that suggests that the sample is between the two classifications. What has happened is that the ANN has learned the original samples so well that instead of identifying the features that differentiate the classifications, it has memorized which samples belong to which class. It would not have identified a pattern that can be applied to other samples but instead would be specifically trained (overfit) to the samples on which it was trained. It has placed an equal balance on all input features as a method of identifying each specific sample and the correct output that correlates to it.

The ANN is not at fault here; rather, the methodology that was used to train the ANN was, when it lacked a testing set to observe when the neural network began overfitting the initial training set. A proper methodology would have trained the neural network on a randomized subset (training set) of the complete sample set and tested after each training iteration to observe an improvement in the classification of a different subset of the sample set, with no intersection of the training set (also known as a testing set). At some point, the neural network would begin overfitting the training set, and the classification accuracy of the testing set would suffer. This should be observed over a few iterations to ensure that the accuracy is indeed diverging, but the neural network that existed at the best classification accuracy of the testing set should be the final neural network returned for analysis. It was the most effective at identifying a pattern that differentiated between the types of classification yet had not overfit the training set as it was still applicable to the testing set.

At this point, the training of the ANN was complete, but it was unclear what the final accuracy of the ANN would be. It was trained using the training set, and it was optimized using the testing set, so the accuracy of these two sets of samples is biased. What is needed is a validation set to identify the expected accuracy of the methodology on an unknown sample. A validation set that has had no impact on the development of the final analysis method will provide an unbiased assessment of the accuracy of the ANN.

The sets need to be selected randomly, but there does need to be a balance of categorization ensured that is consistent throughout the sets. The best approach is to separate all of the samples based on classification and ensure that a percentage is assigned to the sets based on the size of the sets. For example, if the training set was 50% of the samples, the testing set 30%, and the validation set 20%, then 50% of the first classification should be randomly placed in the training set, 30% in the testing, and 20% in the validation. This should be repeated for all of the classifications.

There is a final step necessary for effectively evaluating a methodology, which is repetition. It is still not possible to gauge how effective the nonlinear method will evaluate an unknown new sample without reproducible results. The samples in the training, testing, and validation sets should be shuffled (though still equally dispersed), and the methodology should be repeated many times to understand the strengths and weaknesses of the approach. This step is often ignored in many articles that use nonlinear methods of analysis, and without repeated evaluation, accurate identification of the confusion matrix is not possible. Selection of the method of nonlinear analysis is not nearly as important as the enforcement of a proper training and testing regime that can effectively create and evaluate the method of analysis.

18.5.2 Feature Identification and Extraction

Deconvolution, as described above, identifies the specific peaks that represent specific compounds in the analyzed sample. However, data from certain devices cannot be effectively deconvoluted, and most nonlinear methods are computationally expensive in an exponential relationship to the amount of inputs presented. Therefore, most modern analyses are unable to handle more than a small handful of inputs in a computationally realistic amount of time, but modern devices can return hundreds of compounds in a clinical sample. A method of reducing the number of variables is necessary, grouping the large set of compounds into a smaller set of features.

One popular method is principal component analysis (PCA). The best way to understand PCA is to visualize a set of scattered points on a two-dimensional plot increasing as a positive slope. PCA rotates the data so that the coordinates of the plot align with the primary slope of the data, or the principal component of the data. If the coordinate system is redefined so that the x-axis represents the first principal component, then the data will align principally with the x-axis, and the second principal component will align with the axis orthogonal to the x-axis: the y-axis. It is able to identify these principal components by analyzing the largest variance of the data, creating a *loading* vector that reflects the direction of this variance and a *score* assigned to each sample that reflects the location of the orthogonal value. The variance along the first principal component is removed entirely, therefore ensuring that the remaining components of each point are orthogonal to the first principal component. Applying this example to data gathered by a hyphenated device, the original x- and y-coordinates of each point represent the preprocessed data for each sample, and PCA is able to identify the principal components in a very high-dimensional space, reducing thousands of data points per sample to a range of 10 to 20 components.

The first principal component is the feature with the largest variation in data points between the samples and conveys information from all of the data points, though the importance of each data point varies based on the related coefficient of the loading vector for that principal component. The loading vector for each component indicates a relationship between the data points and therefore suggests a feature between the samples with consistent variability, though the impact of this feature on the desired classification or quantification is uncertain.

The value of PCA for nonlinear methods is that the number of variables can be substantially lowered. PCA provides a *shadow* of the initial data as reflected on the components chosen. In the initial two-dimensional example, two principal components can be found that would fully rotate the data to align with the axes. However, if the second principal component was not calculated, the first principal component would contain the location of the data in relation to the vector of primary variance. If the data were recreated from only one principal component, the points would be located similarly to their respective initial locations, but they would be colinear along the principal component vector and not exhibit the variation along the second principal component. PCA effectively compresses the amount of information necessary to communicate the compounds that define the sample by a substantial amount, containing a large percentage of the complete sample within only a few data points. These compressed samples can then be effectively used as inputs for machine learning algorithms, expressing most of the variance of the samples in very few variables.

As with machine learning methods, PCA uses information from all samples to define the principal components. If a sample was removed, the expectation would be that the principal components should remain similar to those that would have been there if the sample remained, but they will be slightly rotated as the impact of the data points from that specific sample no longer exists. This suggests that the first few components should be relatively unaffected by unknown samples, but as more principal components are documented, they may not effectively reflect unknown samples, having been overfit to the samples used to calculate them. To counter this overfitting, the data should already be separated into training, testing, and validation sets before PCA is calculated. PCA will only be calculated on the training set, and the loading vectors applied to the testing set. If the percentage of variance removed from the testing set is similar to the percentage of variation removed from the training set, then a valid principal component has been found, which should apply in a similar manner to all samples. But if the percentage of variance removed from the testing set strongly diverges from the training set, then a principal component that reflects the training set but is not applicable to other samples has been found. At this point, PCA should stop, and the loading vectors are stored to be applied to future unknowns.

PCA is considered the lowest-level eigenvector-based multivariate analysis. It is primarily found through identifying the eigenvalues of covariance matrices but can also be determined iteratively if too many compounds exist to efficiently calculate the covariance matrices. Other methods for feature selection also exist such as partial least-squares discriminant analysis (PLS-DA), which can contour the result towards features that have a direct impact on the classification desired, and therefore can be a final step in a methodology development, but introduce biases if machine learning algorithms are used afterwards on the found features.

18.6 Clinical Applications of AI Analysis of Metabolite Content

The field of metabolomics has evolved rapidly over the last decade from using simple statistics towards using more complex algorithmic approaches to interpret the massive and complex data sets that are generated [65–67]. A wide array of AI approaches are now used to analyze and interpret these data sets, especially in the context of medical diagnostics. Some of the most widely cited reports in the literature have used the nuclear magnetic resonance (NMR) instrument to generate metabolite profiles of urine, blood, and tissue. A seminal paper in 2007 showed that end-to-end automation of this process was possible for these data, and this process is very amenable to high-throughput clinical diagnostic studies [68]. Data from other instrumentation platforms have also been elucidated using AI for medical diagnostics. Electronic noses in particular produce arrays of data that usually must be interpreted using AI approaches [69,70]. Other examples of technology that have been coupled with AI include the following: capillary GC to identify bacteria relevant to human pulmonary infections [71], bacteria diagnosis using a conducting polymer sensor array [72], and IMS distinction of pulmonary markers associated with chronic obstructive pulmonary disorder (COPD) [73]. Finally, the DMS method has widely benefitted from AI application to its data [74–79], with clinical applications such as identification of various bacterial species and strains via their growth-associated metabolites [75] and also virus [77] and bacteria [74,76,78] identification via pyrolysis-associated breakdown products.

We can also examine the published literature according to the types of human effluents that have been analyzed for their metabolite content using AI methods. In urine, most AI metabolomics modeling methods have been used for small metabolites [68,72,80]; however, proteomic biomarker targets have frequently been the target for clinical diagnostic applications [81–85]. There are also reports of AI methods being used to discriminate skin and body odors [86,87]. However, most research has focused on specific disease targets, and the examples of these studies are wide ranging and include the following: cancer diagnostics [82,85,88–92], identification of active tuberculosis cases [84,93], diagnosing preeclampsia [94], diagnosing inflammatory bowel disease [95], following potential bacterial contamination of chronic wounds [96], determining onset and metabolite association with waterborne pathogens [97], and even identifying strains and species of bacteria in austere environments such as the Antarctic [98].

18.7 Future Outlook on AI in Chemometric Analysis

The interplay between the biological problem and the analytical chemistry and data analysis approaches is illustrated in Figure 18.6. The changes in biological state result in alterations of chemical composition. The differences, for example, an altered proteome or the release of small chemicals, will comprise the body of biomarkers reflective of this biological state. The examples of different biological states are infection, malignancy, senescence, and other clinically meaningful events. Identification, detection, and interpretation of these biomarkers will allow for characterization of the biological state. In order to detect and identify the biomarkers of interest, an appropriate analytical method must be selected. More sophisticated methods may be required for an initial research stage, while specifically tailored sensors may be used when the biomarkers of interest are established and characterized. Finally, a data analysis method needs to be developed to interpret the data off the analytical instrument and identify the biological state based on biomarkers that are measured.

The confluence of all three factors is necessary. In the absence of biomarkers that are indicative of biological state of interest, even the most sophisticated analytical methods in conjunction with data analysis will not result in meaningful outcome. In such cases, medical/biological studies are needed in order to establish how the biological state can be characterized. If the biomarkers of a state exist, but cannot be measured, whether due to low sensitivity or poor reproducibility of the applied instrumentation, positive diagnostic outcomes are not possible. Finally, biomarkers of interest may exist, and can be measured, but the data cannot be interpreted due to excessive complexity and nonoptimal analysis.

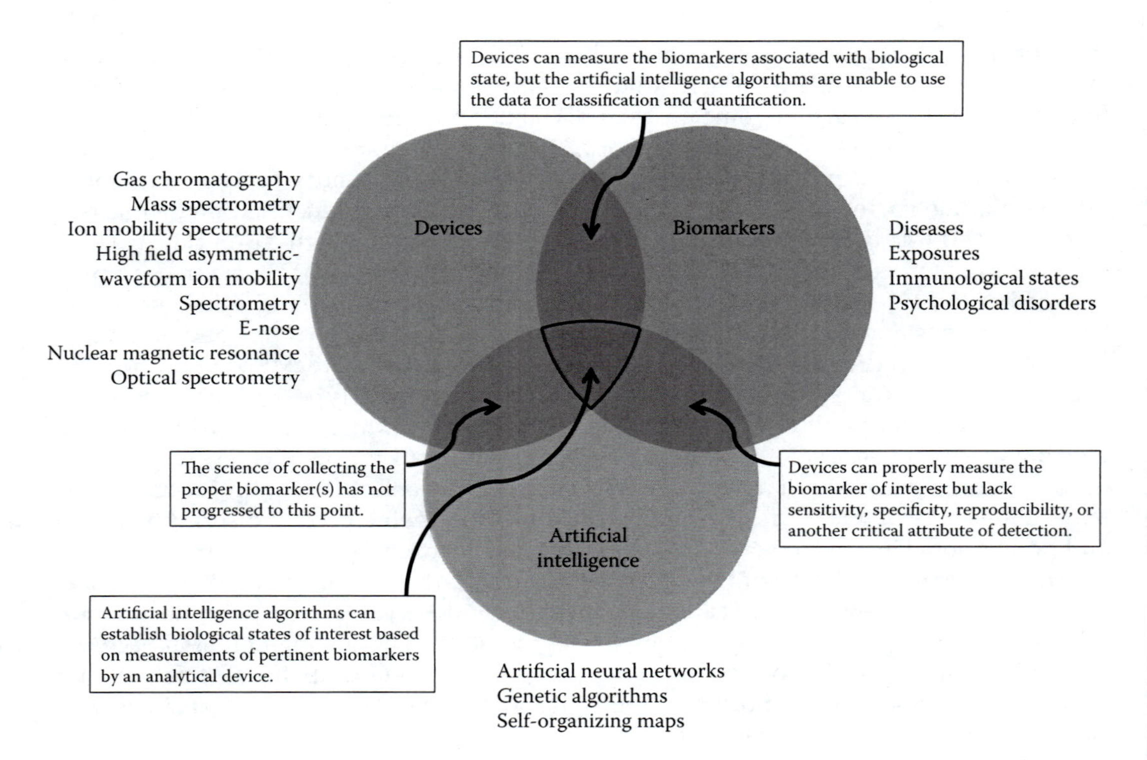

FIGURE 18.6
(See color insert.) A conceptual representation of the interplay between the biological problem, the analytical chemistry instrumentation used to measure metabolic biomarkers of the state, and the data analysis approaches needed to interpret the result. The "Biomarkers" circle represents the body of biomarkers that is indicative of a certain biological state such as infection, malignancy, exposure, or another disorder. The "Devices" circle represents the multitude of analytical methods that are available to detect and measure the biomarkers of interest. The instrumentation options most adequate for the metabolite biomarkers' measurements may differ at the initial research/exploration stage (the biomarkers of interest are being established) versus the targeted analysis stage (the biomarkers of interest are known and documented). The developments in instrumentation, such as incremental improvements in performance, or introduction of conceptually new analysis approaches extends the circle. The "Artificial Intelligence" circle represents the variety of nonlinear data analysis approaches that allow discriminating of the biological states based on the output of the analytical instrumentation employed in the study. Incremental gains in knowledge of mechanistic details of a particular biological process would allow replacing nonlinear AI methods such as ANN with linear approaches. The necessity of a concerted approach required to solve the complex biological/medical challenges is represented by the intersection of all three circles.

This illustrates that in order to achieve a successful outcome, complex biological/medical challenges need to be addressed concertedly through analytical chemistry/instrumentation development in conjunction with mathematical/data analysis approaches.

18.8 Final Words

While the capabilities of nonlinear analysis methods to retrieve small points of discerning data from large samples with convoluted information are extremely powerful, a large section of this chapter was dedicated to preprocessing and training techniques to show the importance of a proper methodology. If the data come to the machine learning algorithm with excessive noise, subdued indicative signals, or biases imprinted by the preprocessing method, it will be impossible for the method of analysis to discern viable information from them.

This lesson applies to the use of machine learning algorithms as well. The choice of which method to use is not nearly as important as the proper application of the method. An imbalance in the number of each classification within the training set will heavily skew the results of analysis in a negative manner, or few repetitions of the technique to varied training, testing, and validation sets will provide very few observations from which to make a well-informed assessment of the method, regardless of whether it is a neural network, SOM, or GA. Every step of a methodology working together in a meaningful way can identify small but important signals to categorize and quantize, but just one incorrectly applied step can cause only noise in the results.

There is a concept called the AI effect, and it perceptively identifies both people's general discomfort with the use of the term *artificial intelligence* and why AI always seems relegated to the fringe of methods of analysis, rather than the heart. The AI effect documents the breaking of new ground by AI that helps to identify the true calculations or systems that were previously unknown. When these systems are identified, linear methods quickly fill the gaps, or the method that was once called AI is renamed to the author of the method. Fred Reed wrote in an editorial on the consistently disdainful opinion of AI,

> "When we know how a machine does something 'intelligent,' it ceases to be regarded as intelligent. If I beat the world's chess champion, I'd be regarded as highly bright. When a computer does it by sorting through all possible moves, people say, 'That's not thinking'" [99].

Nick Bostrom, the director of the Future of Humanity Institute at Oxford, said during an interview,

> "A lot of cutting edge AI has filtered into general applications, often without being called AI because once something becomes useful enough and common enough it's not labeled AI anymore" [100].

This final comment is not to disdain research using AI but, rather, to identify that AI is used on the fringes because what it is attempting to do is *hard*. If there were a linear method, or a better understanding of the system, machine learning algorithms would be put aside for faster and more consistent analyses. A researcher using AI is breaking new ground and trying to identify the patterns that no one has ever seen before. It is not easy, and it follows a path with fits and starts, but AI is the best way to push the boundaries of discovery and create the next generation of analysis methods.

Acknowledgments

This work was generously and partially supported by several funding agencies. The content of this work is solely the responsibility of the authors and does not necessarily represent the official view of these agencies. Partial support is acknowledged from the following:

UL1 RR024146 from the National Center for Research Resources (NCRR), a component of the National Institutes of Health (NIH), and NIH Roadmap for Medical Research [CED]; Gilead Sciences, Inc. [CED]; the Department of the Army [CED]; the Hartwell Foundation [CED]; the Office of Naval Research, Code 34, Warfighter Performance Department [CED]; and the US Department of Veterans Affairs, Post-9/11 GI-Bill [DJP].

References

1. M.M.W.B. Hendriks, F.A. van Eeuwijk, R.H. Jellema, J.A. Westerhuis, T.H. Reijmers, H.C.J. Hoefsloot and A.K. Smilde. Data-processing strategies for metabolomics studies, *Trac-Trend Anal Chem*, vol. 30, no. 10, 2011, pp. 1685–1698; doi 10.1016/j.trac.2011.04.019.
2. D.I. Ellis and R. Goodacre. Metabolic fingerprinting in disease diagnosis: Biomedical applications of infrared and Raman spectroscopy, *Analyst*, vol. 131, no. 8, 2006, pp. 875–885; doi 10.1039/b602376m.
3. G.A.N. Gowda, S.C. Zhang, H.W. Gu, V. Asiago, N. Shanaiah and D. Raftery. Metabolomics-based methods for early disease diagnostics, *Expert Rev Mol Diagn*, vol. 8, no. 5, 2008, pp. 617–633; doi 10.1586/14737159.8.5.617.
4. A. D'Alessandro, B. Giardina, F. Gevi, A.M. Timperio and L. Zolla. Clinical metabolomics: The next stage of clinical biochemistry, *Blood Transfus*, vol. 10, 2012, pp. S19–S24; doi 10.2450/2012.005s.
5. T. Hyotylainen. Novel methodologies in metabolic profiling with a focus on molecular diagnostic applications, *Expert Rev Mol Diagn*, vol. 12, no. 5, 2012, pp. 527–538; doi 10.1586/erm.12.33.
6. R. Kaddurah-Daouk, B.S. Kristal and R.M. Weinshilboum. Metabolomics: A global biochemical approach to drug response and disease, *Ann Rev Pharmacol Toxicol*, 48, 2008, pp. 653–683.
7. H.J. Issaq, S.D. Fox, K.C. Chan and T.D. Veenstra. Global proteomics and metabolomics in cancer biomarker discovery, *J Sep Sci*, vol. 34, no. 24, 2011, pp. 3484–3492; doi 10.1002/jssc.201100528.
8. C.H. Johnson, A.D. Patterson, K.W. Krausz, J.F. Kalinich, J.B. Tyburski, D.W. Kang, H. Luecke, F.J. Gonzalez, W.F. Blakely and J.R. Idle. Radiation metabolomics. 5. Identification of urinary biomarkers of ionizing radiation exposure in nonhuman primates by mass spectrometry-based metabolomics, *Radiat Res*, vol. 178, no. 4, 2012, pp. 328–340; doi 10.1667/rr2950.1.
9. S.M. Rappaport. Biomarkers intersect with the exposome, *Biomarkers*, vol. 17, no. 6, 2012, pp. 483–489; doi 10.3109/1354750x.2012.691553.
10. O. Fiehn. Metabolomics—the link between genotypes and phenotypes, *Plant Mol Biol*, vol. 48, no. 1–2, 2002, pp. 155–171; doi 10.1023/a:1013713905833.
11. R. Goodacre, S. Vaidyanathan, W.B. Dunn, G.G. Harrigan, and D.B. Kell. Metabolomics by numbers: Acquiring and understanding global metabolite data, *Trends Biotechnol*, vol. 22, no. 5, 2004, pp. 245–252; doi 10.1016/j.tibtech.2004.03.007.
12. K. Dettmer, P.A. Aronov, and B.D. Hammock. Mass spectrometry-based metabolomics, *Mass Spectrom Rev*, vol. 26, no. 1, 2007, pp. 51–78; doi 10.1002/mas.20108.
13. M. Katajamaa and M. Oresic. Data processing for mass spectrometry-based metabolomics, *J Chromatogr A*, vol. 1158, no. 1–2, 2007, pp. 318–328; doi 10.1016/j.chroma.2007.04.021.
14. A.T. James and A.J. Martin, Gas-liquid partition chromatography; the separation and micro-estimation of volatile fatty acids from formic acid to dodecanoic acid, *Biochem J*, vol. 50, no. 5, 1952, pp. 679–690.
15. A.J.P. Martin and R.L.M. Synge. A new form of chromatogram employing two liquid phases I. A theory of chromatography 2. Application to the micro-determination of the higher mono-amino-acids in proteins, *Biochem J*, vol. 35, 1941, pp. 1358–1368.
16. L.S. Ettre. Fifty years of gas chromatography–the pioneers I knew, part I, *Lc Gc N Am*, vol. 20, no. 2, 2002, pp. 128–140.

17. F.W. Rowland. What is gas-chromatography, *Hewlett-Packard J*, vol. 34, no. 1, 1983, pp. 32–33.
18. K. Biemann. Four decades of structure determination by mass spectrometry: From alkaloids to heparin, *J Am Soc Mass Spectr*, vol. 13, no. 11, 2002, pp. 1254–1272.
19. F.W. Aston. A positive ray spectrograph, *Philos Mag*, vol. 38, no. 228, 1919, pp. 707–714.
20. F.W. Aston. Neon, *Nature*, vol. 104, 1920, pp. 334–334.
21. K.M. Downard. Cavendish's crocodile and dark horse: The lives of Rutherford and Aston in parallel, *Mass Spectrom Rev*, vol. 26, no. 5, 2007, pp. 713–723; doi 10.1002/Mas.20145.
22. J.S. Grossert. A retrospective view of mass spectrometry and natural products—sixty years of progress, with a focus on contributions by R. Graham Cooks, *Int J Mass Spectrom*, vol. 212, no. 1–3, 2001, pp. 65–79.
23. J.H. Beynon. Qualitative analysis of organic compounds by mass spectrometry, *Nature*, vol. 174, no. 4433, 1954, pp. 735–737.
24. R.A. Brown, W.S. Young and N. Nicolaides. Analysis of high molecular weight alcohols by the mass spectrometer—the wax alcohols of human hair fat, *Anal Chem*, vol. 26, no. 10, 1954, pp. 1653–1654.
25. J.H. Beynon. The use of the mass spectrometer for the identification of organic compounds, *Mikrochim Acta*, vol. 44, no. 1–3, pp. 437–453, 1956.
26. G. Siuzdak. What is mass spectrometry, 2012; http://masspec.scripps.edu/mshistory/whatisms_details.php.
27. G.C. Stafford, Jr., P.E. Kelley, J.E.P. Syka and W.E. Reynolds. Recent improvements in and analytical applications of advanced ion trap technology, *Int J Mass Spectrom Ion Processes*, vol. 60, 1984, pp. 85–98; doi 10.1016/0168-1176(84)80077-4.
28. W.E. Stephens. A pulsed mass spectrometer with time dispersion, *Phys Rev*, vol. 69, 1946, pp. 691.
29. M.B. Comisarow and A.G. Marshall. Fourier transform ion cyclotron resonance spectroscopy, *Chem Phys Lett*, vol. 25, 1974, pp. 282–283; doi 10.1016/0009-2614(74)89137-2.
30. T.W. Burgoyne and G.M. Hleftje. Introduction to ion optics for mass spectrograph, *Mass Spectrom Rev*, vol. 15, 1997, pp. 241–259; doi 10.1002/(sici)1098-2787(1996)15:4<241::aid-mas2>3.0.co;2-i.
31. A. Makarov. Electrostatic axially harmonic orbital trapping: A high-performance technique of mass analysis, *Anal. Chem.*, vol. 72, 2000, pp. 1156–1162; doi 10.1021/ac991131p.
32. R.A. Yost and C.G. Enke. Selected ion fragmentation with a tandem quadrupole mass spectrometer, *J Am Chem Soc*, vol. 100, 1978, pp. 2274–2275; doi 10.1021/ja00475a072.
33. R.B. Cody, J.A. Laramee and H.D. Durst. Versatile new ion source for the analysis of materials in open air under ambient conditions, *Anal Chem*, vol. 77, 2005, pp. 2297–2302; doi 10.1021/ac050162j.
34. A. Hansel, A. Jordan, R. Holzinger, P. Prazeller, W. Vogel and W. Lindinger. Proton transfer reaction mass spectrometry: Online trace gas analysis at the ppb level, *Int J Mass Spectrom Ion Processes*, vol. 149/150, 1995, pp. 609–619; doi 10.1016/0168-1176(95)04294-u.
35. C.L. Wilkins. Hyphenated techniques for analysis of complex organic mixtures, *Science (Washington, D. C., 1883-)*, vol. 222, 1983, pp. 291–296; doi 10.1126/science.6353577.
36. I.A. Buryakov, E.V. Krylov, E.G. Nazarov and U.K. Rasulev. A new method of separation of multi-atomic ions by mobility at atmospheric-pressure using a high-frequency amplitude-asymmetric strong electric-field, *International Journal of Mass Spectrometry and Ion Processes*, vol. 128, no. 3, 1993, pp. 143–148.
37. A.A. Shvartsburg, F.M. Li, K.Q. Tang and R.D. Smith. High-resolution field asymmetric waveform ion mobility spectrometry using new planar geometry analyzers, *Anal Chem*, vol. 78, no. 11, 2006, pp. 3706–3714; doi 10.1021/Ac052020v.
38. B.M. Kolakowski and Z. Mester. Review of applications of high-field asymmetric waveform ion mobility spectrometry (FAIMS) and differential mobility spectrometry (DMS), *Analyst*, vol. 132, no. 9, 2007, pp. 842–864; doi 10.1039/B706039d.
39. A.A. Aksenov, J. Kapron and C.E. Davis. Predicting compensation voltage for singly-charged ions in high-field asymmetric waveform ion mobility spectrometry (FAIMS), *J Am Soc Mass Spectr*, vol. 23, no. 10, 2012, pp. 1794–1798; doi 10.1007/s13361-012-0427-6.

40. A.B. Kanu, P. Dwivedi, M. Tam, L. Matz and H.H. Hill. Ion mobility-mass spectrometry, *J Mass Spectrom*, vol. 43, no. 1, 2008, pp. 1–22; doi 10.1002/Jms.1383.
41. D.A. Barnett, R. Guevremont and R.W. Purves. Determination of parts-per-trillion levels of chlorate, bromate, and iodate by electrospray ionization/high-field asymmetric waveform ion mobility spectrometry/mass spectrometry, *Appl Spectrosc*, vol. 53, no. 11, 1999, pp. 1367–1374.
42. R. Handy, D.A. Barnett, R.W. Purves, G. Horlick and R. Guevremont. Determination of nanomolar levels of perchlorate in water by ESI-FAIMS-MS, *J Anal Atom Spectrom*, vol. 15, no. 8, 2000, pp. 907–911.
43. M.A. McCooeye, Z. Mester, B. Ells, D.A. Barnett, R.W. Purves and R. Guevremont. Quantitation of amphetamine, methamphetamine, and their methylenedioxy derivatives in urine by solid-phase microextraction coupled with electrospray ionization-high-field asymmetric waveform ion mobility spectrometry–mass spectrometry, *Anal Chem*, vol. 74, no. 13, 2002, pp. 3071–3075; doi 10.1021/Ac011296+.
44. M.A. McCooeye, B. Ells, D.A. Barnett, R.W. Purves and R. Guevremont. Quantitation of morphine and codeine in human urine using high-field asymmetric waveform ion mobility spectrometry (FAIMS) with mass spectrometric detection, *J Anal Toxicol*, vol. 25, no. 2, 2001, pp. 81–87.
45. J.T. Kapron, M. Jemal, G. Duncan, B. Kolakowski and R. Purves. Removal of metabolite interference during liquid chromatography/tandem mass spectrometry using high-field asymmetric waveform ion mobility spectrometry, *Rapid Commun Mass Sp*, vol. 19, no. 14, 2005, pp. 1979–1983; doi 10.1002/Rcm.2016.
46. G.A. Eiceman, E.G. Nazarov, R.A. Miller, E.V. Krylov and A.M. Zapata. Micro-machined planar field asymmetric ion mobility spectrometer as a gas chromatographic detector, *Analyst*, vol. 127, no. 4, 2002, pp. 466–471; doi 10.1039/b111547m.
47. G.A. Eiceman, B. Tadjikov, E. Krylov, E.G. Nazarov, R.A. Miller, J. Westbrook and P. Funk. Miniature radio-frequency mobility analyzer as a gas chromatographic detector for oxygen-containing volatile organic compounds, pheromones and other insect attractants, *J Chromatogr A*, vol. 917, no. 1–2, 2001, pp. 205–217.
48. N.P.V. Nielsen, J.M. Carstensen and J. Smedsgaard. Aligning of single and multiple wavelength chromatographic profiles for chemometric data analysis using correlation optimised warping, *J Chromatogr A*, vol. 805, no. 1–2, 1998, pp. 17–35; doi 10.1016/S0021-9673(98)00021-1.
49. R.H. Wallis. An approach for the space variant restoration and enhancement of images, In *Proc. Symposium on Current Mathematical Problems in Image Scenes*, Monterey, CA, 1976.
50. G. Tomasi, F. van den Berg and C. Andersson. Correlation optimized warping and dynamic time warping as preprocessing methods for chromatographic data, *J Chemometr*, vol. 18, no. 5, 2004, pp. 231–241; doi 10.1002/Cem.859.
51. A.M. van Nederkassel, M. Daszykowski, P.H.C. Eilers and Y.V. Heyden. A comparison of three algorithms for chromatograms alignment, *J Chromatogr A*, vol. 1118, no. 2, 2006, pp. 199–210; doi 10.1016/j.chroma.2006.03.114.
52. G. Tomasi, F. Savorani and S.B. Engelsen. icoshift: An effective tool for the alignment of chromatographic data, *J Chromatogr A*, vol. 1218, no. 43, 2011, pp. 7832–7840; doi 10.1016/j.chroma.2011.08.086.
53. R.J.O. Torgrip, M. Aberg, B. Karlberg and S.P. Jacobsson. Peak alignment using reduced set mapping, *J Chemometr*, vol. 17, no. 11, 2003, pp. 573–582; doi 10.1002/Cem.824.
54. A. Savitzky and M.J.E. Golay. Smoothing + differentiation of data by simplified least squares procedures, *Anal Chem*, vol. 36, no. 8, 1964, pp. 1627–1639; doi 10.1021/Ac60214a047.
55. J. Riordon, E. Zubritsky and A. Newman. Top 10 articles – analytical chemistry looks at 10 seminal papers, *Anal Chem*, vol. 72, no. 9, 2000, pp. 324a–329a; doi 10.1021/Ac002801q.
56. P.A. Gorry. General least-squares smoothing and differentiation by the convolution (Savitzky–Golay) method, *Anal Chem*, vol. 62, no. 6, 1990, pp. 570–573; doi 10.1021/Ac00205a007.
57. R.J. Barsanti and J. Gilmore. Comparing noise removal in the wavelet and Fourier domains, In *2011 Proceedings of IEEE 43rd Southeastern Symposium on System Theory (SSST 2011)*, Auburn, AL, 2011, pp. 163–167; doi 10.1109/ssst.2011.5753799.

58. P.H.C. Eilers and H.F.M. Boelens. Baseline correction with asymmetric least squares smoothing, Leiden University Medical Centre Report, 2005.
59. A.C. Sauve and T.P. Speed. Normalization, baseline correction and alignment of high-throughput mass spectrometry data, 2004.
60. R. Perez-Pueyo, M.J. Soneira and S. Ruiz-Moreno. Morphology-based automated baseline removal for Raman spectra of artistic pigments, *Appl Spectrosc*, vol. 64, no. 6, 2010, pp. 595–600.
61. S.E. Stein. An integrated method for spectrum extraction and compound identification from gas chromatography/mass spectrometry data, *J Am Soc Mass Spectr*, vol. 10, no. 8, 1999, pp. 770–781; doi 10.1016/S1044-0305(99)00047-1.
62. J. Biller and K. Biemann. Reconstructed mass spectra, a novel approach for the utilization of gas chromatograph—mass spectrometer data, *Anal Lett*, vol. 7, no. 7, 1974, pp. 515–528.
63. B.N. Colby. Spectral deconvolution for overlapping Gc Ms components, *J Am Soc Mass Spectr*, vol. 3, no. 5, 1992, pp. 558–562; doi 10.1016/1044-0305(92)85033-G.
64. R.G. Dromey, M.J. Stefik, T.C. Rindfleisch and A.M. Duffield. Extraction of mass-spectra free of background and neighboring component contributions from gas chromatography mass spectrometry data, *Anal Chem*, vol. 48, no. 9, 1976, pp. 1368–1375; doi 10.1021/Ac50003a027.
65. D.I. Broadhurst and D.B. Kell. Statistical strategies for avoiding false discoveries in metabolomics and related experiments, *Metabolomics*, vol. 2, no. 4, 2006, pp. 171–196; doi 10.1007/s11306-006-0037-z.
66. D.B. Kell. Metabolomics, modelling and machine learning in systems biology—towards an understanding of the languages of cells, *FEBS J*, vol. 273, no. 5, 2006, pp. 873–894; doi 10.1111/j.1742-4658.2006.05136.x.
67. M. Eliasson, S. Rannar and J. Trygg. From data processing to multivariate validation—essential steps in extracting interpretable information from metabolomics data, *Curr Pharm Biotechnol*, vol. 12, no. 7, 2011, pp. 996–1004.
68. O. Beckonert, H.C. Keun, T.M.D. Ebbels, J.G. Bundy, E. Holmes, J.C. Lindon and J.K. Nicholson. Metabolic profiling, metabolomic and metabonomic procedures for NMR spectroscopy of urine, plasma, serum and tissue extracts, *Nat Protoc*, vol. 2, no. 11, 2007, pp. 2692–2703; doi 10.1038/nprot.2007.376.
69. A.K. Pavlou and A.P.F. Turner. Sniffing out the truth: Clinical diagnosis using the electronic nose, *Clin Chem Lab Med*, vol. 38, no. 2, 2000, pp. 99–112; doi 10.1515/cclm.2000.016.
70. P. Montuschi, M. Santonico, C. Mondino, G. Pennazza, G. Mantini, E. Martinelli, R. Capuano, G. Ciabattoni, R. Paolesse, C. Di Natale, P.J. Barnes and A. D'Amico. Diagnostic performance of an electronic nose, fractional exhaled nitric oxide, and lung function testing in asthma, *Chest*, vol. 137, no. 4, 2010, pp. 790–796; doi 10.1378/chest.09-1836.
71. M.L. McConnell, G. Rhodes, U. Watson and M. Novotny. Application of pattern-recognition and feature extraction techniques to volatile constituent metabolic profiles obtained by capillary gas-chromatography, *J Chromatogr*, vol. 162, no. 4, 1979, pp. 495–506; doi 10.1016/s0378-4347(00)81830-7.
72. S. Aathithan, J.C. Plant, A.N. Chaudry and G.L. French. Diagnosis of bacteriuria by detection of volatile organic compounds in urine using an automated headspace analyzer with multiple conducting polymer sensors, *J Clin Microbiol*, vol. 39, no. 7, 2001, pp. 2590–2593; doi 10.1128/jcm.39.7.2590-2593.2001.
73. A.C. Hauschild, J.I. Baumbach and J. Baumbach. Integrated statistical learning of metabolic ion mobility spectrometry profiles for pulmonary disease identification, *Genet Mol Res*, vol. 11, no. 3, 2012, pp. 2733–2744; doi 10.4238/2012.July.10.17.
74. H. Schmidt, F. Tadjimukhamedov, I.V. Mohrenz, G.B. Smith and G.A. Eiceman. Microfabricated differential mobility spectrometry with pyrolysis gas chromatography for chemical characterization of bacteria, *Anal Chem*, vol. 76, no. 17, 2004, pp. 5208–5217.
75. M. Shnayderman, B. Mansfield, P. Yip, H.A. Clark, M.D. Krebs, S.J. Cohen, J.E. Zeskind, E.T. Ryan, H.L. Dorkin, M.V. Callahan, T.O. Stair, J.A. Gelfand, C.J. Gill, B. Hitt and C.E. Davis. Species-specific bacteria identification using differential mobility spectrometry and bioinformatics pattern recognition, *Anal Chem*, vol. 77, no. 18, 2005, pp. 5930–5937.

76. M.D. Krebs, B. Mansfield, P. Yip, S.J. Cohen, A.L. Sonenshein, B.A. Hitt and C.E. Davis. Novel technology for rapid species-specific detection of Bacillus spores, *Biomol Eng*, vol. 23, no. 2–3, 2006, pp. 119–127; doi 10.1016/j.bioeng.2005.12.003.
77. S. Ayer, W.X. Zhao and C.E. Davis. Differentiation of proteins and viruses using pyrolysis gas chromatography differential mobility spectrometry (PY/GC/DMS) and pattern eecognition, *IEEE Sens J*, vol. 8, no. 9–10, 2008, pp. 1586–1592; doi 10.1109/jsen.2008.923586.
78. W. Cheung, Y. Xu, C.L.P. Thomas and R. Goodacre. Discrimination of bacteria using pyrolysis-gas chromatography-differential mobility spectrometry (Py-GC-DMS) and chemometrics, *Analyst*, vol. 134, no. 3, 2009, pp. 557–563; doi 10.1039/b812666f.
79. W. Zhao, S. Sankaran, A.M. Ibanez, A.M. Dandekar and C.E. Davis. Two-dimensional wavelet analysis based classification of gas chromatogram differential mobility spectrometry signals, *Anal Chim Acta*, vol. 647, no. 1, 2009, pp. 46–53; doi 10.1016/j.aca.2009.05.029.
80. W. Arlt, M. Biehl, A.E. Taylor, S. Hahner, R. Libe, B.A. Hughes, P. Schneider, D.J. Smith, H. Stiekema, N. Krone, E. Porfiri, G. Opocher, J. Bertherat, F. Mantero, B. Allolio, M. Terzolo, P. Nightingale, C.H.L. Shackleton, X. Bertagna, M. Fassnacht and P.M. Stewart. Urine steroid metabolomics as a biomarker tool for detecting malignancy in adrenal tumors, *J Clin Endocrinol Metab*, vol. 96, no. 12, 2011, pp. 3775–3784; doi 10.1210/jc.2011-1565.
81. E.F. Petricoin, C.P. Paweletz and L.A. Liotta. Clinical applications of proteomics: Proteomic pattern diagnostics, *J Mammary Gland Biol Neoplasia*, vol. 7, no. 4, 2002, pp. 433–440; doi 10.1023/a:1024042200521.
82. M. Hilario, A. Kalousis, M. Muller and C. Pellegrini. Machine learning approaches to lung cancer prediction from mass spectra, *Proteomics* vol. 3, no. 9, 2003, pp. 1716–1719; doi 10.1002/pmic.200300523.
83. R.H. Lilien, H. Farid and B.R. Donald. Probabilistic disease classification of expression-dependent proteomic data from mass spectrometry of human serum, *J Comput Biol*, vol. 10, no. 6, 2003, pp. 925–946; doi 10.1089/106652703322756159.
84. D. Agranoff, D. Fernandez-Reyes, M.C. Papadopoulos, S.A. Rojas, M. Herbster, A. Loosemore, E. Tarelli, J. Sheldon, A. Schwenk, R. Pollak, C.F.J. Rayner and S. Krishna. Identification of diagnostic markers for tuberculosis by proteomic fingerprinting of serum, *Lancet*, vol. 368, no. 9540, 2006, pp. 1012–1021; doi 10.1016/s0140-6736(06)69342-2.
85. W.X. Zhao and C.E. Davis. Swarm intelligence based wavelet coefficient feature selection for mass spectral classification: An application to proteomics data, *Anal Chim Acta*, vol. 651, no. 1, 2009, pp. 15–23; doi 10.1016/j.aca.2009.08.008.
86. Z.M. Zhang, J.J. Cai, G.H. Ruan and G.K. Li. The study of fingerprint characteristics of the emanations from human arm skin using the original sampling system by SPME-GC/MS, *J Chromatogr B-Anal Technol Biomed Life Sci*, vol. 822, no. 1–2, 2005, pp. 244–252; doi 10.1016/j.jchromb.2005.06.026.
87. D.J. Penn, E. Oberzaucher, K. Grammer, G. Fischer, H.A. Soini, D. Wiesler, M.V. Novotny, S.J. Dixon, Y. Xu and R.G. Brereton. Individual and gender fingerprints in human body odour, *J R Soc Interface*, vol. 4, no. 13, 2007, pp. 331–340; doi 10.1098/rsif.2006.0182.
88. W. Zhu, X.N. Wang, Y.M. Ma, M.L. Rao, J. Glimm and J.S. Kovach. Detection of cancer-specific markers amid massive mass spectral data, *Proc Natl Acad Sci USA*, vol. 100, no. 25, 200, pp. 14,666–14,671; doi 10.1073/pnas.2532248100.
89. K. Matsumura, M. Opiekun, H. Oka, A. Vachani, S.M. Albelda, K. Yamazaki and G.K. Beauchamp. Urinary volatile compounds as biomarkers for lung cancer: a proof of principle study using odor signatures in mouse models of lung cancer, *Plos One*, vol. 5, no. 1, 2010, e8819; doi 10.1371/journal.pone.0008819.
90. A.D. Patterson, O. Maurhofer, D. Beyoglu, C. Lanz, K.W. Krausz, T. Pabst, F.J. Gonzalez, J.F. Dufour and J.R. Idle. Aberrant lipid metabolism in hepatocellular carcinoma revealed by plasma metabolomics and lipid profiling, *Cancer Res*, vol. 71, no. 21, 2011, pp. 6590–6600; doi 10.1158/0008-5472.can-11-0885.
91. Y. Hanai, K. Shimono, K. Matsumura, A. Vachani, S. Albelda, K. Yamazaki, G.K. Beauchamp and H. Oka. Urinary volatile compounds as biomarkers for lung cancer, *Biosci Biotechnol Biochem*, vol. 76, no. 4, 2012, pp. 679–684; doi 10.1271/bbb.110760.

92. C.L. Silva, M. Passos and J.S. Camara. Solid phase microextraction, mass spectrometry and metabolomic approaches for detection of potential urinary cancer biomarkers—a powerful strategy for breast cancer diagnosis, *Talanta*, vol. 89, 2012, pp. 360–368; doi 10.1016/j.talanta.2011.12.041.
93. M. Phillips, R.N. Cataneo, R. Condos, G.A.R. Erickson, J. Greenberg, V. La Bombardi, M.I. Munawar and O. Tietje. Volatile biomarkers of pulmonary tuberculosis in the breath, *Tuberculosis*, vol. 87, no. 1, 2007, pp. 44–52; doi 10.1016/j.tube.2006.03.004.
94. L.C. Kenny, W.B. Dunn, D.I. Ellis, J. Myers, P.N. Baker, D.B. Kell and the Gopec Consortium. Novel biomarkers for pre-eclampsia detected using metabolomics and machine learning, *Metabolomics*, vol. 1, no. 3, 2005, pp. 227–234; doi 10.1007/s11306-005-0003-1.
95. H.R.T. Williams, I.J. Cox, D.G. Walker, B.V. North, V.M. Patel, S.E. Marshall, D.P. Jewell, S. Ghosh, H.J.W. Thomas, J.P. Teare, S. Jakobovits, S. Zeki, K.I. Welsh, S.D. Taylor-Robinson and T.R. Orchard. Characterization of inflammatory bowel disease with urinary metabolic profiling, *Am J Gastroenterol*, vol. 104, no. 6, 2009, pp. 1435–1444.
96. A.N. Thomas, S. Riazanskaia, W. Cheung, Y. Xu, R. Goodacre, C.L.P. Thomas, M.S. Baguneid and A. Bayat. Novel noninvasive identification of biomarkers by analytical profiling of chronic wounds using volatile organic compounds, *Wound Repair Regen*, vol. 18, no. 4, 2010, pp. 391–400; doi 10.1111/j.1524-475X.2010.00592.x.
97. J.R. Denery, A.A.K. Nunes, M.S. Hixon, T.J. Dickerson and K.D. Janda. Metabolomics-based discovery of diagnostic biomarkers for onchocerciasis, *Plos Negl Trop Dis*, vol. 4, no. 10, 2010, e834; doi 10.1371/journal.pntd.0000834.
98. R. Romoli, M.C. Papaleo, D. de Pascale, M.L. Tutino, L. Michaud, A. LoGiudice, R. Fani and G. Bartolucci. Characterization of the volatile profile of Antarctic bacteria by using solid-phase microextraction-gas chromatography-mass spectrometry, *J Mass Spectrom*, vol. 46, no. 10, 2011, pp. 1051–1059; doi 10.1002/jms.1987.
99. F. Reed. Promise of AI not so bright, *The Washington Times*, 2006.
100. Editorial, *AI Set to Exceed Human Brain Power*, 2006.

19

A Comparison of Seven Discretization Techniques Used for Rule Induction from Data on the Lazy Eye Vision Disorder

Patrick G. Clark, Jerzy W. Grzymala-Busse, and Gerhard W. Cibis

CONTENTS

19.1 Introduction

The human eye is one of the most intricate parts of the human body. It has astounded doctors and scientists and inspired poets and painters with its graceful movement and piercing precision, giving light to life. With this complexity sometimes come problems. Small defects in the development of the eye can lead to a lifelong handicap, where early diagnosis can dramatically improve the outcomes for many young patients. Over decades of ophthalmological research, scientists have developed methods and techniques to provide accurate diagnosis of these defects. However, the scarce number of qualified physicians to help the general population restricts those that can benefit from these developments. In many situations, the underlying causes are minor eye problems and, left untreated, can result in impaired vision or blindness.

In this chapter we look at a specific vision disorder called amblyopia or what is commonly referred to as "lazy eye." This condition is a developmental disorder of the visual system caused by ocular abnormalities early in life. Whether the root cause is strictly physical or is a combination of both physical and neurological problems is still being debated. However, what experts in the field do agree on are the treatments for the disorder. While surgery or optical correction of refractive errors can often address the initial cause of amblyopia, once amblyopia has developed, such interventions cannot restore visual function since amblyopia itself is a cortical deficit, a neurological disorder and not a physical one [1]. Amblyopia has two primary causes, namely, strabismus and anisometropia. Strabismus is a misalignment between the two eyes. Anisometropia is when the refractive error between the two eyes is different [2]. The reason that these two conditions can lead to amblyopia is because they cause the brain to begin ignoring the signals from the weaker or blurrier eye. Corrective action after amblyopia has developed becomes problematic, as the brain will not be able to regenerate the neural pathways. Thus, early detection is essential for the patient to have a healthy visual outcome. Fortunately, amblyopia can be successfully treated if identified when the patient's brain is still in the developmental stages, generally when the patient is younger than 6 years old, with the noncontroversial methods of glasses and patching therapy achieving 80% to 90% effectiveness [3].

The most pervasive challenge to early diagnosis is the accessibility to specialists to screen patients at an early age to identify the condition. Because the current methods of identifying the problem require well-trained operators or even medical doctors, the number of personnel available to evaluate even a small fraction of the worldwide population is not sufficient. It has been reported that amblyopia affects from 2% to 5% of the population [4,5]. Unfortunately, a large percentage of the population lacks access to proper vision care to facilitate treatment of the problem in the optimal years (prior to 6 years of age). In addition, a study by Webber and Wood [4] shows that populations undergoing early intervention have a lower prevalence of amblyopia than those that do not. This implies that the condition does not improve on its own accord and further supports the need for early accurate detection and treatment. The socioeconomic factors of amblyopia are similar to other medical issues that arise when a population does not have reliable access to medical care.

The ideal solution for vision diagnosis of large numbers of the population is a self-contained, low-cost, completely automated system capable of accurate identification of disorders, all with minimal operator training, patient cooperation, and safe to be used with small children. One approach for this sort of solution is based on the work of Cibis [3] and Wang [6], pioneering the science of analyzing images for identifying features that may indicate the development of amblyopia and using artificial intelligence techniques to automate the process [7].

In this chapter, we study the same data sets that have been analyzed with other classifiers and instead measure the utility of different discretization methods for the numeric attributes.

19.2 Background and Related Work

In this section, we will present problems associated with human vision, vision disorders, and vision screening.

19.2.1 Human Vision

A diopter is a unit of measure used for describing the optical power of a lens. When used in the context of the human eye, the measure is used for describing the degree of focusing error, or refractive error, in the eye. For a perfect eye, the refractive error is 0.00 diopters (D), while a hyperopic eye has a positive refractive error (e.g., +3.00 D), and a myopic eye will have negative refractive error (e.g., –1.50 D). Typically, the refractive error is measured to the nearest quarter diopter [7].

Hyperopia is the medical term for a person who is farsighted. Medically speaking, a farsighted eye will focus the light through the lens behind the retina and will cause the image to appear blurry. The hyperopic condition can be dealt with using a positive diopter lens or a convex lens. Myopia is the medical term for a person who is nearsighted. Medically speaking, a nearsighted eye will focus the light through the lens in front of the retina and will cause the image to appear blurry. The myopic condition can be dealt with using a negative diopter lens or a concave lens [7].

19.2.2 Vision Disorders

Amblyopia is the primary vision disorder this research attempts to accurately identify. It has two primary physical causes, anisometropia and strabismus. Anisometropia is a condition where the refractive error in one eye is significantly different than the other. The difference in the refractive errors is difficult to overcome for a developing visual cortex, primarily due to the very different images being presented. Studies have shown that anisometropia is the predisposing condition that leads to amblyopia 50% of the time and that an undiagnosed anisometropia will lead to strabismus [2]. Strabismus can be identified as a misalignment of the focal point between the two eyes, but the condition is not always identifiable with the naked eye and may require a thorough screening. Strabismus typically involves a lack of coordination between the two eyes and the extraocular muscles where the patient is unable to bring both eyes into focus on the same point in space, thus preventing proper binocular vision. One screening test is the Hirschberg test, where a light is reflected off the patient's eyes. The screening involves measuring the point of reflection at each eye. If the reflection is at the same place on both eyes, the eyes are properly aligned [2].

Both of these physical abnormalities in the eye have the potential to cause the development of a patient's visual function to be impaired and cause the image from the amblyopic eye to be disregarded by the visual cortex. If the condition is allowed to persist, the neural pathways become permanently formed, and the use of the amblyopic eye is diminished. The degree to which it is diminished varies based on how early the condition developed in the patient's life and if any remediation treatment was used [2].

19.2.3 Vision Screening

The Hirschberg test is a screening test used in the fields of ophthalmology to identify strabismus. Julius Hirschberg pioneered the procedure in 1886 when he used a candle to observe the reflection of light through the patient's corneas [8]. In a patient with normal eye function, the reflection of the light from the cornea appears in the center of the eye. However, those patients with abnormal eye function will reflect the light off-center, and the degree in which the reflection is off-center can be measured to determine the degree of misalignment [8]. The Hirschberg point is the point on the cornea where light is reflected, and a ratio is calculated based on that point and the edge of the pupil. The difference in the ratio is what determines the degree of misalignment between the two eyes [9]. Today, the procedure is performed with more sophisticated and sensitive tools, but the general idea presented by Hirschberg remains the same.

The red reflex that occurs when a light source shines through the lens of the eye and is reflected off the retina back to the observer is referred to as the Bruchner reflex [9]. It is named for its modern proponent, who used an ophthalmoscope to view and measure the red reflex and compare the difference between the two eyes [9]. When the Bruchner reflex is abnormal, the patient is considered to have positive Bruchner reflex. With a positive Bruchner test, the deviated eye is the brighter of the two [9]. Bruchner reflex is a key feature that is used as a marker to identify a physical abnormality of the eyes that could lead to the amblyopic condition.

Foveation is used to identify the point of true fixation from a slightly off-axis fixation.

Patients will occasionally experience an intermittent deviation of 6° or less from true fixation [9]. An examination of recorded patient video reveals foveation when the focus goes from fixation to slightly off-axis fixation. The change will typically happen in one frame and is approximately 1/30 of a second [7]. Cibis argues that this sort of behavior may hold a large amount of information regarding the true vision of a patient.

19.2.4 Traditional Vision Screening

Traditional vision screening is based on the Snellen E chart or the STYCAR graded-balls test [10]. These involve the patient sitting with a trained specialist (optometrist or ophthalmologist). The Snellen E chart is printed with eleven lines of block letters, where the first line consists of one very large letter and the subsequent rows have increasing numbers of letters that decrease in size. A patient taking the test is positioned 20 ft from the chart, covers one eye, and reads aloud the letters of each row, with the smallest row that can be read accurately indicating the patient's visual acuity in that eye. For example, a patient who needs to stand 20 ft away from a target that could be seen at 40 ft by a standard patient is said to have 20/40 vision. The STYCAR graded-balls test involves using a white ball before a dark background, with a specialist observing the patient. The specialist will hold the white ball approximately 1.2 ft from the patient and measure the time needed for a patient to fixate on the ball. In addition, measurements of peripheral vision are also taken using this method [11]. The specialist uses these two tests to identify if the patient is at risk for developing amblyopia.

19.2.5 Photorefractive Screening

Photorefractive screening is based on a system to interpret the images of the eyes. It does not directly identify amblyopia, but it looks for defects in the eyes that may lead to amblyopia. Examples of these systems include Photoscreener, RetinoMax K-Plus 2, SureSight Vision Screener, Pediatric Vision Screener, MTI Photoscreener (Medical Technology and Innovations, Inc., Lancaster, PA), and Visiscreen [7,12].

The Photoscreener system is based on single images of the patient's eyes. The system is a handheld camera with instructions for the operator on how to identify potential problems. The operator takes a picture of each eye and then analyzes the resulting frame according to the instructions. If the frame shows markers of a vision disorder, the patient is identified for referral. RetinoMax K-Plus 2 and SureSight Vision Screener are two automated screening systems that are commercially available. The benefit of this sort of system is that the subjective nature of an operator's analysis is eliminated, and a more consistent result should be evident [7]. However, the drawback is that it operates on single-frame analysis and not on a video of the patient looking at a light source. In addition, it is more focused on identifying general vision problems than those related to amblyopia. Pediatric Vision Screener, instead of taking pictures with a recording device, measures the frequency of the polarized light off the retina as a light source circles the eye. By comparing the results of the test on both eyes, the system is able to identify strabismus. The system requires the cooperation of the patient, which can be a problem when dealing with children younger than 6 [7]. MTI Photoscreener and Visiscreen systems are not automated and require operation and analysis by skilled professionals [12].

While all of these systems hold promise, they all fall short in one of three ways. First, they require patient cooperation that would typically be beyond the demographic that would be helped by the amblyopic screening. Second, they operate only on single-frame analysis and can miss the subtle foveation that may help to more accurately identify a vision problem. Finally, these systems require a well-trained operator or a specialist to analyze the results.

A system for identifying strabismus and amblyopia using video images, called video vision development assessment (VVDA), has been developed [3]. The method involves a consumer-grade video camera with a light source attached to the base of the camera. The patient sits approximately 52 in. from the camera and looks at the light source while approximately 2 min of video is recorded. The video is then digitized and analyzed by specialists or trained technicians to determine if the patient should be referred to a specialist. This processing is generally divided into frame selection and feature extraction.

19.2.6 Automated Photorefractive Screening

More recently, research has been performed in order to automate the analysis of the frames using artificial intelligence techniques [6,7,13]. These works focused on implementing the image processing *and* case-based reasoning [14] algorithms that constituted the first version of the automated video vision development assessment (AVVDA) system. In a completely automated fashion, they were able to identify the key frames of the video, isolate the pupils, and locate the Hirschberg point [6,13]. Figure 19.1 shows the image output from the AVVDA system, where the Hirschberg reflex and iris diameter are highlighted. The automated photorefractive screening system works by having an operator take a short video of the patient, which is then analyzed automatically by the

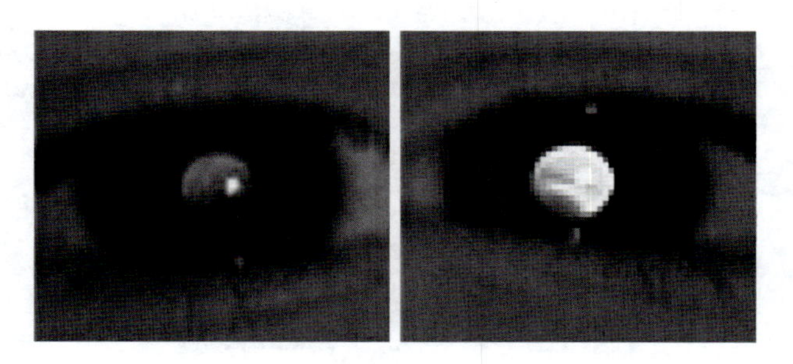

FIGURE 19.1
Key frame output from the AVVDA system.

software in the following manner: Initially, the software identifies the frames where both eyes are open and looking at the light source. These frames are identified as key frames. Next, the software isolates the location of the eyes and pupils in the key frames. Finally, the software uses various techniques to extract the distinguishing features that may be indicators for amblyopia.

In order to further enhance AVVDA, researchers have investigated using the same feature set with a different set of classifiers, with the overall goal of the AVVDA system being to allow an unskilled technician to operate the system and accurately obtain a decision about patient referral to an optometrist or ophthalmologist [7]. In the current form, AVVDA uses case-based reasoning [14], C4.5 decision tree [15], and artificial neural network [16] classifiers to assist in making the decision. Fifty-four features are extracted from the images to train the classifiers.

At this point, various artificial intelligence techniques have been utilized to automatically produce the referral to a specialist [3,7]. The results reported in this chapter expand upon the automated photorefractive screening method, with the goal to evaluate other classifiers, as a means to produce more accurate results from the same feature set.

19.3 Discretization Methods

All three data sets describing lazy eye vision disorder and presented in Table 19.1 have numerical attributes. Typically, for mining such data, a preliminary step called discretization is applied. During discretization, numerical data are converted into symbolic ones (a domain of the numerical attribute is partitioned into intervals).

Discretization methods in which attributes are processed one at a time are called *local* [17,18] (or *static* [19]). If all attributes are taken into account during selection of the best cut point, the method is called *global* [17,18] (or *dynamic* [19]). Additionally, if information about the expert's classification of cases is included in the process of discretization, the method is called *supervised* [19].

The data mining system Learning from Examples based on Rough Sets (LERS) uses a number of discretization algorithms [20]. In our experiments, we used five approaches to discretization based on cluster analysis and two well-known discretization methods based on equal frequency per interval and on minimal conditional entropy. All of these methods are global and supervised. All numerical attributes were normalized [21] (attribute values were divided by the attribute standard deviation).

In our first discretization technique, based on agglomerative cluster analysis [21], initially, each case is a single cluster, and then the clusters are fused together, forming larger

TABLE 19.1

Data Sets Used for Experiments

Cases	Attributes	Data Set
725	54	Middle frame
723	552	Key frame (all pixels)
499	180	Slope frame

and larger clusters. In remaining four cluster analysis discretization methods, where we used divisive techniques, initially, all cases are grouped in one cluster, and then this cluster is gradually divided into smaller and smaller clusters. In both methods, during the first step of discretization, *cluster formation*, cases that exhibit the most similarity are fused into clusters.

Once clusters are formed, the postprocessing starts. Initially, clusters are projected on all attributes. Then the resulting intervals are merged to reduce the number of intervals and, at the same time, to preserve consistency. Merging of intervals begins from *safe merging*, where, for each attribute, their union replaces neighboring intervals labeled by the same decision value. The next step of merging intervals is based on checking every pair of neighboring intervals for whether their merging will result in preserving consistency. If so, intervals are merged permanently. If not, they are marked as unable to be merged. Obviously, the order in which pairs of intervals are selected affects the final outcome. In our experiments we used two criteria:

- Start from an attribute with the most intervals first.
- Start from an attribute with the largest conditional entropy of the decision given the attribute. Conditional entropy is a measure to quantify the amount of information gained by the attribute conditioned on the decision.

In our experiments using divisive cluster analysis, four different discretization methods were used. These methods are named on the basis of two questions asked by the LERS system: first, whether to use divisive cluster analysis (0 means use cluster analysis, 1 means skip cluster analysis and project original, single cases on all attributes and continue with postprocessing), and second, whether to process an attribute with the most intervals first (denoted by 0) or whether to process an attribute with the largest conditional entropy first (denoted by 1). Thus, our methods of discretization are denoted by 00, 01, 10, or 11. For example, 00 denotes using cluster analysis and processing attributes with the most intervals first.

In Tables 19.2 through 19.4, discretization techniques are named as follows:

- **aca**, a discretization technique based on agglomerative cluster analysis [17]
- **dca-00**, a discretization technique based on divisive cluster analysis [22]
- **dca-01**, a discretization technique based on divisive cluster analysis [22]

TABLE 19.2

Error Rates

	Error Rate (%)		
Data Set	**Middle Frame**	**Key Frame (All Pixels)**	**Slope Frame**
aca	39.72	40.25	43.49
dca-00	42.07	42.60	44.49
dca-01	37.93	42.60	44.69
dca-10	37.66	44.67	43.49
dca-11	37.10	44.67	43.49
entropy	35.45	38.45	38.88
equal frequency	36.55	39.56	40.28

TABLE 19.3
Complexity of Rule Sets—Number of Rules

	Number of Rules		
Data Set	**Middle Frame**	**Key Frame (All Pixels)**	**Slope Frame**
aca	193	59	90
dca-00	191	61	92
dca-01	174	61	89
dca-10	186	63	90
dca-11	250	63	90
entropy	137	52	64
equal frequency	136	73	79

TABLE 19.4
Complexity of Rule Sets—Number of Conditions

	Number of Rules		
Data Set	**Middle Frame**	**Key Frame (All Pixels)**	**Slope Frame**
aca	955	1041	707
dca-00	1114	1092	755
dca-01	1015	1092	699
dca-10	1014	1002	707
dca-11	819	1002	707
entropy	810	736	525
equal frequency	826	676	465

- **dca-10**, a discretization technique based on divisive cluster analysis [22]
- **dca-11**, a discretization technique based on divisive cluster analysis [22]
- **entropy**, a globalized discretization technique [17] based on minimal entropy [23]
- **equal frequency**, a globalized discretization technique [17] based on equal frequency of attribute values per interval [18]

19.4 Experiments

In this section, we describe data sets used for experiments and our validation technique.

19.4.1 Data Sets

For our experiments, we used three data sets, presented in Table 19.1.

19.4.1.1 Experimental Data Set 1: Color Density and Hirschberg Reflex

The first experimental data set finds its genesis in previous work from Wang [6,13] and the original AVVDA system. A total of 54 key features of patients' eyes (27 features per eye) were extracted from the video feeds and key frames. The two primary features are the pupil radius and the degree of fixation based on the Hirschberg point. The remaining 25 features for each eye were calculated using the color of pixels within 80% of the radius of the pupil. This includes, for example, the average red, green, and blue values throughout the pupil. The reader is referred to the work of Wang [6] for additional details on how these values are calculated. These 54 features for each of the 725 patients and all of the identified key frames were used as input into the discretization methods and modified learning from examples module, version 2 (MLEM2) classifier.

19.4.1.2 Experimental Data Set 2: Iris and Pupil Color Slope with Middle-Stack Key Frame Selection

The second experimental data set is based on the work of Clark et al. [24]. It utilizes the same 725 patient videos but instead extracts a different set of frame and feature data. The difference begins with the key frame concept. In the previous experiment, a patient will have typically produced many key frames. The software identifies a key frame based on two primary criteria. First, the patient must have both eyes open. Second, the patient must be looking at the light source so that a reflection off of the retina is clear. From the key frame stack identified for a patient, a single frame is then selected. In the second experiment, the research takes a very simple approach to identifying the single frame that will be used for feature extraction. It selects the frame from the middle of the stack of frames. If there are an odd number of frames, it selects the frame that is a whole-number division of the entire frame count by two. The goal is to avoid the fringe frames (the first or last), and analysis of a subset of the data shows that the middle frame has the greatest chance of being one of the best images [24]. Notably, 2 patients did not meet the criteria for frame selection; therefore, 723 patients participated in this experiment.

Now that a frame is selected, the color information is extracted from the image starting at the edges of the iris for both the left and right eyes. The red, blue, and green information is extracted from the eye image in a left-to-right pattern, representing the rate of change of color across they eye, or color slope. This is illustrated in Figure 19.2. The key frame for each patient is then further processed to produce 552 color features.

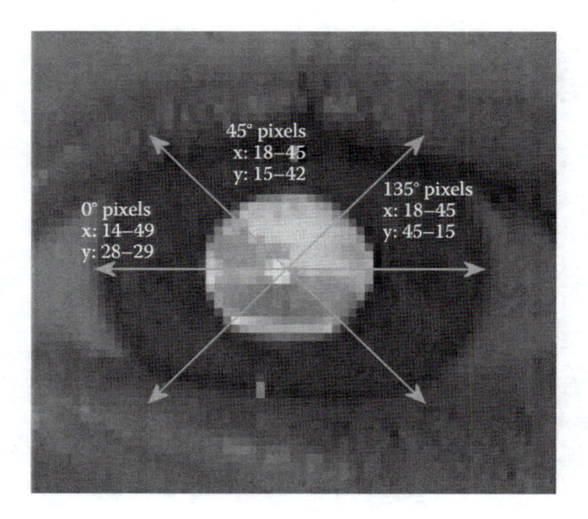

FIGURE 19.2
(See color insert.) Illustration of iris and pupil color slope.

19.4.1.3 Experimental Data Set 3: Pupil Color Slope with Level Key Frame Selection

The third experimental data set is also based on the work of Clark et al. [24]. Utilizing the same initial set of 725 patient videos, the goal is to find the best patient and frame sets for selecting a frame. The set of key frames is still identified as described in Section 19.2. However, the frame selection process was altered so that a method different from selecting the middle frame is utilized. In this case, the frame selected out of all the key frames is that where the patient's head is most level. Since the entire classification problem is centered on the reflection of light off the lens and the retina, keeping the angle of reflection the same for both eyes would, theoretically, further enhance any differences between a healthy eye and the eye of a patient who should be referred to a specialist. The important features extracted around the crescent reflection and the Hirschberg point that is measured from that crescent further support this concept. The frame that is most level is determined by the pixel location of the center of the pupil at the y-axis. The value is compared between the two eyes for each key frame, and the frame corresponding to the closest value is selected. When no key frame could be found where the difference in the height of the eyes was less than five pixels, then the patient data was discarded. The data was discarded so that the classifiers would not be trained with imperfect data and in hopes of further isolating the features that will yield the best results. This process reduced the number of patients in the entire sample from 725 to 499. The key frame is then further processed to extract color data from the pupil only. This process produced 180 features for each patient. Figure 19.3 illustrates the feature vectors extracted.

19.4.2 Validation

In our experiments, first, all three data sets were discretized using seven different approaches to discretization. For rule induction, the Learning from Examples Module, version 2 (LEM2), algorithm of the LERS data mining system was used [25,26]. Z. Pawlak introduced rough set theory in 1982 [27,28]. For validation, we used stratified tenfold cross validation. The LERS data mining system has two options: *variable* and *fixed* tenfold cross validation. In the former, different reordering is used for every experiment; in the latter, random reordering is the same for all experiments. In our experiments, for objective

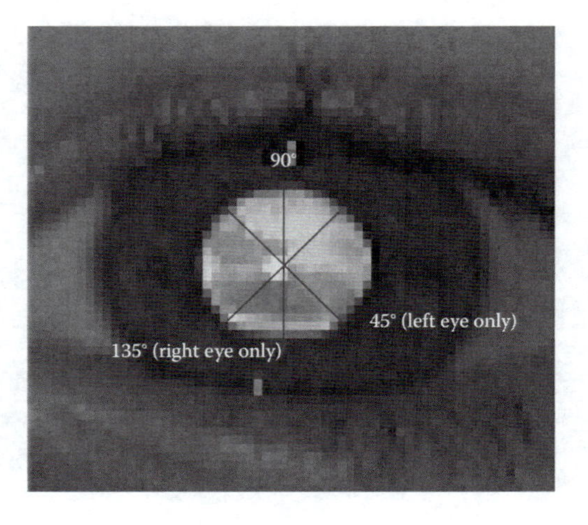

FIGURE 19.3
Illustration of pupil color slope.

comparison of performance of different discretization methods, we used fixed tenfold cross validation.

19.5 Conclusion

As seen in Table 19.2, it is quite clear that the best discretization method was based on conditional entropy. At the same time, rule sets induced from data sets discretized by the conditional entropy method had a smaller number of rules (for two data sets) and had a smaller total number of rules (for one data set). Note that the results of previous research [29] are quite different. (Six of the seven discretization methods—aca, dca-00, dca-01, dca-10, dca-11, and entropy—were incomparable; there was no best or worst method).

References

1. S. J. Anderson, I. E. Holliday, and G. F. Harding. Assessment of cortical dysfunction in human strabismic amblyopia using magnetoencephalography. *Vision Research*, 39:1723–1738, 1999.
2. S. B. Steinman, B. A. Steinman, and R. P. Garzia. *Foundations of Binocular Vision: A Clinical Perspective*. McGraw-Hill, 2000.
3. G. W. Cibis. Video vision development assessment in diagnosis and documentation of microtropia. *Binocular Vision and Strabismus Quarterly*, 20:151–158, 2005.
4. J. L. Webber and J. Wood. Amblyopia: Prevalence, natural history, functional effects and treatment. *Clinical and Experimental Optometry*, 88:365–375, 2005.
5. D. Robaei, E. Ojaimi, A. Kifley, F. Martin, and P. Mitchell. Causes and associations of amblyopia in a population-based sample of 6-year-old Australian children. *Archives of Ophthalmology*, 126:878–884, 2006.
6. T. Wang. Investigation of image processing and computer-assisted diagnosis system for automated video vision development assessment. PhD thesis, Computer Engineering and Computer Science, University of Missouri–Columbia, 2005.
7. J. Van Eenwyk, A. Agah, and G. Cibis. Automated human vision assessment using computer vision and artificial intelligence. In *Proceedings of the IEEE International Conference on System of Systems Engineering*, June 2008.
8. M. Wheeler. Objective strabismometry in young children. *Trans Am Opthalmol Society*, 40: 547–564, 1942.
9. G. W. Cibis. Video vision development assessment (VVDA): Combining the Bruckner test with eccentric photoreflection for dynamic identification of amblyogenic factors. *Transactions of the American Ophthalmological Society*, 84:643–685, 1994.
10. A. R. Kemper, P. A. Margolis, S. M. Downs, and W. C. Bordley. A systematic review of vision screening tests for the detection of amblyopia. *Pediatrics*, 104:1220–1222, 2007.
11. M. D. Sheridan. The STYCAR graded-balls vision test. *Developmental Medicine and Child Neurology*, 15:423–432, 1973.
12. H. L. Freedman and K. L. Preston. Polaroid photoscreening for amblyogenic factors. an improved methodology. *Ophthalmology*, 99:1785–1795, 1992.
13. T. Wang. Eye location and fixation estimation techniques for automated video vision development assessment. Master's thesis, Computer Engineering and Computer Science: University of Missouri–Columbia, 2002.

14. A. Aamodt and E. Plaza. Case-based reasoning: Foundational issues, methodological variations, and system approaches. *AICom—Artificial Intelligence Communications*, 7:39–59, 1994.
15. J. R. Quinlan. *C4.5: Programs for Machine Learning*. Morgan Kaufmann, 1993.
16. Y. Pao. *Adaptive Pattern Recognition and Neural Networks*. Addison-Wesley Publishing Company, 1989.
17. M. R. Chmielewski and J. W. Grzymala-Busse. Global discretization of continuous attributes as preprocessing for machine learning. *International Journal of Approximate Reasoning*, 15(4):319–331, 1996.
18. J. W. Grzymala-Busse. Discretization of numerical attributes. In W. Kloesgen and J. Zytkow, editors, *Handbook of Data Mining and Knowledge Discovery*, pages 218–225. Oxford University Press, New York, 2002.
19. J. Dougherty, R. Kohavi, and M. Sahami. Supervised and unsupervised discretization of continuous features. In *Proceedings of the 12th International Conference on Machine Learning*, pages 194–202, 1995.
20. J. W. Grzymala-Busse. Mining numerical data—a rough set approach. In *Proceedings of the RSEISP'2007, the International Conference of Rough Sets and Emerging Intelligent Systems Paradigms*, pages 12–21, 2007.
21. B. Everitt. *Cluster Analysis*. Heinemann Educational Books, London, 1980.
22. N. Peterson. Discretization using divisive cluster analysis and selected post-processing techniques. Internal Report, Department of Computer Science, University of Kansas, 1993.
23. U. M. Fayyad and K. B. Irani. On the handling of continuous-valued attributes in decision tree generation. *Machine Learning*, 8:87–102, 1992.
24. Patrick G. Clark, Arvin Agah, and Gerhard W. Cibis. Applied artificial intelligence techniques for identifying the lazy eye vision disorder. *Journal of Intelligent Systems*, 20(2):101–127, 2011.
25. J. W. Grzymala-Busse. LERS—a system for learning from examples based on rough sets. In R. Slowinski, editor, *Intelligent Decision Support. Handbook of Applications and Advances of the Rough Set Theory*, pages 3–18. Kluwer Academic Publishers, Dordrecht, 1992.
26. J. W. Grzymala-Busse. A new version of the rule induction system LERS. *Fundamenta Informaticae*, 31:27–39, 1997.
27. Z. Pawlak. Rough sets. *International Journal of Computer and Information Sciences*, 11:341–356, 1982.
28. Z. Pawlak. Rough Sets. *Theoretical Aspects of Reasoning about Data*. Kluwer Academic Publishers, Dordrecht, 1991.
29. P. Blajdo, J. W. Grzymala-Busse, Z. S. Hippe, M. Knap, T. Mroczek, and L. Piatek. A comparison of six approaches to discretization—a rough set perspective. In *Proceedings of the Rough Sets and Knowledge Technology Conference*, pages 31–38, 2008.

20

A Crash Introduction to Ambient Assisted Living

Manuel Fernández-Carmona and Cristina Urdiales

CONTENTS

20.1 From Traditional Domotics to Smart Houses

Ambient intelligence (AmI) was coined as a term by the European Commission, as discussed in the work of (Ducatel et al. 2001), and has received major attention in the last decade. An AmI environment is expected to be sensitive to the needs, personal requirements, and preferences of its inhabitants and capable of anticipating their needs and behavior to interact with people in a user-friendly way (Sadri 2011). There are many similar definitions, but most authors agree that all of them imply the capability of embedding sensors in everyday objects to raise context awareness and processors to learn and adapt to the user's behavior (Aarts et al. 2002; Gaggioli 2005). Indeed, if we plot applications related to AmI against embedding and context awareness, we obtain a diagram like Figure 20.1.

It can be noted that we use the term *smart* instead of *intelligent* because many authors state that we are still a long way off from real artificial intelligence (AI) environments. Similarly, we have chosen *ubiquity* instead of *embedding* to generalize our description. It can be observed that there are very simple systems that have been in use for decades, like *panic buttons* or *smoke alarms*. These systems do not require any intelligence and require only the most basic connectivity. Original air-conditioning systems used to be reactive too, meaning that a splitter would open or close depending uniquely on the local temperature. Current heating, ventilation, and air-conditioning (HVAC) systems, however, are a bit smarter and can be networked, temporized, and even remotely activated. If we go for more complex systems where input signals need to be further processed to choose a

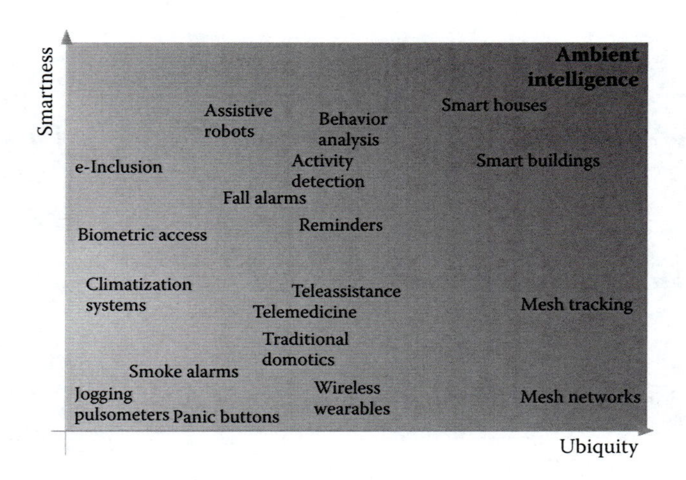

FIGURE 20.1
On the way to ambient intelligence.

proper action, we need to increase the system smartness (e.g., biometric access) or even gather information for a time period in order to build a user profile (e-inclusion). However, in these systems, connectivity is not vital, since no networking is needed in general. A first step in this direction would be wearable systems that are connected to a network during operation, usually via a personal area network (PAN), for example, the bracelets that people with mild cognitive problems wear for tracking. In this case, the device does not need to be too smart, just able to transmit. Please, note that this is not the same as a body sensor where one has to go to a station to download captured data from time to time, like a jogging pulse meter or a dive computer. Nevertheless, many traditional systems of telemedicine were based exactly on this: having the user go to a networking unit connected with a biometrics sensor(s) with some regularity (telemedicine at home) or when a specialist is not locally available (telemedicine at care centers). Communication and sensor technology has improved enough nowadays to actually be able to develop multiparametric wearable sensors that can connect to a wide area network (WAN) using a mobile phone as a gateway (e.g., Equivital). Alternatively, and specially if users are in their preferred environment, sensors can be deployed at their house. The first steps in this direction were traditional domotics systems, where several sensors were embedded in a house to trigger an alarm or activate some automations (e.g., light switching, smoke alarms, person detection, etc.). Most works in this area focused on developing a reliable standard to integrate devices from different manufacturers in a seamless way, like X10, I2C, or KNX. Current installations are also connected to the Internet through a gateway to enable interaction with devices via Web services, from a PC or a mobile phone. Many manufacturers aim at zero installation by deploying their devices in a wireless mesh network, placing them wherever they are supposed to operate and switching them on. Current mesh networks allow communication among a large number of sensors with an average battery life from 1.5 to 2 years, and there has been an effort to work with scavenger devices that can actually recharge from residual energy, like glass vibration due to city traffic. The best-known approach to scavengers would be passive radio-frequency identification (RFID), where tags can send their identification to a transmitter using just energy received from that transmitter. RFID has been widely used for context-based awareness, by tagging everyday objects with passive tags and learning their sequence of activation. This technique indirectly provides behavior and activity analysis without all privacy concerns related to video cameras. If we combine a ubiquitous mesh network with smart devices and processors, we evolve into smart homes, which would be close enough to current AmI environments.

According to Friedewald et al. (2005), smart homes offer one or more of the following features: (1) home automation; (2) communication and socialization; (3) rest, leisure, and sports; and (4) household work and learning. Home automation is the evolution of traditional domotics, whereas communication and socialization include communities and all profile-based e-services, including e-government and e-participation. All biometrics and fitness programs fall within the third category, and household work and learning correspond to smart actuators, like robot vacuum cleaners, smart washing machines, and so forth. These services are often coordinated by an agent architecture [e.g., MavHome (Cook et al. 2006), SHARE-it (Cortés et al. 2010), iDorm (Holmes et al. 2002)]. There are already many existing smart homes used for research around the world, most of them under the Living Lab paradigm. *Massachusetts Institute of Technology (MIT) House_n* (Intille 2002), for example, is used by the Open Source Building Alliance (OSBA) to test new technologies and services—with a focus on advanced sensor networks—under the joined effort of developers, architects, manufacturers, installers, and customers. It includes PlaceLab, a 1000 ft^2 full-service condominium prepared to gather fine-grained data on human behavior. The *Aware Home Research Initiative* has given support to several small-scale projects, including *aging in place*, which was originally planned to support social connections between the elderly and their families and improve cognition aspects like memory and planning skills, but its functionality was extended later to behavior analysis as well, via heavier sensorization (Tran et al. 2010). *Gator Tech Smart House* focuses on elderly people with disabilities, and it is equipped with a large number of smart-labeled furniture and appliances including gadgets like a smart floor for localization and falling detection, a smart closet to advise on clothes depending on the weather, a smart mirror to display information, or a smart bed to monitor sleeping patterns (Helal et al. 2005). Some researchers have taken a step further and, instead of building their own smart home, actually placed their technology in living facilities used by the elderly in a continuous way, like TigerPlace (Skubic et al. 2011). There are even more ambitious approaches to ambient assisted living (AAL), like a full smart city: FUSIONOPOLIS was proposed by the Agency for Science, Technology and Research (A*Star), Singapore, where areas of the city have been built or remodeled according to the AmI paradigm (e.g., Jurong Lake District), including wide-area communication networks (e.g., WiMAX), intelligent energy distribution systems, smart climate control, crowd behavior analysis, and so forth.

It can be noted that most technologies necessary for AmI applications are available nowadays. However, there are still not so many of them deployed. The problem is, of course, the costs of such an installation. While personal computers and smartphones already have a prominent presence in many houses, sensor networks and actuators are still too expensive, for example, for us to afford ourselves homes capable of learning how we like our illumination and HVAC at all times. It is no surprise, consequently, that most state-of-the-art approaches to smart homes are usually related to assisted living, with a special focus on the elderly, as covered in the next section. Cost is not an issue for a person with special needs to achieve a certain degree of autonomy. Even if a given person cannot afford an AmI installation, it is still way cheaper in terms of well-being and cost for social services, health systems, or insurance companies to pay for information and communication technology (ICT)-based AAL than to proceed to institutionalization. Assisted living has the trifecta of *technology, cost*, and *need* (Urdiales 2012).

20.2 AAL

Assistive engineering and design is a field at the intersection between technology, the natural sciences, the humanities, the social sciences, and medicine. Assistive technologies are of special interest, as the average age of the population increases fast: There will be an estimated 74% growth in population aged 65–74 years by the year 2020 (Camarinha and Afsarmanesh 2001). This growth is exceeding that of the under-65 population, which only sees a growth of 24% (Wan et al. 2005). Indeed, the share of the population aged over 80 years (3.6% in 2000) is expected to reach 6% by 2025 and 10% by 2050.

At the same time, the percentage of people with disabilities increases with age: Data show that half of seniors aged 65 years and over have a disability. In the United States alone, there are over 35 million people with disabilities (Pope and Tarlov 1991). The association between old age and disability occurs because more people live longer, and, thanks to the progresses of medicine, many more persons survive acute diseases, resulting in their being affected by chronic conditions and some disabilities. Clearly, human resources will not be sufficient to assist all the elderly or people with disabilities, so ICTs are expected to play a key role in this respect. Indeed, the main goals of assistive technologies are

- To increase self-dependency
- To allow community dwelling: demographic changes have led a large number of elderly people to live alone
- To increase the elderly users' participation in ICT-based assistance
- To provide insightful data to health professionals, caregivers, familiars, psychologists, system designers, and so forth

The alternative to ICT-based assistance is institutionalization, which has a major cost in social, personal, and economic terms. Hence, it is only natural that most efforts in AmI have focused on developing smart homes for people with special needs. Just to mention an example, the Ambient Assisted Living Joint Programme (AAL JP) in Europe had a total budget of 600 M EUR over 6 years, was cofunded by the European Commission and both public and private entities, and has supported 60 projects since 2008.

Smart homes have been around for a while. Indeed, the original concept dates back from the 50s. However, the technology required to make them work has not been available until recently. According to Sadri (2011), a smart home should (1) automatically perform everyday tasks; (2) improve economy of usage of utilities like electricity; (3) improve safety and security; and (4) improve quality of life. Besides, the main challenge in human centered projects is to overcome the digital divide (DD) by following a "design for all" paradigm. Elderly people define a quite heterogeneous group with very different needs and skills, and technology must be available and useful for all of them. The key aspects affecting the DD are reported to be age, technological skills, geographical location, culture, disabilities, and gender. All assisted living projects must keep ergonomics and transparency in mind and make no assumptions regarding the technological skills of potential users. This target implies a high capacity for adaptation, which is typically achieved through experience-based learning.

If we combine the mentioned goals of assistive technology with smart homes, we obtain a functional architecture like the one in Figure 20.2. In brief, both the house and the user are equipped with networking sensors that, in combination with the user's commands and actions, are analyzed by the system to create a user profile. This profile, which could

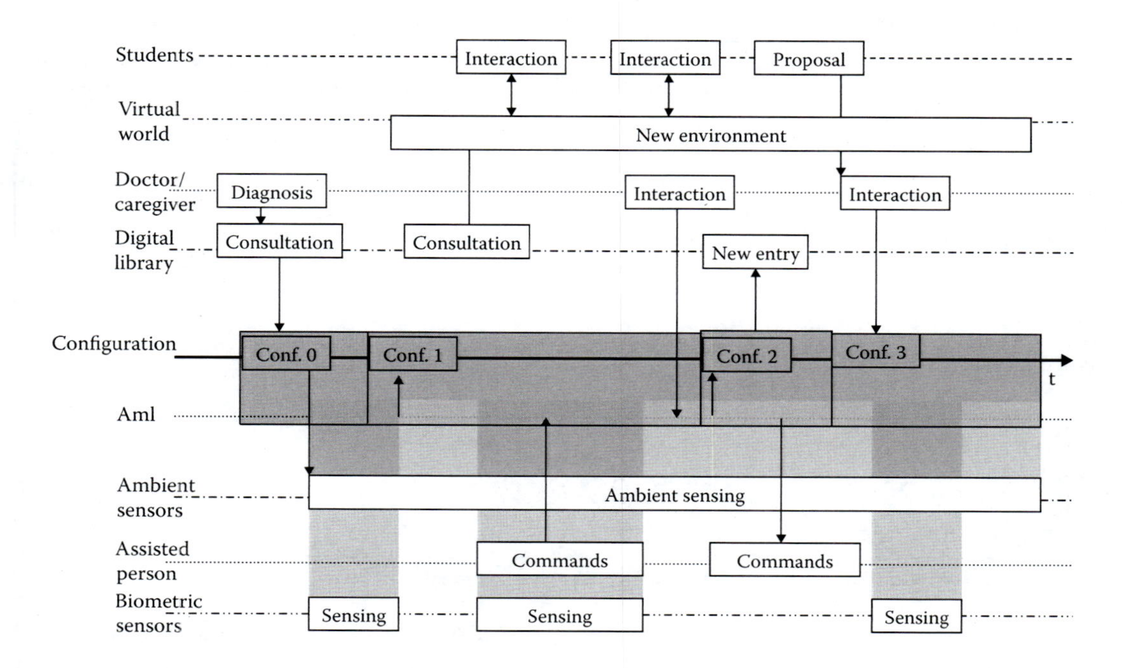

FIGURE 20.2
(See color insert.) Full cycle of a smart assistive home.

be complemented by any medical information available for that particular user, is user primarily to assist the user, personalize services, and generate alarms, if necessary. It can potentially be used as well to provide insightful data to health professionals and system designers, by storing profiles, after they are devoid of sensitive information, into digital libraries. In the extreme, information in these libraries can be used as well to generate virtual environments for teaching and training of future professionals in the area. However, AAL projects usually do not cover the full spectrum of applications. Instead, they focus on a number of the mentioned aspects.

An extensive analysis of the numerous projects on assisted living that have been developed in the last 5 years clearly shows that most of them have similar goals, even though they may be oriented to different target populations. Projects in (approximate) order of smartness can be loosely categorized into the following (Table 20.1):

- *E-Inclusion:* These systems focus on integration of ICT via adapted interfaces that can be personalized using profiles. Research here focuses on integration of communication standards and services, as well as developing or adapting new human/computer interfaces (HCIs) and devices (MyLife, HOST, GoldUI).
- *Social interaction tools:* These systems have been traditionally implemented as call centers offered by social services and health providers to elderly people living alone. These call centers keep periodical communication with the user to prevent isolation

TABLE 20.1

Requirements for Assisted Living Projects

	E-inclusion	Social Interaction	Remote Monitoring	Reminders	Alarm Generation	Behavior Analysis	Autonomous Assistance
New Interfaces	Y	–	N	–	–	–	Y
WPAN	N	N	–	Y	–	Y	–
WLAN	N	N	Y	Y	Y	Y	Y
Web Services	Y	Y	N	–	–	–	–
Internet	Y	Y	N	N	–	–	–
Ambient Sensors	N	N	Y	–	–	Y	–
Biometric Sensors	N	N	Y	–	–	–	–
Wearables	N	N	Y	Y	–	Y	–
Actuators	N	N	N	N	N	N	Y
Reactive AI	N	N	N	Y	Y	–	Y
Deliberative AI	N	N	N	–	N	Y	Y
Profiling	Y	–	N	N	N	Y	Y

Note: WLAN, Wireless Local Area Network; WPAN, Wireless Personal Area Network.

and also whenever the user feels in danger. Most services at the moment include peer-to-peer and broadcast videoconference and sometimes are used to offer loosely supervised cognitive or physical exercises (HERA, CO-LIVING) that can be later evaluated by a professional to check on the user's condition. In recent years, the social networking boom has led to more ambitious tools that also aim at creating communities of the elderly without service supervision. Furthermore, the degree of participation of a person in these networks could be used to assist diagnosis of syndromes like geriatric depression. Some of these projects aim at reutilization of common devices like TV sets (HOMEdotOLD, FOSIBLE), whereas others focus on new ones like smartphones or tablets (GO-MY_LIFE, Express to Connect) or even develop their own platforms, like touch screens (ELISA, ConnectedVitalityNetwork) or even mobile robots (ALIAS, EXCITE). Innovation in these kinds of projects is related to open standards, hardware integration, and development of adapted, user-centered interfaces.

- *Remote monitoring:* Probably one of the simplest approaches in terms of intelligence, remote monitoring consists of periodically evaluating the state of a person—usually chronic—by checking a set of parameters, typically biosignals (AMICA, ALADDIN), but also activity in general (PAMAP). In these, cases employed sensors tend to be noninvasive, either biometric ones or biosensors. The user may be required to perform the measures himself/herself and send them to the monitorization center, or he might be equipped with wearable sensors instead, which could transmit his/her readings automatically when necessary. This solution has been widely used in countries where there are many isolated population nuclei for primary attention, and most effort has been dedicated to development of cheap and reliable communication networks and easy-to-use biometric devices.
- *Reminders:* These systems are based on reminding the user—typically a person with minor to mild cognitive disabilities—to perform a given action at an appropriate time of the day. These actions may range from the very simple, like taking

a medication, to more complex tasks like dressing up. To do so, the system needs an (estimate) agenda of the activities of daily living (ADLs) of the user, and it must be able to provide feedback through an adapted interface, frequently audio. ADLs include activities like eating, getting in and out of bed, using the toilet, bathing or showering, dressing, using the telephone, shopping, preparing meals, housekeeping, doing laundry, or managing medications. The reminder may be triggered simply by the clock, or it also may require additional conditions, like not being in a place of designation at the time. If this is the case, ambient sensors may be needed.

- *Alarm generation:* Users and/or environments are equipped with a set of sensors that can reliably capture isolated and potentially dangerous situations, for example, inertial sensors and fall detection (CARE, eCAALYX, HOPE). When one of these situations is detected, the system automatically informs caregivers or social services, so they can act in consequence. In their simplest version, they do not even require sensors: They have been available for years in the shape of a panic button that the user pushes in a dangerous situation. It needs to be noted that these are purely reactive systems, and most innovations to this respect occur in the employed sensors, often wearable ones (e.g., ISACTIVE) but also external ones (e.g., CARE).
- *Behavior analysis:* These systems are more complex than the previous ones in the sense that they do analyze the user's actions as a whole, create a profile, and check for irregularities. To do so, the user and/or the house must be equipped with a set of sensors that capture his/her activity and correlate it with his/her location, time of the day, and nearby elements. Repeating patterns and sequences shape the so-called user profile, which is expected to be repeated over time. Unlike the agenda for reminders, profiles are not available *a priori* but are learned over time. When these profiles are available, the system may check for deviations in current behaviors with respect to the profile, rank them from neglectful to serious, and, if necessary, generate the proper alarms. This information can also be use for diagnosis support (BEDMOND, REMOTE), as it is gathered through long time periods and, hence, helps to understand better the progress of a given condition and/or treatment (H@H, ROSETA).
- *Autonomous assistance:* It can be observed that most mentioned systems can be defined by the information they capture, but there is usually a human in the loop to make the ultimate decision or provide the required care. A last stage in an assistive system is to include actuators that can provide physical assistance to the system user. These assistive devices may vary greatly, from wearable drug dispensers (HELP) to assistive robots. These robots must be distinguished from social robots (ALIAS, EXCITE) that, in fact, do not support activity. Rather, robots in this group (DOMEO, SHARE-it) provide assistance to mobility (robotic wheelchairs and walkers). Note that most systems provide assistance of some kind. The specifics of these ones are that the amount and type of assistance must be automatically estimated by the system and that some physical interaction with the user is required.

20.3 Localization for Context Awareness

The main challenge in AmI environments is probably *context awareness*. Early definitions of context awareness state that it provides information on "where you are, who you are with, and what resources are nearby" (Schilit et al. 1994). Other authors state that context is the set of variables that may be of interest for an agent and that influence its actions (Bolchini et al. 2009). Chen and Kotz (2000) propose a four-dimensional space composed of computing context, physical context, time context, and user context. Sensors and actuators are specifically related to the so-called physical context and, specifically, to device/user location, which has been reported as key information for context awareness.

Given the importance of the user's location with respect to items for context awareness, it is not surprising that much effort has been focused on estimating it. Localization has been traditionally solved by means of triangulation, and most systems in AmI still rely on this procedure. Many indoor systems rely on using video cameras to localize the user, but recent trends avoid this approach due to privacy-related issues, so they are not going to be included in this chapter. However, while GPS is extensively used outdoors, indoor localization relies on other signal beacons to operate. This procedure is globally known as received signal strength indication (RSSI)-based localization, and it can be based on different wireless protocols that work both indoors and outdoors, like Wi-Fi, Bluetooth (BT), ZigBee, or RFID (Table 20.2). There are also hybrid systems that combine different radio-frequency technologies to combine their strengths (LaMarca et al. 2005). The main issues regarding this approach are power consumption, range, and precision. For example, Wi-Fi is likely to cover a wider area with less emitters than BT or ZigBee, but its power consumption is way higher. If we compare BT to ZigBee, beacons would likely consume less power in the second case, since transmitters would be off most of the time. In applications where device feeding is an issue (typically wide-area applications outdoors, like tsunami detection, volcano alarms, and so forth), it is preferable to use a larger number of low-power-consumption devices, whereas in environments where beacons can be plugged into a power line, it is usually better to rely on just a few wide-range ones. Regarding costs, it is always better to rely on signals that are already deployed in the environment, like global system for mobile communications (GSM) or 3G outdoors (Zhu and Durgin 2005) or Wi-Fi networks indoors (Youssef and Agrawala 2008), if possible.

It also needs to be noted that working with technology that can be used for other purposes makes the system cheaper. For example, the mobile phone network makes it easy to localize any user with a mobile phone without any extra technology. Interferences may be a problem in these cases, though: Wi-Fi and BT networks are quite frequent everywhere and out of the designer's control (laptops and mobile phones of neighbors and bypassers), so frequency-hopping-based devices like BT or ZigBee might find, in the extreme, no room to settle themselves in the spectrum (Trapero et al. 2007). In these cases, BT would overpower ZigBee, but localization beacon packets are so tiny that they may manage to squeeze themselves in any opening in these cases. This problem can obviously be avoided by working with protocols with nonoverlapping operation bandwidths.

The main issue with the mentioned localization techniques is precision. Usually, RSSI-based techniques have an uncertainty between 1 and 2 m at best, depending on the processing algorithms they use. In some application, this is enough, but sometimes, more precision is needed. In these cases, there are commercial solutions based on detection: The user carries a tag, either active or passive, that can be read only at certain distances, so the system knows if the person is nearby one of the readers. Indeed, near field communications (NFC) and, more specifically, RFID-based localization have been widely employed in hospitals to track not only people but also material and equipment (Table 20.2). Torres-Solís et al. (2010) present a deeper survey on indoor localization technologies.

Similarly, passive RFID tags and a portable reader have been used in pervasive computing to detect what the user is actually touching (Larson, 2000), which can intuitively associate

TABLE 20.2

Commercial Indoor Localization Solutions

Product	Use	Notes	URL
RFID			
Identec	User tracking Asset tracking	Proprietary protocol	http://www.identecsolutions.com
Syris Xtive	User tracking Asset tracking	Proprietary protocol	http://xtive.com.tw
UWB			
Ubisense	User tracking Assisted living apps	UWB (TDOA and AOA) techniques	http://www.ubisense.net/en/ http://www.ubisense.net/en/resources/case-studies/assisted-living.html
Bluetooth			
Topaz	User tracking	RSSI techniques	http://www.tadlys.co.il/pages/Product_content.asp?iGlobalId=2
IPCS	User tracking Asset tracking	BT low energy RSSI techniques	http://www.9solutions.com/ipcs-insight/products
ZigBee			
Nanotron products	Asset tracking	RSSI techniques	http://www.nanotron.com/EN/PR_products.php
IDC ZigBee tracking system	Asset tracking	RSSI techniques	http://www.zig-bee.co.uk/tabid/1026/Itemid/10/default.aspx
Netvox	Location—home automation	RSSI techniques	http://www.netvox.com.tw/
Wi-Fi			
Hugs Infant Protection	User tracking	Wi-Fi tag—RSSI techniques	http://www.stanleyhealthcare.com/solutions/patient-security/hugs-infant-protection
Passport Patient Protection	User tracking	Wi-Fi tag—RSSI techniques	http://www.stanleyhealthcare.com/solutions/patient-security/passport-patient-protection
Petz Pediatric Protection	User tracking	Wi-Fi tag—RSSI techniques	http://www.stanleyhealthcare.com/solutions/patient-security/pedz-pediatric-protection
RoamAlert	User tracking	Wi-Fi tag—RSSI techniques	http://www.stanleyhealthcare.com/solutions/patient-security/wander-management/roamalert
Aeroscout	Asset tracking—location—sensoring	Wi-Fi tag—Wi-Fi network—RSSI techniques	http://aeroscout.com/healthcare
Ekahau	User tracking Asset tracking	Wi-Fi tag—Wi-Fi network—RSSI techniques	http://www.ekahau.com/products/products-overview.html
Infrared			
HiBall Tracking System	User tracking	RSSI techniques	http://www.3rdtech.com/HiBall.htm

Note: AOA, Angle of Arrival; IDC, Intelligent Distributed Controls; IPCS, Internet Protocol Communications Security; TDOA, Time Difference of Arrival; UWB, Ultra-Wide-Band.

with the services. Passive tags must be less than 50 cm away from the reader, depending on the technology (active tags can operate within a few meters' distance). Hence, their sequence of activation may be correlated with the user's ADLs or associated with a certain activity (Philipose 2003; Wren and Munguia-Tapia 2006; Tran et al. 2010). Other researchers have also conducted similar studies using domotics networks deployed around the house (presence and motion detection, touch sensors, switches, video cameras, and so forth) (Alwan et al. 2003; Virone et al. 2008) or worn by the user instead (Himberg et al. 2001; Lee and Mase 2002). The main advantage of passive RFID-based systems is that only one reader per user is needed, and cheap sticker-like tags can be distributed around the environment with no need for battery replacement or installation. Unfortunately, commercial readers are still too bulky to wear comfortably, and there are some problems regarding antenna orientation, interferences with materials, and so forth, so this hardware technology is still under development.

Once a person is localized with respect to his/her environment, captured data can be contextualized and modeled into knowledge. Context data can be roughly split into general models, domain-specific models, and no models (Bellavista et al. 2012). The main problem with knowledge extraction in this field is that there is no general model of human activity: Instead, each person has different ADLs depending on his/her preferred environment distribution and state, circumstances, functional profile, and likings. Hence, most AmI applications focus on learning an activity profile of their target user and estimating the importance of deviations from the acquired profile. AmI systems usually mine sequential patterns to detect habits and repetitive actions, since interactions of a person with devices in a smart house as part of his/her routine can be considered as a sequence of events (episode), with some inherent pattern of recurrence (Agrawal and Srikant 1995). There are several techniques to detect these sequences, like the episode detection (ED) algorithm in the work of Heierman and Cook (2003), Bayesian networks (Chen et al. 2012), Markov models (Kautz et al. 2003), and location clustering (Youssef and Agrawala 2008). The sequence set for a given person forms his/her profile, and the system can use it to predict his/her actions, release some system response, detect abnormal behaviors, or trigger alarms.

20.4 Sensors and Actuators

AmI integrates different computational paradigms, as well as many devices that must coexist as seamlessly as possible. Integration is always a challenge, specially when full ad hoc design is unfeasible because not every part of the system is going to be developed by its designers. The safest way to achieve compatibility is to comply with existing standards.

AmI sensor and actuator networks are typically based on the same standards as traditional domotics, and hence, they present similar advantages and drawbacks. It is important to know them in advance when we decide what technology to use in our system. Table 20.3 lists the best-known standards in the field and their applications.

The first issue to consider is whether the standard is open or not. For example, well-known standards like *Lonworks* or *ligthwaveRF* are proprietary standards, meaning that every device inserted in the network must be purchased from its associated company, and interoperability with other networks is very limited or null. This might seem a minor concern, but, in fact, it is very constraining, because some standards do not support some devices. For example, X10 is not reliable, so it does not support alarm devices (fire, smoke, flood, etc.). The only way to include them in an X10-based installation would be to have

TABLE 20.3
Tentative (Real) Budget for a Toy Domotics Installation for Different Standards

Standard	Parts	Price
KNX-EIB	PSTN-EIB gateway (support)	~1700 EUR
	Smoke detector	~170 EUR
	Motion detector (volumetric)	~160 EUR
	Door opening detector	~95 EUR
	Light actuator	~150 EUR
	IP gateway	~1500 EUR
		Total: 3775 EUR
BUSing (wireless)	ETHBUS Web server	~420 EUR
	B-W Wireless gateway	~50 EUR
	MECing-W adapter	~60 EUR
	Smoke detector	~85 EUR
	Motion detector	~105 EUR
	Door opening detector	~85 EUR
	Light actuator	~195 EUR
		Total: 1000 EUR
X10 (wireless)	WLAN X10 interface	~360 EUR
	Smoke detector	Not manufactured
	Motion detector	~62 EUR
	Door opening detector	~34 EUR
	Light RF actuator	~45 EUR
		Total: 501 EUR

Note: ETHBUS, ethernet-busing; PSTN, public swithced, RF, radiofrequency.

two parallel networks in the house. In this sense, open protocols are preferable, since any manufacturer can actually build a standard-compliant device, incorporate a standard-compliant interface to an existing device, or simply tap into the network(s) to extract all required information and forward it to the local processing unit. Casa Agevole, discussed by Annicchiarico and Cortés (2010), for example, combines KNX-European installation bus (EIB) devices with a ZigBee sensor network and an ultrasonic localization system.

In its simplest form, a standard could be simply defined at link/network level in the open system interconnection (OSI) stack, so any compliant device would be fully compatible with OSI-compliant devices and protocols as well (e.g., Wi-Fi or internet protocol (IP) cameras), but isolated devices are usually much more expensive and harder to find.

A second issue is whether the chosen standard requires specialized installation. For example, some standards like KNX-EIB require licensed staff for installation and maintenance. This implies an additional cost in terms of money and time, as well as a dependency or third parties. It also needs to be noted that wired installations require not only structural changes in the house but also additional elements of support to connect to the local bus, feed the different devices, and so forth. In contrast, a completely wireless mesh architecture, like ZigBee working in zero configuration mode, ideally requires only a gateway into the local processing unit to be fully operative.

It could be assumed that wireless systems are always preferable, but, in fact, this is not totally true. Even low-power-consumption standards like ZigBee work on batteries that need to be replaced from time to time. In mesh networks where not all elements spend the same

TABLE 20.4

Standards

Name	Type	Uses	URL
Lonworks	Proprietary wired domotic protocol	Lighting, energy, HVAC, security, building automation	http://www.echelon.com/technology/lonworks/
Bacnet	Open data communication protocol for building automation and control networks	HVAC, interoperability of domotic networks	http://www.bacnet.org/
KNX	Open wired/wireless domotic protocol	HVAC, building automation, alarm monitoring, energy management and electricity/gas/water metering, audio and video distribution	http://www.knx.org/
X10	Proprietary wired/wireless domotic protocol	Building automation	http://www.x10.com
Z-wave	Open wireless domotic protocol	Gateways, hospitality, lighting, locks, remote controllers, security and alarm, sensors, shade control, smart meters, thermostats, window control	http://www.z-wavealliance.com/
ZigBee	Open wireless communication protocol	Low-power communications, building automation, health care, home automation, smart energy	http://www.zigbee.org/Standards/Overview.aspx
ligthwaveRF	Proprietary wireless domotic protocol	Lightning power, heating security, energy, assisted living	http://www.lightwaverf.com/
RFID	Tracking and identification protocols Open and proprietary	Access management; tracking of goods, persons and animals; contactless payment; machine-readable information; massively distributed sensor networks; authentication	http://rfid.net/
ILLUMRA	Open wireless domotic protocol Based on ultralow-power-consumption protocol EnOcean (ISO/IEC 14543-3-10)	HVAC control, house automation	http://www.illumra.com/Products/index.html
Wi-Fi	Open wireless communication protocol	Networking, consumer electronics, computing and peripherals, handsets, transportation management, medical/fitness devices, HVAC, home security and control, embedded sensors	http://www.wi-fi.org/
Bluetooth	Open wireless communication protocol	Automotive, consumer electronics, health and wellness, mobile telephony, PC and peripherals, sports and fitness, smart home	http://www.bluetooth.com
Continua	Industrial interoperability standards between Bluetooth and USB devices, enabling them to more easily share information with caregivers and service providers	Health care	http://www.continuaalliance.org/index.html

time active, each sensor and actuator would consume its battery in its own time, so, in practice, one would spend much time replacing them. Actually, wireless devices that need to send a mild amount of information with some regularity may be wireless in terms of communication but plugged into the electric supply. This arrangement obviously does not require specialized installation. However, it can be observed in Table 20.3 that only a few standards support all kind of devices. This brings us to the third consideration in design: cost.

There are major differences in the cost of an installation depending on the chosen standard. Table 20.3 shows tentative prices for the same toy installation with the basics of being completely functional and having just one sensor of each kind. Prices may change depending on the provider and the time of installation, and we also need to take into account whether equipment needs to be imported to obtain the final cost. However, it is immediate to notice that KNX-EIB is four times more expensive than BUSing and eight times more expensive than X10. As mentioned, X10 is not reliable enough to support fire smoke detectors, but it is fairly cheap to operate via Internet. All in all, the best option could be to install a heterogeneous network where all devices are integrated after data is exported to the processing unit, so that expensive standards only support critical devices, but it can be noted that the enormous price difference is mostly due to functional elements of the network, like the PSTN-EIB and IP gateways. Besides, this approach could imply the need for two different installers, so most systems ultimately choose a unique standard depending on their needs (Cortés et al. 2010) (Table 20.4).

20.5 Biometrics Sensors

Biometrics sensors are well known in telemedicine applications, but in order for them to be used at home, they must be easy, to adjust and their output must be easy to transmit to a processing unit. Monitorization can be either invasive (e.g., needles, probes, etc.) or noninvasive. Even though in extreme cases, invasive devices are acceptable [e.g., electromyography (EMG)-based control, glucose meters, and so forth], biometrics sensors are usually noninvasive, specially if users must attach them on their own.

Ideally, biometrics sensors should be wearable, meaning that the person can be constantly monitorized during his/her daily activity. Wearable sensors must be comfortable, resistant, and wireless. Besides, they must be low-power-consumption devices, so they are based on communication standards that can operate for long time periods without battery replacement.

Most typical commercial sensors matching the discussed criteria for nonspecific conditions include blood pressure, heart rate (HR), frequency rate, blood oxygen saturation, and body temperature, ranging from 2000 to 5000 USD. Some researchers have tried to develop their own ad hoc sensors, for example, Mundt et al. (2005), but most systems rely on market solutions for the following reasons: (1) devices must be safe to avoid any user damage (electric shock, interferences with pacemakers, device warming, etc.); (2) they must be compatible with the usage of other medical devices (i.e., pacemakers) but also with everyday devices such as mobile phones and personal digital assistants (PDAs); and (3) devices should be standard-compliant. One of the main issues with commercial sensors is obtaining their raw readings for processing, as many of them just show processed data in their own displays. As mentioned in the previous section, the best option is to offer data flows via a well-known, popular standard, like BT. However, it is necessary to take into account the bandwidths of the signals to be transmitted if communication is supposed to go live rather than off-line: For example, an electroencephalography (ECG) data flow is within the limit of what BT can actually transmit safely, whereas EMG data would exceed BT bandwidth. Nevertheless, most systems try to achieve compatibility with smartphone-supported protocols, so received signals can be broadcasted using general packet radio service (GPRS) or 3G when there is no local area network (LAN) available (e.g., MOBIHEALTH).

Table 20.5 shows a list of current biometrics devices on the market that present all mentioned features.

TABLE 20.5

Commercial Biometrics Sensors

Name	Company	Notes	URL
ECG			
BT12	Corscience	Bluetooth communication, 12 channels	http://www.corscience.de/en/medical-engineering/products/ecg/bt12-device.html
Pulse oximetry			
4100 Bluetooth Pulse Oximeter	Nonin	Bluetooth	http://www.nonin.com/
Blood pressure			
Boso-Medicus Prestige	Corscience	Bluetooth	http://www.corscience.de/en/medical-engineering/products/blood-pressure/boso-medicus-prestige-bt-device.html
EEG and EMG			
NeXus EEG cap	MindMedia	21 channels	http://www.mindmedia.nl/CMS/en/products/eegaccesories/item/186-nb-eegcap-s/m/l.html
Kine Measurement System (KMS)	KINE	Wireless EMG	http://kine.is/Products/MMS/
Galvanic skin response (GSR)			
Skin Conductance sensor	Thought Technologies		http://www.thoughttechnology.com/
Body and skin temperature			
Temperature sensor	Thought Technologies		http://www.thoughttechnology.com/
HR			
HR-210	Omron	Strapless HR monitor	http://www.omronwebstore.com/detail/OMR+HR-210
Blood pressure			
Upper Arm Blood Pressure Monitor 767PBT	A&D	Bluetooth, Continua-compliant	http://www.andonline.com/medical/products/details.php?catname=&product_num=UA-767PBT-C
Respiratory rate			
Respiratory belt transducer	ADInstruments	DIN or BNC connector	http://www.adinstruments.com/products/mlt1132#overview

TABLE 20.6

Commercial Multiparametric Sensors

Device	Company	Parameters	URL
NIBP2010/ChipOx	Corscience	Blood pressure, oxygen saturation (SpO2) and pulse rate	http://www.corscience.de/en/medical-engineering/products/multiparameter/nibp2010chipox-board.html
ProComp Infinity	Thought Technology Ltd	8 channels for ECG, EEG, EMG, blood volume pulse (BVP), HR, GSR, respiration, or skin temperature	http://www.thoughttechnology.com/
NeXus-10 MKII	MindMedia	4 single-channel inputs, 2 dual-channel inputs, 1 oximetry input, and 1 digital multichannel input; muscle tension, brain waves, HR, relative blood flow, skin conductance, respiration, temperature, oximetry	http://www.mindmedia.nl/CMS/products/nexus-systems/item/175-nexus10mkii.html
Equivital	Hidalgo	ECG, respiratory, tri-axis accelerometer and temperature, core body temperature, GSR, oxygen saturation	http://www.equivital.co.uk/
Sensium Life Platform TZ2050	Toumaz	HR, ECG, accelerometer, temperature	http://www.toumaz.com/uploadsv3/wysiwyg_editor/files/TZ205000-MPB%20Product%20brief%20Sensium%20Life%20Platform%20V1_1.pdf#search=%27sensium%27

It needs to be noted that most chronic patients require supervision of several biometrics parameters at the same time. Since solo devices are still quite bulky, most systems actually rely on multiparametric sensors, which can capture several magnitudes at a time. Most these devices are built around body belts, where capture devices can be plugged in on a need basis. Table 20.6 lists current commercial multiparametric sensors and the parameters they provide.

20.6 Interfaces and Ergonomics

Originally, most interfaces in AAL used to consist of switches and/or a PC-based interface. In these cases, it was assumed that users would need to cope with the system interface, and hence, much effort has focused on ergonomics. There are many questionnaires to evaluate usability in HCIs. One of the best-known general ones is NASA TLX (Hart and Stavenland 1988), but there are several specific ones for persons with disabilities coping with assistive technology, like the Quebec User Evaluation of Satisfaction with Assistive Technology (QUEST) (Gelderblom et al. 2002), Psychosocial Impact of Assistive Devices Scale (PIADS) (Jutai and Bortolussi 2003), assistive technology devices (ATDs) (Scherer et al. 2007), and

Matching Person and Technology (MPT) assessment (Gatti et al. 2004). Systems meant to be operated by persons with potential physical and/or cognitive disabilities should be piloted first to obtain an estimation of usability and, if possible, evaluated by associations of users and caregivers (Annicchiarico and Cortés 2010). Voice control clearly shows the need for these studies. Although voice can be complemented by other physical interfaces, it seems natural that a person can send commands to a smart environment by means of a simple micro, for example, as discussed by Garate et al. (2005) and Noyes et al. (1989). Regarding usability, it is important to note that mild disabilities may affect significantly the vocal skills of the user, so the most frequent voice databases are rendered useless for these people. In some cases, their condition may even change in time, so voice-based systems may require learning capabilities to be useful for these people.

The main problem with voice-based interfaces is how to present information to the user. Simple sentences and status reports are easy enough to provide via an earplug or area speakers, but more complex data are easier to present in a visual format. Visual information was first provided by means of a PDA (Favela et al. 2004; Rodríguez-Losada et al. 2005; Tapia and Corchado 2009), but these devices have been replaced by the latest generation of affordable tablets and smartphones in the market.

Some persons, though, have strong impairments and, consequently, cannot work with the usual controllers. Other physical interfaces can be controlled by head, feet, chin, shoulder switches, and so forth, which are fairly common yet quite expensive in the field of assistive technologies. These interfaces are not as comfortable as the aforementioned ones, but in some cases, there may be no other choice for the user to exert some control on his environment. In these cases, it is extremely important to take into account ergonomics and medical factors. For example, some head trackers might not be advisable for people with spinal cord injury, as they require significant neck motion. Some of these interfaces are presented in Table 20.7. In extreme cases, interfaces can be designed ad hoc. For example,

TABLE 20.7

Control Devices for Persons with Disabilities

Type	Physical Skill Required	Invasiveness	Manufacturers	Price Range (Orientative)
Touch screen	Pointing and clicking	None	CINTIQ, TERN, Honeywell, JUNG	350–1500 USD
Tablet	Pointing and clicking	None	Apple, Samsung, Asus, and so forth	150–500 USD
Smartphones	Pointing and clicking	None	Apple, Samsung, Sony, Nokia, LG, and so forth	80–500 USD
Switches (for disabled users)	Button pressing	None	ASL	2800 USD
Speech recognition	Speaking	Low	Kempf, Dragon Naturally Speaking, IBM Viavoice	500–1600 USD
Head tracker	Head motion	Low–medium	APT Tech. Inc, Peachtree, Magitek	990 USD
Head mouse	Head motion	Low–medium	Smartnav, Tracker-Pro, EagleEyes	500–900 USD
Chin sensor, breath actuators	Partial muscle motion	Medium	Dynamic DX, Systems, Sunrise Medical, Dynamics DX, PG Omni+	1200–1700 USD
Eye tracker	Eye motion	Low	ASL, Eyelink	45,000 USD
EOG	Eye motion	Low–medium	Adinstruments Inc, Biocontrol Systems	1000 USD
BCI	None	Medium	Cyberlink/Brainfingers, Emotiv	2100–3500 USD

the telethesis project (Pino et al. 1998) proposed an on/off switch to choose an option on a screen that is continuously renovated to fit the needs of a patient affected by amyotrophic lateral sclerosis (ALS).

Finally, if mobility is completely out of the question, there are other, less intuitive mechanisms for control. Eye tracking, for example, can be based on tracking the cornea, which is outlined through focal illumination (Li et al. 2007) or using electrodes to measure the electrooculographic potential (EOG), which corresponds to the angle of the eyes in the head (Yanco 1998; Wei et al. 2010). Brain/computer interface (BCI) has also offered promising results, despite its poor signal-to-noise ratio. Some BCIs quantify potential commands into a reduced number of bins (mental states) to choose among a limited number of options (Millán et al. 2004). However, this approach requires the user to be continuously focused to operate the system. Alternatively, other systems rely on the P300 evoked potential, a natural, involuntary response of the brain to infrequent stimuli: If a sequence of stimuli is presented, but only one of these interests the subject, approximately 300 ms after the target flashes, there is a positive potential peak in the EEG signal, which can be reliably detected (Rebsamen et al. 2007). Table 20.7 shows all mentioned commercial devices, with a tentative list of manufacturers and price ranges.

20.7 Robotic Companions

Assistant robots are typically autonomous mobile robots that help the user with minor tasks, like carrying food, drinks, and medication (Hagras et al. 2004; Cesta et al. 2005; Saffiotti and Broxvall 2005). In this kind of system, robots are typically connected to the environment and make decisions depending not only on their onboard sensors but also on the input instance of all sensors in the system as well as any knowledge acquired by the environment sensors. In order to conciliate the behaviors of all the system components, agent architectures are often used. A common requirement in homes equipped with robots is to know the locations of all robots and users, so localization is usually of key importance.

A special category in assistant robots is *socialization robots* (Cauchi et al. 2012; Granata et al. 2012; Tiberio et al. 2012). These mobile robots do not have manipulation capabilities. Instead, their goal is to promote social inclusion by linking users to people and events outside their homes. Socialization robots typically hold a monitor to provide easy access to online services. These services may range from videoconferencing or participation in social networks to cognitive reminders or activity monitoring (telemonitorization and telepresence). It needs to be noted that these robots usually act as a mobile gateway but do not engage the user themselves in any activity: There must be another person on the other extreme, although many of these robots use voice-based interfaces for usability.

Assistive robots are mostly used for physical support for mobility, basically in the shape of walkers or power wheelchairs. Robotic walkers usually improve balance and assist direction (Nejatbakhsh and Kosuge 2005; Rodríguez-Losada et al. 2005; Rentscheler et al. 2007; Wasson et al. 2007). Their input instance typically consists of the force pair exerted by the user in the handles, a front range sensor to detect obstacles, and often, an accelerometer to detect slope changes. Sometimes, these devices are used to keep a constant pushing force when the user loads the walker or faces a slope, specially for frail elderly. In rehabilitation, they may provide haptic feedback to make the user push more or less in a specific way. Robotic wheelchairs are similar to autonomous robots, regarding input instance (controller, encoders, range sensor) and navigation algorithms. These wheelchairs may carry people from one place to another without any actual input, but in fact, physicians have stated that lack of practice leads to loss of residual skills, so it is preferable that their intervention level is very high instead. Systems where user and robot have to cooperate fit the so-called shared-control paradigm. Shared control may range from safeguarded navigation (Miller and Slack 1995)—the wheelchair stops or takes over in case of danger—to full autonomous navigation. Most commercial systems, though, rely on keeping a basic set of primitives like **AvoidObstacle**, **FollowWall**, and **PassDoorway** to assist the person in difficult maneuvers (Connell and Viola 1990; Simpson et al. 1998). Alternatively, other systems rely on combining human control and robot output in a continuous way to force the user to practice any residual skill (Urdiales et al. 2009).

A third, more ambitious goal in domestic robotics is to create a *companion* capable of responding emotionally to situations experienced and to interactions with humans (artificial emotion). Thus far, there has been limited success in robo-pets like Aibo dogs or Ugobe Pleo and, more specifically, the Paro therapeutic robot, mostly based on learned emotive expression in the behavior of the robot for human/machine interactions. Reported effects on persons affected by dementia include stress reduction and improved interaction and socialization. These robots, however, do not understand emotions or interaction; they simply randomize actions from a set and reinforce those that provoke acceptance according to simple sensor readings (e.g., back touch sensors in ERS-7). There have been many works on categorization of expressions. For example, Velásquez (1998) proposed the use of *anger, fear, distress/sadness, enjoyment/happiness, disgust,* and *surprise*, which can be learned via artificial neural networks (ANNs). Further works in the field led to more complex models in robotics. One of the best-known ones is probably the MIT robot head Kismet, which focuses on maintaining the proper level of stimulation during social interaction (Breazeal and Scassellati 1998). Kismet uses computer vision and audio input to determine different simple environment parameters like noise level, color, motion, presence of people in the field of view, and so forth. The system architecture includes several systems: perception, motivation, attention, behavior, and motor. Basically, emotions are triggered by stimuli over a given threshold but also influenced by the so-called drives as well as by other emotions. Drives are partitioned into three regimes—homeostatic, overwhelmed, and underwhelmed—and they are used to modulate emotions. Kismet includes four drives (fatigue, social, security, and stimulation); five emotions (anger, disgust, fear, happiness, and sadness); and up to ten expressive states.

Robots like Kismet aim to prove that emotion plays important functions in social interaction. Ethologists have pointed out that emotional expression has a communicative function and acts as a releaser for the coordination of social behavior (Mook 1996).

20.8 Architecture Design

When a given system becomes very complex, it naturally goes modular, both in software and hardware aspects. Resulting modules need to communicate in an organized way. This decomposition into modules and their interaction is usually called its architecture. Architectures are defined by their structure, which is the number and nature of the modules they are composed of, and their style, which comprises the communication mechanism employed in their interaction (Coste-Maniere and Simmons 2000). Whereas structure is intrinsically linked to each system's functionality and purpose, style is more general, as it is related to efficient computational paradigms for software interaction between modules. Although there are many valid architectures that have been used in AmI projects and different architectures may coexist in most projects, recent ones use multiagent systems (MASs) as part of their structure and Web services as part of their style for integration, since there are already many valid operating third-party Web services that can be integrated into new systems.

Agents are particularly interesting in applications where there are multiple independent computational threads and a task-centered model of computation (Wooldridge and Jennings 1995). Agent technology is most likely popular in AmI because it is efficient at coping with pervasive systems and distributed information. AmI agents can be used as abstractions for devices and functionalities (Misker et al. 2004; Amigone et al. 2005) to coordinate the activities of the lower-level entities (Favela et al. 2004; Rodríguez et al. 2005) or to interface with users (Cortés et al. 2010). Agent communication usually relies on the Common Object Request Broker Architecture (CORBA) model, whereas many implementations rely on the Foundation for Intelligent Physical Agents (FIPA) Java Agent Development Environment (JADE).

In general, these agent-based architectures include four basic functionalities: (1) decision making, (2) information gathering and knowledge extraction, (3) communication between agents, and (4) physical elements. Every agent may yield all four functionalities (e.g., Cook et al. 2006) or just specialize in one (Guralnik and Haigh 2002; Plocher and Kiff 2003). For example, an iDorm embedded agent (Hagras et al. 2004) receives data from the sensors, learns and stores the user behavior, decides what to do, and acts accordingly. ROBOCARE, however, has a dedicated software agent to process camera images, detect users and robots from the background, calculate their 3-D localization, and identify them, whereas a second agent is in charge of controlling the assisted person's schedule, which should fit his location (Bahadori et al. 2005). Independent LifeStyle Assistant (ILSA) agents also specialize in individual behaviors like monitoring medication ingestion, issuing reminders, and so forth (Guralnik and Haigh 2002; Plocher and Kiff 2003). When there might be conflict between agents' decisions, an additional coordination (or manager) agent is sometimes required to work seamlessly. In other approaches, agents are used as abstractions, to act on behalf of users, doctors, caregivers, and sometimes, even robots (Muñoz et al. 2003; Corchado et al. 2008; Cortés et al. 2010).

20.9 Conclusions

This work has presented a review of ambient assistive living systems, with a special focus on assistive technologies and a tentative taxonomy of current projects depending on their demands and target population. We have reviewed as well the most important technologies and computational paradigms employed in these kinds of projects, providing information about off-the-shelf standard-compliant networks and devices and their manufacturers, when possible. We have tried to present a showcase of the most significant ones in each field, since AmI is clearly a multidisciplinary research field, and hence, it usually implies coexistence of many different technologies.

It needs to be noted that major research is being performed at the moment in areas related to AmI, like augmented reality, metamaterials, wearables, nanotechnology, and so forth, but we have tried to focus exclusively on systems that have already been prototyped in smart houses around the world.

Acknowledgments

This work was partially supported by the Spanish Ministerio de Educacion y Ciencia (MEC), project no. TEC2011-06734, and by the Junta de Andalucia, SIAD project no. TIC-3991.

Acronyms

AALJP Ambient Assisted Living Joint Programme
ADLs activities of daily living
ALS amyotrophic lateral sclerosis
AmI ambient intelligence
BCI brain/computer interface
BVP blood volume pulse
DD digital divide
ECG electroencephalography
EMG electromyography
EOG electrooculographic potential
GSR galvanic skin response
HCI human/computer interface
HR heart rate
HVAC heating, ventilation, and air-conditioning
ICT information and communication technology
PAN personal area network
RFID radio-frequency identification
WAN wide area network

References

Aarts, E., Harwig, R., and Schuurmans, M. 2002. Ambient intelligence. In Denning J. (ed). *The Invisible Future*. McGraw-Hill, New York, pp. 235–250.

Agrawal, R., and Srikant, R. 1995. Mining sequential patterns. In *Proceedings of the 11th International Conference on Data Engineering*. Taiwan, pp. 3–14.

Alwan, M., Kell, S., Dalal, S., Turner, B., Mack, D., and Felder, R. 2003. In-home monitoring system and objective ADL assessment: Validation study. In *Proceedings of the International Conference on Independence, Aging and Disability*. Washington, DC.

Amigoni, F., Gatti, N., Pinciroli, C., and Roveri, M. 2005. What planner for ambient intelligence applications? *IEEE Transactions on Systems, Man and Cybernetics, Part A: Systems and Humans*, 35(1), 7–21.

Annicchiarico, R., and Cortés, U. 2010. To share or not to share SHARE-it: Lessons learnt. In *Electronic Healthcare—Third International Conference, eHealth 2010*, Revised Selected Papers. Lecture Notes of the Institute for Computer Sciences, Social Informatics and Telecommunications Engineering, vol. 69. Springer, Casablanca, Morocco, pp. 295–302.

Bahadori, S., Cesta, A., Iocchi, L., Leone, G.R., Nardi, D., Pecora, F., Rasconi, R., and Scozzafava, L. 2005. Towards ambient intelligence for the domestic care of the elderly. In *Ambient Intelligence*, Remagnino, P., Foresti, G., and Ellis, T. (eds). Springer, New York, pp. 15–38.

Bellavista, P., Corradi, A., Fanelli, M., and Foschini, L. 2012. A survey of context data distribution for mobile ubiquitous systems. *ACM Computing Surveys*, 44(4), 24:1–24:45.

Bolchini, C., Curino, C., Quintarelli, E., Schreiber, F., and Tanca, L. 2009. Context information for knowledge reshaping. *International Journal of Web Engineering and Technology*, 5(1), 88–103.

Breazeal, C., and Scassellati, B. 1998. Infant-like social interactions between a robot and a human caretaker. *Adaptive Behavior, Special Issue on Simulation Models of Social Agents*, 8, pp. 49–74. MIT Press Cambridge, MA.

Camarinha-Matos, L., and Afsarmanesh, H. 2001. Virtual communities and elderly support. In *Advances in Automation, Multimedia and Video Systems, and Modern Computer Science*, Kluev, V.V., D'Attellis, C.E., Mastorakis, N.E. (eds.). WSES Press, Greece, pp. 279–284.

Cauchi, B., Goetze, S., and Doclo, S. 2012. Reduction of non-stationary noise for a robotic living assistant using sparse non-negative matrix factorization. In *Proceedings of the Speech and Multimodal Interaction in Assistive Environments Workshop (SMIAE 2012)*. Jeju Island, Republic of Korea.

Cesta, A., Cortellessa, G., Pecora, F., and Rasconi, R. 2005. Exploiting scheduling techniques to monitor the execution of domestic activities. *Intelligenza Artificiale (Italian Journal of Artificial Intelligence)*, 2(4), 74–81.

Chen, G., and Kotz, D. 2000. A survey of context-aware mobile computing research. Tech. Rept. TR2000-381. Dartmouth Computer Science.

Chen, L., Pei, L., Kuusniemi, H., Chen, Y., Kröger, T., and Chen, R. 2012. Bayesian fusion for indoor positioning using bluetooth fingerprints. *Wireless Personal Communications*. Springer US, New York, pp. 1–11.

Connell, J., and Viola, P. 1990. Cooperative control of a semi-autonomous mobile robot. In *Proceedings of the IEEE International Conference on Robotics and Automation*. Cincinnati, OH, pp. 1118–1121.

Cook, D., Youngblood, G.M., and Das, S.K. 2006. A multi-agent approach to controlling a smart environment. In *Designing Smart Homes*, Augusto, J.C., and Nugent, C.D. (eds). Lecture Notes in Computer Science, vol. 4008. Springer-Verlag, pp. 165–182.

Corchado, J.M., Bajo, J., and Abraham, A. 2008. GerAmi: Improving healthcare delivery in geriatric residences. *Intelligent Systems (Special Issue on Ambient Intelligence), IEEE*, 23(2), 19–25.

Cortés, U., Barrué, C., Martínez, A.B., Urdiales, C., Campana, F., Annicchiarico, R., and Caltagirone, C. 2010. Assistive technologies for the new generation of senior citizens: The SHARE-it approach. *International Journal of Computers in Healthcare*, 1(1), 35–65.

Coste-Maniere, E., and Simmons, R. 2000. Architecture, the backbone of robotic systems. In *Proceedings of the IEEE International Conference on Robotics and Automation, 2000 (ICRA '00)*, vol. 1. San Francisco, CA, pp. 67–72.

Ducatel, K., Bogdanowicz, M., Scapolo, F., Leijten, J., and Burgelman, J.C. 2001. Scenarios for ambient intelligence in 2010. Tech. Rept. IST Advisory Group. Seville, Spain, Feb.

Favela, J., Rodríguez, M., Preciado, A., and González, V. 2004. Integrating context-aware public displays into a mobile hospital information system. *IEEE Transactions on Information Technology in Biomedicine*, 8(3), 279–286.

Friedewald, M., Da Costa, O., Punie, Y., Petteri, A., and Heinonen, S. 2005. Perspectives of ambient intelligence in the home environment. *Telematics and Informatics*, 22(3), 221–238.

Gaggioli, A. 2005. Optimal experience in ambient intelligence. In *Ambient Intelligence*, Riva, G., Vatalaro, F., Davide, F., and Alcaniz, M. (eds). IOS Press, Amsterdam, pp. 35–43.

Gárate, A., Herrasti, N., and López, A. 2005. GENIO: An ambient intelligence application in home automation and entertainment environment. In *sOc-EUSAI '05: Proceedings of the 2005 Joint Conference on Smart Objects and Ambient Intelligence*. ACM Press, New York, pp. 241–245.

Gatti, N., Matteucci, M., and Sbattella, L. 2004. An adaptive and predictive environment to support augmentative and alternative communication. In *Proceedings of the 9th International Conference on Computers Helping People With Special Needs (ICCHP 2004)*, Klaus, J., Miesenberger, K., Zagler, W.L., and Burger, D. (eds). Paris, France, July 7–9, 2004. Lecture Notes in Computer Science, vol. 3118. Springer-Verlag, pp. 983–990.

Gelderblom, G.J., de Witte, L.P., Demers, L., Weiss-Lambrou, R., and Ska, B. 2002. The Quebec User Evaluation of Satisfaction with Assistive Technology (QUEST 2.0): An overview and recent progress. *Technology and Disability*, 14(3), 101–106.

Granata, C., Bidaud, P., Beck, C., and Mayer, P. 2012. Experimental analysis of interactive behaviors for a personal mobile robot. In *Fifteenth International Conference on Climbing and Walking Robots and the Support Technologies for Mobile Machines*. Baltimore, MD, July 23–26.

Guralnik, V., and Haigh, K. 2002. Learning models of human behaviour with sequential patterns. In *Proceedings of Fourteenth Conference on Innovative Applications of Artificial Intelligence Workshop "Automation as Caregiver"*. Edmonton, Alberta, Canada, pp. 24–30.

Hagras, H., Callaghan, V., Colley, M., Clarke, G., Pounds-Cornish, A., and Duman, H. 2004. Creating an ambient-intelligence environment using embedded agents. *IEEE Intelligent Systems*, 19(6), 12–20. New York.

Haigh, K., Kiff, L., Myers, J., Guralnik, V., Geib, C., Phelps, J., and Wagner, T. 2004. The Independent LifeStyle Assistant (I.L.S.A.): AI lessons learned. In *Proceedings of the Nineteenth National Conference on Artificial Intelligence, Sixteenth Conference on Innovative Applications of Artificial Intelligence*, McGuinness, D.L., and Ferguson, G. (eds). AAAI Press/The MIT Press, San Jose, CA, pp. 852–857.

Hart, S., and Staveland, L. 1988. Development of NASA-TLX (Task Load Index): Results of empirical and theoretical research. In *Human Mental Workload*, Hancock, P., and Meshkati, N. (eds), Chap. 7. Elsevier, Amsterdam, North Holland, pp. 139–183.

Heierman, III, E., and Cook, D. 2003. Improving home automation by discovering regularly occurring device usage patterns. In *Proceedings of the Third IEEE International Conference on Data Mining (ICDM '03)*. IEEE Computer Society, Washington, DC, pp. 537–540.

Helal, S., Mann, W., El-Zabadani, H., King, J., Kaddoura Y., and Jansen, E. 2005. The gator tech smart house: A programmable pervasive space. *IEEE Computer Magazine*, 38(3), 50–60.

Himberg, J., Mantyjarvi, J., and Seppanen, T. 2001. Recognizing human motion with multiple acceleration sensors. In *IEEE International Conference on Systems, Man and Cybernetics*. pp. 747–752.

Holmes, A., Duman, H., and Pounds-Cornish, A. 2002. iDorm: Gateway to heterogeneous networking environments. In *International ITEA Workshop on Virtual Home Environments*, vol. 1. February 20–21, pp. 85–93.

Infield, G. 1968. A computer in the basement? *Popular Mechanics*, 1, 77–79.

Intille, S. 2002. Designing a home of the future. *IEEE Pervasive Computing, Integrated Environments*, 1(2), 76–82.

Jutai, J., and Bortolussi, J. 2003. *Psychosocial Impact of Assistive Technology: Development of a Measure for Children. Assistive Technology Research Series*. IOS Press, pp. 936–940.

Kautz, H., Etziono, O., Fox, D., and Weld D. 2003. Foundations of assisted cognition systems. Tech. Rept. CSE-02-AC-01. University of Washington, Department of Computer Science and Engineering.

Kiff, L., and Plocher, T. 2003. Mobility Monitoring with the Independent LifeStyle Assistant (ILSA). In *The Proceedings of the International Conference on Aging, Disability and Independence*. Washington, DC, December, pp. 170–171.

LaMarca, A., Chawathe, Y., Consolvo, S., Hightower, J., Smith, I., Scott, J., Sohn, T., Howard, J., Hughes, J., Potter, F., Tabert, J., Powledge, P., Borriello, G., and Schilit, B. 2005. Place lab: Device location. Positioning using radio beacons in the wild. In *Proceedings of 3rd International Conference on Pervasive Computing*. Munich, Germany.

Larson, K. 2000. The home of the future. *A+U Architecture and Urbanism*, (361).

Lee, S.W., and Mase, K. 2002. Activity and location recognition using wearable sensors. *IEEE Pervasive Computing*, 1(3), 24–32.

Li, F., Kolakowski, S.M., and Pelz, J.B. 2007. Using structured illumination to enhance video-based eye tracking. In *Proceedings of the International Conference on Image Processing (ICIP '07)*, vol. 1. San Antonio, TX, pp. 373–376.

Millán, J., Renkens, F., Mourino, J., and Gerstner, W. 2004. Noninvasive brain-actuated control of a mobile robot by human EEG. *IEEE Transactions on Biomedical Engineering*, 51, 1026–1033.

Miller, D., and Slack, M. 1995. Design and testing of a low-cost robotic wheelchair prototype. *Autonomous Robots*, 3(1), 77–88.

Misker, J., Veenman, J., and Rothkrantz, L. 2004. L.J. groups of collaborating users and agents in ambient intelligent environments. In *Proceedings of the 3rd International Joint Conference on Autonomous Agents and Multi-Agent Systems (AAMAS '04)*, vol. 3. ACM Press, New York, pp. 1320–1321.

Mook, D.G. 1996. *Motivation: The Organization of Action*. W.W. Norton & Company Incorporated, Virginia.

Mundt, C., Montgomery, K., Udoh, U., Barker, V., Thonier, G., Tellier, A., Ricks, R., Darling, R., Cagle, Y., Cabrol, N., Ruoss, S., Swain, J., Hines, J., and Kovacs, G. 2005. A multiparameter wearable physiologic monitoring system for space and terrestrial applications. *IEEE Transactions on Information Technology in Biomedicine*, 9, 382–391.

Muñoz, M.A., Rodríguez, M., Favela, J., Martínez-García, A.I., and González, V.M. 2003. Context-aware mobile communication in hospitals. *Computer*, 36(9), 38–46.

Nejatbakhsh, N., and Kosuge, K. 2005. User-environment based navigation algorithm for an omnidirectional passive walking aid system. In *IEEE International Conference on Rehabilitation Robotics*. Chicago, IL, pp. 178–181.

Noyes, J.M., Haigh, R., and Starr, A.F. 1989. Automatic speech recognition for disabled people. *Applied Ergonomics*, 20(4), 293–298.

Philipose, M., Fishkin, K.P., Perkowitz, M., Patterson, D., and Hähne, D. 2003. The probabilistic activity toolkit: Towards enabling activity-aware computer interfaces. Intel Research Seattle Tech. Rept.

Pino, P., Amoud, P., and Brangier, E. 1998. A more efficient man machine interface: Fusion of the interacting telethesis and smart wheelchair projects. In *Proceedings of the 1998 Second International Conference on Knowledge-Based Intelligent Electronic System*. Adelaide, South Australia, April, pp. 21–23.

Pope, A.M., and Tarlov, A.R. 1991. *Disability in America: Toward a National Agenda for Prevention*. The National Academies Press, Washington, DC.

Rebsamen, B., Burdet, E., Guan, C., Zhang, H., Teo, C.L., Zeng, Q., Ang, M., and Laugier, C. 2007. Controlling a wheelchair indoors using thought. *IEEE Intelligent Systems*, 22, 18–24.

Rentscheler, A.J., Cooper, R.A., Blash, B., and Boninger, L.M. 2007. Intelligent walkers for the elderly: Performance and safety testing of VAPAMAID robotic walker. *Journal of Rehabilitation Research and Development*, 40, 423–432.

Rodríguez, M.D., Favela, J., Preciado, A., and Vizcaíno, A. 2005. Agent-based ambient intelligence for healthcare. *AI Communications*, 18(3), 201–216.

Rodríguez-Losada, D., Matía, F., Jiménez, A., Galán, R., and Lacey, G. 2005. Implementing map based navigation in Guido, the robotic smartwalker. In *2005 IEEE International Conference on Robotics and Automation*. Barcelona, Spain, pp. 3390–3395.

Sadri, F. 2011. Ambient intelligence: A survey. *ACM Computing Surveys*, 43(4), 1–66.

Saffiotti, A., and Broxvall, M. 2005. PEIS ecologies: Ambient intelligence meets autonomous robotics. In *Proceedings of the 2005 Joint Conference on Smart Objects and Ambient Intelligence: Innovative Context-Aware Services: Usages and Technologies*, vol. 121. ACM, New York, pp. 277–281.

Scherer, M., Jutai, J., Fuhrer, M., Demers, L., and Deruyter, F. 2007. *A Framework for Modeling the Selection of Assistive Technology Devices (ATDs)*, vol. 2. Taylor and Francis, Limited, Abingdon, Oxfordshire, UK, pp. 1–8.

Schilit, W., Adams, N., and Want, R. 1994. Context-aware computing applications. In *Proceedings of the Workshop on Mobile Computing Systems and Applications*. IEEE Computer Society, Santa Cruz, CA, pp. 85–90.

Simpson, R., Levine, S., Bell, D., Jaros, L., Koren, Y., and Borenstein, J. 1998. NavChair: An assistive wheelchair navigation system with automatic adaptation. In *Assistive Technology and Artificial Intelligence*, Mittal, V., Yanco, H., Aronis, J., and Simpson, R. (eds). Lecture Notes in Computer Science, vol. 1458. Springer Berlin/Heidelberg, pp. 235–255.

Skubic, M., Koopman, R., Phillips, L., Alexander, G.L., Miller, S.J., and Guevara, R.D. 2011. Using sensor networks to detect urinary tract infections in older adults. In *Proceedings of the 13th IEEE International Conference on e-Health Networking Applications and Services*. pp. 142–149.

Tapia, D.I., and Corchado, J.M. 2009. An ambient intelligence based multi-agent system for alzheimer health care. *International Journal of Ambient Computing and Intelligence*, 1(1), 15–26.

Tiberio, L., Cesta, A., Cortellessa, G., Pauda, L., and Pellegrino, A.R. 2012. Assessing affective response of older users to a telepresence robot using a combination of psychophysiological measures. In *21st IEEE International Symposium on Robot and Human Interactive Communication*. Paris, France.

Torres-Solís, J.H., Falk, T., and Chau, T. 2010. A review of indoor localization technologies: Towards navigational assistance for topographical disorientation. *InTech*, 1, 51–84.

Tran, A.C., Marsland, S., Dietrich, J., Guesgen, H., and Lyons, P. 2010. Use cases for abnormal behavior detection in smart homes. In *Proceedings of the Aging Friendly Technology for Health and Independence (ICOST '10)*. Tran & Mynatt, pp. 144–151.

Trapero, R., Urdiales, C., Sigler, F., Domínguez-Durán, M., La Torre, J., Coslado, F.J., Perez-Parras, J., and Sandoval, F. 2007. On practical issues about interference in Telecare applications based on different wireless technologies. *Telemedicine and e-Health*, 13(5), 519–534.

Urdiales, C. 2012. *Collaborative Assistive Robot for Mobility Enhancement (CARMEN): The Bare Necessities Assisted Wheelchair Navigation and Beyond*. Springer Publishing Company, Incorporated, Berlin, Heidelberg.

Urdiales, C., Peula, J.M., Fernández-Carmona, M., Annicchiarico, R., Sandoval, F., and Caltagirone, C. 2009. Adaptive collaborative assistance for wheelchair driving via CBR learning. In *IEEE 11th International Conference on Rehabilitation Robotics (ICORR 2009)*. Kyoto, Japan, June.

Velásquez, J. 1998. A computational framework for emotion-based control. In *Workshop on Grounding Emotions in Adaptive Systems, Conference on Simulation of Adaptive Behavior*.

Virone, G., Alwan, M., Dalal, S., Kell, S., Turner, B., Stankovic, J., and Felder, R. 2008. Behavioural patterns of older adults in assisted living. *IEEE Transactions on Information Technology in Biomedicine*, 12(3), 387–398.

Wan, H., Sengupta, M., Velkoff, V., and DeBarros, K. 2005. *65+ in the United States, 2005*. U.S. Dept. of Commerce, Economics and Statistics Administration, U.S. Census Bureau, Washington, DC.

Wasson, G., Huang, C. A., Majd, S.P., and Ledoux, A. 2007. Shared navigational control and user intent detection in an intelligent walker. *Aging Medicine*, 53–76.

Wei, L., Hu, H., Lu, T., and Yuan, K. 2010. Evaluating the performance of a face movement based wheelchair control interface in an indoor environment. In *2010 IEEE International Conference on Robotics and Biomimetics (ROBIO)*. Tianjin, China, Dec., pp. 387–392.

Wooldridge, M., and Jennings, N.R. 1995. Intelligent agents: Theory and practice. *Knowledge Engineering Review*, 10(2), 115–152. Cambridge Univ. Press, Cambridge, UK.

Wren, C.R., and Munguia-Tapia, E. 2006. Toward scalable activity recognition for sensor networks. In *Proceedings of the 2nd International Workshop in Location and Context-Awareness (LoCA '06)*, 3987. Dublin, Ireland, pp. 168–185.

Yanco, H.A. 1998. Wheelesley: A robotic wheelchair system: Indoor navigation and user interface, assistive technology and artificial intelligence. *Applications in Robotics, User Interfaces and Natural Language Processing*. Springer, Berlin Heidelberg New York, pp. 256–268.

Youssef, M., and Agrawala, A. 2008. The Horus location determination system. *Wireless Networks*, 14, 357–374.

Zhu, J., and Durgin, G.D. 2005. Indoor/outdoor location of cellular handsets based on received signal strength. In *Vehicular Technology Conference, 2005*. VTC 2005-Spring. 2005 IEEE 61st, vol. 1, no., pp. 92, 96 Vol. 1, May 30–June 1 2005, Stockholm, Sweden.

21

Intelligent Light Therapy for Older Adults: Ambient Assisted Living

Joost van Hoof, Eveline J. M. Wouters, Björn Schräder, Harold T. G. Weffers, Mariëlle P. J. Aarts, Myriam B. C. Aries, and Adriana C. Westerlaken

CONTENTS

21.1 Introduction

Light therapy is increasingly administered and studied as a nonpharmacologic treatment for a variety of health-related problems, including treatment of people with dementia. It is applied in a variety of ways, ranging from being exposed to daylight (in sanatoria) to being exposed to light emitted from electrical sources. These include light boxes, light showers, and ambient bright light. Light therapy covers an area in medicine where medical sciences meet the realms of physics, engineering, and technology (van Hoof et al. 2012).

One of the areas within medicine in which light therapy is administered is geriatric psychiatry, which includes the care of older adults with dementia. Dementia can be caused by a number of progressive disorders that affect memory, thinking, behavior, and the ability to perform everyday activities. Alzheimer's disease is the most common cause of dementia. Other types include vascular dementia, dementia with Lewy bodies, and frontotemporal dementia. Dementia mainly affects older people, although there is a growing awareness of a substantial amount of cases that start before the age of 65. After age 65, the likelihood of developing dementia roughly doubles every 5 years.

In the 2009 World Alzheimer Report, Alzheimer's Disease International estimated that there would be 35.6 million people living with dementia worldwide in 2010, increasing to 65.7 million by 2030 and 115.4 million by 2050. Within Europe, nearly two-thirds live in low- and middle-income countries, where the sharpest increases in numbers will occur.

The societal cost of dementia is already enormous. Dementia is significantly affecting every health and social care system in the world. The economic impact on families is insufficiently appreciated. New technological services and government policies may help to address this problem.

Low-income countries accounted for just under 1% of the total worldwide costs (but 14% of the prevalence), middle-income countries for 10% of the costs (but 40% of the prevalence), and high-income countries for 89% of the costs (but 46% of the prevalence). About 70% of the global costs occurred in just two regions: Western Europe and North America.

Recent research indicates that dementia could be slowed down significantly by treatments that reset the body's biological clock. This kind of research started with the work of van Someren et al. (1997), who conducted a study on the effects of ambient bright light emitted from ceiling-mounted luminaires. In a randomized controlled study by Riemersma-van der Lek et al. (2008), brighter daytime lighting was applied to improve the sleep of persons with dementia (PwDs) and slow down cognitive decline. Applying bright light techniques is expected to lengthen the period PwDs can continue to live in their own home. It can also reduce the speed of health decline once the PwD has been moved to a care facility. In these situations, the quality of life will be positively impacted, and the workload of the care professionals and assisting relatives is expected to decrease. At the same time, Forbes et al. (2009) concluded that due to the lack of randomized controlled trials, there are no clear beneficial outcomes of light therapy for older persons. Similar lighting systems have also been tested in school environments; however, these systems did not have any effect on school children in a controlled setting (Sleegers et al. 2013).

PwDs do not venture outdoors as much as healthy younger adults, due to mobility impairments, and inside their homes, they are exposed to light levels that are not sufficient for proper vision, let alone yielding positive outcomes to circadian rhythmicity and mood (Aarts and Westerlaken 2005; Sinoo et al. 2011). Using light as a care instrument does not only apply to people with dementia, it also applies to ageing in general. Ageing impacts the circadian rhythm of people and increases the gap between level of sleep required and level of sleep achieved (older people do not get enough deep sleep). Moreover, it impacts the vision, since the eyes become affected due to biological ageing. These ageing effects include, among others, the yellowing of the lens and vitreous.

Although the evidence regarding the positive impact of light on well-being, especially of older people and PwDs or persons with other neurological diseases, is hopeful but not convincingly and scientifically affirmed (Forbes et al. 2009), these insights are already being converted to implementable solutions. Applying light as an instrument for care has tremendous benefits. It is noninvasive, it is cheap in implementation and maintenance, and it has a high level of intuitive use, creating a low threshold for acceptance.

Applying bright light techniques can slow down cognitive decline and reduce the speed of health decline. There is, however, no conclusive evidence on which lighting conditions are most favorable for yielding positive health outcomes. We also lack a clear definition of what technicians and product developers call healthy lighting. It is also not known how to design such healthy lighting systems, for instance, in relation to the emergence of new energy-friendly light sources such as LED (van Hoof et al. 2012). There is no validated set of algorithms as used in current lighting systems, including the effects of static versus dynamic lighting protocols (Barroso and den Brinker 2013), the contribution of the dynamic component of daylight, vertical and horizontal illuminance levels, and color temperature. The available knowledge has not yet been converted into widespread implementable lighting solutions, and the solutions available are often technologically unsophisticated, uneducated guesses and poorly evaluated from the perspective of end users. New validated approaches in terms of ceiling-mounted luminaires, the inclusion of low-energy light sources, and integration of computerized controls are needed.

This chapter will focus first on the effects of biological ageing and dementia on our lighting needs and second on the application of intelligent light therapies for older adults with dementia.

21.2 The Effects of Biological Ageing and Dementia

The age-related sensory changes, involving sensory receptors in the eyes, ears, nose, buccal cavity, and peripheral afferent nerves, frequently affect the way we perceive the environment. Apart from the sensory changes, incorrect or malfunctioning visual aids and hearing aids may have negative effects, too. Sensory losses or impairments, together with cognitive deficits, make it difficult for the individual to interpret and understand the environment (perception and comprehension phase) (van Hoof et al. 2010).

21.2.1 Ageing-Related Changes in Vision

Ageing negatively affects vision. In general, the performance of the human eye deteriorates already at a relatively early age. Many people aged 45 and over wear glasses to compensate for impaired vision due to presbyopia, caused by reduced elasticity of the lens of the eye resulting in significant loss of focusing power. Older people are known to have vision impairments stemming from the normal ageing process, which include an impaired ability to quickly adapt to changes in light levels, extreme sensitivity to glare, reduced visual acuity, restricted field of vision and depth perception, reduced contrast sensitivity, and restricted color recognition. Changes in vision do not happen overnight and depend on the progress of age. After the age of 50, glare and low levels of light become increasingly problematic. People require more contrast for proper vision and have difficulty perceiving patterns. After the age of 70, fine details become even harder to see, and color and depth perception may be affected. Apart from the influence of ageing, there are pathological changes leading to low vision and eventual blindness, such as cataracts, macular degeneration, glaucoma, and diabetic retinopathy (van Hoof et al. 2010; Sinoo et al. 2011). In Table 21.1 an overview of age-related eye pathology is given.

21.2.2 Ageing and Nonvisual Effects of Light

Apart from being indispensable for proper vision, light plays a role in regulating important biochemical processes, immunologic mechanisms, and neuroendocrine control (for instance, melatonin and cortisol pathways), via the skin and via the eye (Hughes and Neer 1981). Light exposure ($\lambda \sim 460$–480 nm) is the most important stimulus for synchronizing the biological clock, suppressing pineal melatonin production, elevating core body temperature, and enhancing alertness (van Hoof et al. 2010, 2012). The circadian system, which is orchestrated by the hypothalamic suprachiasmatic nuclei (SCN), influences virtually all tissues in the human body. In the eye, light activates intrinsically photosensitive retinal ganglion cells (Brainard et al. 2001; Thapan et al. 2001), which discharge nerve impulses that are transmitted directly to the SCN and, together with the photoreceptors for scotopic and photopic vision, participate in mammalian circadian phototransduction.

In older adults, the orchestration by the SCN requires ocular light levels that are significantly higher than those required for proper vision, but the exact thresholds are unknown

TABLE 21.1

Age-Related Sensory Changes to Vision

- Lid elasticity diminished, leading to pouches under the eyes.
- Loss of orbital fat, leading to excessive dryness of eyes.
- (1) Decreased tears; (2) arcus senilis becomes visible; (3) sclera yellows and becomes less elastic; (4) yellowing and increased opacity of cornea, which may lead to a lack of corneal luster.
- (1) Increased sclerosis and rigidity of the iris and (2) a decrease in elasticity and convergence ability of the lens, leading to presbyopia.
- Decline in light accommodation response leads to lessened acuity.
- Diminished pupillary size leads to a decline in depth perception.
- Atrophy of the ciliary muscles (holding the lens) leads to a diminished recovery from glare.
- Night vision diminishes leading to night blindness.
- Yellowing of the lens may lead to a diminished color perception (blues and greens).
- Lens opacity may develop, leading to cataract.
- Increased ocular pressure may lead to seeing rainbows around lights.
- Shrinkage of gelatinous substance in the vitreous, which may lead to altered peripheral vision.
- Vitreous floaters appear.
- Ability to gaze upward decreases.
- Thinning and sclerosis of retinal blood vessels.
- Atrophy of photoreceptor cells.
- Degeneration of neurons in visual cortex.

Sources: Hughes, P.C., and Neer, R.M., *Hum. Factors*, 23(1), 65–85, 1981. Ebersole, P. et al., editors, *Toward Healthy Aging*, sixth edition, Mosby, St. Louis, MO, 2004.

to date. Research by Aarts and Westerlaken (2005) in the Netherlands has shown that light levels, even during daytime, are too low both to allow for proper vision and, consequently, also for non–image-forming effects, even though the semi-independently living older persons were satisfied with their lighting conditions. A similar study was carried out among 40 community-dwelling older people in New York City by Bakker et al. (2004). Even though nearly all of them had inadequate light levels for both image-forming and non-image-forming effects, subjects rated their lighting conditions as adequate.

An additional problem is formed by the ageing of the eye, which leads to opacification and yellowing of the vitreous and the lens, limiting the amount of bluish light reaching the retinal ganglion cells. This can be as much as a 50% reduction in 60-year olds compared to 20-year olds.

Many older adults are not exposed to high-enough illuminance levels, due to decreased lens transmittance, poorly lit homes (up to 400 lx), and the short periods of time spent outdoors. The indoor illuminance levels are too low for any non-image-forming effects to take place.

21.2.3 Dementia-Related Changes in Vision

Dementia has a severe impact on the human visual system (Guo et al. 2010), and the effects of biological ageing often aggravate the visual dysfunctions stemming from dementia. Persons with Alzheimer's disease frequently show a number of visual dysfunctions, even in the early stages of the disease (Kergoat et al. 2001; Redel et al. 2012). These dysfunctions include impaired spatial contrast sensitivity, motion discrimination, and color vision, as well as blurred vision. Altered visual function may even be present if people with dementia have normal visual acuity and have no ocular diseases (Kergoat et al. 2001). Another dysfunction is diminished contrast sensitivity, which may exacerbate the effects of other cognitive losses and increase confusion and social isolation (Boyce 2003). Impaired visual acuity may be associated with visual hallucinations (Desai and Grossberg 2001). According to Mendez et al. (1996), persons with Alzheimer's disease have disturbed interpretation of monocular as well as binocular depth cues, which contributes to visuospatial deficits. The impairment is largely attributed to disturbances in local stereopsis and in the interpretation of depth from perspective, independent of other visuospatial functions.

21.2.4 Dementia and Nonvisual Effects of Light

In people with Alzheimer's disease, the SCN is affected by the general atrophy of the brain, leading to nocturnal restlessness due to a disturbed sleep–wake rhythm and wandering (van Someren 2000; Waterhouse et al. 2002). The timing of the sleep–wake cycle can show a far wider variation; times of sleep and activity can vary substantially from day to day or can be temporarily inverted (Waterhouse et al. 2002), which has great implications for both the PwD and his/her family carer. Restlessness and wandering form a high burden for carers and are among the main reasons for institutionalization (Health Council of the Netherlands 2002; Abbott 2003; Harper et al. 2005). Marshall (1995) stated that lighting technology deserves more attention as a means to help with managing problem behavior. Hopkins et al. (1992) have suggested a relation between illuminance levels and this type of behavior before, and today, light therapy is used as a treatment to improve sleep in people experiencing sundowning behavior (Brawley 2006). Sundowning is associated with increased confusion and restlessness in PwDs in the evening.

It is hypothesized that high-intensity lighting, with vertical (instead of horizontal, as our eye is located in a vertical plain) illuminance levels of well over 1000 lx (eye height), may play a role in the management of dementia. Bright light treatment with the use of light boxes is applied to entrain the biological clock, to modify behavioral symptoms, and to improve cognitive functions, by exposing people with dementia to high levels of ocular light (see, for instance, Lovell et al. 1995; Thorpe et al. 2000; Yamadera et al. 2000; Graf et al. 2001; Dowling et al. 2005). This intervention requires supervision to make PwDs follow the total protocol and may cause a bias in the outcomes of the therapy, for instance, as the level of personal attention is higher. The results of bright light therapy on managing sleep and behavioral, mood, and cognitive disturbances show preliminary positive signs, but there is a lack of adequate evidence obtained via randomized controlled trials to allow for widespread implementation in the field (Kim et al. 2003; Terman 2007; Forbes et al. 2009).

Another approach that is gaining popularity, from a research, ethical, and practical point of view, is to increase the general illuminance level in rooms where people with dementia spend their days in order for non-image-forming effects of light to take place (Boyce 2003). Studies by Rheaume et al. (1998), van Someren et al. (1997), Riemersma-van der Lek et al. (2008), and van Hoof et al. (2009a,b) that exposed institutionalized PwDs to ambient bright light through ceiling-mounted luminaires showed short-term and long-term effects, such as lessened nocturnal restlessness, a more stable sleep–wake cycle, possible improvement to restless and agitated behavior as well as better sleep quality, increased amplitude of the circadian body temperature cycle, and a lessening of cognitive decline.

The occurrence of nonvisual effects of light does not only depend on light intensity. As stated before, certain parts of the light spectrum (specific short wavelengths) are more effective than others. The human circadian photoreception sensitivity peaks at approximately 480 nm, which is associated with the neuroendocrine and neurobiological systems. This sensitivity is graphically represented in the so-called $C(\lambda)$ curve, which is used

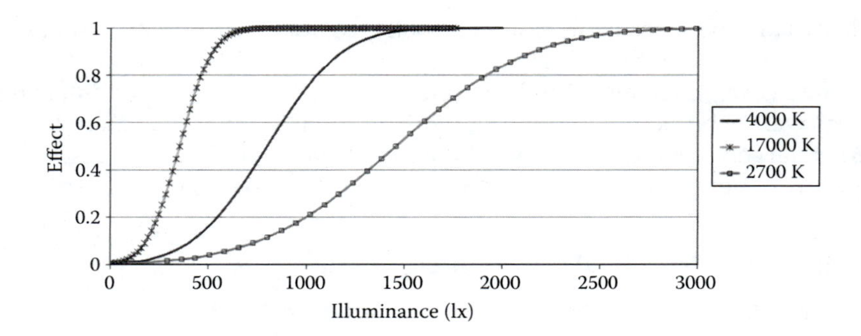

FIGURE 21.1
Hypothesized size of nonvisual effects of light during daytime for different illuminance levels at the eye and different color temperatures. (Adapted from Górnicka, G.B., Lighting at Work. Environmental Study of Direct Effects of Lighting Level and Spectrum on Psychophysiological Variables. Dissertation, Eindhoven University of Technology, Eindhoven, 2008. Reprinted from *Building and Environment*, 44(9), van Hoof, J. et al., High colour temperature lighting for institutionalised older people with dementia, 1959–1969, Copyright 2009, with permission from Elsevier. License number: 3087171431049.)

in lighting research and practice (Pechacek et al. 2008). Generally, required illuminance levels are higher than average, and so is the (correlated) color temperature of the light (Górnicka 2008) (Figure 21.1). The (correlated) color temperature is one of the measures for the amount of short-wavelength light present in the spectrum. As stated, there are several short-term and long-term effects (van Hoof et al. 2012). A cluster-unit crossover intervention trial by Sloane et al. (2007) on the effects of high-intensity light found that nighttime sleep of older adults with dementia improved when exposed to morning and all-day light, with the increase most prominent in participants with severe or very severe dementia. Hickman et al. (2007) studied the effects on depressive symptoms in the same setting as Sloane et al. (2007). Their findings did not support the use of ambient bright light therapy as a treatment for depressive symptoms. To date, it is unknown if the light therapy is effective, how long effects of bright light last, and how to predict which persons (may) respond favorably to light treatment. These points were already made by the Health Council of the Netherlands in 2002.

21.3 Technological Solutions: Design and Practice

The increasing numbers of older people with dementia in combination with the lack of available care professionals go together with a need for technological solutions and services to support activities of daily living and reduce the burden on carers. Light therapy is hypothesized to play a role in improving the well-being and quality of life of PwDs. To date, the administration of light therapy via ceiling-mounted luminaires is a relatively new area of study and innovation. As the current state of science permits us to design and model healthy lighting solutions, it is time to improve the quality of life of PwDs. In order to do so, we need to investigate the recent innovations in the field of lighting technology in relation to the underlying algorithms of the lighting equipment's steering mechanism.

New dynamic lighting protocols are being implemented in the lighting solutions offered to older adults (Figures 21.2 and 21.3). The underlying assumption of such systems is that

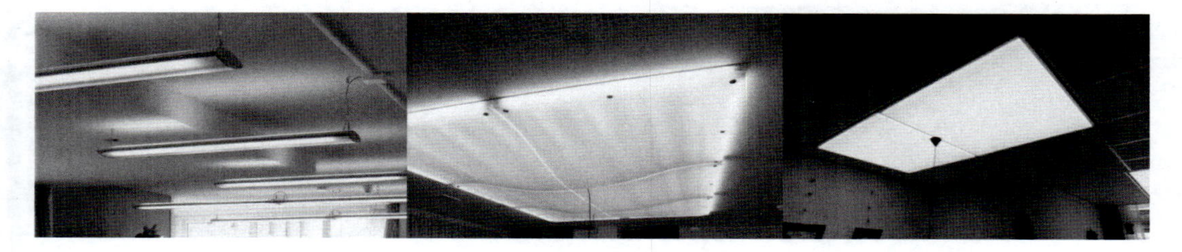

FIGURE 21.2
(See color insert.) Examples of dynamic lighting installed in Dutch nursing homes: Amadea by Derungs (left), Biosun by Van Doorn (middle), and Strato by Philips (right).

FIGURE 21.3
Examples of dynamic lighting systems in Dutch nursing homes. Figure upper right. (Reprinted from *Building and Environment*, 44(9), van Hoof, J. et al., High colour temperature lighting for institutionalised older people with dementia, 1959–1969, Copyright 2009, with permission from Elsevier. License number: 3087171431049.)

human beings evolved in daylight conditions and that the dynamic component further contributes to the positive effects of the lighting systems. As there is no validated set of algorithms as used in current lighting systems, including the relative effects of static versus dynamic lighting protocols (Rea et al. 2002; Figueiro 2008; Barroso and den Brinker 2012), there is still plenty of room for innovation and research. Figure 21.4 shows the rationale behind a dynamic lighting protocol and the way it has been shaped in practice. As can be seen in the figure, both illuminance and color temperature are controlled through dedicated software. In most projects, only the main luminaire in the living room is steered via a dynamic protocol.

In all the studies concerning light therapy, the exposure to daylight is often an ill-described aspect. We therefore do not fully understand the effects of these interactions. Daylight has a dynamic character, which is mimicked by dynamic lighting systems. With new technologies, lighting can be supplemented to the available daylight, which also has positive effects on energy consumption. This requires that new lighting solutions are to

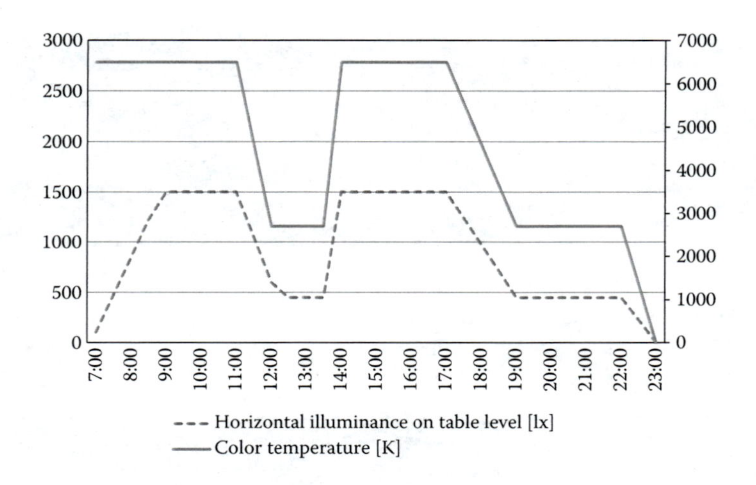

FIGURE 21.4
Example of a protocol for dynamic lighting. The horizontal illuminance on the table level varies between 0 and 1500 lx. The color temperature of the light varies between 0 and 6500 K. The so-called post-lunch dip can be found in the first hour of noon.

be integrated within a framework of the digitalization of the home (including sensor-based networks, ambient assisted living, and e-health) (van Hoof et al. 2011a,b) and smart façade systems (measuring incoming daylight and its spectral composition), which can go together with an optimization of exposure to certain light levels and, hence, energy use. Moreover, the exposure to light can be part of a digital patient or care file and form an additional source of information to medication protocols. Ambient intelligence in the home environment can thus lead to new improvements in the administration of ambient bright light, starting with institutional settings.

New lighting solutions should combine all sources of light (including daylight); novel lighting technologies such as LED technology (with energy-efficient options for a high output in the blue region of the spectrum) or ceiling- or wall-mounted lighting applications (also referred to as ambient bright light); smart sensors that measure the quality and quantity of light needed at a specific moment for a specific individual; and the platform that both controls the supply of light and, in parallel, measures the effectiveness of the light. The intricate balance between daylight and electrical light sources calls for smart sensors in the dwelling and dedicated software to steer the lighting exposure of persons. These sensors should also measure the presence of occupants and the type of activity they are engaged in. The dip in lighting protocols (Figure 21.4) is related to the length of the lunch break. If this break takes longer, perhaps the dip in the protocol should be longer too. The protocol should ideally follow the care regime. Apart from manual controls, a sensor-based network can help achieve accounting for the regime.

In addition, light therapy also calls for an attractive design of the lighting equipment. Indoor lighting conditions may be perceived as unfavorable from the perspective of personal preferences and taste, particularly with higher color temperature lighting or light with a dedicated, spectrum which accounts for the human circadian photoreception sensitivity that peaks at approximately 480 nm [following from the C(λ) curve (Pechacek et al. 2008)]. This can be solved by adding an additional decorative luminaire underneath the luminaires used for light therapy.

One other factor that makes it hard to implement innovations in the field is that to date, there are no extensive documents on lighting for older people, which can serve as an underlying basis when conducting research and for product and service innovation. There is, however, a growing interest within the building community in the nonvisual aspects of light (Webb 2006; McNair et al. 2010). CEN EN 12464-1 (2011) summarizes recommendations of lighting of indoor work places. The standard specifies horizontal illuminances for health care facilities, such as waiting rooms, corridors, examination rooms, and spaces for diagnostics in hospitals. Nursing homes are not included in this standard. The standard does not specifically include color temperature for health care facilities in general, too. There need to be more efforts to improve current standards and guidelines to account for light therapy and ambient bright light. We need to look at how science can find its way into practice in order to improve the quality of life of PwDs and the work of their carers and the installers.

At the same time, we need to be critical. There is still a lot we do not know. To date, we do not know if dynamic systems have better outcomes than static lighting systems and how such systems (should) interact with available daylight. We do know that vision can be improved by raising general illuminance levels and glare control; the nonvisual benefits cannot be quantified yet. Therefore, the economic and well-being benefits of accounting for these parameters are not yet clear. As long as there are many uncertainties, maybe we should suggest exposing our older citizens to plenty of daylight, for instance, by taking them out for a stroll. The innovations in the realm of sensors and ambient intelligence in the home environment hold a promise that, in the future, the nursing home can administer the right amount of lighting to its residents, which allows them to enjoy the highest degree of quality of life.

References

Aarts MPJ, Westerlaken AC. Field study of visual and biological light conditions of independently-living elderly people. *Gerontechnology* 2005;(3):41–152.

Abbott A. Restless nights, listless days. *Nature* 2003;245(6961):896–898.

Alzheimer's Disease International (2009). World Alzheimer report. Alzheimer's Disease International, London.

Bakker R, Iofel Y, Mark S, Lachs MS. Lighting levels in the dwellings of homebound older adults. *Journal of Housing for the Elderly* 2004;18(2):17–27.

Barroso A, den Brinker B. Boosting circadian rhythms with lighting: a model driven approach. *Lighting Research and Technology* 2013;45(2):197–216 doi: 10.1177/1477153512453667.

Boyce PR (2003). *Human Factors in Lighting. Lighting Research and Technology,* 2013;45(2):197–216.

Brainard GC, Hanifin JR, Greeson JM, Byrne B, Glickman G, Gerner E, Rollag MD. Action spectrum for melatonin regulation in humans: Evidence for a novel circadian photoreceptor. *Journal of Neuroscience* 2001;21(16):6405–6412.

Brawley EC (2006). *Design Innovations for Aging and Alzheimer's. Creating Caring Environments*. John Wiley & Sons, Hoboken, NJ.

CEN. EN 12464-1 (2011). *Light and Lighting—Lighting of Work Places—Part 1: Indoor Work Places.* Comité Européen de Normalisation CEN, Brussels, Belgium.

Desai AK, Grossberg GT. Recognition and management of behavioral disturbances in dementia. *Primary Care Companion to the Journal of Clinical Psychiatry* 2001;3(3):93–109.

Dowling GA, Hubbard EM, Mastick J, Luxenberg JS, Burr RL, van Someren EJW. Effect of morning bright light treatment for rest-activity disruption in institutionalized patients with severe Alzheimer's disease. *International Psychogeriatrics* 2005;17(2):221–236.

Ebersole P, Hess P, Schmidt-Luggen A, editors (2004). *Toward Healthy Aging*. Sixth edition. Mosby, St. Louis, MO.

Figueiro MG. A proposed 24 h lighting scheme for older adults. *Lighting Research and Technology* 2008;40(2):153–160.

Forbes D, Culum I, Lischka AR, Morgan DG, Peacock S, Forbes J, Forbes S. Light therapy for managing cognitive, sleep, behavioural, or psychiatric disturbances in dementia. *Cochrane Database of Systematic Reviews* 2009;(4):CD003946.

Górnicka GB (2008). Lighting at work. Environmental study of direct effects of lighting level and spectrum on psychophysiological variables. Dissertation. Eindhoven: Eindhoven University of Technology.

Graf A, Wallner C, Schubert V, Willeit M, Wlk W, Fischer P, Kasper S, Neumeister A. The effects of light therapy on Mini-Mental State Examination in demented patients. *Biological Psychiatry* 2001;50(9):725–727.

Guo L, Duggan J, Cordeiro MF. Alzheimer's disease and retinal neurodegeneration. *Current Alzheimer Research* 2010;7(1):3–14.

Harper DG, Volicer L, Stopa EG, McKee AC, Nitta M, Satlin A. Disturbance of endogenous circadian rhythm in aging and Alzheimer disease. *The American Journal of Geriatric Psychiatry* 2005;13(5):359–368.

Health Council of The Netherlands (2002). Dementia. Publication no. 2002/04. Health Council of The Netherlands, The Hague, The Netherlands [in Dutch].

Hickman SE, Barrick AL, Williams CS, Zimmerman S, Connell BR, Preisser JS, Mitchell CM, Sloane PD. The effect of ambient bright light therapy on depressive symptoms in persons with dementia. *Journal of the American Geriatrics Society* 2007;55(11):1817–1824.

Hopkins RW, Rindlisbacher P, Grant NT. An investigation of the sundowning syndrome and ambient light. *The American Journal of Alzheimer's Care and Related Disorders and Research* 1992;7(2): 22–27.

Hughes PC, Neer RM. Lighting for the elderly: a psychobiological approach to lighting. *Human Factors* 1981;23(1):65–85.

Kergoat H, Kergoat M-J, Justino L, Robillard A, Bergman H, Chertkow H. Normal optic nerve head topography in the early stages of dementia of the Alzheimer type. *Dementia and Geriatric Cognitive Disorders* 2001;12(6):359–363.

Kim S, Song HH, Yoo SJ. The effect of bright light on sleep and behavior in dementia: an analytic review. *Geriatric Nursing* 2003;24(4):239–243.

Lovell BB, Ancoli-Israel S, Gevirtz R. Effect of bright light treatment on agitated behavior in institutionalized elderly subjects. *Psychiatric Research* 1995;57(1):7–12.

Marshall M. Technology is the shape of the future. *Journal of Dementia Care* 1995;3(3):12–14.

McNair D, Cunningham C, Pollock R, McGuire B (2010). *Light and Lighting Design for People With Dementia*. Stirling, Dementia Services Development Centre, University of Stirling, UK.

Mendez MF, Cherrier MM, Meadows RS. Depth perception in Alzheimer's disease. *Perceptual and Motor Skills* 1996;83(3 Pt 1):987–995.

Pechacek CS, Andersen M, Lockley SW. Preliminary method for prospective analysis of the circadian efficacy of (day)light with applications to healthcare architecture. *LEUKOS* 2008;5(1):1–26.

Rea MS, Figueiro MG, Bullough JD. Circadian photobiology: an emerging framework for lighting practice and research. *Lighting Research and Technology* 2002;34(3):177–187.

Redel P, Bublak P, Sorg C, Kurz A, Förstl H, Müller HJ, Schneider WX, Perneczky R, Finke K. Deficits of spatial and task-related attentional selection in mild cognitive impairment and Alzheimer's disease. *Neurobiology of Aging* 2012;33(1):195.e27–195.e42.

Rheaume YL, Manning BC, Harper DG, Volicer L. Effect of light therapy upon disturbed behaviors in Alzheimer patients. *American Journal of Alzheimer's Disease* 1998;13(6):291–295.

Riemersma-van der Lek RF, Swaab DF, Twisk J, Hol EM, Hoogendijk WJG, van Someren EJW. Effect of bright light and melatonin on cognitive and noncognitive function in elderly residents of group care facilities. A randomized controlled trial. *The Journal of the American Medical Association* 2008;299(22):2642–2655.

Sinoo MM, van Hoof J, Kort HSM. Lighting conditions for older adults in the nursing home: assessment of environmental illuminances and colour temperature. *Building and Environment* 2011;46(10):1917–1927.

Sleegers PJC, Moolenaar NM, Galetzka M, Pruyn A, Sarroukh BE, van der Zande B. Lighting affects students' concentration positively: findings from three Dutch studies. *Lighting Research and Technology* 2013;45(2):159–175, doi:10.1177/1477153512446099

Sloane PD, Williams CS, Mitchell CM, Preisser JS, Wood W, Barrick AL, Hickman SE, Gill KS, Connell BR, Edinger J, Zimmerman S. High-intensity environmental light in dementia: effect on sleep and activity. *Journal of the American Geriatrics Society* 2007;55(10):1524–1533.

Terman M. Evolving applications of light therapy. *Sleep Medicine Reviews* 2007;11(6):497–507.

Thapan K, Arendt J, Skene DJ. An action spectrum for melatonin suppression: evidence for a novel non-rod, non-cone photoreceptor system in humans. *Journal of Physiology* 2001;535(1):261–267.

Thorpe L, Middleton J, Russell G, Stewart N. Bright light therapy for demented nursing home patients with behavioral disturbance. *American Journal of Alzheimer's Disease* 2000;15(1):18–26.

van Hoof J, Aarts MPJ, Rense CG, Schoutens AMC. Ambient bright light in dementia: effects on behavior and circadian rhythmicity. *Building and Environment* 2009a;44(1):146–155.

van Hoof J, Schoutens AMC, Aarts MPJ. High colour temperature lighting for institutionalised older people with dementia. *Building and Environment* 2009b;44(9):1959–1969.

van Hoof J, Kort HSM, Duijnstee MSH, Rutten PGS, Hensen JLM. The indoor environment and the integrated building design of homes for older people with dementia. *Building and Environment* 2010;45(5):1244–1261.

van Hoof J, Kort HSM, Rutten PGS, Duijnstee MSH. Ageing-in-place with the use of ambient intelligence technology: perspectives of older users. *International Journal of Medical Informatics* 2011a;80(5):310–331.

van Hoof J, Wouters EJM, Marston HR, Vanrumste B, Overdiep RA. Ambient assisted living and care in The Netherlands: the voice of the user. *International Journal of Ambient Computing and Intelligence* 2011b;3(4):25–40.

van Hoof J, Westerlaken AC, Aarts MPJ, Wouters EJM, Schoutens AMC, Sinoo MM, Aries MBC. Light therapy: methodological issues from an engineering perspective. *Health Care and Technology* 2012;20(1):11–23.

van Someren EJW. Circadian and sleep disturbances in the elderly. *Experimental Gerontology* 2000;35(9–10):1229–1237.

van Someren EJW, Kessler A, Mirmiran M, Swaab DF. Indirect bright light improves circadian rest-activity rhythm disturbances in demented patients. *Biological Psychiatry* 1997;41(9):955–963.

Waterhouse JM, Minors DS, Waterhouse ME, Reilly T, Atkinson G (2002). *Keeping in Time With Your Body Clock.* Oxford University Press, Oxford.

Webb AR. Considerations for lighting in the built environment: non-visual effects of light. *Energy and Buildings* 2006;38(7):721–727.

Yamadera H, Ito T, Suzuki H, Asayama K, Ito R, Endo S. Effects of bright light on cognitive and sleep–wake (circadian) rhythm disturbances in Alzheimer-type dementia. *Psychiatry and Clinical Neurosciences* 2000;54(3):352–353.

22

Context Awareness for Medical Applications

Nathalie Bricon-Souf and Emmanuel Conchon

CONTENTS

22.1 Introduction

Context-aware computing has been defined as "an application's ability to adapt to changing circumstances and respond according to the context of use" (Dey 2001). It can be considered as one of the main aspects among five key technology features able to provide ambient intelligence, identified by Aarts (2004) as embedded, context-aware, personalized, adaptive, and anticipatory.

New technologies using wireless networks, mobile tools, sensors, wearable instruments, intelligent artifacts, and handheld computers are now pretty mature and available to support the development of context-aware applications. The medical domain, due to its complexity and its variability, is a major candidate for the use of such technologies, not only for the mobile perspective but also to address in a more efficient way the problem of care, as underlined by Montani (2011): "The issue of reaching context awareness and of realizing context-based reasoning are being routinely addressed in mobile and ubiquitous computing [...], but are now recognized as key aspects also for a wide range of other areas, among which is patient care."

Context has been studied since the 1990s; this chapter does not pretend to present an exhaustive view of what has been done in this area in health care, but it would help to understand the issues, the place, and the importance of context awareness for future care.

22.1.1 Problem Statement

Schilit et al. (1994) introduced the notion of context awareness in a paper entitled "Context Aware Computing Application." This paper is very important for the context-awareness community as it launched most of the research on context awareness. In this paper, context-aware computing was defined, and four categories of context-aware applications were described: *proximate selection, automatic contextual reconfiguration, contextual information and commands,* and *context-triggered actions.*

Despite a lot of other research that has been performed until now in this area, context is still difficult to define and to represent, and context-aware applications or services are consequently still difficult to define too. Moreover, actual context-aware applications are not so numerous.

In order to present the recent trends in context-awareness applications in health care, we have organized our chapter into three sections. In the first part of this chapter, we aim at clarifying the notion of context by highlighting some important research topics on context. The second part of this chapter analyzes some of the surveys that have been performed on context-aware applications. The third part of the chapter presents some of the recent works that are proposing context-aware health care applications.

22.2 Context

22.2.1 Context Definition

Dey (2001), in his paper entitled "Understanding and Using Context," proposed a definition of context that is still predominantly used: "any information that can be used to characterize the situation of an entity (i.e., whether a person, place or object) that are considered relevant to the interaction between a user and an application, including the user and the application themselves." He also described a context-aware system named the Context Toolkit. He illustrated his context definition as follow: "Context is typically the location identity and state of people, groups and computational and physical objects." This definition is not satisfying for everyone working in the context area. Dourish (2001), for example, argued that from a phenomenological point of view, context cannot be described independently of the activity. Winograd (2001) remarked that "something is context because of the way it is used in interpretation." Souza et al. (2008) proposed the following definition for context: "a set of elements surrounding a domain entity of interest which are considered relevant in a specific situation during some time interval." Chen and Kotz (2000) propose their definition: "Context is the set of environmental states and settings that either determines an application's behavior or in which an application event occurs and is interesting to the user." Many other definitions for context are proposed, and Bazire and Brezillon (2005) picked up more than 150 definitions of context, from various disciplines and underlined a lack of consensus on the following questions: "Is context external or internal? Is context a set of information or processes? Is context static or dynamic? Is context a simple set of phenomenon or an organized network?" Pomerol and Brezillon (1999) introduced the notion of proceduralized context: Context is seen as linked to a focus, and it allows distinguishing between external knowledge, which is not relevant for the situation, and context knowledge, which is pertinent according to the situation. Proceduralized context is then built into a given situation, as an instantiation of contextual knowledge. This definition was refined by Vieira et al. (2011). They propose a context meta-model in order to design context-sensitive systems. This meta-model distinguishes *contextual elements*, which are defined during design; *contextual knowledge*; and *context*, which appears during enactment. Some other authors focus on the different kind of context entities. Chen and Kotz (2000) propose a four-dimensional space for context organized into *computing context*, *physical context*, *time context*, and *user context*. Zimmerman et al. (2007) propose an operational definition of context, introducing a formal extension to the definition through five fundamental categories of context (*individuality*, *time*, *location*, *activity*, *relations*) and an operational extension through the description of different uses of context. Soylu et al. (2009), who work on pervasive learning with a context-awareness system, propose a hierarchical representation of context. Two main roots are defined: *user* and *environment*; different entities are modeled: *user*, *external devices*, *application*, *environment*, *time*, *history*, and *relation*. They also propose a formal representation of context according to its type (the entities), its dynamism, and its level of adaptation.

22.2.1.1 Context Representation

One of the first works about context representation is that of McCarthy (1993), who proposed in 1993 in his "Notes for Formalizing Context" an axiomatization of contexts described as formal objects. Different approaches have then been performed. One could also refer to the recent survey of Bellavista et al. (2012), who propose a classification with three kinds of context data representation: *no model for data representation*, *domain-specific model*, or *general model*. It refers to the Strang and Linnhoff-Popien classification, which proposes six different main ways to represent the context: *key-value* (attribute–value pairs allow the description of context information or context meta-information); *markup scheme* (based on XML representation); *graphical*; *object-oriented* (defining context types with object classes); *logic-based*; and *ontology-based* (allowing expressiveness and reasoning through the strengths of an ontology) models. We can cite contextual graphs proposed by Brezillon (2007) as an example of graphical models; the work performed by Serafini and Homola (2011) allowing a contextualized knowledge repository as a representation of description logic; and C-OWL contextualizing ontologies (Bouquet et al. 2003) as ontology-based models.

22.2.2 Dealing with Context: Other Important Points

Two other issues have to be taken into account to propose efficient context-awareness applications: context elicitation and quality of context (QoC).

22.2.2.1 Context Elicitation

This deals with the selection of the items that are used to identify and to keep track of context setting (it is particularly pertinent during design phases of a project). Usually, determining what should be used as context is a major problem. Contextual situations are frequent, and for each situation, it is difficult to determine what the contextual elements are. "In any case, let us remark that context modeling is still a complex, human-based, and time-consuming task very prone to errors. Even the usage of powerful tools that help designers during the modeling phase does not succeed in guaranteeing that the final model well represents context" (Bellavista et al. 2012). In the medical domain, a lot of information can be used by the application, and context is very often needed in health care when it is surrounding the application domain of interest and is relevant for it. An alert system, for example, can be different according to the medical units specialties (e.g., a geriatric unit does not have the same needs as a pediatric one). A care recommendation can vary according to the availability of a specific device (for instance, magnetic resonance imaging does not exist in every small hospital or is rare in developing countries). Elicitation of context should be used to decide on what information entities to follow up on as contextual elements.

Souza et al. (2008) propose an elicitation process to get the taxonomy of domain entities and the contextual elements of a contextual ontology for data integration from an empirical methodology based on face-to-face meetings with data integration experts and literature reviews. The process of elicitation and modeling is thus based on a spiral lifecycle, which involves four phases: identification of technical constraint, stakeholder, and end users; building of a cognitive context model; analyzing and optimizing context space; and detection of context change. Yang et al. (2008) propose a Java Expert System Shell (JESS)–enabled context elicitation system. It involves three phases—form-filling to acquire directly contextual information from requesters' inputs, context detection (such as positioning), and context extraction to derive contextual information from ontologies. Bricon-Souf et al. (2011) propose an elicitation cycle, in order to state what the context is used for, to determine the useful knowledge needed to compute the application, to gather different areas of expertise in order to focus on the information used to build knowledge, and to perform the acquisition of the pertinent information.

22.2.2.2 QoC

QoC is linked to the quality of the contextual information (it is particularly relevant during enactment phases of an application). The quality of the contextual information that is collected is of major importance. Different parameters have to be considered to assume the quality of context-aware applications. Buchholz (2003) proposes a focus on the quality of the physical sensors, on the quality of the contextual information, and on the quality of the delivery process. Manzoor et al. (2008) define several parameters such as up-to-dateness, trustworthiness, completeness significance, precision, and resolution in order to evaluate the provided QoC.

22.2.3 Context Architecture

Such context-aware systems are dedicated to context-aware applications, which are relying on contextual knowledge to adapt their behaviors. Figure 22.1 presents an architectural representation of such systems, which can be found in most existing systems with slight variations. A context-aware system can be decomposed into four main components (layers): an acquisition layer, a management layer, a distribution layer, and an adaptation layer.

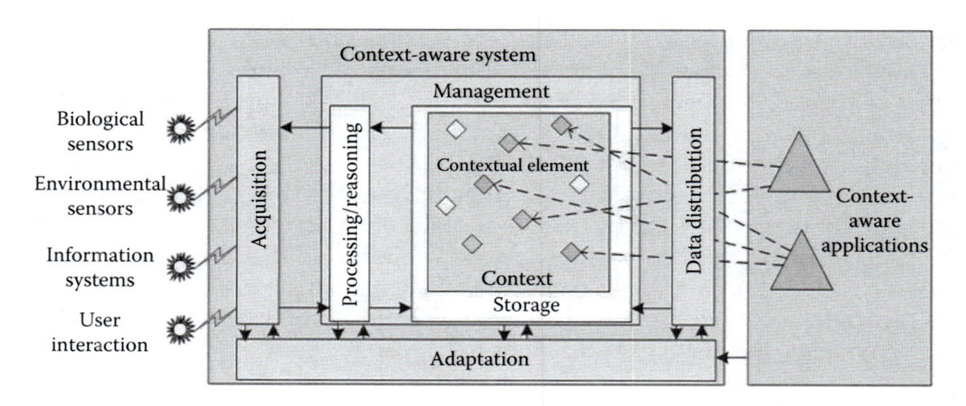

FIGURE 22.1
(See color insert.) Architectural representation of context-aware systems.

22.2.3.1 Acquisition (Capture)

The *context acquisition* layer is in charge of the raw data capture to acquire data from heterogeneous devices and data sources and to provide them to the management layer. It also has to deal with specific communication settings and has to provide mechanisms to ensure that the information is available. This is particularly critical in mobile environments, which need to focus on the ability to continuously capture external data (i.e., information) and to integrate this knowledge into the systems. External data can be acquired from physical sensors, information systems, or even user interaction (graphical interface, type of device in use, voice, camera, and so on). The heterogeneity of devices can lead to the provision of information on the quality of the captured data. For instance, the localization can be detected with GPS, but it can also be inferred from the activation of a force sensor positioned on a chair or in a smart pillow. Both methods give information about localization, but the force sensor accuracy is higher than that of GPS; therefore, the quality of the context is better with the force sensor. The acquisition layer should provide such information to help the processing layer.

22.2.3.2 Management

Management of context aims to process and to store contextual information. The management layer is an intermediate layer between context sources and context consumers. It integrates sometimes the data distribution layer. Zhang et al. (2005), for instance, propose a layered context management framework that provides contextual information to context-aware systems. In this framework, different modules are used: context source, context query engine, context knowledge base, context discoverer, and context application. In this work, we choose to limit the definition of the management layer to the main functionalities of processing and storage.

The processing sublayer receives raw data from the acquisition layer (with quality indicators in the best scenario) and is in charge of the reasoning part of the system. In this layer, classical artificial intelligence methods can be performed, ranging from Bayesian networks to Markovian approaches, for instance. Other solutions propose the use of rules and inference engines. Korel (2010) provides a good overview of these different techniques. The aim of this reasoning process is to provide new context information with a higher level of abstraction than those provided by sensors. Indeed, it is possible to make a differentiation between low-level contextual elements (position, temperature, and so on) and high-level contextual elements ("the person is eating"). Wang et al. (2004) and Soylu et al. (2009) are some of the authors who highlight how to manage low-level context (issued from raw data) towards high-level context (more related to the activities).

Once the relevant elements are selected, they have to be stored. The "context storage" sublayer commonly proposes the use of a dedicated contextual database. This storage sublayer helps to constitute a history of the context evolution so that the reasoning sublayer improves the inference of high-level contextual elements. This is critical for context-aware systems especially for activity recognition, where the notion of habits is very important. In monitoring systems, the history record helps to trigger alarms.

22.2.3.3 Distribution

The *data distribution* layer makes the gateway between the context management system and the context-aware applications and has to take into account the underlying network (fixed or mobile) and the way the data are sent. The underlying network has a huge impact on the overall quality of the system. For instance, specific problems appear when dealing with context-aware mobile applications such as the node density or the interaction caused by user mobility. Several solutions are then used in context-aware system, ranging from publish/subscribe mechanisms using CORBA to dedicated multicast protocols or gossip-based approaches for routing the data from the system to the context-aware applications. Moreover, security and privacy issues are of major concern especially for health context-aware applications. Furthermore, the dissemination the layer should guarantee the QoC and should ensure that the dissemination of the data does not introduce some gaps in the measured QoC at the management layer. Therefore, the underlying network quality of service (QoS) has to be considered to ensure the QoC in the overall system. Bellavista et al. (2012) provide a deep survey about these distribution aspects and point out different issues and possible solutions.

22.2.3.4 Adaptation

The adaptation layer controls the way the context is captured and processed according to the needs of the application. For instance, an application informs the adaptation layer on the required QoC so that this layer can adapt the behavior of the reasoning component (context processing layer). The adaptation also takes place at the management layer triggered by the adaptation layer when specific information is needed to build the relevant contextual elements, especially when the notion of QoC is provided by the system. As previously stated, the distribution layer also may have to adapt its behavior regarding to the QoS of the underlying layer.

22.3 Analyses of Surveys on Context Awareness

22.3.1 Introduction

In order to present the different trends of research, we propose reporting and commenting on the different surveys on the theme of context awareness. The aim is not to provide a deep presentation of every existing context-aware solution but, rather, to provide to the reader some indications on where to find the relevant information among the large number of existing surveys.

We propose the following classification, based on the focus of the reviews. We will first report on the papers dedicated to the general aspects of the context-aware systems such as the middleware architecture proposed. Then we will present some reviews reporting on some specific aspect of context awareness such as the representation model proposed or the targeted domain of the paper: We are interested in health care.

22.3.2 Global Context-Awareness Surveys

Over the past decade, lots of research efforts have focused on middleware solutions to enable context-aware services, and general surveys often present these middleware propositions.

- Chen and Kotz (2000): To our knowledge, this is the first survey about context-aware middleware solutions. In this survey, the authors focus on mobile-aware applications. They explore different definitions of context, and they propose their own, as previously cited. They distinguish active context and passive context according to how context influences the behavior of the application. They consequently propose definitions for context-aware computing: "Active context awareness: an application automatically adapts to discovered context, by changing the applications' behavior" and "Passive context awareness: an application presents the new or adapted context to an interested user and makes the context persistent for the user to retrieve later." They present some of the context-aware applications developed in the 90s that propose location-based services. They also focus on how to propose sensors (location, high-level context, context changes); describe models; and finally, present different architectures for context.
- The European project WASP (2003) published a report on existing context-aware solutions. The concepts of context and context awareness are investigated, referring to the Dey and Lieberman definitions. A three-step architecture for context-aware application management is proposed:
  - Sensing the context makes a differentiation between low-level context (location, network conditions, time, etc.) and a higher level of context (activity, social interactions, etc.).
  - Modeling the context focuses on ontological approaches based on resource description framework (RDF) formalism to highlight the semantic meaning of sensed information and their impact on the context model.
  - Monitoring the context introduces a new generic model to keep the observed context up-to-date.

  A survey of existing architectures in 2003 is presented. Most of the presented architectures are related to ubiquitous computing: the Cooltown project, the Context Toolkit designed by Dey, an adaptive solution based on a service platform (M3 43), a universal information appliance (UIA), Ektara, Portolano, and PIMA.

  These different solutions are compared according to eight criteria: support for device heterogeneity, support for device mobility, management of application mobility and distribution, support for context-aware issues, support for adaptation, support for rapid deployment, management of context, and support for user mobility.

They conclude that, in 2003, the management of application mobility and distribution and the support for user mobility were not addressed in the reviewed solutions, and they point out that the privacy and security aspects are not included in the existing solutions.

Based on this observation, they propose several solutions that could be investigated for a middleware solution.

- Henricksen et al. (2005) focus on middleware solutions for context-aware systems. For once, the localization systems are not taken into account even if they are considered as the first context-aware solutions. The authors propose a five-layered architecture to describe context-aware systems ignoring the data distribution component.
  - Sensors
  - Context processing component
  - Context repository
  - Decision support tools
  - Application component

  They define a seven-paradigm taxonomy to compare the surveyed middleware: support for heterogeneity, support for mobility, scalability, support for privacy, traceability and control, tolerance for component failures, and ease of deployment and configuration. They investigate five middlewares—the Context Toolkit, context fusion networks (CFNs), the Context Fabric (Confab), Gaia, and the reconfigurable context-sensitive middleware (RCSM)—through a short description of each of them.

  Their main conclusion is that none of these systems are providing decision support tools. They present their own solution—the PACE middleware—to tackle this problem.
- Singh (2006) starts his survey with a brief definition of context. He defines a three-layer architecture for software context-aware agents:
  - An interface with the outside world (environment, other agents, etc.)
  - A context described with an ontology and an intelligence that uses rules
  - An inference engine for the reasoning part

  This solution is rather close to the proposed architecture in the WASP project and is quite restrictive.

  He makes a brief presentation of four middlewares—Context Toolkit, CoBrA, Context Mediated Framework (CMF), and SOCAM—but does not provide any taxonomy or a clear comparison of them. Finally, he presents his own solution: the STU21 context-aware framework.
- Baldauf et al. (2007) published what will become one of the most cited surveys on context-aware middlewares. After a short overview of location-aware systems (ActiveBadge, for instance), the authors present different definitions of context and highlight the distinction of two context dimensions: external/physical (hardware related) and internal/logical (specified by the user or by his/her interactions) (Prekop et al. 2003; Gustavsen 2002; Hofer et al. 2002).

  They present a brief survey of existing architectures for context-aware systems highlighting the work of Chen and Kotz (2000) on the acquisition of contextual information, which defines three main techniques: direct sensing, middleware, and context server. They also point out Winograd's (2001) proposition based on widgets, which work as an interface to a hardware sensor; network services; and a blackboard model, which is data oriented and based on a producer/consumer paradigm.

A six-layered architecture ranging from sensors to application based on a middleware and on a context provider approach is provided.

They present research on context modeling introducing the classification work of Strang and Linnhoff-Popien, showing that the ontology-based approach is the most effective one.

They present a small survey of existing framework based on middleware focusing on location-aware systems, where they present and compare several context-aware frameworks: CASS, CoBrA, context management framework, Context Toolkit, CORTEX, Gaia, Hydrogen, and SOCAM with several indicators—*the way the context is sensed, context model, context processing method, resource discovery, historical context data (supported or not), and security and privacy issues.*

They conclude that these solutions do not provide good mechanisms for resource discovery especially for dynamic environments, except for SOCAM. They also show that the historical context data are not used as they could be, as, for instance, there is no learning algorithm implemented that could use the past data to proactively adapt the system behavior. They also point out that only CoBrA and Gaia provide both security and privacy support.

- Kjær (2007) introduced a taxonomy to compare context-aware middlewares. This taxonomy is composed of seven elements:
  - Environment (infrastructure based or self-contained)
  - Storage (to store context data)
  - Reflection (meta-data of the application, the context, or the middleware)
  - Quality
  - Composition (of components)
  - Migration
  - Adaptation (system can adapt itself to a change in the context)

  The security and privacy aspects are not addressed by Kjær; neither is the data distribution.

  With this taxonomy, several middlewares are presented and compared: Aura, CARMEN, CARISMA, Cooltown, CORTEX, Gaia, MiddleWhere, MobiPADS, and SOCAM.

  Kjær remarks that the QoC is not taken into account except for MiddleWhere and that the adaptation is not always provided, which seems rather problematic for context-aware applications/services.

- Miraoui et al. (2008) aimed to provide a guide to context-aware systems' developers and architecture designers. They start with a small overview of existing surveys, and they conclude that none of them provide a good comparison of existing systems.

  For their own comparison, they chose five main criteria: level of context abstraction, communication model, reasoning system, extensibility, and reusability.

The presented systems are Active Badge, ParcTab, Stick-e-notes, Cyberguide and Guide, CASS, CORTEX, context management framework, JCAF, Context Toolkit, Hydrogen, SOCAM, and finally, CoBrA.

The paper ends on a discussion on context-aware architectures, defining a composition of five layers: *sensing, interpreting, reasoning, storage and management, and adaptation*. The authors state that most of the presented architectures do not provide every layer and that the model used to represent the context is the main key to designing a context-aware architecture. As in previous surveys, their conclusion is that no generic solution can be provided for context-aware systems.

- Romero (2008) compared 10 different context-aware middlewares with regard to seven functionalities:
  - Context management
  - Adaptation
  - Communication
  - Service discovery
  - Persistence
  - Application building support
  - The different paradigms used

  He briefly describes the following middlewares: Gaia, Gaia microserver, CORTEX, Aura, CARISMA, MiddleWhere, MobiPADS, SOCAM, RCSM, and CAPNET.

  The author points out that every middleware is focused on a specific problem and that no generic solution is proposed. Moreover, the comparison is only based on architectural design, as in previous surveys.
- Hong et al. (2009) introduce their work in reference to the Chen and Kotz (2000) and Baldauf et al. (2007) surveys on context-aware systems. They performed a review of 237 articles from the literature issued from 2000 to 2007 and report on the general characteristics of these articles, according to where the context-aware systems papers were found: online databases or journals.

  They propose a classification framework of context-aware systems in five layers:
  - Concept and research
  - Network infrastructure layer
  - Middleware layer
  - Application layer
  - User infrastructure layer

  According to this classification, they perform a distribution of their literature selection. For example, hospital applications were evaluated in 2.9% of the selected papers. This classification allows them to represent the distribution of articles (for example, in 2007, articles about concepts are the most numerous, followed by articles about applications).

  Discussion and suggestions are then proposed, and the following questions are discussed: how to extract and use the cognitive context in context-aware application; what are the design patterns of context-aware systems; which is the best inferring algorithm to extract user context and to provide service to user; how to deal with concurrently enormous data; how to manage information and knowledge having different formats and offering suitable service to users; how to extract the best solution when the context of users is conflicted, how to reflect the preference of users for satisfying user needs; how to save users' information in context-aware systems; and how to evaluate performance of context-aware systems.

- Saeed and Waheed (2010) describe their survey as "the most extensively conducted survey on context-aware middleware architectures." After a presentation of some related works on context-aware middleware system overviews, 15 different systems are listed and described: Aura, CAPNET, CARISMA, CARMEN, Cooltown, CORTEX, Gaia, MiddleWhere, FlexiNet, NEXUS, One.World, AspectIX, MobiPADS, HOMEROS, and SOCAM.

  An analysis of these middlewares according to architectural style, location transparency, aspect-oriented decomposition, fault tolerance, interoperability, service discovery, and adaptability is then proposed in order to highlight weaknesses and strengths of the different systems.

22.3.3 Focused Context-Awareness Surveys

22.3.3.1 Representation-Focused

- Strang and Linnhoff-Popien (2004) dedicated a survey on context modeling approaches for ubiquitous computing based on five criteria: *distributed composition, partial validation, richness and quality of information, incompleteness and ambiguity, level of formality,* and *applicability to existing environments.* Then, they present the most relevant context modeling approaches, which are key-values models, markup scheme models, graphical models, object-oriented models, logic-based models, and ontology-based models. An evaluation of these modeling approaches is then performed based on the proposed criteria, showing that ontological modeling is the best suited for context representation.
- Bolchini et al. (2007) focus on context modeling rather than on context-aware architecture. First, a new taxonomy is defined to compare the different solutions. This taxonomy can be split into three big areas: modeled aspects, representation features, and context management and usage. Each of these three areas has a direct impact on the way context is modeled. The authors propose five categories of context: *context as a matter of channel-device presentation; context as a matter of location and environment; context as a matter of user activity; context as a matter of agreement and sharing;* and *context as a matter of selecting relevant data, functionality, and services.* They investigate 16 context-aware systems regarding their model representation: ACTIVITY, CASS, CoBrA, CoDaMoS and SOCAM, COMANTO, Context-ADDICT, Conceptual-CM, CSCP, EXPDOC, FAWIS, Graphical-CM, HIPS/HyperAudio, MAIS, SCOPES, SOCAM, and U-Learn.

22.3.3.2 Health Care-Focused

- Bricon-Souf and Newman (2007) attempted to derive an objective view of the dynamism of context awareness in health care and to identify strengths and weaknesses in this field, through a survey of the research literature. The paper highlights that the majority of laboratories identified as working on context awareness are mainly involved in computer sciences research. Health care, if it exists, appears as one of their interests for a potential testing area for the proposed frameworks or tools. Only a few laboratories were identified in the literature as being involved in context awareness dedicated to health care research.

A more detailed analysis of actual characteristics of context-awareness application at the hospital is proposed, identifying the following characteristics:

- Type of work done (proposed model, prototype, application, or scenario)
- Use of context (presentation of information, execution of services, storage of contextual information)
- Description of the different features of context manipulated by the system (i.e., activity, people, and environment)
- Organization or categorization of context
- Objectives of use of context awareness

According to this classification, an overview of the features of context actually under use is proposed. The gap between actual context-awareness prototypes in health care and the modeling of the contextual applications proposed in the literature is highlighted through this analysis.

- Orwat et al. (2008) give an overview of developments and implementations of pervasive computing systems from 2002 to 2006 in health care based on the identification and the analysis of 69 relevant papers. Different systems are found and categorized according to the following:
    - Project status
    - Health care settings
    - User groups
    - Improvement aims
    - System features

  A large majority of them are prototypes; more than half of the systems deal with home care or mobility. They are mainly dedicated to doctors or nurses if they aim to be addressed to health care professionals or dedicated to the patient if they aim to be addressed to laypersons. They usually seek to enhance prevention and care. Diverse medical domains are covered. Monitoring of data is frequently taken into account, and the system often proposes the transmission of data, data analysis performance, and diagnostic support.

- Silva et al. (2011) propose a health care monitoring systems survey. They highlight the importance of ambient assisting living and therefore of wireless health monitoring. They introduce the issues with context awareness based on mobility. They present several systems, focusing on different levels of context-awareness treatment, such as the following:
    - Making better decisions on radio interface commutation
    - Extracting context information from specific scenarios where a mobile device may gather information from its actual state, based on the device and radio access technology state
    - Modeling context with ontologies
    - Using Bayesian networks for context-reasoning algorithms

22.3.4 Conclusion of the Survey Analysis

Even if context awareness leads to numerous research works, it is still difficult to get a precise idea of what could be the actual applications with such systems. A lot of issues are mentioned. Numerous solutions in context representation as well as in context architecture are proposed. For dedicated applications, prototypes are proposed, but it is still difficult to know how contextual elements are chosen (context elicitation) or how QoC can be taken into account.

Furthermore, context should help to focus on pertinent information. In the above studied surveys, information retrieval, for instance, is not addressed. But, context can also be used to reduce the amount of information to deal with in health care applications/systems. Similarly, context-aware decision support systems are not considered even if they are investigated in many research works.

22.4 Context-Aware Applications in Health Care

22.4.1 Research Trends

In order to highlight the evolution of the research on the topic of context awareness, we decided to do a quantitative view of the published papers about context-aware applications in health care. We have performed queries both on IEEE Xplore and on MEDLINE. IEEE was the top source of papers in the health care survey by Hong et al. (2009), and MEDLINE was used because of its medical purpose. We think that it could show us the actual trends even if it does not provide an exhaustive view.

To compose a first set of queries, we used the key word "context-aware" associated with one of the identified topics: modeling, middleware, reasoning, monitoring, information retrieval, and decision support. A second set of queries was then performed, adding "health care" to each previous query to highlight the impact of health in the area of context awareness. As the health area is a large domain with very specific issues and lots of medical specialties, finding relevant key words remains difficult. We retain the "health care" key word, which of course conducts us to a restrictive subset of results but ensures that they actually will be focused on the health application target.

Queries have been performed on the two reference databases from 2000 to 2012, at 2-year intervals. The results in terms of number of publications are presented in Table 22.1.

The different research topics for context awareness have been particularly dynamic since 2006 and seem to have slowly decreased since 2010 while remaining very active. The health care subset also started to grow in 2006 but shows a continuous progression over the years until now.

The use of IEEE Xplore and MEDLINE shows a difference in the nature of context-aware topics, as "modeling," "middleware," and "reasoning" do not produce any results on MEDLINE.

22.4.2 Description of the Applications

Two main sets of context-aware systems can be distinguished:

- The first one is dedicated to the management of context-aware systems. It deals with modeling of context, middleware specifications, and reasoning with contextual elements. Such works are not specific to health care and are well described by the previously cited surveys.

TABLE 22.1
Number of Publications Issued from Different Context-Aware Research Topics

Topic		Context-Aware							Context-Aware Health Care						
		[2000–2001]	[2002–2003]	[2004–2005]	[2006–2007]	[2008–2009]	[2010–2011]	[2012–present]	[2000–2001]	[2002–2003]	[2004–2005]	[2006–2007]	[2008–2009]	[2010–2011]	[2012–present]
Modeling	IEEE Xplore	49	111	286	654	802	511	137	1	3	2	13	18	19	5
	MEDLINE	0	0	1	0	1	2	1	0	0	0	0	0	0	0
Middleware	IEEE Xplore	10	40	117	233	224	131	19	0	0	1	10	13	7	0
	MEDLINE	0	0	0	1	0	2	3	0	0	0	1	0	0	0
Reasoning	IEEE Xplore	8	13	40	126	141	110	23	0	0	1	4	5	3	1
	MEDLINE	0	0	1	1	1	1	1	0	0	0	0	0	0	0
Monitoring	IEEE Xplore	6	20	49	146	184	158	39	1	2	1	17	18	21	6
	MEDLINE	0	0	4	5	7	8	7	0	0	1	2	3	2	3
Information retrieval	IEEE Xplore	5	24	49	66	102	38	16	0	1	0	0	1	2	2
	MEDLINE	0	0	2	2	3	4	3	0	0	0	0	1	2	1
Decision support	IEEE Xplore	4	8	25	44	64	60	16	0	0	5	1	4	5	2
	MEDLINE	0	0	2	1	1	2	3	0	0	1	0	1	1	1

Note: Raw retrieval per 2 years.

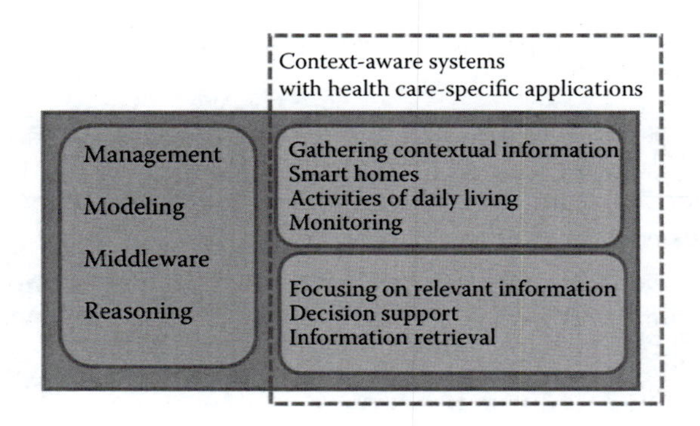

FIGURE 22.2
The broad categories for context-aware systems.

- The second one includes context-aware systems that can be more specifically dedicated to health care applications. Such applications can be split into two categories depending on what the context is used for.
  - The first category contains context-aware applications that gather contextual elements in order to build a clever representation of the applications settings. These systems are often system-centric. Many health care applications are concerned with this kind of application, such as smart homes, activity of daily living (ADL), and monitoring.
  - The second category contains context-aware systems that use context in order to focus on interesting information. These applications are more data-centric. Context-aware decision support systems as well as context retrieval information systems use contextual information to adapt and restrict their feedback to users.

In order to clarify our presentation of health care context-aware applications, we will use these two categories (in the dashed square in Figure 22.2).

22.4.2.1 Gathering Contextual Information

Context should help us to better understand and use a complex environment. Introducing context-aware systems at home is very promising, for instance. Allowing people to stay longer at home helps to improve the quality of life as well as to reduce hospitalization cost. Elderly people or patients with chronic diseases could benefit from such sophisticated environments. Different kinds of context-aware applications are proposed:

- Smart homes can provide health care facilities with the use of sensors or portable devices to generate health reports, which can then be used by health professionals or health care service providers, such as in the work of Virone et al. (2002). More details on smart homes can be found in the work of Chan et al. (2008) and Alam et al. (2012).
- Sophisticated knowledge engineering coupled with captors helps to provide ADL detection (see Ye et al. 2012) and improvement of quality of life (wellness).

In recent years, several solutions have been investigated, for instance:

- A context-aware system to support elderly patients undergoing total hip replacement recovery at home (Jimenez et al. 2011)
- A system for increased data quality in home blood pressure monitoring through context awareness (Wagner et al. 2011)
- A sensor platform and a machine learning approach to sense and detect phases of a surgical operation (Bardram et al. 2011)
- An ambient intelligent nurse call system that uses gathered context data to find the most appropriate caregivers to handle a call of a patient and generate new calls based on sensor data (Ongenae et al. 2012)

22.4.2.2 Focusing on Relevant Information

In the domain of medical decisions, context-aware applications should help to focus on pertinent information according to the context.

Context seems to be important in building efficient alert systems as well as in proposing adapted guidelines. Alert systems exist, but overalerting is a real problem. Adapting alerts will reduce alert fatigue and help the clinicians not to be reluctant to adopt decision support tools. Anya et al. (2010), for instance, propose a context-aware decision support system for collaborative e-work and propose the provision of context-specific guidelines according to the activity. In the European patient safety through intelligent procedures (PSIP) project, which deals with adverse drug events due to product safety problems and medication errors due to human factors, a taxonomy integrating contextualization of elements issued from an activity analysis to guide the design of a contextualized clinical decision supported system (CDSS) is proposed (Bernonville et al. 2011).

Contextualization improves information retrieval, for example, the work performed by Dinh and Tamine (2012) in order to use knowledge-specific domain sources (as health ontologies) to perform a context-sensitive retrieval of documents.

22.4.2.3 Conclusion on Context-Aware Applications

Many context-aware applications are dedicated to specific medical problems with a large diversity of application domains. The medical domain is, on the one hand, particularly interesting for context-aware applications due to its inherent complexity, but on the other hand, it is also particularly difficult to get a proper application. Technical problems (equipment interoperability in smart homes, for instance); security and privacy problems (privacy of medical information, for instance); and medical constraints (hygiene issues, availability of data) are some of the difficulties that cause a gap between prototypes and real applications.

22.5 Conclusion

Nowadays, technologies of communication and information are omnipresent. They can be found at home with sensors, in the street, as well as in mobile equipment that is carried by individuals, such as smartphones. This massive introduction coupled with the high complexity of medical care enables an obvious need for context-aware applications dedicated to health. In this chapter, the complexity of context-aware systems and the difficulty in defining the notions of *context* and *context awareness* have been shown. The difficulty in defining context and, more specifically, defining what information is relevant for a contextual situation induces a need for context elicitation. Based on existing surveys on context awareness, a description of context-aware systems architecture has been presented to guide researchers. Most surveys focus on middleware solutions or on very specific points such as modeling or reasoning techniques. These surveys are system-centric, and the work centered on the relevant data selection is not considered even if it represents several research topics such as information retrieval or decision support systems.

An analysis of the recent works performed in the health care domain has been carried out from these observations and considering both aspects (system-centric and data-centric). Several propositions have been presented in order to show the diversity of the research effort and the difficulty in coming up with a generic solution for health care applications.

References

Aarts, Emile. 2004. Ambient intelligence: A multimedia perspective. *IEEE Multimedia* 11 (1) (January): 12–19.

Alam, Muhammad Raisul, Student Member, Mamun Bin, Ibne Reaz, Mohd Alauddin, and Mohd Ali. 2012. A review of smart homes—past, present, and future. *Systems, Man, and Cybernetics, Part C: Applications and Reviews, IEEE Transactions* 42 (6): 1–14.

Anya, Obinna, Hissam Tawfik, Atulya Nagar, and Saad Amin. 2010. Context-aware decision support in knowledge-intensive collaborative e-work. *Procedia Computer Science* 1 (1) (May): 2281–2290.

Baldauf, Matthias, Schahram Dustdar, and Florian Rosenberg. 2007. A survey on context-aware systems. *International Journal of Ad Hoc and Ubiquitous Computing* 2 (4): 263.

Bardram, Jakob E., Afsaneh Doryab, Rune M. Jensen, Poul M. Lange, Kristian L. G. Nielsen, and Soren T. Petersen. 2011. Phase recognition during surgical procedures using embedded and body-worn sensors. In *The 2011 IEEE International Conference on Pervasive Computing and Communications (PerCom)*. (March): 45–53. Seattle.

Bazire, Mary, and Patrick Brézillon. 2005. Understanding context before using it. In *Modeling and Using Context*, ed. Anind Dey, Boicho Kokinov, David Leake, and Roy Turner, (3554): 29–40. Springer Berlin/Heidelberg.

Bellavista, Paolo, Antonio Corradi, Mario Fanelli, and Luca Foschini. 2012. A survey of context data distribution for mobile ubiquitous systems. *ACM Computing Surveys* 44 (4) (August 1): 1–45.

Bernonville, Stéphanie, Romaric Marcilly, Radja Messai, Nicolas Leroy, Emma Przewozny, Nathalie Souf, and Marie-Catherine Beuscart-Zéphir. 2011. Implementation of a taxonomy aiming to support the design of a contextualised clinical decision support system. *Studies in Health Technology and Informatics* 166 (January): 74–83.

Bolchini, Cristiana, Carlo A. Curino, Elisa Quintarelli, Fabio A. Schreiber, and Letizia Tanca. 2007. A data-oriented survey of context models. *ACM SIGMOD Record* 36 (4): 19–26.

Bouquet, Paolo, Fausto Giunchiglia, and Frank Van Harmelen. 2003. C-OWL: Contextualizing ontologies. *Journal of Web Semantics* 1 (4) (October): 164–179.

Bricon-Souf, Nathalie, and Conrad R. Newman. 2007. Context awareness in health care: A review. *International Journal of Medical Informatics* 76 (1) (January): 2–12.

Bricon-Souf, Nathalie, Radja Messai, and Stephanie Bernonville. 2011. A framework for context elicitation in medicine. In *The 23rd International Conference of the European Federation for Medical Informatics*. MIE 2011/CD/*Short Communications*, 7–9.

Brézillon, Patrick. 2007. Context modeling: task model and practice model. In *Modeling and Using Context*, ed. Boicho Kokinov, Daniel C. Richardson, Thomas R. Roth-Berghofer, and Laure Vieu, 4635: 122–135. Springer Berlin Heidelberg.

Buchholz, Thomas. 2003. Quality of context: What it is and why we need it. In *Proceedings of the 10th Workshop of the OpenView University Association: OVUA'03*, 1–14. Geneva, Switzerland.

Chan, Marie, Daniel Estève, Christophe Escriba, and Eric Campo. 2008. A review of smart homes—Present state and future challenges. *Computer Methods and Programs in Biomedicine* 91 (1) (July): 55–81.

Chen, Guanling, and David Kotz. 2000. *A Survey of Context-Aware Mobile Computing Research*. Time. Hanover, USA.

Dey, Anind K. 2001. Understanding and using context. *Personal and Ubiquitous Computing* 5 (1) (February 28): 4–7.

Dinh, Duy, and Lynda Tamine. 2012. Towards a context sensitive approach to searching information based on domain specific knowledge sources. *Web Semantics: Science, Services and Agents on the World Wide Web* 12–13 (null) (April): 41–52.

Dourish, Paul. 2001. Seeking a foundation for context-aware computing. *Human-Computer Interaction* 16 (2): 229–241.

Gustavsen, Richard Moe. 2002. Condor – an application framework for mobility-based context-aware applications. In *Proceedings of the Workshop on Concepts and Models for Ubiquitous Computing*, 1–6.

Henricksen, Karen, Jadwiga Indulska, Ted McFadden, Sasitharan Balasubramaniam. 2005. On the Move to Meaningful Internet Systems 2005: CoopIS, DOA, and ODBASE Lecture Notes in Computer Science Volume 3760, pp 846–863.

Hofer, Thomas, Wieland Schwinger, Mario Pichler, Gerhard Leonhartsberger, Josef Altmann, A.-Hagenberg, Werner Retschitzegger, and Johannes Kepler. 2002. Context-awareness on mobile devices—the hydrogen approach. In *Proceedings of the 36th Annual Hawaii International Conference on System Sciences*, Vol. 43.

Hong, Jong-yi, Eui-ho Suh, and Sung-Jin Kim. 2009. Context-aware systems: a literature review and classification. *Expert Systems With Applications* 36 (4) (May): 8509–8522.

Jimenez, Juan, Natalia Romero, and David Keyson. 2011. Capturing patients' daily life experiences after total hip replacement. In *Proceedings of the 5th International ICST Conference on Pervasive Computing Technologies for Healthcare* 226–229.

Kjær, Kristian Ellebæk. 2007. A survey of context-aware middleware. In *Proceedings of the 25th Conference on IASTED International Multi-Conference: Software Engineering*, 148–155. Innsbruck, Austria.

Korel, Barbara T. 2010. A survey on context-aware sensing for body sensor networks. *Wireless Sensor Network* 02 (08): 571–583.

Manzoor, Atif, Truong, Hong-Ling, and Dustdar, Schahram. 2008. On the evaluation of quality of context. In *Proceedings of the 3rd European Conference on Smart Sensing and Context*, Volume 5279, 140–153.

McCarthy, John. 1993. Notes on formalizing context. In *IJCAI*, ed. Ruzena Bajcsy, 555–562. Morgan Kaufmann.

Miraoui, Moeiz, Chakib Tadj, and Route De Soukra. 2008. Architectural survey of context-aware systems in pervasive computing environment. *Ubiquitous Computing and Communication Journal* 3: 1–9.

Montani, Stefania. 2011. How to use contextual knowledge in medical case-based reasoning systems: a survey on very recent trends. *Artificial Intelligence in Medicine* 51 (2) (February): 125–31.

Ongenae, Femke, Pieter Duysburgh, Mathijs Verstraete, Nicky Sulmon, Lizzy Bleumers, An Jacobs, Ann Ackaert, Saar De Zutter, Stijn Verstichel, and Filip De Turck. 2012. User-driven design of a context-aware application: An ambient-intelligent nurse call system. In *Proceedings of the 6th International Conference on Pervasive Computing Technologies for Healthcare (PervasiveHealth) and Workshops*, 205–210.

Orwat, Carsten, Andreas Graefe, and Timm Faulwasser. 2008. Towards pervasive computing in health care—A literature review. *BMC Medical Informatics and Decision Making* 8 (January): 26.

Pomerol, Jean-Charles, and Patrick Brezillon. 1999. Dynamics between contextual knowledge and proceduralized context. In *Modeling and Using Context*, ed. Paolo Bouquet, Massimo Benerecetti, Luciano Serafini, Patrick Brézillon, and Francesca Castellani, 1688: 284–295. Springer Berlin Heidelberg.

Prekop, Paul, Mark Burnett, and Fern Hill Park. 2003. Activities, context and ubiquitous computing 2. Existing context-awareness approaches. *Special Issue on Ubiquitous Computing Computer Communications* 26 (11): 1168–1176.

Romero, Daniel. 2008. Context-aware middleware: An overview. *Paradigma* 2, 3 (2008): 1–11.

Saeed, Aamna, and Tabinda Waheed. 2010. An extensive survey of context-aware middleware architectures. In *The 2010 IEEE International Conference on Electro/Information Technology*, (May): 1–6. Normal, IL.

Schilit, Bill, Norman Adams, and Roy Want. 1994. Context-aware computing applications. *Workshop on Mobile Computing Systems and Applications* 85–90.

Serafini, Luciano, and Martin Homola. 2011. Contextualized knowledge repositories for the semantic web. *Web Semantics: Science, Services and Agents on the World Wide Web*, 12–13, April 2012: 64–87.

Silva, Joao C., Artur M. Arsenio, and Nuno M. Garcia. 2011. Context-awareness for mobility management: A systems survey for healthcare monitoring. *7th International Conference on Broadband Communications and Biomedical Applications* (November): 18–23.

Singh Abtishek, 2006. Survey of context aware frameworks—Analysis and criticism. Its2.unc.edu 1–23.

Souza Damires, Belian Rosalie, Salgado Ana Carolina, and Tedesco Patricia Azevedo 2008. Towards a context ontology to enhance data integration processes. Proceedings of the 4th International VLDB Workshop on Ontolog-based Techniques for DataBases in Information Systems and Knowledge Systems, ODBIS 2008, Auckland, New Zealand, August 23, 2008, Co-located with the 34th International Conference on Very Large Data Bases. 49–56.

Soylu, Ahmet, Patrick De Causmaecker, and Piet Desmet. 2009. Context and adaptivity in pervasive computing environments: Links with software engineering and ontological engineering. *Journal of Software* 4 (9): 992–1013.

Strang, Thomas, and Claudia Linnhoff-Popien. 2004. A context modeling survey. In *Workshop on Advanced Context Modelling, Reasoning and Management, UbiComp 2004—The Sixth International Conference on Ubiquitous Computing (September 2004)*. Nottingham, England.

Vieira, Vaninha, Patricia Tedesco, and Ana Carolina. 2011. Expert systems with applications designing context-sensitive systems: An integrated approach. *Expert Systems With Applications* 38 (2): 1119–1138.

Virone, Gilles, Norbert Noury, and Jacques Demongeot. 2002. A system for automatic measurement of circadian activity deviations in telemedicine. *IEEE Transactions on Biomedical Engineering* 49 (12) (December): 1463–1469.

WASP. 2003. Context-aware services—state of the art. *Telematica*.

Wagner, Stefan, Thomas Toftegaard, and Olav Bertelsen. 2011. Increased data quality in home blood pressure monitoring through context awareness. In *Proceedings of the 5th International ICST Conference on Pervasive Computing Technologies for Healthcare* (3): 234–237. Vienna, Austria.

Wang, Xiao Hang, Da Qing Zhang, Tao Gu, and Hung Keng Pung. 2004. Ontology based context modelling and reasoning using OWL. In *CoMoRea, the Second IEEE International Conference on Pervasive Computing and Communications*. Orlando, Florida.

Winograd, Terry. 2001. Architectures for context. *Human-Computer Interaction* 16 (2): 401–419.

Yang, Stephen J. H., Jia Zhang, and Irene Y. L. Chen. 2008. A JESS-enabled context elicitation system for providing context-aware web services. *Expert Systems with Applications* 34 (4) (May): 2254–2266.

Ye, Juan, Simon Dobson, and Susan McKeever. 2012. Situation identification techniques in pervasive computing: a review. *Pervasive and Mobile Computing* 8 (1) (February): 36–66.

Zhang, Daqing, Zhiwen Yu, and Chung-Yau Chin. 2005. Context-aware infrastructure for personalized healthcare. *Studies in Health Technology and Informatics* 117 (January): 154–63.

Zimmermann, Andreas, Andreas Lorenz, Reinhard Oppermann, and Sankt Augustin. 2007. An operational definition of context. CONTEXT'07 Proceedings of the 6th International and Interdisciplinary Conference on Modeling and using Context 558–571.

23

Natural Language Processing in Medicine

Rui Zhang, Yan Wang, and Genevieve B. Melton

CONTENTS

23.1 Introduction

A natural language refers to any language used by people for communication, other than machine or computerized language such as C++ or Java. Natural language processing (NLP), a field of artificial intelligence and computational linguistics, is the automated analysis of natural language. Within biomedicine, the biomedical literature includes a large number of publications written in text format to which NLP techniques are applied. In the clinical domain, there has been a surge of interest in the secondary use of electronic health record (EHR) system data, including electronic clinical notes to improve health care quality through disease surveillance, decision support, and evidence-based medicine. To improve the use of textual information in EHR systems, the development of effective NLP methods for clinical texts is an important and challenging task for effectively using EHR data more reliably.

The application of NLP to process medical literatures and documents has rapidly attracted researchers, especially with the surge of EHR system adoption. However, NLP algorithms require special development for medical tasks because medical sublanguages differ largely from general English across several linguistic dimensions introduced by Harris. For instance, clinical notes are often entered by physicians who have limited time, and therefore, they frequently use domain-specific abbreviations, omit information that can be assumed by context, and have language problems such as misspellings or incorrect word usage. As a result, out-of-the-box existing NLP applications for general English usually did not perform well for medical text. Moreover, domain terminologies and local dialects are prevalent in medical documents. For example, it is not uncommon for physicians at different hospital sites to develop their own local jargon for devices, techniques, or other items. System performance is also challenged in that the outputs of medical NLP systems are frequently used in health care systems or clinical research that requires reliable, high-quality NLP performance and modular, flexible, fast systems. One other major challenge for medical NLP systems are barriers faced from data availability and confidentiality. Many medical NLP systems need to access medical documents from EHR clinic information systems. This can often be problematic because access to patient records is confidential, requires the approval of institutional review boards (IRBs), and may require data de-identification. Also, it is difficult to share data across institutions, which creates another challenge for system interoperability and interinstitutional validation of systems.

23.2 NLP Tasks in Medicine

In the section below, we enumerate some actively researched medical NLP system components from both a fundamental and a higher level, as well as issues that complicate these tasks in the medical domain.

23.2.1 Low-Level NLP Components

23.2.1.1 Tokenization

Tokenization is an initial step of automated processing of a text. It is the task of identifying boundaries that separate semantic units, which include morphemes, words, dates, and symbols within a text. The primary indication of such semantic units, also called tokens, in general English is white space that occurs before and after a word. A token may also be separated by punctuation marks instead of a word space, such as by a period, comma, semicolon, or question mark. Some of the difficulties that occur with tokenization stem from ambiguous punctuation, such as the colon in "2:30am" or the periods in "M.D." In the medical literature, in addition to typical ambiguous punctuations often seen in general English, the biomedical literature will also have certain technical terms and heterogeneous orthographics, such as "Adams Stokes" and "Adams-Stokes," which add additional difficulties in tokenization (Arens 2004; Barrett and Weber-Jahnke 2011; Jiang and Zhai 2007; Wrenn et al. 2007). For this reason, a simple tokenizer for general English text will typically not work well in biomedical text. Therefore, tokenization algorithms often need new heuristics and domain-specific training corpora to accommodate the distinct features of medical sublanguages.

23.2.1.2 Sentence Boundary Detection

Sentence boundary detection (SBD), also called sentence boundary disambiguation or sentence breaking, can also be a challenging NLP component, particularly for clinical documents. This task aims to detect where sentences start and end. A simple SBD system can identify sentence boundaries using a small set of rules. However, the task can be complicated by the fact that punctuation marks such as question marks, semicolons, and periods are often ambiguous and need more complex logic in special cases. In addition to rule-based systems, AI methods such as decision trees, neural networks, and hidden Markov models (HMMs) are frequently used for SBD. Also, the biomedical literature and clinical documents are full of abbreviations (e.g., "q.i.d.," "p.r.n."), acronyms (e.g., "OD," "OS"), and symbolic constructions (e.g., "blood pressure: 130/67") that add difficulty to SBD (Barrows et al. 2000; Friedman 1997; Huang et al. 2005). For medical NLP systems, one frequent approach for SBD includes the use of domain lexical resources such as the National Library of Medicine's (NLM's) SPECIALIST Lexicon (McCray et al. 1994) and annotated domain corpora to ensure satisfactory SBD performance.

23.2.1.3 Part-of-Speech Tagging

Part-of-speech (POS) tagging is the process for determining the part of speech of words in a piece of text, based on both definition as well as local context. The example below shows the tagging output of the following sentence using the Penn Treebank tag set (Marcus et al. 1994): "The cystic duct was triply clipped distally and singly proximally and transected."

> "The/DT cystic/JJ duct/NN was/VBD triply/RB clipped/VBN distally/RB and/CC singly/RB proximally/RB and/CC transected/VBN."

POS tagging is an essential step of NLP systems where errors can propagate upward to the syntactic processing level and produce more errors in the syntactic output, which provides important information necessary for text understanding. Therefore, having reliable POS information is critical to successful implementation of various NLP applications. POS taggers trained merely on general English do not usually achieve state-of-the-art performance on medical text. A number of POS taggers (Fan et al. 2011; Pakhomov et al. 2006; Smith et al. 2004) have been developed specifically for the medical domain, such as the adapted Trigrams'n'Tags (TnT) tagger (Pakhomov et al. 2006), which is a TnT tagger (http://www.coli.uni-saarland.de/~thorsten/tnt/) trained on a relatively small set of clinical notes, and the MedPost tagger (Smith et al. 2004), a POS tagger based on an HMM and trained on manually tagged sentences in medical text.

23.2.1.4 Shallow Parsing

Shallow parsing, also called chunking, is the process of identifying constituents (syntactically correlated parts of words like noun groups, verb groups, etc.) in a sentence. As an intermediate step toward deep parsing, shallow parsing produces a limited amount of syntactic information from sentences and does not specify internal structures or roles of each constituent in the main sentence. The sentence below exemplifies shallow parsing output:

> "[NP The cystic duct] [VP was triply clipped] [ADVP distally and singly] [ADVP proximally] and [UCP transected]"

In the medical domain, shallow parsing is used in a wide range of tasks such as drug–drug interaction (DDI) detection, medical problem assertion detection, biological entity relation extraction, and medical information extraction (IE). Several shallow parsers have been built for medical text processing, such as the SPECIALIST minimal commitment parser (McCray et al. 1993), which produces high-level syntactic information rather than the traditional full syntactic information for better noun phrase discovery in medical text.

23.2.1.5 Deep Parsing

Deep parsing is the process to produce an ordered, rooted tree that represents the syntactic structure of a string according to some formal grammar such as constituency grammars (Sipser 1996) and dependency grammars (Mel'Čuk 1988). Figure 23.1 shows the constituency parse tree of the sentence "The dressing was removed from it."

Full syntactic parsing of text can provide a large amount of deep linguistic information such as sentence voice, phrase type, and POS tags, which are shown to perform considerably better than surface-oriented features (e.g., pattern matching) for many NLP tasks. Because of the special features of medical sublanguage (e.g., domain vocabulary, telegraphic text,

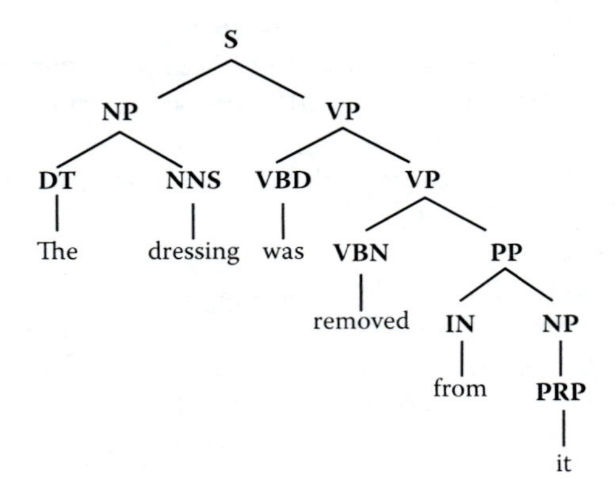

FIGURE 23.1
A constituent (phrase structure) tree for "The dressing was removed from it."

special grammar), parsers trained on general English corpus like the *Wall Street Journal* (http://www.ldc.upenn.edu/Catalog/CatalogEntry.jsp?catalogId=LDC2000T43) only have limited performance on medical text. NLP experts have investigated several methods to adapt parsers trained on general English to new target domains. New entries can be imported from domain resources to existing parser lexicons using morphological clues, heuristic mapping, and direct expansion (Szolovits 2003). POS tag information of domain-specific lexical elements can also be provided to a parser to avoid inconsistencies between domain POS tags and parser lexicon POS tags (Rimell and Clark 2009). Moreover, better parsing performance can also be acquired by adjusting the syntactical category statistics for important domain lexical elements like verbs and other lexical elements that have unusual usage in a particular domain (Huang et al. 2005).

23.2.2 High-Level NLP Components

23.2.2.1 Negation Detection

Many medical documents such as discharge summaries and radiology reports contain large amounts of important information of patients, like conditions, findings, and diseases, that can be used for a wide range of secondary applications. In these reports, about half of the described findings and diseases are actually absent in a given patient. For example: "They have not noticed any abnormal behaviors, movements, or rash anywhere else on his body" or "no significant complications of bleeding." As a result, negation detection is a critical component in medical NLP systems. In medical text, negation detection is not an easy task as negation can be explicit (e.g., "Patient denies any fevers, emesis, or diarrhea") or implied (e.g., "Chest x-ray is clear upon my read").

The scope of negation is another challenge for the negation detection process. Consider two sentences: "The child is not tired" and "The child is not very tired." In the first sentence, the word "not" scopes over "tired," while in the second sentence, the word "very" redirects the scope of "not" to itself and away from "tired."

Negation detection systems can detect negation solely by rules developed on a handcrafted list of negation phrases that appear before or after a term of interest or through the use of an ontology of negative medical concepts generated from a standard medical dictionary, such as the Systematized Nomenclature of Medicine—Clinical Terms (SNOMED CT), to decide if a term is negated (Chapman et al. 2001; Elkin et al. 2005; Mutalik et al. 2001). In more complex systems, machine learning techniques such as naïve Bayes and decision trees are applied with more clues, such as deeper syntactic structure and lexical cues (Goryachev et al. 2006; Sarafraz and Nenadic 2010; Yang et al. 2009).

23.2.2.2 Relationship Extraction

Relation extraction aims to determine or discover relationships between entities (e.g., drugs, diseases, findings, genes) in medical texts. Relations among these entities, in their simplest form, are binary, involving only two entities. Other relations can involve more than two entities. A large variety of relations have been investigated, such as interactions between drugs, genes, associations between diseases and symptoms, and relations between patient problems and treatments. Co-occurrence statistics are effective methods that are frequently used for identifying relations between medical entities by collecting instances where the entities co-occur (Cao et al. 2005; Chen et al. 2008; Wang et al. 2009). The hypothesis behind this approach is that an entity and its related entities are more likely to appear together than random combinations of entities. Thus, if entities are repeatedly mentioned together, then there is a good chance that they may be related. However, the nature of the relationship between these associated entities usually cannot be determined by the method alone.

Rule-based approaches for relation extraction work by exploiting the particular linguistic patterns exhibited by relations. Rules used can be manually defined by domain experts or derived from annotated corpora. Machine learning-based systems rely on machine learning techniques along with a variety of features based on the nature of the relationship, such as lexical, syntactic, semantic, and dependency features (Barnickel et al. 2009; Katrenko and Adriaans 2007).

Several important challenges are associated with relation extraction in the medical domain. First, in the medical domain, annotation of relations can be complicated because relations are often expressed across discontinuous spans of text. Secondly, there can be a lack of consensus on how to best annotate a particular type of relation. As a result, annotation resources between research groups can be largely incompatible and the quality of systems constructed based upon these resources can be difficult to evaluate.

23.2.2.3 Named Entity Recognition

Named entity recognition (NER) aims to identify and classify elements into named entities, which are predefined categories such as names (e.g., drugs, genes, person), findings, diseases, and medications. The biomedical literature is full of terms particular to the biomedical domain that are typically not detected by conventional general English NLP systems. In order to extract relations between entities, it is crucial for the system to be able to detect unknown nouns or named entities. Some named entities can be effectively identified solely through surface patterns (e.g., phone number: xxx-xxx-xxxx, person: Carole Green MD). However, rule-based systems require a significant manual effort, and these rules may not easily extend to a new domain. Machine learning is another effective approach for NER. Compared with rule-based systems, it requires less human intuition with rules, and this approach can easily be adapted to new domains. The disadvantage of machine learning approaches is the large annotation corpus required for model training.

23.2.2.4 Word Sense Disambiguation

Ambiguity is a problem inherent to natural language, where a term can have more than one meaning depending upon the context or use of the term in a particular text. Word sense disambiguation (WSD) is the process of understanding which sense of a term, including single words, abbreviations, or acronyms, is being used in a particular context among a list of predefined sense candidates. In the medical domain, researchers have suggest that the problem might be more restricted compared to general English based upon the idea that since medicine is scientific, it might be more specific than general English. Instead, the problem is more extensive in the medical domain due to the high use of abbreviations and acronyms in medical documents and the biomedical literature.

Approaches to WSD (McInnes et al. 2011; Pakhomov et al. 2005; Stevenson et al. 2012; Xu et al. 2007) generally rely on a particular domain knowledge source, such as the Unified Medical Language System (UMLS) (Humphreys et al. 1998), MEDLINE (http://www.nlm.nih.gov/bsd/pmresources.html), and Entrez Gene Database (http://jura.wi.mit.edu/entrez_gene/), and domain corpora such as GENIA (Kim et al. 2003) or the Biosemantics test collection (Weeber et al. 2001) for sense collection, sample collection, and model training.

23.2.2.5 Semantic Role Labeling

Semantic role labeling (SRL) is the task of detecting semantic roles associated with predicates, which are mainly verbs, such as "hit" and "move," in a sentence. For example, in the sentence "He placed the ball beside the couch," the predicate is the verb "place." The semantic roles associated with "place" include "placer"—who placed; "thing placed"—what is placed; and "location"—where it (the ball) is placed. Labeling semantic roles like above for predicates in a given text answers questions such as "who," "when," "what," "where," and "why." SRL can be used for IE, question answering (QA), text summarization, and other NLP tasks that require some kind of semantic interpretation.

Example semantic roles include agent, patient, instrument, and adjunctive arguments indicating other meanings such as locative and temporal. In general, SRL can be accomplished using supervised machine learning approaches. Given a predicate and each constituent in a syntactic parsed output, an SRL system assigns a semantic role from a predefined set of roles for the predicate. A typical supervised SRL system can be designed by extracting machine learning features for each constituent, training a machine learning classifier on an annotated, training set and then labeling each unlabeled constituent in a new set of text with a given set of features. In order to build an SRL system to process medical text, domain resources such as semantic annotated corpus and semantic frames (a semantic frame is a collection of semantic roles of a predicate) often must be created (Dahlmeier and Ng 2010; Kogan et al. 2005). Also, domain-specific features often boost the SRL performance in medical domain (Tsai et al. 2006).

23.2.2.6 IE

IE is a task that involves extracting problem-specific information from the text of interest and then transforming this information into structured form. For example, vaccination reactions can be extracted from medical reports, and relationships between genes and diseases from the biomedical literature are all cases of IE. Most early and straightforward IE systems were built mostly using pattern matching techniques such as regular expressions over features such as text strings, syntactic structure, semantic type, and dictionary entries.

Recent systems are mostly based on machine learning methods. State-of-the-art lower-level components and high-level components introduced before such as deep parsing, NER, and WSD are often part of an IE system. In the medical domain, a variety of IE systems have been built for various tasks (Dang et al. 2008; Denecke and Bernauer 2007; Hripcsak et al. 1998; Lakhani and Langlotz 2010; Long 2005) as well as many NLP tools for IE, such the Medical Language Extraction and Encoding System (MedLEE) (Hripcsak et al. 1998) and the clinical Text Analysis and Knowledge Extraction System (cTAKES) (Savova et al. 2010).

23.3 NLP Methods

NLP methods include symbolic (linguistics-based) methods, statistics-focused methods, and machine learning methods. Symbolic methods are built based on linguistic rules, while statistical methods and machine learning methods require training to build models. In this section, we cover a few of these methods briefly.

23.3.1 Support Vector Machine

A support vector machine (SVM) is a discriminative machine learning approach that belongs to the family of margin-based classifiers. It works by looking for an optimal hyperplane that maximizes the distance between the hyperplane and the nearest samples from each of the two classes. In the simplest two-feature case, a straight line would separate samples in an X–Y plot. In a general *N*-feature case, the separator will be a hyperplane with *N*–1 dimensions. For an *N*-feature case, the input may be transformed mathematically using a kernel function (e.g., Gaussian), to allow linear separation of the data points. During the learning process, the separation process selects a subset of the training data, the "support vectors," that best differentiates the classes. The resulting hyperplane maximizes the distance between each class and the support vectors, as shown in Figure 23.2.

As a binary classification approach, SVM is only directly applicable for two-class tasks. However, it can be easily extended to multiclass classification problems by the one-versus-all (OVA) approach or pairwise approach. The prediction accuracy of SVM is generally high because of the sound mathematical theory behind it and the robustness of the method. It generally works well when training examples contain errors, as well, because of its use of a separation process. On the downside, SVM is computationally expensive. Its training process is a convex optimization problem that requires at least quadratic time with respect to the number of training examples. In the medical domain, SVM has been shown to perform well on many classification tasks such as smoking status classification (Cohen 2008), SRL for protein transport predicates (Bethard et al. 2008), and disease comorbidity status classification (Ambert and Cohen 2009).

23.3.2 Maximum Entropy Modeling

The principal of maximum entropy modeling (MEM) is simple and realistic. It is a statistical learning method that models all that is known and assumes nothing about that which is unknown. The modeling contains a set of predefined features or constraints as shown below.

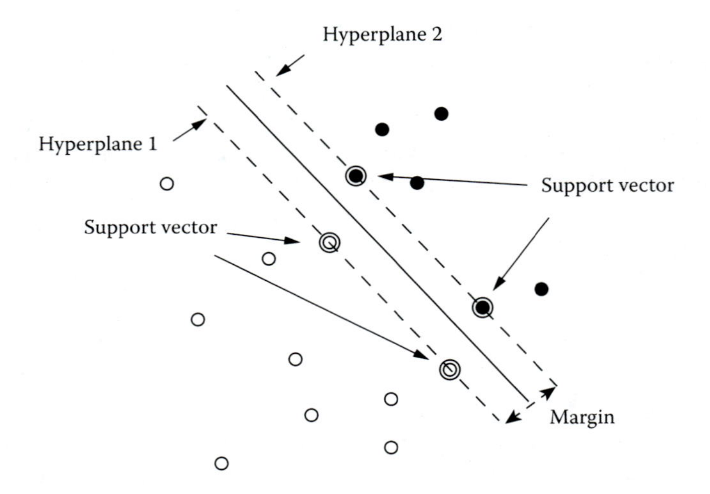

FIGURE 23.2
Linear separating hyperplanes.

$$\sum_{x,y} \tilde{p}(x)p(y|x)f(x,y) = \sum_{x,y} \tilde{p}(x,y)f(x,y)$$

Here, x is a random variable representing some context information, y is the output, and $f(x, y)$ is the feature function, as shown below.

$$f(x,y) = \begin{cases} 1 & \text{if } x \text{ denotes a context and } y \text{ indicates an output} \\ 0 & \text{otherwise} \end{cases}$$

$\tilde{p}(x,y)$ is the joint empirical distribution that is derived from the training data expressing some relationship between features and outcome, as shown below.

$$\tilde{p}(x,y) = \frac{1}{N} \times \text{number of } (x,y)$$

$p(x, y)$ is the conditional probability model for predicting the output y given a context x. Among many conditional probability models, the best model p^* is the one that maximizing the conditional entropy $H(p)$, which is shown below, on $p(x, y)$, as it has a more uniform probability distribution on unseen x in the training set, consequently allowing less bias for unseen contexts.

$$H(p) = -\sum_{x,y} \tilde{p}(x)p(y|x)\log p(y|x)$$

For details of how MEM works and why it works for NLP, readers may refer to an article by Berger (1996).

One advantage of MEM is that heterogeneous information sources such as lexical, syntactical information and bigrams can be modeled easily as features in an integrated model. Another advantage of MEM is that it handles overlapping features very well. It is sometimes more effective to use a combined feature together with its component features, compared with using simple features alone. Because of its ability to incorporate heterogeneous features, MEM has been used in the medical domain for a diverse set of NLP tasks, such as patient medication status mining (Pakhomov et al. 2002), SRL for biomedical verbs (Tsai et al. 2006), and noun phrase identification in radiology reports (Huang et al. 2005).

23.3.3 *n*-Gram Model

Statistical language modeling (SLM) is widely used for many NLP tasks, such as POS tagging, parsing, information retrieval, and machine translation. Generally speaking, SLM assigns a probability to a set of n words based on a probability distribution.

An n-gram model is a typical language method used in the field of computational linguistics and NLP (Manning and Schütze 2003). The word "n-gram" means consecutive items, such as words or terms. n-gram models (n = 2, 3, 4 …) are used to estimate the probability of the existence of the n-gram. To simplify the calculation of the probability of the word, the Markov assumption states that the probability of the word is only based on the prior few words instead of all previous words. The probability of a word is then simplified as follows:

$$P(w_1^n) \approx \prod_{k=1}^{n} P(w_k \mid w_{k-1})$$

An n-gram model (which checks the n-1 previous words) is an (n-1)th order Markov model.

Probability of a given word can be estimated by its relative frequency. One commonly used estimate is called maximum likelihood estimate (MLE).

$$P_{MLE}(w_1 \dots w_n) = \frac{C(w_1 \dots w_n)}{N}$$

$$P_{MLE}(w_n \mid w_1 \dots w_{n-1}) = \frac{C(w_1 \dots w_n)}{C(w_1 \dots w_{n-1})}$$

where N is the number of training instances and C is the count.

In practice, MLE is not an ideal estimator due to sparseness, which occurs in many data sets. MLE assigns a value of zero to the probability of unseen events (in the training set), which will propagate to the whole sentence. To avoid issues related to the sparseness of data sets, which always exist if the data set is large, discounting methods are commonly used, such as the Good–Turing (GT) method and that by Ney and Essen.

GT discounting is based on the assumption that the probability of items follows a binomial distribution. It is suitable for a large data set.

$$\text{If } C(w_1 \dots w_n) = r > 0,\ P_{GT}(w_1 \dots w_n) = \frac{r^*}{N}$$

where

$$r^* = \frac{(r+1)S(r+1)}{S(r)}$$

$$\text{If } C(w_1 \dots w_n) = 0,\ P_{GT}(w_1 \dots w_n) = \frac{1 - \sum_{r=1}^{\infty} N_r \frac{r^*}{N}}{N_0} \approx \frac{N_1}{N_0 N}$$

where s is the function that fits the observed values of (r, Nr), and $S(r)$ is the expectation of the frequency.

Ney and Essen proposed two discounting models for estimating frequencies of n-grams. One is absolute discounting:

$$\text{If } C(w_1 \dots w_n) = r,\ P_{abs}(w_1 \dots w_n) = \begin{cases} (r - \delta/N) & \text{if } r > 0 \\ (B - N_0\delta/N_0 N) & \text{otherwise} \end{cases}$$

where δ is a small constant number for all nonzero MLE frequencies and B is the number of target feature values.

Another is linear discounting:

$$\text{If } C(w_1 \ldots w_n) = r,\ P_{abs}(w_1 \ldots w_n) = \begin{cases} (1-\alpha)r/N & \text{if } r > 0 \\ \alpha/N_0 & \text{otherwise} \end{cases}$$

where α is a constant slightly less than 1.

These discounting methods make the probability of unseen events a small number instead of zero and rescale the other probabilities. The absolute discounting approach is very successful, while the linear discounting is less so since it does not approximate even higher frequencies.

n-Grams have been used for low-level NLP tasks such as spelling correction, speech recognition, and word disambiguation. They have been used in a longitudinal analysis of clinical notes to identify new information (Zhang et al. 2012). *n*-Grams have also been used to improve the efficiency of systematic reviews (SRs) for evidence-based medicine (Munshaw and Kepler 2010).

23.3.4 HMM

HMM (Figure 23.3) is another statistical NLP method. Markov models are built on the Markov assumption that the current state occurs based upon on the previous state(s). For the simplest first-order Markov model, there are M^2 transitions between M states. Unlike deterministic models, where each state is dependent on another state, Markov models assign probability to each transition between two states. In a visible Markov model, the state is visible, and state transition probabilities are the only parameters to calculate. In an HMM, hidden states have a probability contribution to the outputs. For example, in a speech recognition system, the sound we hear is the output of hidden states, such as vocal chords, the size of the person's throat, the position of the person's tongue, and many other factors. Each sound of a word is generated from changes of these hidden factors.

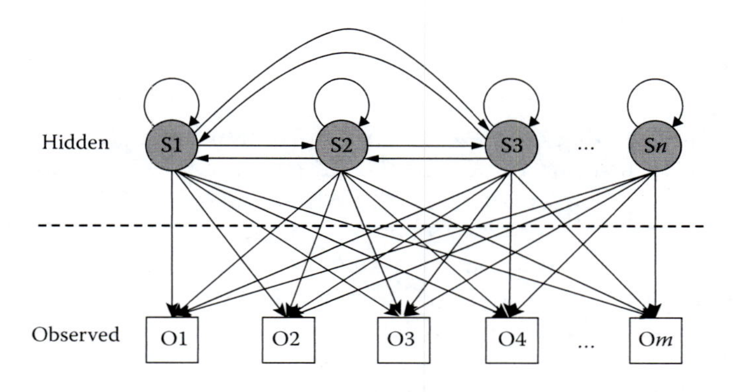

FIGURE 23.3
Hidden Markov models. S1, S2, S3...S*n* are hidden states; O1, O2, O3, O4...O*m* are outputs. Each state can transit to other states or itself, shown in the lines between states. Transitions between the state S*n* and other states are not shown. Each observed output is generated from hidden states with probabilities, indicated as darker lines.

HMMs are defined by the following quintuple:

$$\lambda = (N, M, A, B, \pi)$$

where N is the number of states for the model; M is the number of distinct observation symbols per state; A is the $N \times N$ state transition probability distribution given in the form of a matrix $A = \{a_{ij}\}$; B is the $N \times M$ observation symbol probability distribution given in the form of a matrix $B = \{b_j(k)\}$; and π is the initial state distribution vector $\pi = \{\pi_i\}$.

Three canonical problems are associated with HMM:

1. Evaluation: computing the probability of a particular output sequence based on the given model parameters. This is typically implemented using Viterbi–forward or forward algorithms (Viterbi 1967).
2. Decoding: finding the state candidates that can generate a particular output sequence based on the given model parameters, which is typically computed by using the Viterbi algorithm.
3. Learning: finding the set of state transition and output probabilities that fit the given output sequence(s) the best.

HMM has been widely used in speech recognition and bioinformatics (Drawid et al. 2009; Munshaw and Kepler 2010). It has been used in medicine to describe the effect of alcoholism treatment on the likelihood of healthy/unhealthy populations (Wall and Li 2009), to estimate the transition probabilities between states of liver cirrhosis (Bartolomeo et al. 2011), and for disease surveillance with public health data (Watkins et al. 2009).

23.4 Clinical NLP Resources and Tools

In this section, we mainly focus on several NLP resources and tools available in the biomedical domain. Specifically, we introduce the resources and tools supported by the US NLM at the National Institutes of Health (NIH).

23.4.1 UMLS

UMLS (http://www.nlm.nih.gov/research/umls/) was developed by and is maintained by the NLM to provide health care professionals and researchers with a biomedical domain knowledge resource (Humphreys et al. 1998). UMLS is a structured knowledge base that connects different biomedical sources and enables biomedical research application development. UMLS contains three knowledge sources: Metathesaurus, Semantic Network (McCray 2003), and SPECIALIST Lexicon (McCray et al. 1994) and lexical tools.

Metathesaurus is created based on over 100 vocabularies, code sets, and thesauri. It covers several major categories, including comprehensive vocabularies [e.g., SNOMED CT, http://www.nlm.nih.gov/research/umls/Snomed/snomed_main.html, Medical Subject Headings (MeSH, http://www.nlm.nih.gov/mesh/)]; laboratory and observational data [e.g., Logical Observation Identifier Names and Codes (LOINC, http://loinc.org/) (Forrey et al. 1996; McDonald et al. 2003)]; diseases [e.g., International Classification of Diseases and Related Health Problems (ICD, http://www.who.int/classifications/icd/en/)]; and procedures and supplies [e.g., Current Procedural Terminology (CPT, http://www.ama-assn.org/ama/pub/physician-resources/solutions-managing-your-practice/coding-billing-insurance/cpt.page)]. While 60% of the Metathesaurus terms are in English, it also contains terms in 17 other languages, including Spanish, French, Dutch, Italian, Japanese, and Portuguese. Each equivalent biomedical term (same meaning, various names) from different sources is assigned a concept unique identifier (CUI).

The Semantic Network is an upper-level ontology of biomedical knowledge. It contains 135 semantic types and 54 relationships between semantic types. Each concept is assigned at least 1 of 135 defined semantic types. All semantic types are hierarchically organized under two main topics: Entity and Event. The top level of semantic types in the UMLS Semantic Network is depicted in Figure 23.4.

The relationship "ISA" is used to link most concepts. For example, "Carbohydrate" ISA "Chemical." There are also five major, nonhierarchical relationships: physical (e.g., PART_OF, BRANCH_OF); spatial (e.g., LOCATION_OF, ADJACENT_TO); temporal (e.g., CO-OCCURS_TO, PRECEDES); functional (e.g., TREATS, CAUSES); and conceptual (e.g., EVALUATION_OF, DIAGNOSES).

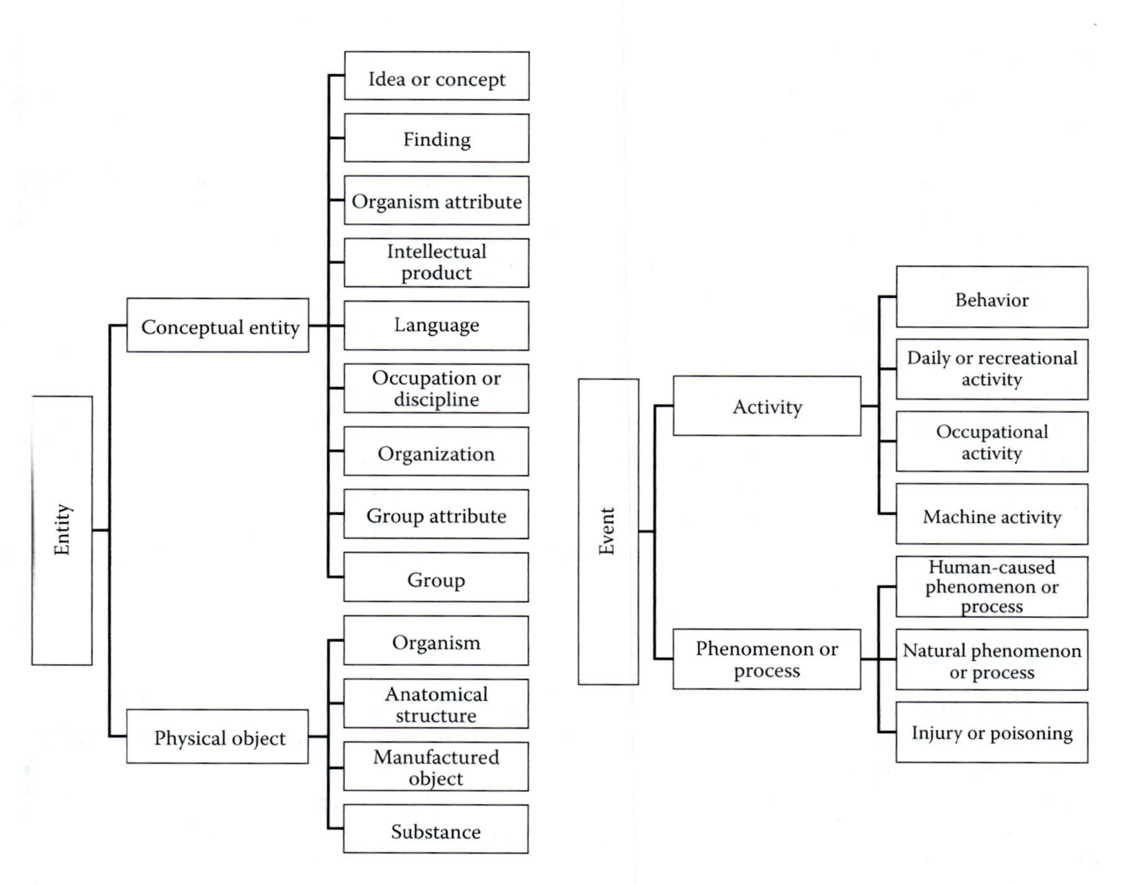

FIGURE 23.4
Hierarchy structure of UMLS semantic types.

The SPECIALIST Lexicon is an English dictionary including over 200,000 biomedical terms as well as common English words. It also contains syntactic, morphological, and orthographic information of each term or word. For example, each lexical record contains base forms of the term; the part of speech; a unified identifier; spelling variants; and inflection for nouns, verbs, and adjectives.

23.4.2 Corpora

Development of NLP systems requires large volumes of biomedical and clinical texts. The MEDLINE database is a collection of biomedical abstracts. It is maintained by the NLM and contains over 21 million reference from 1946 to the present. The GENIA corpus (http://www.nactem.ac.uk/genia/genia-corpus) collects 1999 MEDLINE abstracts, selected from a PubMed query for MeSH terms "human," "blood cells," and "transcription factors." The corpus has been annotated with various levels of linguistic and semantic information covering POS, syntactic, term, event, relation, and coreference annotation (Kim et al. 2003). In the clinical domain, there are a few collections of clinical texts, including the Pittsburgh collection of clinical reports (http://www.dbmi.pitt.edu/nlpfront), Multiparameter Intelligent Monitoring in Intensive Care (MIMIC II) database (Saeed et al. 2011), and Informatics for Integrating Biology & the Bedside (i2b2) NLP research data sets (https://http://www.i2b2.org/NLP/DataSets/Main.php). Most research groups created their own clinical text corpus and annotations for specific NLP tasks locally.

23.4.3 SPECIALIST NLP Tools

SPECIALIST NLP tools (http://lexsrv3.nlm.nih.gov/Specialist/Home/index.html) are computer programs developed by the NLM to aid in dealing with different biomedical NLP tasks. Tools include lexical tools such as lexical variant generator (LVG), normalized string generator (Norm), word index generator (WordInd), dTagger POS tagger, subterm mapping tools (STMTs), and others.

LVG contains a series of commands to perform lexical transformation of text. Norm provides a normalization process for those terms included in the SPECIALIST Lexicon. Norm can help to find similar terms and to map terms to UMLS concepts. WordInd creates a sequence of alphanumeric characters by reading the text, helping UMLS to produce the word index for the Metathesaurus. dTagger is a POS tagger built specifically with SPCIALIST Lexicon. dTagger was trained on MedPost corpus, a set of annotated MEDLINE abstracts, and tokenizes text into multiword terms. STMT was built to provide comprehensive subterm-related features, including all subterms, the longest prefix subterm, and synonymous subterm substitutions.

23.4.4 MetaMap

MetaMap (http://metamap.nlm.nih.gov/) is a program developed by the NLM to map biomedical text to the UMLS Metathesaurus (Aronson 2001; Aronson and Lang 2010). MetaMap provides various options, including data option (choose specific vocabularies and data model); processing options (such as author-defined acronyms/abbreviations, negation detection, WSD) and output options (human readable, machine output, and XML). Released application programming interfaces (APIs) provide options to integrate MetaMap into other programs. MetaMap was originally developed for information retrieval from bibliographic data such as MEDLINE citations. As it is an effective tool to map biomedical terms, MetaMap has been widely used in applications of the clinical domain, such as detection of clinical findings.

23.4.5 SemRep

SemRep is a rule-based, symbolic NLP program developed by NLM for semantic knowledge representation from biomedical literatures, mainly from titles and abstracts in MEDLINE (Fiszman et al. 2003; Rindflesch and Aronson 1993; Rindflesch and Fiszman 2003; Srinivasan and Rindflesch 2002). SemRep uses underspecified syntactic analysis and structured domain knowledge from UMLS. SemRep relies on syntactic analysis based on the SPECIALIST Lexicon and the MedPost POS tagger (Smith et al. 2004). MetaMap helps to map noun phrases in the sentences to UMLS Metathesaurus concepts. SemRep interpreted the semantic relationships (syntactic indicators in the sentence, such as verbs, nominalizations, prepositions, etc.) between two concepts in the sentences based on dependency grammar rules and ontology (i.e., an extended version of the UMLS Semantic Network). SemRep represents semantic knowledge from each sentence in citations as the format of semantic predications (a subject–predicate–object triplet). Both subjects and objects are Metathesaurus concepts and predicates that correspond to a relation type in SemRep ontology. For example, SemRep interprets sentence 1 as semantic predications in sentence 2.

1. We used hemofiltration to treat a patient with digoxin overdose that was complicated by refractory hyperkalemia
2. Hemofiltration-TREATS-Patients
   Digoxin overdose-PROCESS_OF-Patients
   Hyperkalemia-COMPLICATES-Digoxin overdose
   Hemofiltration-TREATS-Digoxin overdose

23.5 Current Clinical NLP Systems

In this section, we introduce several existing clinical NLP systems. We summarize some current clinical NLP systems in Table 23.1 and then discuss a few of them.

23.5.1 MedLEE

Medical language extraction and encoding system (MedLEE) is an NLP system that extracts information from clinical texts into a structured format and translates the information to terms in a controlled dictionary. MedLEE has been used to process various types of clinical records, including radiology reports, discharge summaries, sign-out notes, pathology reports, electrocardiogram reports, and echocardiogram reports (Cao et al. 2004; Chen et al. 2008; Chun et al. 2005; Friedman et al. 1994, 2004; Xu et al. 2004). The MedLEE preprocessor first transforms reports into a structure for the core NLP engine to process, for example, adding or changing section headers and expanding abbreviations (e.g., "hx" to "history"). The core MedLEE engine maps medical terms to semantic types and uses grammar rules to extract their semantic relationships. A structured output in

TABLE 23.1
Current Clinical NLP Systems

System	Description	Institution (Principle Investigator)	References
BioMedICUS[a]	A UIMA pipeline system designed for researchers for extracting and summarizing information from unstructured text of clinical reports	University of Minnesota (Pakhomov)	http://code.google.com/p/biomedicus/
cTAKES[a]	A UIMA pipeline built around OpenNLP, Lucene, and LVG for extracting disorder, drug, anatomical site, and procedure information from clinical notes	Mayo Clinic (Chute)	Savova et al. 2010
HITEx[a]	An NLP system distributed through i2b2	Harvard (Zeng)	Goryachev et al. 2006
MedEx[a]	A semantic-based medication extraction system designed to extract medication names and prescription information	Vanderbilt (Xu)	Xu et al. 2010 Doan et al. 2010
MedLEE	An expert-based NLP system for unlocking clinical information from narratives	Columbia (Friedman)	Friedman and Hripcsak 1998 Friedman 2000
MedTagger[a]	A machine learning–based name entity detection system utilizing existing terminologies	Mayo Clinic (Liu)	Torii et al. 2011
MetaMap[a]	An expert-based system for mapping text to the UMLS	NLM (Aronson)	Aronson and Lang 2010
SecTag[a]	A system to tag clinical note section headers	Vanderbilt (Denny)	Denny et al. 2009 Denny et al. 2008

Note: Systems are listed alphabetically.
[a] Publicly available systems.

XML format is then generated for each sentence. The data are finally transformed and stored in a clinical repository.

23.5.2 cTAKES

Clinical text analysis and knowledge extraction system (cTAKES) is an NLP system developed at Mayo Clinic for IE (specifically disorders, drugs, anatomical sites, and procedures) from free texts in clinical notes (Savova et al. 2010). cTAKES was built on a pipeline framework called the Unstructured Information Management Architecture (UIMA, IBM), which allows components in the system to be implemented sequentially. UIMA enables NLP systems to be decomposed into components, each of which is responsible for different tasks in analyzing the unstructured information. In cTAKES, components include basic NLP tasks such as sentence boundary detector, tokenizer, morphologic normalizer, part-of-speech tagger, dependency parser, NER annotator, and negation detector. It also contains clinical-specific tasks including the patient's smoking status identifier and drug mention annotator.

23.5.3 HITEx

Health Information Text Extraction (HITEx) is an open-source NLP system developed at the National Center for Biomedical Computing, i2b2 (Goryachev et al. 2006). HITEx was built on General Architecture for Text Engineering (GATE) framework and assembles GATE pipeline application and standard NLP components (such as POS tagger, parser). Each pipeline was developed to extract different clinical information, including diagnoses, discharge medications, smoking status, negation finding, and so forth. For example, to find principal diagnoses, the pipeline searches UMLS concepts in specific note sections and filters semantic types of the concepts that are either findings or symptoms (Zeng et al. 2006). To assign various diagnoses to the correct patient family member, from discharge summary to outpatient notes, HITEx mapped the family member concept and eight diagnosis semantic types from notes and associated diagnosis with the most relevant family member by using a set of rules (Goryachev et al. 2008).

23.6 Medical Applications of NLP

In a time-constrained clinical practice environment, clinicians may have limited time to review and synthesize all clinical notes. Thus, successful processing of large volumes of clinical narratives is the key component for improving health care. In this section, we will provide some research studies to discuss the role of NLP in health care.

23.6.1 NLP for Surveillance

Surveillance is a fundamental and important task in health care, especially surveillance of adverse events (AEs) based on the clinical texts. Hripcsak et al. (2003) developed a framework to discover AEs from clinical notes. They used MedLEE to parse the clinical narratives and generate a coded database, followed by query generation to detect and classify events. Penz and colleagues (2007) also used MedLEE to detect AEs related to the placement of central venous catheters (CVCs) from clinical texts in the Veterans Administration database. They used both an NLP program and a phrase-matching algorithm to achieve a sensitivity of 72% and specificity of 80.1%. Melton and Hripcsak (2005) constructed an AE detection system using MedLEE to identify 45 AE types from discharge summaries and obtained better results than traditional methods, with sensitivity of 28% and specificity of 98.5%. Recently, Friedman et al. reported adverse drug reactions (ADRs) using EHR and an automated method that combines MedLEE with an expert-generated disease identifier. They applied the method to identify two serious ADRs from almost 0.2 million records and reached a good performance (sensitivity of 93.8% and specificity of 91.8%) (Haerian et al. 2012).

23.6.2 NLP for Clinical Decision Support

An NLP system can transfer clinical texts to encoded information, which meets the needs of clinical decision support (CDS) (Demner-Fushman et al. 2009). For example, NLP systems can help to find patients who match certain criteria based on the information extracted from clinical texts. Jain et al. (1996) used MedLEE to encode the information in chest radiograph and mammogram reports and identified patients at risk of having tuberculosis (TB). Fiszman et al. (2000) found that an NLP system for automatic detection of acute bacterial pneumonia from chest x-ray reports performed similarly to physicians and better than lay persons and keyword searching. Day et al. (2007) have developed a daily program using the MPLUS NLP system and decision support technologies to automatically identify trauma patients. Compared with results with clinicians' judgments, the system performed well, with sensitivity of 71% and specificity of 99%.

References

Ambert, K. H., and Cohen, A. M. (2009). A system for classifying disease comorbidity status from medical discharge summaries using automated hotspot and negated concept detection. *J Am Med Inform Assoc*, 16(4), 590–595.

Arens, R. (2004). A preliminary look into the use of named entity information for bioscience text tokenization. In *Proceedings of the Student Research Workshop at HLT-NAACL*, 37–42.

Aronson, A. R. (2001). Effective mapping of biomedical text to the UMLS Metathesaurus: the MetaMap program. In *AMIA Annual Symposium Proceedings*, Washington, DC, 17–21.

Aronson, A. R., and Lang, F. M. (2010). An overview of MetaMap: historical perspective and recent advances. *J Am Med Inform Assoc*, 17(3), 229–236.

Barnickel, T., Weston, J., Collobert, R., Mewes, H.-W., and Stümpflen, V. (2009). Large scale application of neural network based semantic role labeling for automated relation extraction from biomedical texts. *PLoS One* 2009, 4(7), e6393.

Barrett, N., and Weber-Jahnke, J. (2011). Building a biomedical tokenizer using the token lattice design pattern and the adapted Viterbi algorithm. *BMC Bioinformatics*, 12(Suppl 3), S1.

Barrows, R. C., Busuioc, M., and Friedman, C. (2000). Limited parsing of notational text visit notes: ad-hoc vs. NLP approaches. In *AMIA Annual Symposium Proceedings*, Los Angeles, 51–55.

Bartolomeo, N., Trerotoli, P., and Serio, G. (2011). Progression of liver cirrhosis to HCC: an application of hidden Markov model. *BMC Med Res Methodol*, 11, 38.

Berger, A. L., Pietra, V. J. D., and Pietra, S. A. D. (1996). A maximum entropy approach to natural language processing. *Comput Linguist*, 22(1), 39–71.

Bethard, S., Lu, Z., Martin, J., and Hunter, L. (2008). Semantic role labeling for protein transport predicates. *BMC Bioinformatics*, 9(1), 277.

Cao, H., Chiang, M. F., Cimino, J. J., Friedman, C., and Hripcsak, G. (2004). Automatic summarization of patient discharge summaries to create problem lists using medical language processing. *Stud Health Technol Inform*, 107, 1540.

Cao, H., Markatou, M., Melton, G. B., Chiang, M. F., and Hripcsak, G. (2005). Mining a clinical data warehouse to discover disease-finding associations using co-occurrence statistics. In *AMIA Annual Symposium Proceedings*, Washington, DC, 106–110.

Chapman, W. W., Bridewell, W., Hanbury, P., Cooper, G. F., and Buchanan, B. G. (2001). A simple algorithm for identifying negated findings and diseases in discharge summaries. *J Biomed Inform*, 34(5), 301–310.

Chen, E. S., Hripcsak, G., Xu, H., Markatou, M., and Friedman, C. (2008). Automated acquisition of disease drug knowledge from biomedical and clinical documents: an initial study. *J Am Med Inform Assoc*, 15(1), 87–98.

Chen, H., Fuller, S. S., Friedman, C., and Hersh, W. (2005). *Medical Informatics: Knowledge Management and Data Mining in Biomedicine*. Springer, New York.

Cohen, A. (2008). Five-way smoking status classification using text hot-spot identification and error-correcting output codes. *J Am Med Inform Assoc*, 15(1), 32–35.

Dahlmeier, D., and Ng, H. T. (2010). Domain adaptation for semantic role labeling in the biomedical domain. *Bioinformatics*, 26(8), 1098–1104.

Dang, P. A., Kalra, M. K., Blake, M. A., Schultz, T. J., Halpern, E. F., and Dreyer, K. J. (2008). Extraction of recommendation features in radiology with natural language processing: exploratory study. *Am J Roentgenol*, 191(2), 313–320.

Day, S., Christensen, L. M., Dalto, J., and Haug, P. (2007). Identification of trauma patients at a level 1 trauma center utilizing natural language processing. *J Trauma Nurs*, 14(2), 79–83.

Demner-Fushman, D., Chapman, W. W., and McDonald, C. J. (2009). What can natural language processing do for clinical decision support. *J Biomed Inform*, 42(5), 760–772.

Denecke, K., and Bernauer, J. (2007). Extracting specific medical data using semantic structure. *Artif Intel Med*, 4594, 257–264.

Denny, J. C., Miller, R. A., Johnson, K. B., and Spickard, A. 3rd. (2008). Development and evaluation of a clinical note section header terminology. In *AMIA Annual Symposium Proceedings*, Washington, DC, 156–160.

Denny, J. C., Spickard, A. 3rd, Johnson, K. B., Peterson, N. B., Peterson, J. F., and Miller, R. A. (2009). Evaluation of a method to identify and categorize section headers in clinical documents. *J Am Med Inform Assoc*, 16(6), 806–815.

Doan, S., Bastarache, L., Kilmkowski, S., Denny, J. C., and Xu, H. (2010) Integrating existing natural language processing tools for medication extraction from discharge summaries. *J Am Med Inform Assoc*, 17, 528–531.

Drawid, A., Gupta, N., Nagaraj, V. H., Gelinas, C., and Sengupta, A. M. (2009). OHMM: a Hidden Markov Model accurately predicting the occupancy of a transcription factor with a self-overlapping binding motif. *BMC Bioinformatics*, 10, 208.

Elkin, P. L., Brown, S. H., Bauer, B. A., Husser, C. S., Carruth, W., Bergstrom, L. R., and Wahner-Roedler, D. L. (2005). A controlled trial of automated classification of negation from clinical notes. *BMC Med Inform Decis Mak*, 5, 13.

Fan, J.-W., Prasad, R., Yabut, R. M., Loomis, R. M., Zisook, D. S., Mattison, J. E., and Huang, Y. (2011). Part-of-speech tagging for clinical text: wall or bridge between institutions. In *AMIA Annual Symposium Proceedings*, Washington, DC, 2011, 382–391.

Fiszman, M., Chapman, W. W., Aronsky, D., Evans, R. S., and Haug, P. J. (2000). Automatic detection of acute bacterial pneumonia from chest x-ray reports. *J Am Med Inform Assoc*, 7(6), 593–604.

Fiszman, M., Rindflesch, T. C., and Kilicoglu, H. (2003). Integrating a hypernymic proposition interpreter into a semantic processor for biomedical texts. In *AMIA Annual Symposium Proceedings*, Washington, DC, 239–243.

Forrey, A. W., McDonald, C. J., DeMoor, G., Huff, S. M., Leavelle, D., Leland, D., Fiers, T., Charles, L., Griffin, B., Stalling, F., Tullis, A., Hutchins, K., and Baenziger, J. (1996). Logical Observation Identifier Names and Codes (LOINC) database: a public use set of codes and names for electronic reporting of clinical laboratory test results. *Clin Chem*, 42(1), 81–90.

Friedman, C. (1997). Towards a comprehensive medical language processing system: methods and issues. In *AMIA Annual Symposium Proceedings*, 595–599.

Friedman, C. (2000). A broad-coverage natural language processing system. In *AMIA Annual Symposium Proceedings*, Los Angeles, 270–274.

Friedman, C., Alderson, P. O., Austin, J. H., Cimino, J. J., and Johnson, S. B. (1994). A general natural-language text processor for clinical radiology. *J Am Med Inform Assoc*, 1(2), 161–174.

Friedman, C., and Hripcsak, G. (1998). Evaluating natural language processors in the clinical domain. *Methods Inf Med*, 37, 334–344.

Friedman, C., Shagina, L., Lussier, Y., and Hripcsak, G. (2004). Automated encoding of clinical documents based on natural language processing. *J Am Med Inform Assoc*, 11(5), 392–402.

Goryachev, S., Kim, H., and Zeng-Treitler, Q. (2008). Identification and extraction of family history information from clinical reports. In *AMIA Annual Symposium Proceedings*, Washington, DC, 247–251.

Goryachev, S., Sordo, M., and Zeng, Q. T. (2006). A suite of natural language processing tools developed for the I2B2 project. In *AMIA Annual Symposium Proceedings*, Washington, DC, 931.

Haerian, K., Varn, D., Vaidya, S., Ena, L., Chase, H. S., and Friedman, C. (2012). Detection of pharmacovigilance-related adverse events using electronic health records and automated methods. *Clin Pharmacol Ther*, 92(2), 228–234.

Hripcsak, G., Bakken, S., Stetson, P. D., and Patel, V. L. (2003). Mining complex clinical data for patient safety research: a framework for event discovery. *J Biomed Inform*, 36(1–2), 120–130.

Hripcsak, G., Kuperman, G. J., and Friedman, C. (1998). Extracting findings from narrative reports: software transferability and sources of physician disagreement. *Methods Inform Med*, 37, 1–7.

Huang, Y., Lowe, H. J., Klein, D., and Cucina, R. J. (2005). Improved identification of noun phrases in clinical radiology reports using a high-performance statistical natural language parser augmented with the UMLS SPECIALIST Lexicon. *J Am Med Inform Assoc*, 12(3), 275–285.

Humphreys, B. L., Lindberg, D. A., Schoolman, H. M., and Barnett, G. O. (1998). The Unified Medical Language System: an informatics research collaboration. *J Am Med Inform Assoc*, 5(1), 1–11.

Jain, N. L., Knirsch, C. A., Friedman, C., and Hripcsak, G. (1996). Identification of suspected tuberculosis patients based on natural language processing of chest radiograph reports. In *AMIA Annual Symposium Proceedings*, Washington, DC, 542–546.

Jiang, J., and Zhai, C. (2007). An empirical study of tokenization strategies for biomedical information retrieval. *Inf Retr*, 10(4–5), 341–363.

Katrenko, S., and Adriaans, P. (2007). Learning relations from biomedical corpora using dependency trees. *KDECB*. 4366, 61–80.

Kim, J., Ohta, T., Tateisi, Y., and Tsujii, J. (2003). GENIA corpus—a semantically annotated corpus for bio-text mining. *Bioinformatics*, 19(90001), 180–182.

Kogan, Y., Collier, N., Pakhomov, S., and Krauthammer, M. (2005). Towards semantic role labeling and IE in the medical literature. In *AMIA Annual Symposium Proceedings*, Washington, DC, 410–414.

Lakhani, P., and Langlotz, C. (2010). Automated detection of radiology reports that document non-routine communication of critical or significant results. *J Digit Imaging*, 23(6), 647–657.

Long, W. (2005). Extracting diagnoses from discharge summaries. In *AMIA Annual Symposium Proceedings*, 2005, Washington, DC, 470–474.

Manning, C. D., and Schütze, H. (2003). *Foundations of Statistical Natural Language Processing*. The MIT Press, Cambridge, MA.

Marcus, M., Santorini, B., and Marcinkiewicz, M. (1994). Building a large annotated corpus of English: The Penn Treebank. *Comput Linguist*, 19(2), 313–330.

McCray, A. T. (2003). An upper-level ontology for the biomedical domain. *Comp Funct Genomics*, 4(1), 80–84.

McCray, A., Aronson, A., Browne, A., Rindflesch, T., Razi, A., and Srinivasan, S. (1993). UMLS knowledge for biomedical language processing. *Bull Med Libr Assoc*, 81(2), 184–194.

McCray, A. T., Srinivasan, S., and Browne, A. C. (1994). Lexical methods for managing variation in biomedical terminologies. In *Proceedings of the Annual Symposium on Computer Applications in Medical Care*, Washington, DC, 235–239.

McDonald, C. J., Huff, S. M., Suico, J. G., Hill, G., Leavelle, D., Aller, R., Forry, A., Mercer, L., DeMoor, G., Hook, J., Williams, W., Case, J., and Maloney, P. (2003). LOINC, a universal standard for identifying laboratory observations: a 5-year update. *Clin Chem*, 49(4), 624–633.

McInnes, B. T., Pedersen, T., Liu, Y., Pakhomov, S. V., and Melton, G. B. (2011). Using second-order vectors in a knowledge-based method for acronym disambiguation. In *Proceedings of the International Conference on Computational Natural Language Learning*, Portland, OR, 145–153.

Mel'Čuk, I. A. (1988). *Dependency Syntax: Theory and Practice*. Albany, NY, State University of New York Press.

Melton, G. B., and Hripcsak, G. (2005). Automated detection of adverse events using natural language processing of discharge summaries. *J Am Med Inform Assoc*, 12(4), 448–457.

Munshaw, S., and Kepler, T. B. (2010). SoDA2: a Hidden Markov Model approach for identification of immunoglobulin rearrangements. *Bioinformatics*, 26(7), 867–872.

Mutalik, P. G., Deshpande, A., and Nadkarni, P. M. (2001). Use of general-purpose negation detection to augment concept indexing of medical documents. *J Am Med Inform Assoc*, 8(6), 598–609.

Pakhomov, S. V., Coden, A., and Chute, C. G. (2006). Developing a corpus of clinical notes manually annotated for part-of-speech. *Int J Med Inform*, 75(6), 418–429.

Pakhomov, S., Pedersen, T., and Chute, C. (2005). Abbreviation and acronym disambiguation in clinical discourse. In *AMIA Annual Symposium Proceedings*, Washington, DC, 589–593.

Pakhomov, S. V., Ruggieri, A., and Chute, C. G. (2002). Maximum entropy modeling for mining patient medication status from free text. In *AMIA Annual Symposium Proceedings*, San Antonio, TX, 587–591.

Penz, J. F., Wilcox, A. B., and Hurdle, J. F. (2007). Automated identification of adverse events related to central venous catheters. *J Biomed Inform*, 40(2), 174–182.

Rimell, L., and Clark, S. (2009). Porting a lexicalized-grammar parser to the biomedical domain. *J Biomed Inform*, 42(5), 852–865.

Rindflesch, T. C., and Aronson, A. R. (1993). Semantic processing in information retrieval. In *Proceedings of the Annual Symposium on Computer Applications in Medical Care*, Washington, DC, 611–615.

Rindflesch, T. C., and Fiszman, M. (2003). The interaction of domain knowledge and linguistic structure in natural language processing: interpreting hypernymic propositions in biomedical text. *J Biomed Inform*, 36(6), 462–477.

Saeed, M., Villarroel, M., Reisner, A. T., Clifford, G., Lehman, L. W., Moody, G., Mark, R. G. (2011). Multiparameter intelligent monitoring in intensive care II: a public-access intensive care unit database. *Crit Care Med*, 39(5), 952–960.

Sarafraz, F., and Nenadic, G. (2010). Using SVMs with the command relation features to identify negated events in biomedical literature. In *Proceedings of the Workshop on Negation and Speculation in Natural Language Processing*, Uppsala, Sweden, 78–85.

Savova, G. K., Masanz, J. J., Ogren, P. V., Zheng, J., Sohn, S., Kipper-Schuler, K. C., and Chute, C. G. (2010). Mayo clinical Text Analysis and Knowledge Extraction System (cTAKES): architecture, component evaluation and applications. *J Am Med Inform Assoc*, 17(5), 507–513.

Sipser, M. (1996). *Introduction to the Theory of Computation*. PWS Pub. Co., Boston.

Smith, L., Rindflesch, T., and Wilbur, W. J. (2004). MedPost: a part-of-speech tagger for bioMedical text. *Bioinformatics*, 20(14), 2320–2321.

Srinivasan, P., and Rindflesch, T. (2002). Exploring text mining from MEDLINE. In *AMIA Annual Symposium Proceedings*, San Antonio, TX, 722–726.

Stevenson, M., Agirre, E., and Soroa, A. (2012). Exploiting domain information for Word Sense Disambiguation of medical documents. *J Am Med Inform Assoc*, 19(2), 235–240.

Szolovits, P. (2003). Adding a medical lexicon to an English parser. In *AMIA Annual Symposium Proceedings*, Washington, DC, 2003:639–643.

Torii, M., Wagholikar, K., and Liu, H. (2011). Using machine learning for concept extraction on clinical documents from multiple data sources. *J Am Med Inform Assoc*, 18(5), 850–857.

Tsai, R. T. H., Chou, W. C., Su, Y. S., Lin, Y. C., Sung, C. L., Dai, H. T., Yeh, I. T., Ku, W., Sung, T. Y., and Hsu, W.-L. (2006). BIOSMILE: adapting semantic role labeling for biomedical verbs: an exponential model coupled with automatically generated template features. *BMC Bioinformatics*, 2007(8), 325–339.

Viterbi, A. (1967). Error bounds for convolutional codes and an asymptotically optimum decoding algorithm. *IEEE Trans Inform Theor*, 13, 260–269.

Wall, M. M., and Li, R. (2009). Multiple indicator hidden Markov model with an application to medical utilization data. *Stat Med*, 28(2), 293–310.

Wang, X., Hripcsak, G., Markatou, M., and Carol, F. (2009). Active computerized pharmacovigilance using natural language processing, statistics, and electronic health records: a feasibility study. *J AM Med Inform Assoc*, 16(3), 328–337.

Watkins, R. E., Eagleson, S., Veenendaal, B., Wright, G., and Plant, A. J. (2009). Disease surveillance using a hidden Markov model. *BMC Med Inform Decis Mak*, 9, 39.

Weeber, M., Weeber, M., Mork, J. G., and Aronson, A. R. (2001). Developing a test collection for biomedical Word Sense Disambiguation. In *AMIA Annual Symposium Proceedings*, Washington, DC, 746–750.

Wrenn, J. O., Stetson, P. D., and Johnson, S. B. (2007). An unsupervised machine learning approach to segmentation of clinician-entered free text. In *AMIA Annual Symposium Proceedings*, Chicago, 811–815.

Xu, H., Anderson, K., Grann, V. R., and Friedman, C. (2004). Facilitating cancer research using natural language processing of pathology reports. *Stud Health Technol Inform*, 107, 565–572.

Xu, H., Fan, J.-W., Hripcsak, G., Mendonça, E. A., Markatou, M., and Friedman, C. (2007). Gene symbol disambiguation using knowledge-based profiles. *Bioinformatics*, 23(8), 1015–1022.

Xu, H., Stenner, S. P., Doan, S., Johnson, K. B., Waitman, L. R., and Denny, J. C. (2010). MedEx: a medication information extraction system for clinical narratives. *J Am Med Inform Assoc*, 17(1), 19–24.

Yang, H., Spasic, I., Keane, J. A., and Nenadic, G. (2009). A text mining approach to the prediction of disease status from clinical discharge summaries. *J Am Med Inform Assoc*, 16(4), 596–600.

Zeng, Q. T., Goryachev, S., Weiss, S., Sordo, M., Murphy, S. N., and Lazarus, R. (2006). Extracting principal diagnosis, co-morbidity and smoking status for asthma research: evaluation of a natural language processing system. *BMC Med Inform Decis Mak*, 6, 30.

Zhang, R., Pakhomov, S., and Melton, G. B. (2012). Automated identification of relevant new information in clinical narrative. In *IHI'12 2nd ACM SIGHIT International Health Informatics Symposium*, Miami, FL, 837–841.

24

Intelligent Personal Health Record

Gang Luo, Selena B. Thomas, and Chunqiang Tang

CONTENTS

24.1 Introduction

As a result of the deployment of several major Internet companies including Microsoft [1], WebMD [2], and Office Ally [3] over the past few years, Web-based personal health records (PHRs) have now become widely available to ordinary consumers. These PHR systems enable consumers to actively manage their health records and, subsequently, their health through a Web interface but have limited intelligence and can fulfill only a small portion of users' health care needs. To improve PHR's capability and usability, we previously proposed the concept of an intelligent PHR (iPHR) [4–6] by introducing and extending expert system technology, Web search technology, natural language generation technology, database trigger technology, and signal processing technology into the PHR domain.

iPHR serves as a centralized portal for automatically providing users with comprehensive and personalized health care information to facilitate their activities of daily living. Due to a lack of health knowledge, consumers often are unaware of their health care needs and/or unable to identify proper keywords to search health care information [7]. To address this problem, iPHR extensively uses health knowledge to (1) anticipate users' needs, (2) guide users to provide the most important information about their health condition, (3) automatically form queries, and (4) proactively push relevant health care information to users whenever their potential need for it is detected. In this chapter, we use the term "health knowledge" to refer to all categories or types of knowledge related to health care, for example, disease diagnosis knowledge and nursing knowledge.

iPHR provides its intelligent functions to users via a Web interface. On the right side of the main Web page of iPHR, there are multiple buttons, one for each intelligent function. After clicking a button, the user is directed to a Web page for executing the corresponding

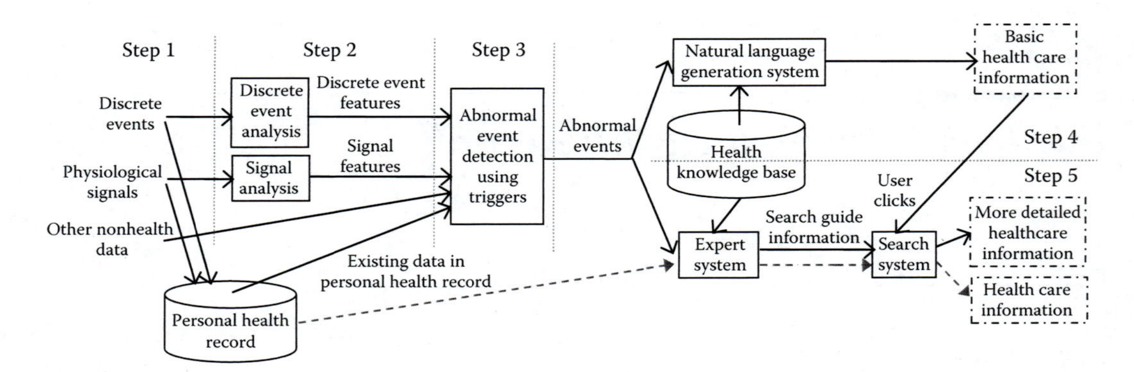

FIGURE 24.1
Architecture of the intelligent personal health record system.

function. At present, our iPHR system provides four functions covering almost 1000 health issues:

1. Guided search for disease information
2. Recommending self-care activities (SCAs)
3. Recommending home health products (HHPs)
4. Continuous user monitoring

Each of the first three functions can be implemented as a standalone health search engine outside of a PHR system. However, with the support of a PHR system, these three functions can be performed more effectively. In comparison, the fourth function of continuous user monitoring depends on the PHR system.

As shown in Figure 24.1, in addition to a standard PHR system, iPHR has a number of other components including a discrete event analysis system, a signal processing system, a trigger system, a natural language generation system, a health knowledge base, an expert system, and a search system. At a high level, the dashed arrow path near the bottom of Figure 24.1 illustrates the work flow of the first three functions of iPHR. The expert system uses health knowledge to convert information in the PHR into a set of keyword phrases termed "search guide information," which reflects the user's health condition and health care needs. Then the Web search engine uses this search guide information as seeds to retrieve personalized health care information. The solid arrow paths in Figure 24.1 illustrate the work flow of the fourth function of continuous user monitoring. The description of that work flow is provided in Section 24.5.

In the rest of this chapter, we present an overview of the four functions of iPHR one by one. Interested readers can find the details of our design rationale and implementation techniques of these four functions in our previous publications [4–6,8–16].

24.2 Guided Search for Disease Information

iPHR includes our intelligent medical Web search engine called iMed [8–10,13,14] to help users find disease information related to their health condition. The main idea of iMed is to use disease diagnosis knowledge and an interactive questionnaire to guide users to provide the most important information about their health condition and to automatically form queries. This eliminates the challenge for users to come up with appropriate medical keyword queries on their own.

iMed uses diagnostic decision trees written by medical professionals [17] as its built-in disease diagnosis knowledge. As shown in Figure 24.2, each diagnostic decision tree corresponds to either an objective sign (e.g., low blood pressure) or a subjective symptom (e.g., headache). In a diagnostic decision tree, each node that is neither a leaf node nor the parent of a leaf node represents a question that iMed can ask. Different child nodes of this node correspond to different answers to this question. The medical phrases in a leaf node are the topics (typically diseases) potentially relevant to the user's health condition.

At a high level, iMed works in the following way. The user is first presented with a list of signs and symptoms, from which he/she selects the ones that he/she is currently having. Then iMed asks questions related to these selected signs and symptoms and lists possible answers to these questions. Based on the answers selected by the user, iMed navigates the corresponding diagnostic decision trees and eventually reaches multiple topics potentially relevant to the user's health condition. For each of these topics, iMed automatically uses the topic name to form a query to retrieve some related Web pages. Moreover, iMed presents a set of predetermined aspects (e.g., symptom, diagnosis, treatment, and risk factor). If the user clicks a particular aspect of the topic, iMed automatically forms a query by combining the aspect name and the topic name and uses this query to retrieve Web pages related to this aspect of the topic. In this way, without the need to form any medical keyword query by himself/herself, the user can find disease information that is potentially related to his/her health condition.

For example, Figure 24.2 shows the diagnostic decision tree in Collins [17] for the symptom "face pain." If "face pain" is the only symptom chosen by the user, iMed's first question is "How often do you have facial pain?" If the user selects the answer "occasionally, from time to time" to this question, iMed's next question is "Is your facial pain increased by chewing?" If the user selects the answer "yes" to the second question, iMed reaches multiple topics including dental caries.

A typical user has little health knowledge and frequently encounters challenges during the entire disease information search process. To address this problem, iMed offers

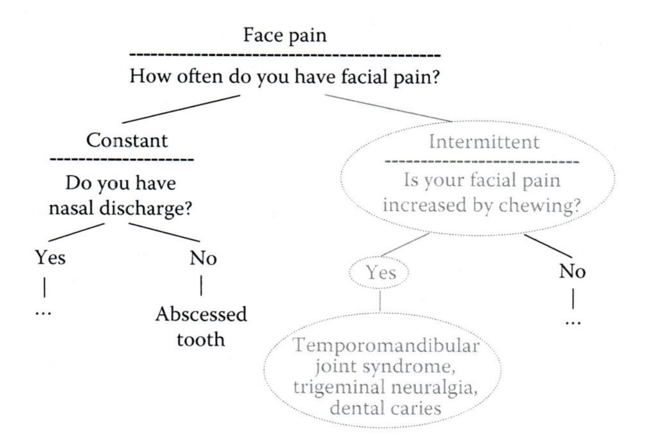

FIGURE 24.2
The diagnostic decision tree for the symptom "face pain."

various kinds of suggestions to provide help. If the user suspects that he answered questions incorrectly, he can access alternative answers to the questions suggested by iMed. To facilitate the quick digestion of search results by the user and to refine his/her inputs, iMed suggests diversified and related medical phrases. iMed also suggests signs and symptoms related to the user's health condition.

24.3 Recommending SCAs

iPHR can automatically recommend SCAs based on the user's health issues [5,15]. The user can click each nontrivial SCA to find various detailed implementation procedures for it on the Web. The main idea of this SCA recommendation function is to use nursing knowledge presented in standardized nursing languages [18].

The nursing informatics community has systematically organized nursing knowledge into multiple standardized nursing languages [18]. At present, iPHR's knowledge base includes two such standardized nursing languages covering the entire nursing domain: the North American Nursing Diagnosis Association International (NANDA-I) nursing diagnoses and the Nursing Interventions Classification (NIC) nursing interventions. A *NANDA-I nursing diagnosis* is a clinical judgment about individual, family, or community responses to actual or potential health problems [19]. A *NIC nursing intervention* is a treatment that can be performed to enhance patient/client outcomes [20].

As shown in Figure 24.3, each health issue links to one or more NANDA-I nursing diagnoses [19]. Every nursing diagnosis usually links to 10 or more NIC nursing interventions [18]. Each nursing intervention includes multiple *care activities* that are used to implement it [20]. In this way, each health issue is connected to multiple care activities via the linkage provided by nursing diagnoses and nursing interventions. Nurses, patients, and/or caregivers can perform these care activities to achieve desirable outcomes for this health issue. For iPHR, we focus on SCAs that patients and caregivers can perform at home or in the community because iPHR is designed to be used by consumers.

At a high level, iPHR's SCA recommendation function works in the following way. iPHR automatically extracts from PHR the user's current health issues (e.g., diseases), uses the linkage method mentioned above to find all of his/her linked SCAs, and then displays

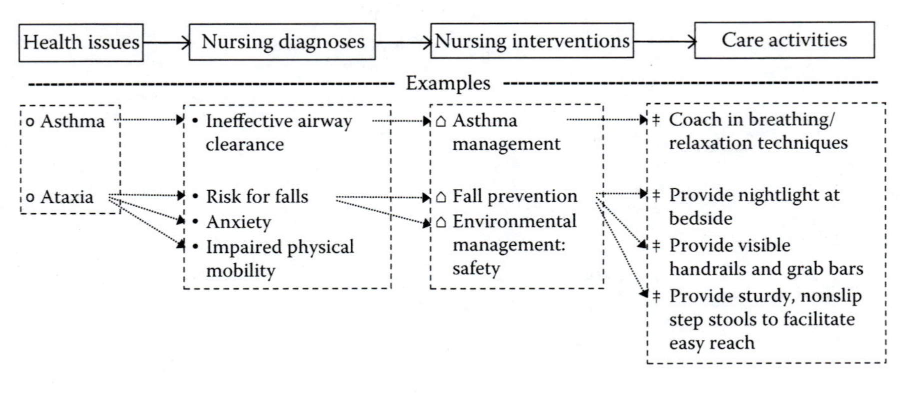

FIGURE 24.3
Linking health issues to care activities.

these SCAs as a prioritized hierarchy. An SCA can have one or more aspects. For each aspect of a nontrivial SCA, a hyperlink is added to the displayed Web page. Also, a precompiled phrase is stored in iPHR's knowledge base as the SCA search guide information of this aspect. If the user clicks this hyperlink, iPHR will submit this phrase as a query to a large-scale health Web search engine. Then by reading the search results returned by this search engine, the user can find various, detailed implementation procedures for this aspect.

For example, as shown in Figure 24.3, the health issue "asthma" links to the SCA "coach in breathing/relaxation techniques." For the breathing aspect of this SCA, the precompiled phrase "asthma breathing techniques" retrieves the following top results: (1) the Buteyko method for breathing (http://www.correctbreathing.com/), (2) two new breathing exercises for asthma (http://www.sciencedaily.com/releases/2008/05/080528095853.htm), (3) a video teaching the pranayama breathing method for asthma (http://www.youtube.com/watch?v=vplrJtp3zB4), and (4) the book *Reversing Asthma: Breathe Easier with this Revolutionary New Program*, with a chapter on teaching breathing techniques (http://www.amazon.com/Reversing-Asthma-Breathe-Revolutionary-Program/dp/0446673633).

It is not uncommon for a person to have multiple health issues simultaneously (e.g., comorbidities). In this case, an SCA that is suitable for a single health issue can become undesirable in the presence of another health issue. This is called contraindication in health care [21]. For instance, the health issue cancer is a contraindication for the SCA massage because massage increases lymphatic circulation and hence may potentially facilitate the spread of cancer through the lymphatic system. In the case where the user has multiple health issues simultaneously, iPHR uses a hierarchical propagation method based on the medical terminology of the International Classification of Diseases (ICD-10) [22] to automatically detect contraindicated SCAs so that they will not be recommended to the user [5].

24.4 Recommending HHPs

iPHR can automatically recommend HHPs based on the user's health issues [11,12,16]. The main idea of this HHP recommendation function is to use both nursing knowledge and treatment knowledge and to extend the language modeling method [23] in information retrieval to combine and rank HHPs retrieved by multiple queries. During this process, various relevant factors are taken into account.

iPHR uses both nursing knowledge and treatment knowledge to obtain HHP search guide information. For each SCA, a set of phrases is precompiled as its HHP search guide information and stored in iPHR's knowledge base. Each such phrase provides one way of retrieving HHPs related to this SCA. For each health issue (e.g., disease, symptom, surgery), a set of HHP search guide phrases is precompiled using disease/symptom treatment knowledge and stored in iPHR's knowledge base. These treatment-based HHP search guide phrases can bridge the semantic gap between the literal meaning and the underlying medical meaning of the health issue.

At a high level, iPHR's HHP recommendation function works in the following way. iPHR automatically extracts from the PHR the user's current health issues and uses the linkage method described in Section 24.3 to find all of his/her linked SCAs. Combining together the HHP search guide information for these SCAs and the treatment-based HHP search guide phrases precompiled for these health issues, we obtain the complete set of

< Medical supplies and equipment
  < Daily living aids
    * Bath and body aids (797)
    * Low vision aids (633)
    * Medication aids (178)
    * Ramps (157)
    * Low strength aids (137)
    * Eating and drinking aids (77)
    * Dressing aids (42)
    * Hearing aids (28)
    * Telephones (27)
    * Hearing aid accessories (12)
    * Others (46)

FIGURE 24.4
A sample navigation hierarchy constructed for the health issue ataxia.

search guide information. iPHR submits each search guide phrase in this set as a query to a vertical search engine to retrieve relevant HHPs. Then all retrieved HHPs are combined together and returned to the user through a navigation interface. On each search result Web page, a navigation hierarchy based on product categories is displayed on the left side, as shown in Figure 24.4. Recommended HHPs are displayed sequentially on the right side.

For example, as shown in Figure 24.3, the health issue ataxia (cannot coordinate muscle movement) links to the following SCAs:

1. Provide sturdy, nonslip *step stools* to facilitate easy reach.
2. Provide *nightlight* at bedside.
3. Provide visible *handrails* and *grab bars*.

Step stools, nightlights, handrails, and grab bars are HHPs relevant to ataxia. iPHR can recommend these HHPs to ataxia patients using nursing knowledge. Nevertheless, for both ataxia and its symptoms, it is likely that neither their names nor their treatment methods appear in the Web pages describing these HHPs. Consequently, without resorting to iPHR, the average consumer would encounter difficulty in finding these HHPs on his/her own.

24.5 Continuous User Monitoring

iPHR can perform continuous user monitoring and proactively push personalized, relevant health care information to users whenever their potential need for it is detected [4]. The main idea of this continuous user monitoring function is to combine techniques from multiple computing areas, including expert systems, Web search, natural language generation, database triggers, and signal processing, to make iPHR active.

More specifically, triggers are precompiled by health care professionals and stored in iPHR's knowledge base. The concept of triggers was originally developed in the database field [24] and is extended here to fit the purpose of our specific iPHR application. Each trigger corresponds to a unique abnormal event that may have a potential health impact. Based on the user's health condition, iPHR automatically determines which triggers will be used. iPHR keeps collecting, processing, and analyzing the user's health data from various sources and detecting abnormal events from it. Whenever a trigger fires, signaling the occurrence of an abnormal event, iPHR recognizes that the user needs to be aware of the related, personalized health care information and automatically pushes this information to the user.

At trigger compilation time, for each abnormal event, health care professionals use their health knowledge to perform the following four actions:

1. Compile the corresponding trigger
2. Compile a template of basic health care information related to this abnormal event
3. For the content included in this template, mark one or more items that they anticipate some iPHR users would want to know more about
4. For each such marked item, compile a set of phrases that can be used to retrieve its detailed information as its search guide information

As described below, the continuous user monitoring function uses all of the materials compiled from these four actions.

The solid arrow paths in Figure 24.1 illustrate the overall work flow of continuous user monitoring in iPHR. At a high level, this work flow consists of the following five steps.

In step 1, the user's health data are collected from multiple sources, such as wearable sensors and passive sensors [25]. These data include both discrete events (e.g., sporadically measured body weight) and continuous time series physiological signals (e.g., electrocardiogram).

In step 2, the user's health data are preprocessed to filter out artifacts. Their essential information is then extracted as various kinds of features.

In step 3, abnormal events are detected from the user's health data using triggers. In iPHR, each trigger corresponds to an abnormal event $E_{abnormal}$ and is of the form

$$\text{firing condition } C_f \rightarrow \text{action } A \text{ (triggering event } E_{triggering}, \text{ applicable condition } C_a\text{)},$$

meaning that if the applicable condition C_a applies to the user, then the action A will be taken when the triggering event $E_{triggering}$ occurs and the features extracted from the user's health data satisfy the firing condition C_f describing the abnormal event $E_{abnormal}$.

In step 4, when one or more abnormal events are detected from the user's health data and their corresponding triggers fire, they are collected together and sent to the natural language generation [26] system of iPHR. Based on how the firing conditions of their corresponding triggers are satisfied, iPHR uses their precompiled templates of basic health care information to generate basic personalized health care information related to them and presents this information to the user on a Web page. Moreover, for each item in this basic health care information that health care professionals mark during trigger compilation time, iPHR automatically adds a hyperlink to this Web page.

In step 5, the user views this Web page. If he/she is interested in knowing more details about a specific item in this basic health care information, he/she can click the hyperlink of this item. In this case, iPHR will use the search guide information precompiled for this item to automatically form one or more queries to retrieve detailed information about this item and then present this retrieved information to the user.

For example, chronic obstructive pulmonary disease (COPD) patients often experience weight loss, an abnormal event that is associated with increased risks of mortality, disability, and handicap. The defining criteria of weight loss are losing >5% weight in the past month or losing >10% weight in the past 6 months [27]. For a user with COPD, iPHR

Significant weight loss is detected, as you have lost >5% of your weight in the past month. This is particularly problematic as you also have COPD.

COPD patients often experience weight loss, which is associated with increased risks of mortality, disability, and handicap.

COPD patients experiencing weight loss may need nutritional therapy. (Click here to view related food and nutritional supplements.) Since weight loss in COPD patients is often accompanied by muscle wasting, nutritional therapy may only be effective if it is combined with anabolic stimuli such as exercise.

FIGURE 24.5
An example of basic personalized health care information provided by iPHR.

will automatically monitor his/her body weight measures. When iPHR detects that he/she has lost >5% of his/her weight in the past month, iPHR will present to him/her the basic personalized health care information shown in Figure 24.5.

For the "nutritional therapy" item in this basic health care information, the precompiled search guide information contains three phrases: *COPD nutritional therapy*, *COPD nutritional supplement*, and *COPD nutrition*. If the user clicks this item, iPHR will present the following top results retrieved by these phrases as detailed information about this item: (1) nutritional guidelines for people with COPD (http://my.clevelandclinic.org/disorders/chronic_obstructive_pulmonary_disease_copd/hic_nutritional_guidelines_for_people_with_copd.aspx), (2) nutrition and COPD—dietary considerations for better breathing (http://www.todaysdietitian.com/newarchives/td_020909p54.shtml), and (3) nutritional therapy for COPD (http://www.lef.org/protocols/respiratory/copd_01.htm).

24.6 Conclusions

iPHR is a new and rapidly moving field. This chapter presents an overview of our iPHR system. As described in the work of Luo et al. [5], there are many open issues in iPHR, and much research work is needed to address them to a satisfactory degree. To improve the existing functions of iPHR as well as to add new functions into iPHR, we expect that more computer science technology will be introduced into the PHR domain and more health knowledge will be incorporated into iPHR in the near future. Moreover, we expect that many techniques originally developed for iPHR, possibly after certain domain-specific extensions, could be applied to other domains for the purpose of using domain knowledge to facilitate the finding of desired information by users.

References

1. Microsoft HealthVault homepage. http://www.healthvault.com, 2012.
2. WebMD personal health record homepage. http://www.webmd.com/phr, 2012.
3. Office Ally personal health record homepage. https://www.patientally.com/Main, 2012.
4. Luo, G., Triggers and monitoring in intelligent personal health record. *Journal of Medical Systems (JMS)*, Vol. 36, No. 5, Oct. 2012, pp. 2993–3009.
5. Luo, G., Tang, C., and Thomas, S.B., Intelligent personal health record: experience and open issues. *Journal of Medical Systems (JMS)*, Vol. 36, No. 4, Aug. 2012, pp. 2111–2128.

6. Luo, G., Thomas, S.B., and Tang, C., Intelligent consumer-centric electronic medical record. In *Proceedings of the 2009 International* Conference of the European Federation for Medical Informatics *(MIE'09)*, Sarajevo, Bosnia and Herzegovina, Sep. 2009, pp. 120–124.
7. Luo, G., Tang, C., Yang, H., and Wei, X., MedSearch: a specialized search engine for medical information retrieval. In *Proceedings of the 2008 ACM Conference on Information and Knowledge Management (CIKM'08)*, Napa Valley, CA, Oct. 2008, pp. 143–152.
8. Luo, G., Intelligent output interface for intelligent medical search engine. In *Proceedings of the 2008 AAAI Conference on Artificial Intelligence (AAAI'08)*, Chicago, July 2008, pp. 1201–1206.
9. Luo, G., Design and evaluation of the iMed intelligent medical search engine. In *Proceedings of the 2009 International Conference on Data Engineering (ICDE'09)*, Shanghai, China, Apr. 2009, pp. 1379–1390.
10. Luo, G., Lessons learned from building the iMed intelligent medical search engine. In *Proceedings of the 2009 Annual International Conference of the IEEE Engineering in Medicine and Biology Society (EMBC'09)*, Minneapolis, MN, Sep. 2009, pp. 5138–5142.
11. Luo, G., On search guide phrase compilation for recommending home medical products. In *Proceedings of the 2010 Annual International Conference of the IEEE Engineering in Medicine and Biology Society (EMBC'10)*, Buenos Aires, Argentina, Sep. 2010, pp. 2167–2171.
12. Luo, G., Navigation interface for recommending home medical products. *Journal of Medical Systems (JMS)*, Vol. 36, No. 2, Apr. 2012, pp. 699–705.
13. Luo G. and Tang, C., Challenging issues in iterative intelligent medical search. In *Proceedings of the 2008 International Conference on Pattern Recognition (ICPR'08)*, Tampa, FL, Dec. 2008, pp. 1–4.
14. Luo, G. and Tang, C., On Iterative intelligent medical search. In *Proceedings of the 2008 International ACM SIGIR Conference on Research and Development in Information Retrieval (SIGIR'08)*, Singapore, July 2008, pp. 3–10.
15. Luo, G. and Tang, C., Automatic home nursing activity recommendation. In *Proceedings of the 2009 American Medical Informatics Association Annual Symposium (AMIA'09)*, San Francisco, Nov. 2009, pp. 401–405.
16. Luo, G., Thomas, S.B., and Tang, C., Automatic home medical product recommendation. *Journal of Medical Systems (JMS)*, Vol. 36, No. 2, Apr. 2012, pp. 383–398.
17. Collins, R.D., *Algorithmic Diagnosis of Symptoms and Signs: Cost-Effective Approach*, 2nd ed. Lippincott Williams & Wilkins, Philadelphia, PA, 2002.
18. Johnson, M., Moorhead, S., Bulechek, G.M., Dochterman, J.M., Butcher, H.K., Maas, M.L., and Swanson, E., *NOC and NIC Linkages to NANDA-I and Clinical Conditions: Supporting Critical Reasoning and Quality Care (NANDA, NOC, and NIC Linkages)*, 3rd ed. Mosby, Maryland Heights, MO, 2011.
19. Ackley, B.J. and Ladwig, G.B., *Nursing Diagnosis Handbook: An Evidence-Based Guide to Planning Care*, 9th ed. Mosby, Maryland Heights, MO, 2010.
20. Bulechek, G.M., Butcher, H.K., and Dochterman, J.M., *Nursing Interventions Classification (NIC)*, 5th ed. Mosby, St. Louis, MO, 2007.
21. Batavia, M., *Contraindications in Physical Rehabilitation: Doing No Harm*. Saunders, St. Louis, MO, 2006.
22. International Classification of Diseases (ICD-10) homepage. http://www.who.int/classifications/icd/en/, 2012.
23. Ponte, J.M. and Croft, B.W., A language modeling approach to information retrieval. In *Proceedings of the 1998 International ACM SIGIR Conference on Research and Development in Information Retrieval (SIGIR'98)*, Melbourne, Australia, Aug. 1998, pp. 275–281.
24. Paton, N.W. and Díaz, O., Active database systems. *ACM Computing Surveys*, Vol. 31, No. 1, 1999, 63–103.
25. Skubic, M., Alexander, G., Popescu, M., Rantz, M., and Keller, J., A smart home application to eldercare: current status and lessons learned. *Technology and Health Care* Vol. 17, No. 3, 2009, 183–201.
26. Reiter, E. and Dale, R., *Building Natural Language Generation Systems*. Cambridge University Press, Cambridge, UK, 2000.
27. Celli, B.R., MacNee, W., and ATS/ERS Task Force, Standards for the diagnosis and treatment of patients with COPD: a summary of the ATS/ERS position paper. *European Respiratory Journal* Vol. 23, No. 6, 2004, 932–946.

25

Application of Artificial Intelligence in Minimally Invasive Surgery and Artificial Palpation

Siamak Najarian and Pedram Pahlavan

CONTENTS

25.1 Introduction

The last century was definitely a milestone in improvements in the art and science of surgery. Introduction of different surgical and antiseptic techniques, as well as various post-operational surveillance improvements, has resulted in a substantial decrease in death rates. However, the core task of "surgery," that is, "cutting and sewing," remained basically unchanged; surgeons were still using basic hand tools and instruments, and the only way to access the surgery site was direct contact and visualization of the organ or tissue.

Nevertheless, during the last decades, a paradigm shift has occurred in the methods of performing surgery, especially in access to the surgical site. Reducing the "invasiveness" of the access approach to organs or tissues has resulted in superior outcomes manifested as improved survival, fewer complications, and quicker return to functional health and a productive life [1]. This focus on less or "minimally" invasive approaches has increased dramatically and has been the subject of a tremendous number of investigations in recent years.

In the following sections, we will first have a brief look at minimally invasive surgery (MIS) and then explore the ways in which this class of surgery has benefited from artificial intelligence.

25.2 Minimally Invasive Surgeries versus Open Surgeries

Open surgery is the traditional form of performing a surgery where large incisions (15 to 30 cm) are required to insert the instruments. These large incisions provide the surgeon with a large, direct visual field; direct tactile contact with the surgical field; and significant freedom of motion.

In this type of surgery, surgeons heavily rely on their sense of touch; guidance to manipulate instruments and analysis of a wide variety of anatomical structures and pathologies are achieved through the sensations perceived from their fingertips. Moreover, it is mainly by means of this sense that surgeons can distinguish between different tissues and organs during a surgery; this way, they can determine how much force they need to exert with their hands and how to avoid injuring nontargeted tissues.

However, despite all the benefits this type of surgery provides for the surgeons, patients suffer from significant blood loss, long hospital stays, long recovery times, and greater probability of infections [2].

To reduce the disadvantages associated with open surgeries, a new revolutionary surgical technique known as MIS has been introduced. During this type of surgery, the operation is performed through small incisions or ports (about 1 cm) in the patient's body to minimize the trauma resulting from open surgeries' large incisions. Through these ports, cannulas are introduced, permitting the insertion of long surgical instruments (including camera, light source, graspers, scissors, etc.) into the patient's body.

It is well known that MIS has many advantages for the patient, such as less postoperative pain, reduced morbidity, shorter hospitalization, decreased rates of infection, better cosmetic results, earlier return to normal activity, and overall reduction in costs. Performing MIS, however, coerces the surgeon to acquire unique psychomotor skills that are different from those needed to perform conventional open surgical procedures [3]. The main reasons for these shifts are that tactile feedback is totally diminished in comparison to open surgery, and the visual feedback available to the surgeon is greatly changed from a conventional 3-D operating field to a 2-D monitor display [4–8].

To solve these drawbacks, a considerable number of research activities are being undertaken, which have focused on developing and using tactile sensory systems to provide the surgeon with tactile feedback information received from the surgical site via the MIS instruments [2]. These tactile sensors try to simulate the same tactile stimulus on the surgeons' fingertips that they would receive if they could touch the tissue or organ directly. These remote palpation devices, generally, have three basic parts: a *tactile sensor* to receive the tactile data through contact, a *tactile data processing unit* to processes the transduced data and extract the useable information from the obtained tactile data, and finally, a *tactile display interface* to present the processed data to the surgeon in a perceivable way [9]. What artificial intelligence does is to enhance the function of the second part of a remote palpation unit, the data processing unit.

Among the different artificial intelligence approaches, artificial neural network is probably the most well-known technique. It is a computational model that tries to mimic the function of the brain. And like the brain, which consists of huge number of neurons, an artificial neuron is the fundamental element of an artificial neural network. Each artificial neuron is a simple mathematical model in the form of $y = f\left(\sum_{i=1}^{n} w_i x_i\right)$ where x_i is the input; w_i is the weight of each input identifying its value; f is the activation function, supposed to produce an output according to the inputs; and y is the output. A schematic of an artificial neuron is depicted in Figure 25.1.

In an artificial neural network, many artificial neurons and a huge number of interconnections between them are configured by the network *topology* or *architecture*. There are a number of different topologies in an artificial neural network, but the most common configuration is the layered network approach. In a layered neural network, neurons are organized in three different types of layers: input, output, and middle layers. The input layer

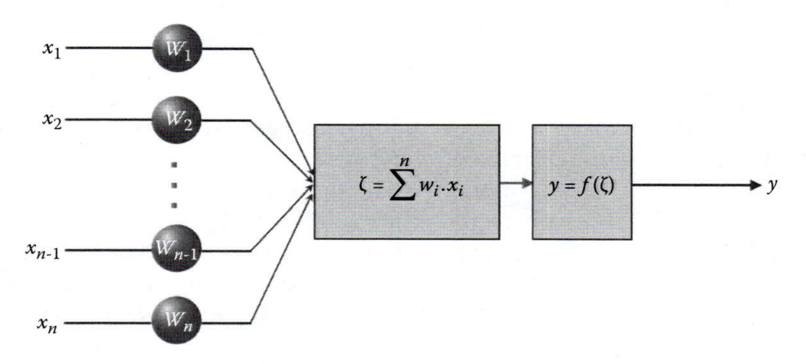

FIGURE 25.1
Schematic of an artificial neuron.

is the layer whose neurons receive the input information. A neural network is allowed to have only one input layer, and each neuron in the input layer is allowed to receive only one input value. The last layer, whose neurons produce the output data, is called the output layer. Like the input layer, each neuron in the output layer is responsible for just one single output value. The layers between these two layers (if they exist) are the middle layers, or hidden layers. This is because they have no connection with the external world and cannot directly interface between the user and the input/outputs.

Like the brain, it is possible for an artificial neuron to change its output (behavior) according to a specific input (environment). Since changing the input and, usually, the activation function is not possible, the weights of the input/output need to be changed for acquiring a desired output with an input. When we are working with one single artificial neuron, finding an appropriate weight value in a way that we can obtain the desired output according to a piece of input information is quite easy. But when we have an artificial neural network consisting of thousands of neurons, it is not usually possible to find all the necessary weights by hand. However, some algorithms have been developed that are able to adjust the weights of an artificial neural network according to the given input and the desired output. This process of adjusting the weights is called *learning* or *training.*

In a neural network, there are two general forms of learning: supervised, in which the network is guided by a "teacher," and unsupervised, in which no external teacher is present. There is also a third, less popular type of learning, called reinforced learning, in which the teacher is present but does not guide the network actively. These algorithms themselves are subcategorized according to their governing rule.

In *supervised learning,* the training process is observed by an external agent (supervisor, teacher). The supervisor compares the output of the network with the desired one and determines the differences between these two. The objective in training is to decrease this difference as much as possible.

In *unsupervised learning,* no desired output is provided for the network. The neural network is supposed to discover the structural features of the input data and adapt itself to them.

In *reinforced learning,* the desired or expected output is not given to the network, but the network is informed of whether its obtained output is correct or incorrect. It will receive either a reward or a penalty according to its computed output.

Now let us consider a tactile sensory system used for breast mass detection as an example of a minimally invasive, palpation-based medical assessment system to explain the theoretical basis of how an artificial neural network could enhance the interpretation of tactile information.

Breast cancer is one of the leading causes of death; as a result, early detection of the disease can save thousands of lives each year. Clinical breast examination (CBE), mammography, magnetic resonance imaging (MRI), and ultrasound imaging (sonography) are the most well-known breast screening modalities being used today [10]. CBE is the primary screening method, in which the physician palpates different points on the breast in order to detect any abnormality [11]; however it suffers from low sensitivity and accuracy of the physician's sense of touch in finding small, deep lesions. Mammography is probably the best screening test currently being applied; however, it has a false positive rate of over 49% in mass detection, and it is not very sensitive for dense breasts. MRI is an important tool in detecting and screening breast cancers and is based on the principles of the nuclear magnetic resonance (NMR) technique. But breast MRI is a long and costly approach, and it is difficult to differentiate between benign and cancerous lesions [12]. Ultrasound sonography uses sound waves directed as a beam and reflected by objects; however, it is not very sensitive to calcifications of small size in the tissue or to breasts of low density.

Moreover, there are some major arguments against localization of breast lesions by imaging techniques, including the inability of each technique to detect calcifications without the help of other imaging modalities [13], the requirement for a well-trained and experienced operator, difficulties in imaging areas located deep inside the breast tissue, and the existence of considerable differences between the patient's breast position during imaging procedures and the operational situation [14]. These drawbacks make these techniques strongly operator dependent [15].

As mentioned above, artificial tactile sensing is a novel, noninvasive technique that mimics CBE and, while enhancing sensitivity and specificity for detection and visualization of breast abnormalities, has no harmful effects of x-ray accumulation in the body or biopsy invasiveness [16,17]. Artificial tactile sensory systems are based on reconstructing the internal structure of soft tissues using the data obtained by a pressure sensor array pressed against the examined site [18,19]. As a result, it is essential to have reference mechanical data for the tissue being palpated. If we consider the breast tissue from outside to inside as layers of skin, fat, gland, and mass, it has been shown that each layer has an elastic modulus (E) of 12, 4.8, 19.2, and 57.6 kPa, respectively, with an overall Poisson ratio of $\nu_{\text{skin-fat-gland}} = 0.49$ for the three most topside layers and $\nu_{\mathbf{mass}} = 0.3$ for the innermost one [20,21]. A tactile sensory system, to be able to detect any mass calcifications inside the breast tissue, needs to palpate different points on the tissue, calculate different parameters according to the reference mechanical properties, and decide whether there are any abnormalities. These mechanical parameters include normal stress, normal strain, hysteresis, and some innovative stiffness-related measurements, including the following [22]:

- Start point stiffness (SPS), defined as the slope of the loading curve at the start point. This is a valuable parameter in detecting masses located near the surface of the tissue (Equation 25.1).

$$SPS_i = \left(\frac{\partial F_i}{\partial z}\right)\Bigg|_{z=0} \tag{25.1}$$

where F_i is the force–displacement curve of loading and unloading processes for each testing point and has a general form of $F = ae(bz)$.

- Average stiffness (AvS), defined as the average stiffness during compression (Equation 25.2). This parameter improves the detection capability of a deeply located mass, which does not have a noticeable effect on the surface of the tissue.

$$AvS_i = \frac{\sum_{j=1}^{n} \left(\frac{\partial F_i}{\partial z} \right)_j}{n}$$
$$AvS_{tot} = \frac{\sum_{i=1}^{N} AvS_i}{N} \tag{25.2}$$

where n is the number of data acquisitions for each test point during the course of compression, and N is the number of test points on each breast of each patient.

- Normalized stiffness (normS), defined as the ratio of the SPS to the average stiffness (Equation 25.3). This parameter reduces the effects of the anatomical differences in different parts of an individual breast and is the one that would be used to compare mechanical parameters of different regions, such as medial and lateral parts.

$$normS_i = \frac{AvS_i}{AvS_{tot}} \tag{25.3}$$

The three major states for this ratio can be considered as follows [20]:

- Ratio ≈ 1: The mechanical properties of the point are very similar to other regions of the breast, usually suggestive of normal tissue.
- Ratio << 1: The point includes much softer tissue than the normal texture of the breast, usually happening in the lateral parts of the breast or in the existence of a fatty mass.
- Ratio >> 1: This case is suggestive of much stiffer tissue than the normal texture of the breast.

In Table 25.1, these stiffness parameters are shown for some sample points. It should be mentioned that in these data, "o'clock" shows the angle of the point with respect to the nipple. By combination of above-mentioned three parameters, detection of an abnormal mass would be possible.

Now that a lesion has been detected, the next step is to localize it more exactly, since identifying the position of the mass abnormality as deep or superficial could give important data about cancerous or benign calcifications. To identify the probability of a lesion depth, logistic regression can be used. It is a generalized linear model used for binomial regression that uses fitting depth estimation data of the study population to a logistic function and has a form of

$$f(z) = \frac{e^z}{1+e^z} \quad \text{where} \quad z = \beta_0 + \beta_1 x_1 + \beta_2 x_2 + \ldots + \beta_5 x_5 \tag{25.4}$$

TABLE 25.1

Stiffness Parameters Measured for Equivalent Symmetric Test Points for One Patient

	Point Number	1	2	3	4	5	6
Right Breast	O'clock	Nipple	2	3	6	9	12
	SPS_i	0.12	1.32	0.56	1.41	0.58	0.88
	AvS_i	1.73	4.85	2.64	7.52	4.25	4.68
	$NormS_i$	0.41	1.13	0.62	1.76	0.99	1.09
Left Breast	O'clock	Nipple	10	9	6	3	12
	SPS_i	0.39	2.08	5.03	0.34	1.04	1.46
	AvS_i	2.77	17.28	26.29	1.74	3.20	6.85
	$NormS_i$	0.29	1.78	2.71	0.18	0.33	0.71
Ratio (Left/Right)	SPS_i	3.38	1.57	8.93	0.24	1.80	1.66
	AvS_i	1.60	3.57	9.96	0.23	0.75	1.46
	$NormS_i$	0.70	1.57	4.40	0.10	0.33	0.65

Source: Mojra, A., Design and fabrication of an intelligent, artificial-tactile-based system for breast masses detection, PhD thesis, Amirkabir University of Technology, Tehran, Iran, 2011.

where $f(z)$ is the probability of the mass depth prediction (deep or superficial); the variable z is a measure of the total contribution of all the independent variables used in the model; β_0 is the intercept; $\beta_1, \beta_2, \ldots, \beta_5$ are the regression coefficients of $x_1, x_2, \ldots, x_5$, respectively; x_1 is SPS_i; x_2 is AvS_i; x_3 is $normS_i$; x_4 is the amount of tissue compression; and x_5 is the age of the patient [23].

As described earlier, during the process of training a neural network, the weights of the input data are the parameters that adjust according to desired output. Here, the neural network is trained under supervision to select the best amount of $\beta_1, \beta_2, \ldots, \beta_5$ to show the importance of each input value. The desired outputs are the data acquired from sonography.

Table 25.2 reports the values of β_i for the logistic model of the mass depth estimation. Each of the regression coefficients describes the size of the contribution of that risk factor. The superficial mass was considered as code 1; as a result, the equation predicts the probability of a lesion to be superficial. As shown in the table, β_1, which is the coefficient of SPS_i, has a positive regression coefficient. This means that this variable increases the probability of being superficial, while the negative regression coefficient of β_2 related to AvS_i means

TABLE 25.2

Values of β_i (Regression Coefficients) for the Logistic Model of Mass Depth Prediction

Stiffness Parameters	β_i	Standard Error	P Value	Exp (β_i)
SPS_i	1.307	0.269	.000	3.6996
AvS_i	−0.465	0.094	.000	0.628
$NormS_i$	1.145	0.848	.177	3.143
Amount of Compression	−0.182	0.087	.036	0.833
Age	−1.08	0.037	.004	0.897
Constant	7.547	2.782	.007	1895.429

Source: Mojra, A., Design and fabrication of an intelligent, artificial-tactile-based system for breast masses detection, PhD thesis, Amirkabir University of Technology, Tehran, Iran, 2011.

that the variable decreases the probability. Moreover, the large coefficient β_1 in comparison to the small amount of β_2 means that the risk factor SPS_i strongly influences the probability of that outcome.

References

1. Mack MJ, Minimally invasive and robotic surgery, *Journal of American Medical Association (JAMA)*, 2001, 285(5): 568–572.
2. Najarian S, Dargahi J, and Mehrizi AA, *Artificial Tactile Sensing in Biomedical Engineering*. McGraw-Hill, New York, 2009.
3. Chmarra MK, TrEndo tracking system; motion analysis in minimally invasive surgery, PhD thesis, Magdalena K. Chmarra, Delft, The Netherlands, 2009.
4. Peine WJ, Remote palpation instruments for minimally invasive surgery, PhD dissertation, Harvard University, United States, 1999. (Dissertations & Theses: A&I database, Publication No. AAT 9921526.)
5. Yen PL, Palpation sensitivity analysis of exploring hard objects under soft tissue, In *International Conference on Advanced Intelligent Mechatronics*, 2003, 2: 1102–1106.
6. Breedveld P, Stassen HG, Meijer DW, Jakimowicz JJ, Observation in laparoscopic surgery: overview of impending effects and supporting aids, *Journal of Laparoendoscopic and Advanced Surgical Techniques*, 2000, 10: 231–241.
7. Bholat OS, Haluck RS, Murray WB, Gorman PJ, and Krummel TM, Tactile feedback is present during minimally invasive surgery. *Journal of the American College of Surgeons*, 1999, 189: 349–355.
8. Dion YM and Gaillard F, Visual integration of data and basic motor skills under laparoscopy. Influence of 2-D and 3-D video-camera systems. *Surgical Endoscopy*, 1997, 11: 995–1000.
9. Eltaib MEH and Hewit JR, Tactile sensing technology for minimal access surgery—a review, *Mechatronics*, 2003, 13: 1163–1177.
10. Yegingil HO, Breast cancer detection and differentiation using piezoelectric fingers, PhD thesis, Drexel University, Philadelphia, PA, 2009.
11. Mittra I, Baum M, Thornton, H, and Houghton, J, Is clinical breast examination an acceptable alternative to mammographic screening? *British Medical Journal*, 2000, 321: 1071.
12. Davis PS, Wechsler RJ, Feig SA, and March DE, Migration of breast biopsy localization wire. *American Journal of Roentgenology*, 1988, 150: 787–788.
13. Malur S, Wurdinger S, Morizt A, Michels W, and Schneider W, A comparison of written reports of mammography, sonography, and magnetic resonance mammography for preoperative evaluation of breast lesions, with special emphasis on magnetic resonance mammography. *Breast Cancer Research*, 2001, 3: 55–60.
14. Rajagopal V, Nielsen PMF, and Nash MP, Modeling breast biomechanics for multi-modal image analysis—successes and challenges. *WIREs Systems Biology and Medicine*, 20102, 3: 293–304.
15. Tanter M, Bercoff J, and Athanasiou A, Quantitative assessment of breast lesion viscoelasticity: initial clinical results using supersonic shear imaging. *Ultrasound in Medicine and Biology*, 34, 1373–1386.
16. Najarian S, Dargahi J, and Mirjalili V, Detecting embedded objects using haptic with applications in artificial palpation of masses, *Sensors and Materials*, 2006, 18: 215–229.
17. Qiu Y, Sridhar M, Tsou JK, Lindfors KK, and Insana MF, Ultrasonic viscoelasticity imaging of nonpalpable breast tumors: preliminary results1. *Academic Radiology*, 2008, 15: 1526–1533.
18. Najarian S, Fallahnezhad M, and Afshari E, Advances in medical robotic systems with specific applications in surgery—a review, *Journal of Medical Engineering and Technology*, 2011, 35: 19–33.

19. Mojra A, Najarian S, Hosseini SM, Towliat Kashani SM, and Panahi F, Abnormal mass detection in a real breast model: a computational tactile sensing approach. *IFMBE Proceedings*, 2009, 25(12): 115–118.
20. Mojra A, Najarian S, Towliat Kashani SM, Panahi F, and Tehrani MA, A novel robotic tactile mass detector with application in clinical breast examination, *Minimally Invasive Therapy and Allied Technology*, 2012, 21(3): 210–221.
21. Wellman PS and Howe RD, Mechanical properties of breast tissue in compression. Harvard BioRobotics Laboratory Technical Report #99003, 1999.
22. Mojra A, Design and fabrication of an intelligent, artificial-tactile-based system for breast masses detection, PhD thesis, Amirkabir University of Technology, Tehran, Tehran, Iran, 2011.
23. Mojra A, Najarian S, Kashani SM, and Panahi F, A novel tactile-guided detection and three-dimensional localization of clinically significant breast masses, *Journal of Medical Engineering and Technology*, 2012, 36(1): 8–16.

26

Wearable Behavior Navigation Systems for First-Aid Assistance

Eimei Oyama, Norifumi Watanabe, Naoji Shiroma, and Takashi Omori

CONTENTS

26.1 Introduction

In this chapter, we present novel wearable systems for first-aid assistance based on video conferencing and augmented reality (AR) technology. The concept of a wearable computing system is under active research since components such as computers, sensors, and actuators are becoming smaller. Similarly, wearable virtual reality (VR) and wearable robotic systems have become a popular research topic [1–5]. Most mobile phones are now equipped with navigation functionality based on the global positioning system (GPS). However, the demand for guidance not only for transportation but also for other activities in daily life is expected to arise from the development of wearable technology [5–12]. Although people rarely need guidance from other people with respect to everyday activities, if, for example, one goes to the aid of an injured or ill person, receiving proper instructions for first aid from an expert might be necessary. Potentially, there are a large number of tasks for which receiving guidance from the experts would be useful. In this context, we have proposed and developed wearable guidance systems for the purpose of providing general or specific guidance in various situations [9–12]. Hereafter, BNS stands for behavior navigation system and WBNS stands for wearable BNS, where the latter is expected to enable the wearer to perform first aid with the accuracy of an expert.

In this chapter, we present the concept and operating principle of WBNSs mainly for first-aid assistance. It should be noted that current WBNSs utilize human intelligence rather than artificial intelligence (AI). However, in the future, AI is expected to be able to provide assistance for first aid if assistance provided through a worldwide communication network becomes popular and the associated data are accumulated and utilized for the construction of such specialized AI.

26.2 Behavior Navigation

We have proposed and developed a basic mechanism for realizing efficient behavior navigation. Remote assistance based on voice instructions issued by an expert at an ambulance service center is the most common method for providing guidance in situations where first aid must be provided immediately. However, the utilization of visual information for remote assistance provided by an expert is not common even in developed countries. In this respect, guidance without visual feedback is limited; an analogy is the process of guiding a person to create a shadowgraph by using only verbal instructions.

The development of communication and wearable technologies enables us to use visual information for behavior navigation. Since 2007, we have proposed and developed WBNSs under the CREST "Multi-sensory Communication, Sensing the Environment, and Behavioral Navigation with Networking of Parasitic Humanoids" project of the Japan Science and Technology Agency (JST) [13]. The leader of the project is Prof. Taro Maeda of Osaka University.

In this chapter, we introduce two representative types of WBNSs for assisting non-experts performing first aid. The first type of WBNS is based on commodity devices and is intended for more general cases of first-aid assistance [10,12]. The second type is based on AR technology and provides behavior navigation not only for first-aid assistance but also for everyday activities [10,11]. Hereafter, WBNSAR stands for WBNS using AR technology.

All WBNSs utilize visual information feedback from a cooperator located next to the patient to an expert located at a remote site, such as an emergency room or a critical care center. The expert shows the cooperator how to perform first aid in accordance with the status of the patient.

The first type of WBNS utilizes a mannequin or a staff member assuming the same posture as the patient in order to allow the expert to demonstrate first-aid techniques to the cooperator. In contrast, WBNSAR has a display system that superimposes the computer-generated (CG) video of the expert's hands onto the actual footage captured by the cameras on the head-mounted display (HMD) of the cooperator. On the screens of their respective HMDs, both the expert and the cooperator see the CG video of the expert's hands as well as the actual camera footage. This allows the cooperator to follow the movements of the expert directly and accurately, providing first aid in the same way as the expert. Such WBNSAR systems are described in detail in later sections.

26.3 Commodity-Based BNSs

In this section, we present the configuration of commodity-based BNSs (CBNSs) for providing assistance with general first aid [10,12].

26.3.1 Configuration of CBNSs

Figure 26.1 shows a conceptual diagram of a CBNS. The cooperator in Figure 26.1 is the nonexpert standing next to an injured or ill person and preparing to provide first aid.

In many cases, it is not possible to acquire video of the hand movements of the cooperator by using one or even several fixed cameras. Furthermore, the position of cameras installed on laptop PCs or smartphones placed on the floor is too low for providing proper feedback for first aid, even if the cooperator and the patient themselves are on the floor. As it is not always possible to fix the camera at the appropriate height, movable or wearable cameras are necessary.

To determine the status of the patient, the expert must acquire an abundance of sensory information from the cooperator, and at the same time, the sensory system of the cooperator must be light and compact. Therefore, in WBNSs, the cooperator wears a head-mounted camera (HMC), which acquires visual information on the patient from the perspective of the cooperator. The camera footage is sent to the expert together with an audio signal, and this is one of the central features of BNSs. The cooperator can see the expert's movements on the HMD for performing first aid.

Figure 26.2 shows the configuration of a CBNS. In order to provide adequate assistance in various cases where first aid is needed, CBNS requires at least two additional staff members to be present at the remote site in addition to the expert. One staff member (referred to as the "actor") plays the role of the patient by assuming the same posture and performing the same movements. The expert and the actor demonstrate a first-aid procedure in accordance with the status of the patient. However, some aspects of first aid, such as cardiopulmonary resuscitation (CPR), can cause injury to the patient. In such cases, a CPR mannequin is used instead of an actor. In the future, it is conceivable that humanoid robots will serve as actors. We assume that the expert and the operator have access to the same tools for providing first aid. The second staff member is a camera operator who acquires video of the movements of the expert and the actor. CBNS transmits the video

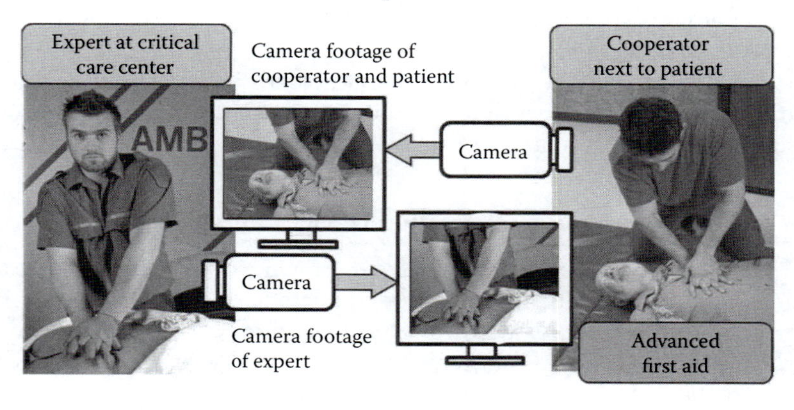

FIGURE 26.1
Conceptual diagram of CBNS for first-aid assistance.

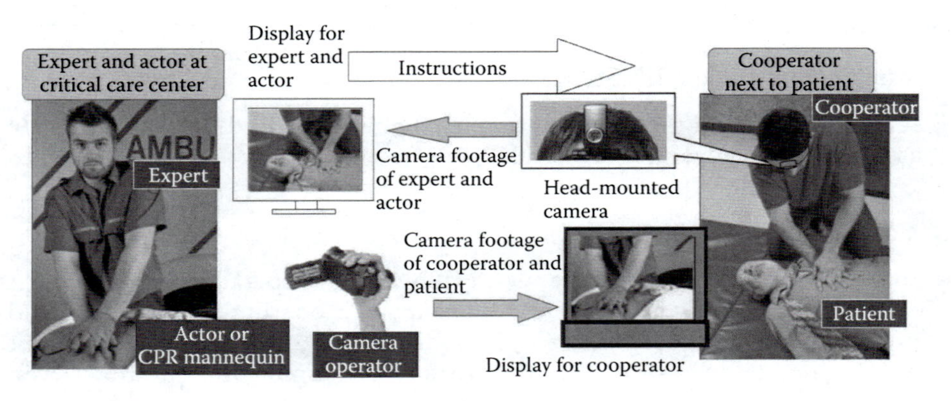

FIGURE 26.2
Configuration and principle of CBNS.

to the cooperator over a communication network, and the video captured by the camera operator at the remote site is shown on the display of the cooperator. CBNS becomes a wearable system when a laptop PC display is utilized by the cooperator.

The cooperator can follow the expert's movements shown on the display. Also, the cooperator can request that the expert advance, stop, or repeat the steps of the first-aid procedure, or request that the camera operator move the camera to capture a particular aspect of the procedure. In turn, the expert monitors the procedure performed by the cooperator by viewing the footage from the cooperator's HMC. The expert can also ask the cooperator to advance, stop, or repeat the steps of the procedure to ensure that it is performed accurately.

The described method is intended for assistance targeting the provision of first aid rather than general behavior navigation. Furthermore, CBNSs are useful only when the expert and the cooperator have access to the same first-aid equipment. Nevertheless, we believe that the proposed method can be applied in the majority of cases where first aid must be provided.

26.3.2 Experimental Prototype of CBNS

We developed a simple experimental prototype of a CBNS consisting of two displays, two PCs, and two cameras, as shown in Figure 26.3 [10].

We conducted experiments to evaluate the effectiveness of the prototype CBNS. Eight participants who played the role of the cooperator were instructed to use a triangular bandage to make an arm sling for a patient with an injured forearm. The results of one evaluation experiment are presented in Section 26.4.

26.3.3 BNSs Using Common Communication Devices for CPR

Most individuals have or will soon have Internet-connected TVs, laptop PCs, and smartphones. Widespread adoption of BNSs can be expected if such devices can be utilized for behavior navigation. We developed three types of simple BNSs using common communication devices (BNSCCDs), as shown in Figure 26.4. The first type of BNS utilizes a large display and the video conferencing system at the cooperator's end. Although this BNS is not a wearable system, Internet-connected TVs have the potential to become central devices for receiving remote medical assistance at home. Therefore, BNSs should be

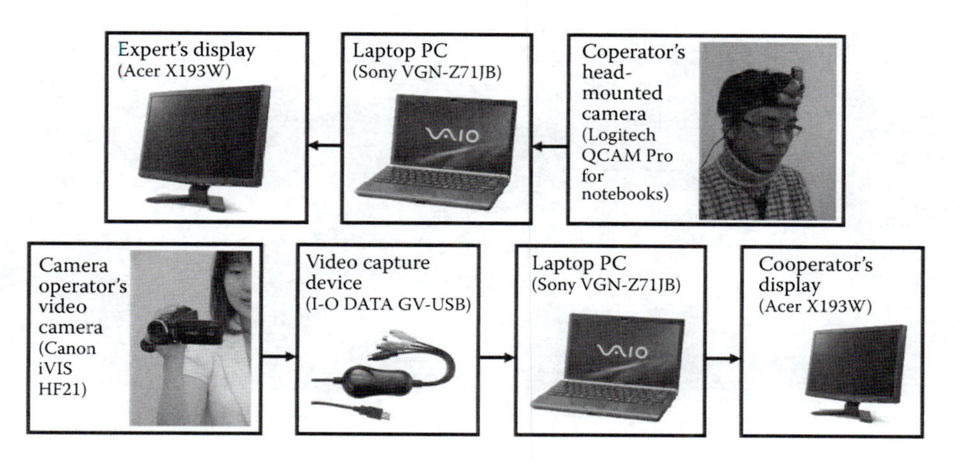

FIGURE 26.3
Hardware configuration of prototype CBNS,

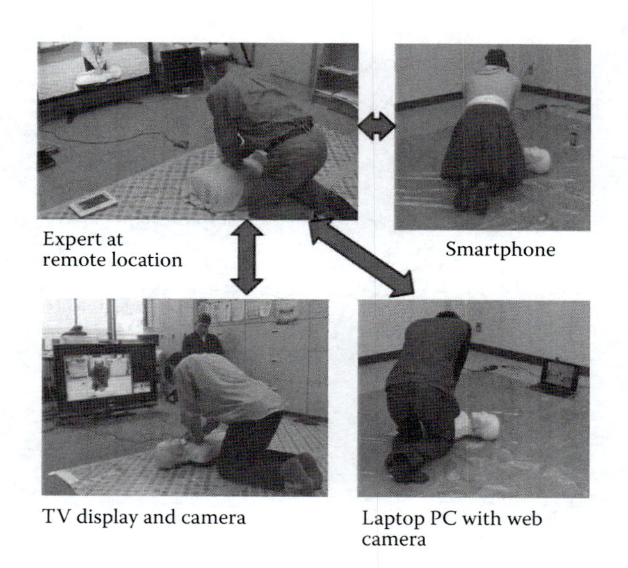

FIGURE 26.4
Concept of BNSCCDs.

investigated from the perspective of their ability to use everyday communication devices. The second type of BNS utilizes a video conferencing system on a laptop PC at the cooperator's end, and the third type of BNS utilizes the cooperator's smartphone, as shown in Figure 26.5. The second and third BNSs can be regarded as wearable systems. They utilize video conferencing software, such as Skype [14] or FaceTime [15]. The expert and the cooperator use the video conferencing function of the software for acquiring and providing visual feedback. The expert utilizes a large display and a high-quality audio system.

It should be noted that these BNSs do not provide behavior navigation for the provision of first aid since the camera fixed on the display cannot always capture the movements of the expert or the cooperator. However, they are sufficient for providing behavior navigation for bystanders performing CPR (Figure 26.6), which is the most critically important

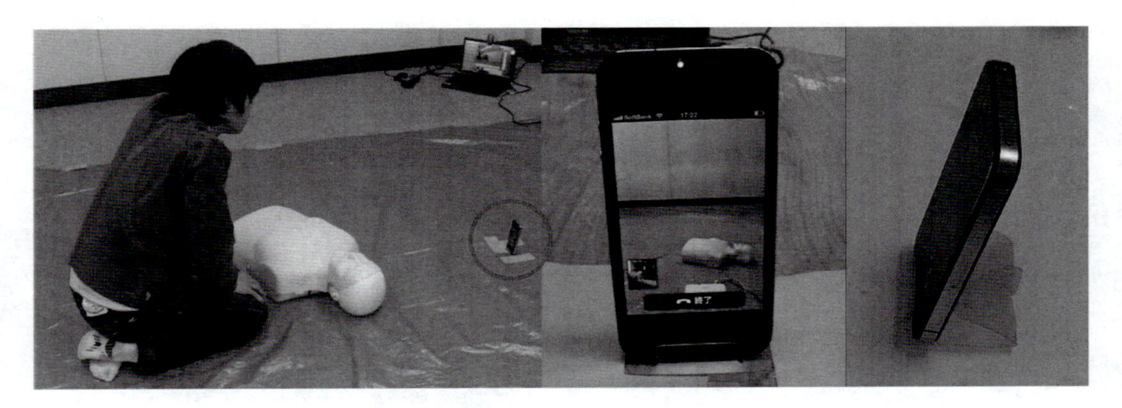

FIGURE 26.5
Example of BNS with a cooperator using a smartphone.

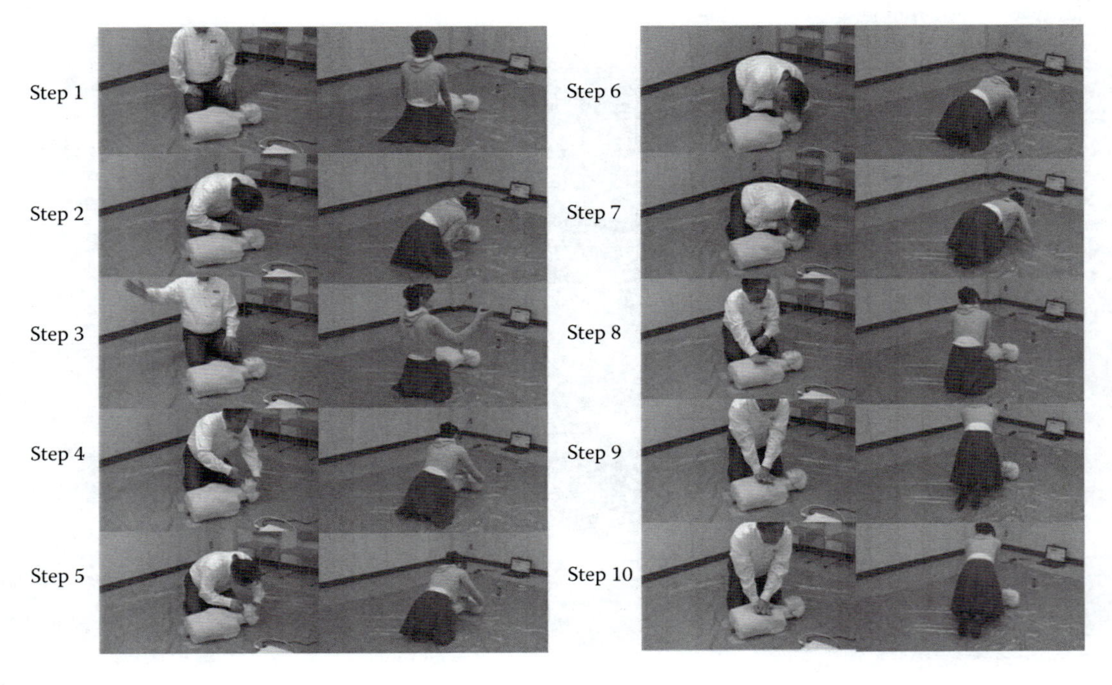

FIGURE 26.6
(See color insert.) Setup of experiment using a smartphone.

procedure of basic life support. The potential usability of BNSs is confirmed by the experiments [12]. The effectiveness and reliability of BNSs is expected to be improved by the optimization of the instruction procedure, the optimization of the position/orientation of the devices, the utilization of figures illustrating the CPR procedure, the utilization of multiple cameras or camera operators as described in the work of Oyama et al. [10], and so forth.

We hope that BNSCCDs will soon find application in providing assistance with CPR. To popularize the use of BNSs for CPR, evidence of their effectiveness should be obtained by a wide range of medical experiments.

26.3.4 Interim Conclusions

We proposed and developed CBNSs for more general first-aid assistance. Nowadays, videos illustrating most aspects of first aid are readily available, and CBNSs are useful for the first-aid procedures presented in such videos. Although this method is useful only when the target of the procedure is a person and the expert has access to the same tools as the cooperator, CBNSs can nevertheless be used in the majority of cases where first aid must be provided urgently.

26.4 WBNSAR

As illustrated in Section 26.3.4, CBNSs have limitations. In order to realize behavior navigation applicable in more general situations, we illustrate the concept of WBNSAR. The origin of WBNSAR is the SharedView System proposed by Kuzuoka [6]. WBNSAR is expected to be useful even when the expert does not have access to the same tools as the cooperator. Furthermore, the first-person view of WBNSAR is likely to increase the effectiveness and accuracy of behavior navigation. In this section, we describe the concept and operation principle of WBNSAR in detail.

26.4.1 Basic Operation Mechanism of WBNSAR

A conceptual diagram of WBNSAR, where the key technology is the sharing of sensory information between the expert and the cooperator, is presented in Figure 26.7. WBNSAR provides rich visual feedback in the form of footage acquired by the cooperator and then provided to the expert, who gathers information about the status of the patient. The camera

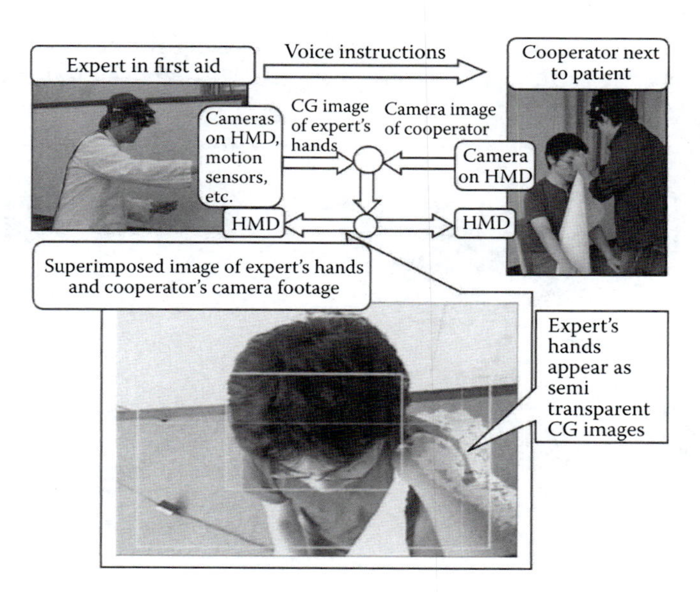

FIGURE 26.7
Conceptual diagram of WBNSAR.

mounted on the HMD of the cooperator can capture visual information from the perspective of the cooperator. In addition to visual information, auditory information acquired by microphones installed on the HMD of the cooperator is highly important for ensuring smooth communication between the expert and the cooperator and presenting the status of the patient accurately.

In addition to feedback in the form of rich sensory information from the cooperator to the expert, a technique for showing the cooperator how to perform the same operations as the expert is essential. The most distinctive technique of our WBNS is the AR display system, which superimposes the CG video of the expert's actions onto the footage captured by the cameras mounted on the HMD of the cooperator, as shown in Figure 26.7. Both the expert and the cooperator see the CG image of the expert's hands as well as the actual footage captured by the camera of the cooperator, allowing the cooperator to copy the movements of the expert directly. The first-person view is the key to the effectiveness of WBNSAR.

Ideally, more extensive sensory feedback, including the sense of touch, should be transmitted to the expert. However, WBNS ignores many aspects of sensory feedback since the wearable systems are limited in size and weight. According to research on telepresence robots, which are capable of transmitting and reproducing the sense of touch, the techniques for presenting the sense of touch are useful in a large number of practical teleoperation tasks [16,17]. However, these technologies are still under active study, and it is still unknown whether these technologies can be integrated into practical wearable systems in the foreseeable future.

26.4.2 Information Processing in a WBNSAR Prototype

In this section, the outline of information processing adopted in WBNSAR is illustrated. As stated above, WBNSAR utilizes CG video of the expert's hands. However, it is difficult to create such images of the expert's hands by using wearable sensors. Therefore, the present WBNSAR directly uses the camera footage of the expert's hands.

Figure 26.8 illustrates the flow of information processing in the WBNSAR prototype. The system worn by the expert extracts the portion corresponding to the expert's hands

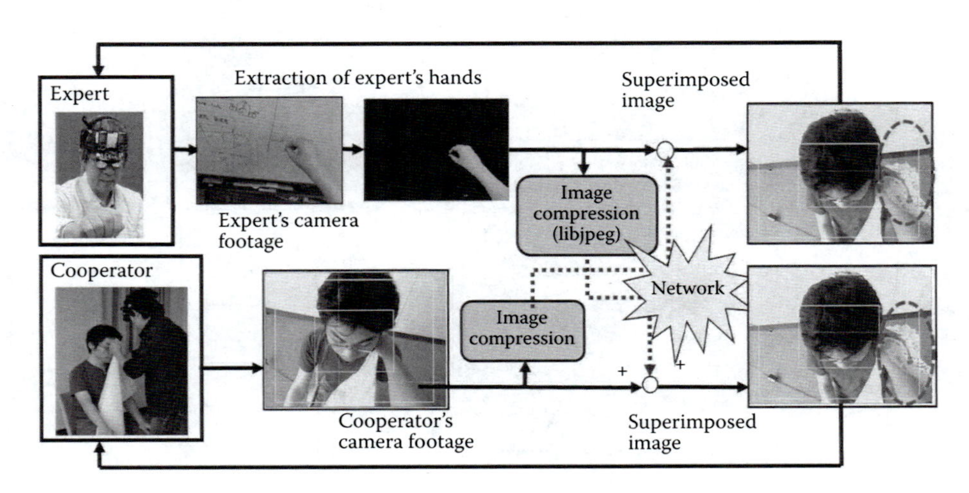

FIGURE 26.8
Information processing in WBNSAR.

from the image obtained by the camera mounted on the expert's HMD by performing relatively simple image processing, described in the work of Oyama et al. [9]. The system converts the input image from an RGB (red, green, and blue) color space to an HSV/HSL (hue, saturation, and value/lightness) color space and extracts the image of the hands as a manually specified region in HSV/HSL color space. The image of the expert's hands is the logical conjunction of the value of the captured image and the image of the extracted region.

The extracted image of the expert's hands and the image captured by the cooperator are compressed using the JPEG algorithm and exchanged via a wireless network. The laptop PCs of the expert and the cooperator superimpose the extracted image of the expert's hands over the actual images captured by the camera of the cooperator. Both the expert and the cooperator see the overlaid images on the displays of their respective HMDs.

Although the processing of the extracted images of the hands is rudimentary and should be improved, it is sufficient for demonstrating the effectiveness of WBNSAR. The extracted images of the expert's hands and the image captured by the camera mounted on the cooperator's HMD are overlaid as shown in Figure 26.8. Furthermore, a rectangular CG image indicating the field of view (FOV) of the expert is overlaid onto the image. Both the expert and the cooperator see the overlaid image on the displays of their respective HMDs.

Since the image extraction procedure used in the WBNSAR prototype is rather primitive, the extracted image occasionally contains regions that do not correspond to the image of the cooperator's body, and the image extraction sometimes misses the cooperator's hands. However, in most cases, the expert and the cooperator can obtain visual information about the expert's hands and the cooperator's FOV due to the high performance of human visual perception. To increase the reliability and robustness of extraction of the images of the expert's hands, we are developing the extraction procedure so as to utilize a body image constructed by using the outputs of motion sensors.

26.4.3 Prototypes of WBNSAR

We developed a number of prototypes of WBNSs. In 2009, we developed the first simple prototype of a WBNS (presented in the work of Oyama et al. [9]), which consisted of two HMDs (eMagin Z800), two cameras (Logitech WebCam Pro 9000), and a custom-made data acquisition suite equipped with 3-D motion sensors (NEC/TOKIN).

The outputs of the 3-D motion sensors are used to move the rectangular image on the display to communicate the FOV of the expert to the cooperator. The expert's HMD is connected to the cooperator's HMD with a laptop PC. Since this simple system was utilized to evaluate the feasibility of the system, network communication between the expert's HMD and the cooperator's HMD was not implemented. The image processing procedures described in Section 26.4.2 are implemented using OpenCV 1.0 [18].

We constructed a stereo HMD for behavior navigation, as shown in Figure 26.9. The HMD consists of an eMagin Z800 dual-input display system, two Logitech Portable Webcam C905 cameras, and a ZMP IMU-Z2 3-D motion sensor. Furthermore, a camera with a wide FOV is attached to the HMD of the cooperator.

In order to obtain a wide FOV and a natural simultaneous correspondence between the generated and actual images, the display system for the expert consists of the HMD and multiple projection screens, as shown in Figure 26.10. The images of two stereo cameras of the cooperator's HMD are directly transmitted to the HMD of the expert. The image from the camera with a wide FOV mounted on the cooperator's HMD is displayed on the screen after subjecting it to image stabilization.

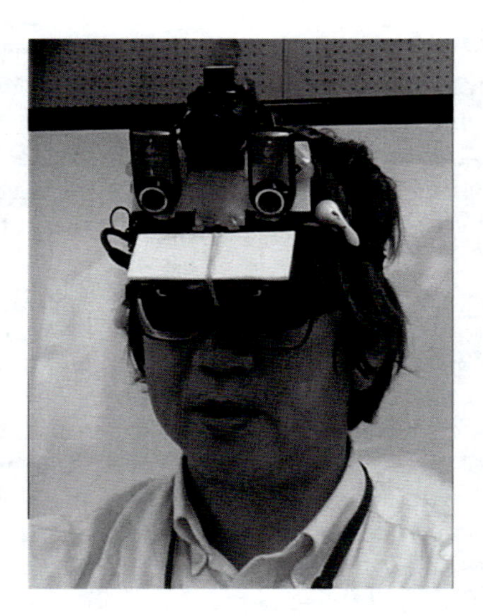

FIGURE 26.9
Advanced HMD of WBNSAR.

FIGURE 26.10
Hybrid display system for the expert.

The captured images of the cooperator and the extracted image of the expert's hands are compressed using the JPEG algorithm by using libjpeg-turbo [19] and sent through a wireless network. An analysis of the performance of the advanced system will be conducted in future work.

26.4.4 Evaluation Experiments

In order to evaluate the effectiveness of the BNSs, we have conducted basic experiments in which a total of 32 participants wearing BNSs were instructed to use a triangular bandage to make an arm sling for a person with an injury in the forearm.

26.4.4.1 Making an Arm Sling without Instruction

Before the experiments using verbal instructions, a CBNS, and a WBNSAR, we conducted a preliminary experiment in which eight participants (different from the 32 participants above) were instructed to make an arm sling by using a triangular bandage. The participants, who had no experience or training on the use of a triangular bandage, were provided only with an image showing the goal of the task (Figure 26.11).

26.4.4.2 Experimental Procedure

A total of 24 participants were instructed to wear an HMC, CBNS, and WBNSAR to make an arm sling by using a triangular bandage. Unlike the previous preliminary experiment, a staff member playing the role of the expert instructed the participants on how to make an arm sling by using the BNS.

We asked the remaining 8 of the 32 participants to wear an HMC and make an arm sling by following guidance provided only in the form of verbal instructions from the expert. We refer to this simple system as verbal instruction. In this experiment, the expert wore an HMD, which displayed video acquired from the perspective of the cooperator; however, the feedback to the participants was only verbal. Hereafter, this group of participants is referred to as "verbal instruction" (VI). Furthermore, eight participants were instructed to wear an HMC and make an arm sling by following the displayed image of the expert and the actor, as described in Section 26.3.1.

In order to present the first-aid procedure in a clear manner, next we described the triangular bandage. The longest side of the triangular bandage was referred to as the base, the corner directly opposite the middle of the base was referred to as the point, and the other two corners were referred to as the ends. The expert indicated the start and end times of the procedure to the staff in order to measure the duration of the procedure, which had to satisfy the following two criteria to be considered successful: (1) the arm sling supported the injured arm, and (2) the elbow of the injured arm was covered by the triangular bandage for stable support. Since various methods can be used to make an arm sling, we evaluated only the final result of the procedure because no researchers involved in this experiment were actual experts in first aid. The duration of the procedure was measured as the time between the start and the end of the procedure, which were indicated to the staff by the expert. Figure 26.12 shows the setup of the experiment using CBNS.

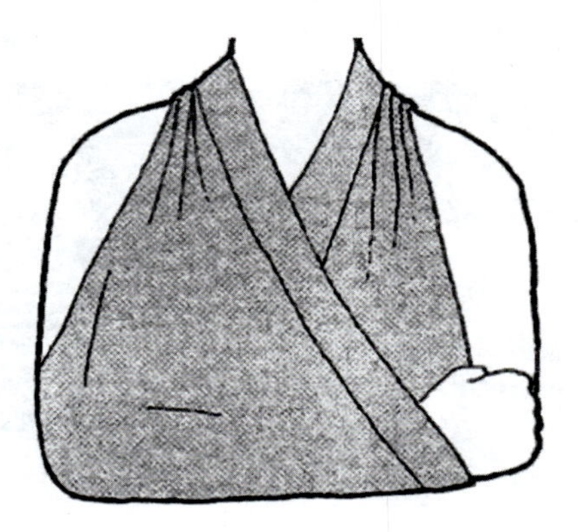

FIGURE 26.11
Goal of experimental task.

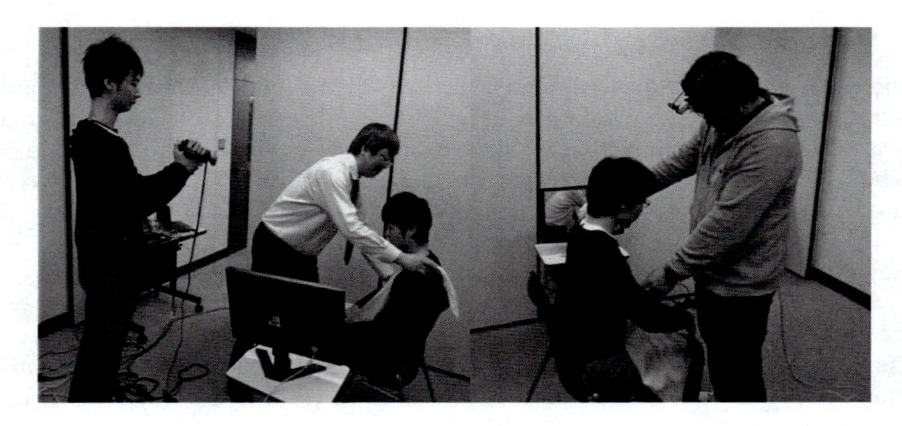

FIGURE 26.12
Setup of experiment using CBNS.

Figure 26.13 shows images displayed to the cooperator, which were captured by the camera operator (right), and images captured by the HMC of the cooperator, which were displayed to the expert and actor (left).

Figure 26.14 shows the setup of the experiment using WBNSAR. Figure 26.15 shows the AR image displayed on the HMD of WBNSAR, which corresponds to the above procedural steps. Although WBNSAR includes a custom-made data acquisition suite equipped with 3-D motion sensors and the HMD is equipped with a single 3-D motion sensor, the

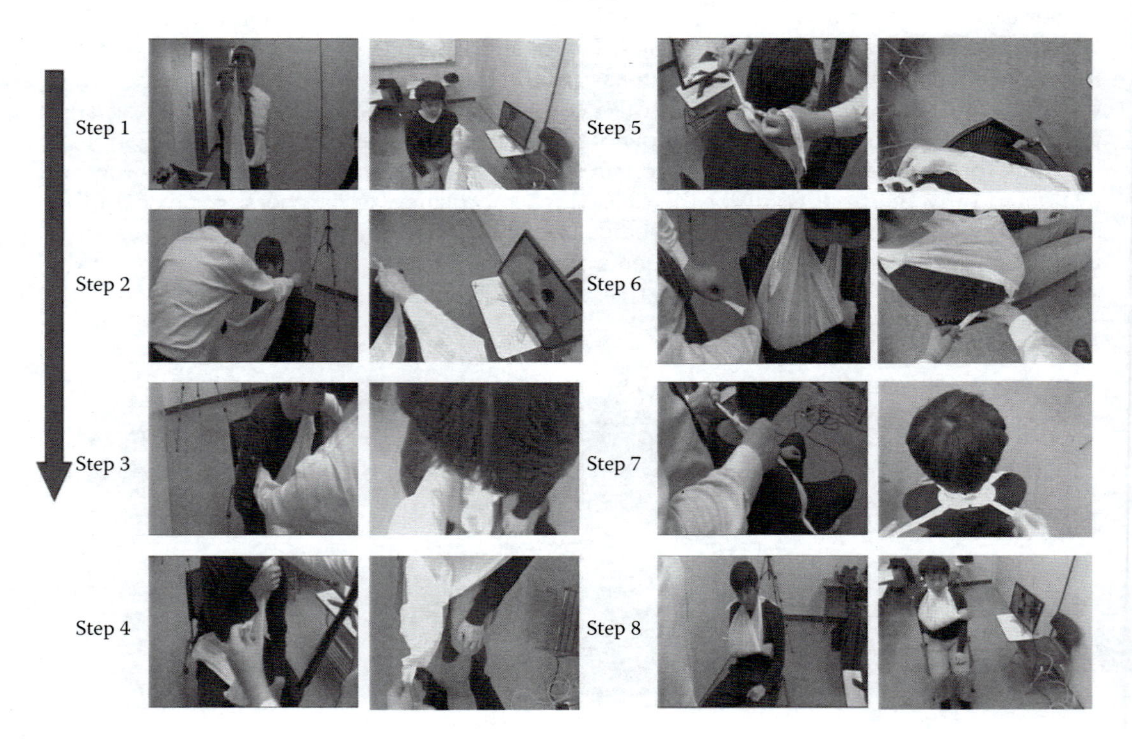

FIGURE 26.13
Images displayed to cooperator (left) and images displayed to expert and actor (right) using CBNS.

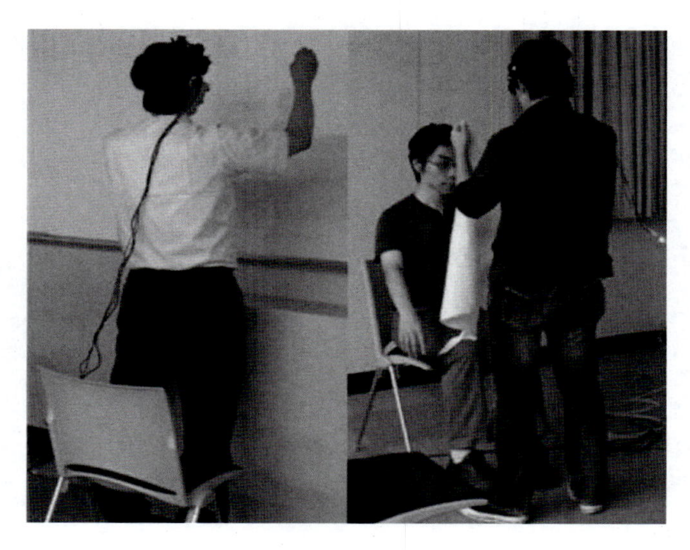

FIGURE 26.14
Setup of experiment using WBNSAR.

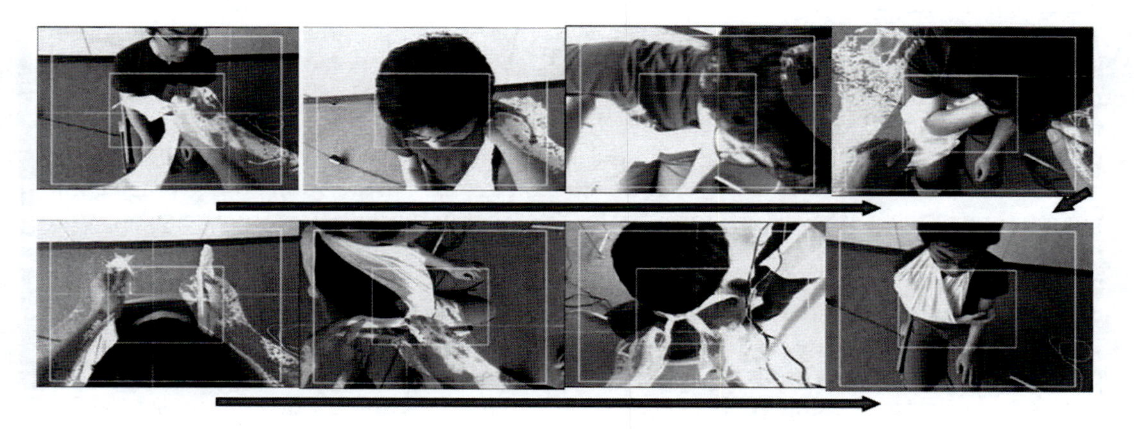

FIGURE 26.15
Images displayed during experiment using WBNSAR.

outputs of the 3-D motion sensors are not used in the experiment since the performance of the minimal system should be investigated at the beginning of the development of WBNS.

26.4.4.3 Experimental Results

The average and standard deviation of the time required to complete the procedure and the percentage of successful procedures are shown in Figure 26.16. The vertical black bars show the standard deviations.

Following the expert's instructions, all participants successfully completed the procedure, including in the cases of verbal instruction, CBNS, and WBNSAR. The participants receiving no instruction used less time overall but occasionally failed to make a proper sling. Since the average person who does not know how to make an arm sling usually

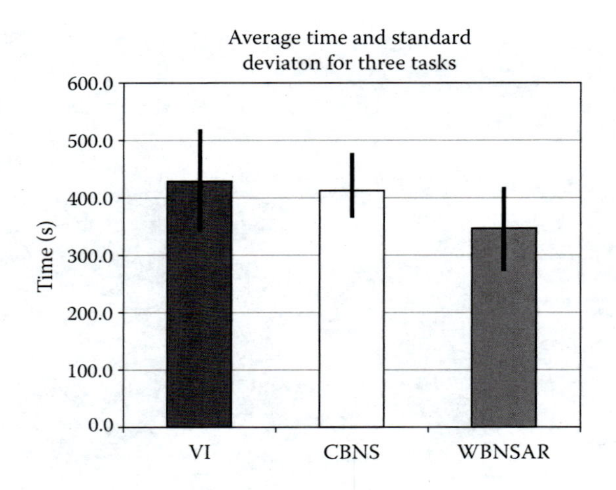

FIGURE 26.16
Experimental results.

cannot complete the task without expert assistance, the success of the participants receiving assistance demonstrates the effectiveness of the BNSs introduced above.

VI in Figure 26.16 indicates the experimental results for the eight participants who followed guidance provided only in the form of verbal instructions from the expert. The participants in the verbal instruction group completed the treatment as fast as those using the CBNSs. However, the standard deviation of the results for the CBNS group is smaller that of the verbal instruction group, and the longest completion time in the CBNS group is notably shorter than that in the verbal instruction group. Furthermore, the average completion time in the WBNSAR group is shorter than that in the verbal instruction group and the CBNS group. WBNSAR in Figure 26.16 indicates the experimental results for the seven participants who used the WBNSAR effectively. The expert was unable to see the hands of one participant clearly because of incorrect initial settings of the camera on the HMD. The participant did not successfully complete the task in the first trial, after which the participant performed the procedure again and completed it successfully. The average time for the eight participants using WBNSAR was 387.2 s, and the standard deviation was 136.3 s.

Participants in the verbal instruction group took longer to perform certain steps because the steps were difficult to explain without visual feedback. For example, verbal instructions such as "make a knot near the point of the bandage to fasten the bandage in order to support the patient's elbow" were difficult to understand, as a result of which participants in the verbal instruction group required a longer time to process the instructions as compared with the participants using CBNS or WBNSAR.

26.4.4.4 Problems Associated with WBNSs

Because the cooperator and the expert in the case of CBNSs must look alternately at the display and down at the patient to perform the first-aid procedure, the process is time consuming. Usually, CBNSs require more time than WBNSAR. However, this drawback is expected to be eliminated by the implementation of image stabilization techniques [20,21].

The present AR technology is also associated with a number of issues due to the following limitations of HMDs: (1) ordinary users must wear the HMD for a few minutes before their vision adjusts to the displays; (2) some individuals may experience motion sickness while wearing am HMD; (3) the FOV of low-cost HMDs is usually narrow; and (4) the HMDs are relatively expensive. However, we believe that these drawbacks of current AR technology will be solved in the near future.

26.5 Conclusions

This chapter presented basic mechanisms for realizing behavior navigation, the configurations of various WBNSs, and the results of an experimental evaluation of the performance of several WBNSs. The presented WBNSs are only rudimentary systems, and thus, they are still far from finding practical application. More advanced and refined WBNSs must be developed and evaluated experimentally. However, the potential advantage of WBNSs has been confirmed by a number of experiments.

In addition, WBNSs are expected to integrate more types of sensors and corresponding sensory information display functionality, such as the sense of touch. We will report on the advancement of development of WBNSs integrating these functions in the near future.

We believe that the services utilizing BNSs for first-aid assistance should be implemented soon. For this purpose, the effectiveness of BNSs must be verified through large-scale medical experiments. We plan to conduct such experiments and report their results in the near future.

Acknowledgments

Research and development of the WBNSs was supported through the CREST "Multi-sensory Communication, Sensing the Environment, and Behavioral Navigation with Networking of Parasitic Humanoids" program by the JST. The leader of the project is Prof. Taro Maeda (Osaka University). Research and development of first-aid assistance using WBNSs was supported by MEDIC First Aid Japan Co., Ltd. We thank the following individuals at MEDIC First Aid Japan: Chikako Uramoto, CEO; Tetsushi Suzuki, a master instructor; and Masatoshi Yamada, an instructor. We also thank Yasuo Kunimi, an associate professor at Tamagawa University and an instructor at MEDIC First Aid Japan.

References

1. T. Starner, S. Mann, B. Rhodes, J. Levine, J. Healey, D. Kirsch, R. Picard, and A. Pentland, Augmented reality through wearable computing, *Presence*, vol. 6, no. 4, Winter, pp. 386–398, 1997.
2. T. Starner, D. Kirsch, and S. Assefa, The locust swarm: an environmentally-powered, networkless location and messaging system, in *The First Int'l Symposium on Wearable Computers*, Boston, pp. 169–170, 1997.
3. M. Hirose, K. Hirota, T. Ogi, H. Yano, N. Kakehi, M. Saito, and M. Nakashige, Hapticgear: the development of a wearable force display system, in *Proc. of IEEE Virtual Reality (IEEE VR) 2001*, Yokohama, Japan, pp. 123–129, 2001.

4. Y. Seo, H. Park, and H.S. Yang, Wearable telepresence system using multi-modal communication with humanoid robot, in *The 13th International Conference on Artificial Reality and Telexistence (ICAT)*, Tokyo, Japan, 2003.
5. T. Kurata, M. Kourogi, N. Sakata, U. Kawamoto, and T. Okuma, Recent progress on augmented-reality interaction in AIST, in The *2nd International Digital Image Forum: The Future Direction and Current Development of User-Centered Digital Imaging Technology and Art*, Seoul, Korea, 2007.
6. H. Kuzuoka, Spatial workspace collaboration: a SharedView video support system for remote collaboration capability, in *Proc. of ACM CHI'92*, Monterey, CA, pp. 533–540, 1992.
7. T. Maeda, H. Ando, M. Sugimoto, J. Watanabe, and T. Miki, Wearable robotics as a behavioral interface—the study of the parasitic humanoid, in *Proc of 6th International Symposium on Wearable Computers*, Seattle, USA, pp. 145–151, 2002.
8. N. Sakata, T. Kurata, T. Kato, M. Kourogi, and H. Kuzuoka, WACL: supporting telecommunications using wearable active camera with laser pointer, in *7th IEEE International Symposium on Wearable Computers ISWC2003*, NY, USA, pp. 53–56, 2003.
9. E. Oyama, N. Watanabe, H. Mikado, H. Araoka, J. Uchida, T. Omori, K. Shinoda, I. Noda, N. Shiroma, A. Agah, K. Hamada, T. Yonemura, H. Ando, D. Kondo, and T. Maeda, A study on wearable behavior navigation system—development of simple parasitic humanoid system, in *IEEE ICRA 2010*, Anchorage, USA, 2010.
10. E. Oyama, N. Watanabe, H. Mikado, H. Araoka, J. Uchida, T. Omori, K. Shinoda, I. Noda, N. Shiroma, A. Agah, T. Yonemura, H. Ando, D. Kondo, and T. Maeda, A study on wearable behavior navigation system (II)—a comparative study on remote behavior navigation systems for first-aid treatment, in *IEEE RO-MAN 2010*, Viareggio, Italy, pp. 808–814, 2010.
11. E. Oyama and N. Shiroma, Behavior navigation system for use in harsh environment, in *2011 IEEE International Symposium on Safety, Security and Rescue Robotics*, Kyoto, Japan, pp. 272–277, 2011.
12. E. Oyama, N. Watanabe, H. Mikado, H. Araoka, J. Uchida, T. Omori, I. Noda, N. Shiroma, and A. Agah, Behavior navigation using common communication devices for CPR, in *2012 IEEE/SICE International Symposium on System Integration (SII2012)*, Fukuoka, Japan, 2012.
13. T. Maeda, Immersive tele-collaboration with parasitic humanoid: how to assist behavior directly in mutual telepresence, in *ICAT2011*, Osaka, Japan, 2011.
14. Skype Technologies S.A., http://www.skype.com/, 2012.
15. Apple Inc., http://www.apple.com/mac/facetime/, 2012.
16. M. S. Shimamoto, TeleOperator/telePresence System (TOPS) Concept Verification Model (CVM) development, in *Recent Advances in Marine Science and Technology*, N.K. Saxena (ed.), (Pacon International, 1992), pp. 97–104, 1992.
17. D. G. Caldwell, A. Wardle, O. Kocak, and M. Goodwin, Telepresence feedback and input systems for a twin armed mobile robot, *IEEE Robotics and Automation Magazine*, vol. 3, no. 3, pp. 29–38, 1996.
18. G. Gilin et al., http://opencv.willowgarage.com/wiki/, 2009.
19. libjpeg-turbo, http://libjpeg-turbo.virtualgl.org/, 2012.
20. N. Shiroma and E. Oyama, Asynchronous visual information sharing system with image stabilization, in *Proc. of IROS 2010*, Taipei, Taiwan, pp. 2501–2506, 2010.
21. N. Shiroma and E. Oyama, Development of virtual viewing direction operation system with image stabilization for asynchronous visual information sharing, in *Proc. of RO-MAN 2010*, Viareggio, Italy, pp. 76–81, 2010.

27

Artificial Intelligence Approaches for Drug Safety Surveillance and Analysis

Mei Liu, Yong Hu, Michael E. Matheny, Lian Duan, and Hua Xu

CONTENTS

27.1 Introduction

From the earliest moments in the modern history of computers, scientists have conceptualized the potential of artificial intelligence (AI) in medicine with the hope to create AI systems that can assist clinicians in the complex medical diagnosis processes [1,2]. The earliest work in medical AI dates back to the early 1970s with applications primarily focused on constructing AI programs that perform diagnoses and make therapy recommendations [3]. Much has changed since then. A wide array of AI-inspired methods have been developed to solve a broad range of important clinical and biological problems such as computer-based knowledge generation, decision support systems, and clinical data mining [4]. This chapter focuses on one specific application of AI methodologies to medical research—medication safety surveillance.

Every year, the US public spends billions of dollars on prescription medications. However, an awareness of the potential side effects that can occur with the use of medications is important for patients, providers, health care systems, payers, and regulatory agencies. Some side effects are minor, but others can turn out to be severe adverse drug reactions (ADRs) leading to patient morbidity. ADRs are defined as "any unintended and undesirable effects of a drug beyond its anticipated therapeutic effects occurring during clinical use at normal dose" [5]. According to the national surveillance study conducted by Budnitz et al. [6] on outpatient emergency visits, a total of 21,298 adverse drug events occurred from January 1, 2004, through December 31, 2005, which would yield a weighted annual estimate of 701,547 individuals or 2.4 individuals per 1000 population treated in the emergency departments experiencing an adverse event. Moreover, Lazarou et al. [7] estimated that 6%–7% of hospitalized patients would experience severe ADRs, causing 100,000 deaths annually in the United States. Additionally, both reported ADRs and related deaths have increased ~2.6 times over the past decade, and numerous drugs withdrew from the US market after presenting unexpected severe ADRs [8,9]. Consequently, ADRs presents a huge financial burden on the national economy, with an estimated $136 billion annual cost in the United States [10,11].

Drug discovery and development is a long and expensive process (Figure 27.1). To bring a new drug from discovery to market, it can take 10–17 years and millions of dollars [12]. And yet the majority of drug discovery endeavors are not successful due to the high failure rate of drug candidates in clinical trials, and approximately 30% of the failures are linked to unacceptable toxicities [13]. Hence, predicting potential ADRs at early stages is essential to reduce the risks of costly failures. Furthermore, even after a drug is approved to market, undiscovered ADRs may occur and lead to withdrawals, which can be financially detrimental for the manufacturers. Therefore, it is critical to predict and monitor a drug's ADRs throughout its life cycle, from preclinical screening phases to postmarket surveillance. Pharmacovigilance (PhV) is the science to address this problem.

In this chapter, we will cover a broad spectrum of the current AI approaches for PhV at both premarketing and postmarketing stages. The methodologies are presented along

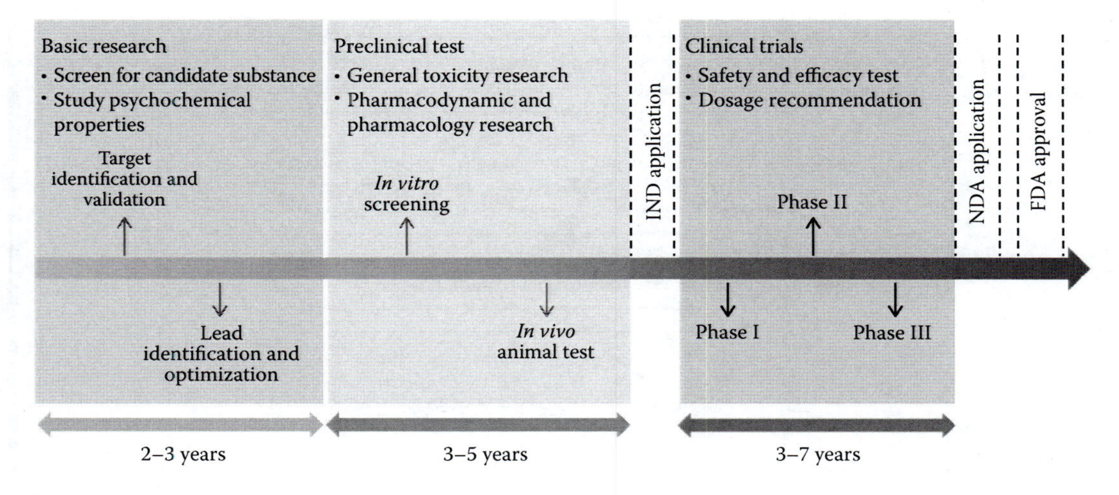

FIGURE 27.1
(See color insert.) Overview of the drug development process. Once a new chemical compound passes preclinical test, pharmaceutical company files an IND with the US Food and Drug Administration (FDA) to obtain approval for testing the new drug in humans. After the drug passes all phases of clinical trials, a pharmaceutical company formally submits a proposal to the FDA to approve the new drug for sale and marketing. IND, investigational new drug; NDA = new drug application.

different axes according to the data sources utilized with respect to each PhV stage. We will present a general overview of PhV in Section 27.2, followed by applications of AI methods in the premarketing surveillance stage (Section 27.3) and postmarketing surveillance stage (Section 27.4), and conclude with useful resources for future research needs (Section 27.5) and future perspectives and challenges (Section 27.6).

27.2 PhV—Drug Safety Surveillance: An Overview

PhV, also known as drug safety surveillance, is to enhance patient care and patient safety with regard to the use of medicines through the collection, monitoring, assessment, and evaluation of information from health care providers and patients. Generally speaking, PhV is conducted throughout the drug development process and market life and can be divided into two major stages: (1) premarketing surveillance—analyzing information collected from preclinical screening and phase I to III clinical trials; and (2) postmarketing surveillance—evaluating data accumulated in the postapproval stage and throughout a drug's market life (Figure 27.2).

Historical PhV efforts have relied on biological experiments or manual review of case reports; however, due to the vast quantities and complexity of the data to be analyzed, computational methods that can accurately and efficiently identify ADR signals have become a critical component in PhV. A variety of enabling resources are available for the computerized ADR detection methods, which include large-scale compound databases containing structure, bioassay, and genomic information, such as the National Institutes of Health's (NIH's) Molecular Libraries Initiative (MLI) [14], as well as comprehensive clinical observational data like electronic medical record (EMR) databases.

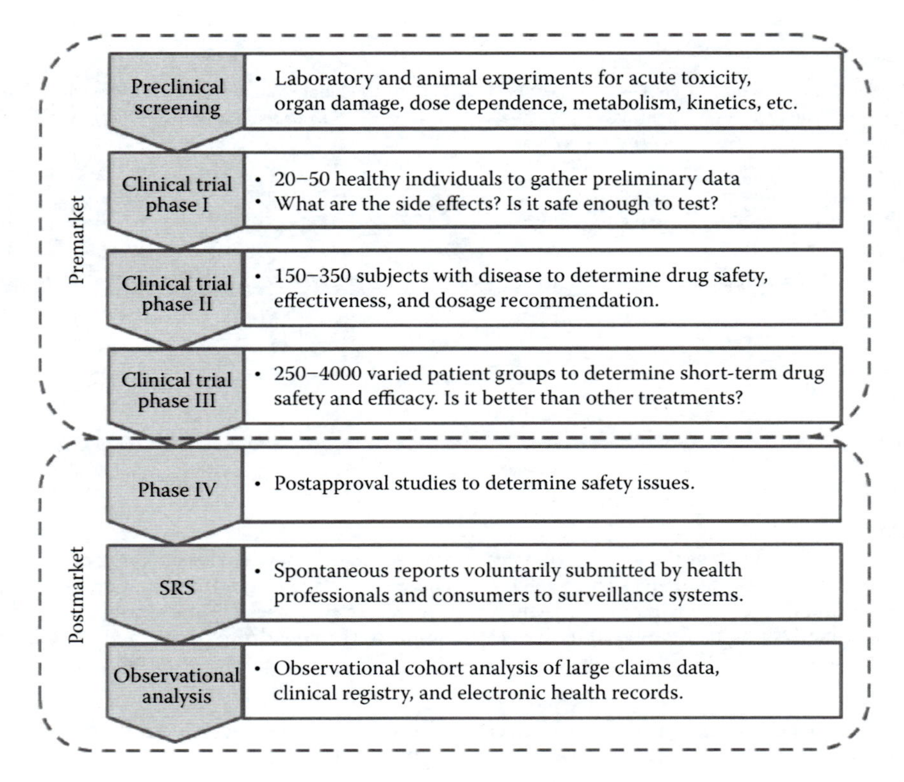

FIGURE 27.2
Pharmacovigilance at different stages of drug development. Different types of data are generated at each stage and can be used for drug safety surveillance. SRS, spontaneous reporting system.

At the premarketing stages of a drug, computational PhV efforts primarily focus on predicting potential ADRs using preclinical characteristics of the compounds (e.g., drug targets, chemical structure), or screening data (e.g., bioassay data). At the postmarketing stage, PhV has traditionally been involved in mining spontaneous reports submitted to national surveillance systems. In recent years, the postmarketing PhV research efforts have shifted toward the use of data generated from other platforms outside the conventional framework such as EMRs, clinical registries, biomedical literature, and patient-reported data in online health forums. Furthermore, an emerging trend of PhV is to link preclinical screening data from the experimental platform with human safety information observed in the postmarketing phase. The following sections provide a general overview of the current computational methodologies applied for PhV utilizing data accumulated at different stages of drug development.

27.3 Premarketing Surveillance

Many PhV efforts at the premarketing stage have been devoted to predicting or assessing potential ADRs during preclinical tests. Before a candidate compound can be approved to be tested on humans in clinical trials, it must pass a preclinical safety screening. One of the fundamental screening methods is to apply broad-scale *in vitro* pharmacology profiling to test new chemical compounds with biochemical and cellular assays [15]. The hypothesis is that if a compound binds to a certain target, then its effect may translate into a possible occurrence of an ADR in humans. However, experimental detection of ADRs remains challenging in terms of cost and efficiency [15]. A large amount of academic research activities have been devoted to developing computational approaches to predict potential ADRs using pharmacological properties of the compounds or bioassay screening data. The existing research can be roughly categorized into protein target-based and chemical structure-based approaches, while others have also explored an integrative approach.

27.3.1 Protein Target-Based Approach

Drugs typically work by activating or inhibiting the function of a protein that plays an important role in the disease mechanism, which in turn results in therapeutic benefits to a patient. Thus, drug design essentially involves the design of small molecules that have complementary shapes and charges to the protein target with which they can bind and interact. ADRs are complex phenomenological observations on drugs that have been attributed to a variety of molecular scenarios, such as unexpected interaction with the primary targets or off-targets, downstream pathway perturbations, and kinetics [16]. Many researchers believe that direct interaction with proteins is one of the most important scenarios [15,17]. Table 27.1 provides a summary of the key articles.

Through hierarchical clustering of biological activity spectra and adverse event data on 1045 prescription drugs and 92 ligand-binding assays, Fliri et al. [18] have shown that drugs with similar *in vitro* protein binding profiles tend to exhibit a similar array of side effects. This concept was further illustrated by Campillos et al. [19], who extrapolated new drug targets by analyzing the likelihood of protein target sharing for 277,885 pairs of 746 marketed drugs using drug side-effect similarities. Furthermore, Scheiber et al. [20] demonstrated the concept by comparing pathways affected by toxic compounds versus those affected by nontoxic compounds. Fukuzaki et al. [21] proposed a method to predict ADRs

TABLE 27.1

Protein Target-Based Approaches to PhV for Preclinical Safety Screening

Concept	Method	Article
Similar *in vitro* protein binding profiles tend to exhibit similar ADRs.	Clustering	Fliri et al. [18] Campillos et al. [19]
Compounds with similar toxicity profiles may have common pathway activation conditions.	Bayesian model Network analysis	Scheiber et al. [20] Fukuzaki et al. [21]
ADRs may be explained by a drug's protein–ligand binding network where many off-targets can bind to the drug and lead to unexpected ADRs.	Statistical models for binding site alignment	Xie et al. [22]
ADR similarity could be caused by their target proteins being close in the protein interaction network.	Statistical measure to assess the distance between two drugs	Brouwers et al. [23]
Molecular actors of ADRs may involve interactions detectable using compound bioactivity screening data.	Logistic regression	Pouliot et al. [24]
In vitro drug target activities are associated with ADRs.	Disproportionality analysis	Lounkine et al. [25]

using subpathways that share correlated modifications of gene-expression profiles in the presence of the drug of interest. In order to find the "cooperative pathways" (pathways that function together), they developed an algorithm called Cooperative Pathway Enumerator (CREPE) to select combinations of subpathways that have common activation conditions.

Xie et al. [22] developed a chemical systems biology approach to identify off-targets of a drug by docking the drug into binding pockets of proteins that are similar to its primary target. Then the drug–protein interaction pair with the best docking score was mapped to known biological pathways to identify potential off-target binding networks of the drug. Unfortunately, scalability of the method is hindered by its requirement for protein 3-D structures and known biological pathways.

Brouwers et al. [23] quantified the contribution of a protein interaction network neighborhood on the observed side-effect similarity of drugs. Their fundamental idea is that side-effect similarity of drugs can be attributed to their target proteins being close in a molecular network. They proposed a pathway neighborhood measure to assess the closest distance of drug pairs according to their target proteins in the human protein–protein interaction network. The authors observed that network neighborhoods only account for 5.8% of the side-effect similarities, compared to 64% from shared drug targets.

Pouliot et al. [24] applied logistic regression (LR) models to identify potential ADRs manifesting in 19 specific system organ classes (SOCs), as defined by the Medical Dictionary for Regulatory Activities [26], across 485 compounds in 508 bioassays in the PubChem database [27,28]. The models were evaluated using leave-one-out cross-validation. The mean area under the receiver operating characteristic curve (AUC) ranged from 0.60 to 0.92 across different SOCs.

Lounkine et al. [25] recently systematically evaluated the potential clinical relevance of protein targets with ADRs for 2760 drugs. The authors used disproportionality analysis (DPA) in conjunction with a chi-squared test to assess the associations for all target–ADR pairs and identified a total of 3257 significant target–ADR association pairs.

27.3.2 Chemical Structure-Based Approach

The chemical structure-based approach attempts to link ADRs to their chemical structures. Table 27.2 summarizes the key articles.

As a proof-of-concept, Bender et al. [29] explored the chemical space of drugs and established its correlation for ADR prediction; however, the positive predictive value was quite low, under 0.5. Thereafter, Scheiber et al. [30] presented a global analysis that identified

TABLE 27.2

Chemical Structure-Based Approaches to PhV for Preclinical Safety Screening

Concept	Method	Article
Investigate correlation between chemical structures and ADRs	Bayesian model	Bender et al. [29]
	Build individual ADR model using Bayesian model and use Pearson correlation to assess similarity between a pair of ADR models	Scheiber et al. [30]
	Support vector machine (SVM)	Yamanishi et al. [31]
	Decision tree	Hammann et al. [32]
	Compared nearest neighbor (NN), SVM, ordinary canonical correlation analysis (OCCA), and sparse canonical correlation analysis (SCCA)	Pauwels et al. [33]

chemical substructures associated with ADRs, but the method was not designed to predict ADRs for any specific drug molecule. Yamanishi et al. [31] proposed a method that predicted pharmacological effects or adverse effects from chemical structures and then used the effect similarity to infer drug–target interactions.

Hammann et al. [32] employed the decision tree model to determine the chemical, physical, and structural properties of compounds that predispose them to causing ADRs in different organ systems. In the study, the authors focused on ADRs in the central nervous system (CNS), liver, and kidney as well as allergic ADRs for 507 compounds. The features used were numerical attributes computed from a compound's structure, which included elemental analysis (e.g., atom count), charge analysis (e.g., polarizability, ion charge, topological polar surface area), and geometry (e.g., number of aromatic rings, rotable bonds), as well as partitioning coefficients and miscellaneous other characteristics (e.g., indicators of hydrogen bonding) [32]. Their decision tree model was shown to produce predictive accuracies ranging from 78.9 to 90.2% for allergic, renal, CNS, and hepatic ADRs.

Pauwels et al. [33] developed a sparse canonical correlation analysis (SCCA) method to predict high-dimensional side-effect profiles of drug molecules based on the chemical structures. The authors demonstrated the usefulness of SCCA by predicting 1385 side effects in the side effect resource (SIDER) database [34] from the chemical structures of 888 approved drugs. They compared five algorithms: random assignment (random) as a baseline, nearest neighbor (NN), support vector machine (SVM), ordinary canonical correlation analysis (OCCA), and SCCA for their abilities to predict known side-effect profiles through five-fold cross-validation. The best resulting AUC scores were 0.6088, 0.8917, 0.8930, 0.8651, and 0.8932 for random, NN, SVM, OCCA, and SCCA, respectively. Their results suggest that the proposed method, SCCA, outperforms OCCA, and its performance is comparable to SVM and NN. The main advantage of OCCA and SCCA over other algorithms is their biological interpretability to understand relationships between the chemical substructures and ADRs.

27.3.3 Integrative Approach

Recently, computational approaches that integrate various types of data relating to drugs for ADR prediction have gained increasing interest. Huang et al. [35] proposed a new computational framework to predict ADRs by integrating systems biology data that include protein targets, protein–protein interaction networks, gene ontology (GO) annotation [36], and reported side effects. SVM was applied as the predictive model for heart-related ADRs (i.e., cardiotoxicity), which resulted in the highest AUC of 0.771. Soon after, Cami et al. [37] developed another ADR prediction framework by combining the network structure formed by drug–ADR relationships (809 drugs and 852 ADRs) and information regarding specific drugs and adverse events. The LR model was used as the predictive model and achieved an AUC of 0.87.

Despite the success of using chemical and biological information on drugs for ADR prediction, few studies have investigated the use of phenotypic information (e.g., indication and other known ADRs). Existing resources, such as the SIDER database [34], contain comprehensive drug phenotypic information, which has been demonstrated to be useful for other drug-related studies [19]. Liu et al. [38] investigated the use of phenotypic information, together with chemical and biological properties of drugs, to predict ADRs. Similar to the work by Pauwels et al. [33], Liu et al. conducted a large-scale study to develop and validate the ADR prediction model on 1,385 known ADRs for 832 US FDA-approved drugs in SIDER with five machine learning algorithms: LR, naïve Bayes (NB), *k*-nearest neighbor (KNN), random forest (RF), and SVM. Evaluation results showed that the integration of chemical, biological, and phenotypic properties outperforms the chemical structured-based method (from 0.9054 to 0.9524 with SVM) and has the potential to detect clinically important ADRs at both the preclinical and postmarket phases for drug safety surveillance.

27.4 Postmarketing Surveillance

Although a drug undergoes extensive screening (Figure 27.2) before its approval by the FDA, many ADRs may still be missed because the clinical trials are often small, short, and biased by excluding patients with comorbid diseases. In general, for any potentially rare ADR with an occurrence rate less than 0.1%, it will be extremely difficult for the premarketing trials to identify the related rare ADRs due to the limitation in size [2]. In addition, premarketing trials do not mirror actual clinical use situations for diverse (e.g., inpatient) populations; therefore, it is important to continue the surveillance postmarket. Several unique data sources are available for postmarketing PhV.

27.4.1 Spontaneous Reports

Spontaneous reporting systems (SRSs) have served as the core data collection system for postmarketing drug surveillance since 1960. Some of the prominent SRSs are the Adverse Event Reporting System (AERS) maintained by the US FDA and the VigiBase managed by the World Health Organization (WHO). Although the SRSs may differ in structure and content, most of them rely on health care professionals and consumers to identify and report suspected cases of ADRs. Information collected usually includes the drugs suspected to cause the ADR, concomitant drugs, indications, suspected events, and limited demographic information. Many postmarketing surveillance analyses are based on these reports voluntarily submitted to the national SRSs, which include DPA for identifying single-drug and ADR associations and data mining algorithms for discovering combinatorial adverse effect of multiple drugs.

These systems are quite useful and have been in existence for many years but are subject to significant limitations in conducting postmarketing surveillance. SRSs suffer from severe underreporting. The extent of underreporting of ADRs to SRSs was investigated across 37 studies, and the median underreporting rate was estimated to be 94% [39]. Additionally, these systems do not have "denominator"-level information, which limits the types of statistical methods that can be applied to them, and reporting rates are influenced by media reports and publicity as much as by clinical and safety concerns [40]. SRSs are the most effective at detecting completely unexpected events, in which any reports of that adverse event profile are notable and suggest the need for further evaluation of the medication.

27.4.1.1 DPA for Single-Drug and ADR Associations

DPA has been the driving force behind most PhV methods involving SRS data. The first use of DPA for drug safety dates back to the early 1980s [41]. It is not our intention to exhaustively list and examine all relevant DPA work in this chapter. Rather, we aim to present the basic concepts and highlight some representative work here. DPA mainly

TABLE 27.3

Contingency Table Commonly Used in DPA Methods

	ADR Occurred	No ADR Occurred
Exposed to drug	A	B
Exposed to other drugs	C	D

TABLE 27.4

Definitions of the Frequentist Measures of Association

Common Frequentist Association Measures	Definition
Relative reporting ratio (RRR)	((a + b + c + d) * a)/((a + c) * (a + b))
Proportional reporting ratio (PRR)	(a/(a + b))/(c/(c + d))
Reporting odds ratio (ROR)	(a * d)/(c * b)

involves frequency analyses of 2 × 2 contingency tables to quantify the degree to which a drug and ADR co-occurs "disproportionally" compared with what would be expected if there were no association (Table 27.3) [42].

Straightforward DPA methods involve the calculation of frequentist metrics. Some of the widely applied frequentist measures (Table 27.4) include the relative reporting ratio (RRR) [43], proportional reporting ratio (PRR) [44] adopted by the Medicines and Healthcare Products Regulatory Agency (MHRA) in the United Kingdom, and reporting odds ratio (ROR) [45] adopted by the Netherlands Pharmacovigilance Center. Hypothesis tests of independence (i.e., chi-square test or Fisher's exact test) are typically used in conjunction with the above association estimates as extra precautionary measures.

In addition to the frequentist approaches, more complex algorithms based on Bayesian statistics were developed such as the gamma-Poisson shrinker (GPS) [46], the multi-item gamma-Poisson shrinker (MGPS) [47,48], and the empirical Bayesian geometric means (EBGMs) [49,50]. The GPS and MGPS methods are currently utilized by the FDA. Moreover, Bayesian Confidence Propagation Neural Network (BCPNN) [51–53] analysis was proposed based on Bayesian logic where the relation between the prior and posterior probability was expressed as the "information component" (IC). The IC given by the BCPNN is adopted by the WHO Uppsala Monitoring Center (UMC) to monitor safety signals in their SRSs.

Other groups have also investigated the James–Stein type of shrinkage estimation strategies in a Bayesian LR model to analyze SRSs data [54]. More recently, Ahmed et al. [55,56] proposed false discovery rate (FDR) estimation for the frequentist methods to address the limitation of arbitrary thresholds. Caster et al. [57] proposed the first implementation of shrinkage LR for large-scale pattern discovery of ADRs using the WHO global individual case safety reports database, VigiBase. Furthermore, Hopstadius and Noren [58] proposed a framework to detect local associations by employing shrinkage observed-to-expected ratios and multiple stratifications. As of now, there is no consensus on which DPA method is better due to the lack of a gold-standard data set for systematic evaluation.

27.4.1.2 Data Mining Algorithms for Multidrug and ADR Associations

The above-mentioned DPA methods are effective in detecting single-drug and ADR associations, but multidrug and ADR associations are also important because they can suggest possible drug–drug interactions. A typical SRS database contains thousands of drugs and ADRs, so it is impractical to enumerate all combinations for statistical analysis. Thus, data mining algorithms have been employed to address this problem.

Harpaz et al. [59] applied the association rule mining algorithm to identify multi-item ADRs. Using a set of 162,744 reports submitted to the FDA in 2008, 1,167 multi-item ADR associations were identified, and among those, 67% were validated by a domain expert. Later, Harpaz et al. [60] applied the biclustering algorithm to identify drug groups that share a common set of ADRs in SRSs data. Tatonetti et al. [61] proposed an algorithm to mine drug–drug interactions from the adverse event reports by analyzing latent signals that indirectly provide evidence for ADRs. Interestingly, they discovered that coadministration of pravastatin and paroxetine had a synergistic effect on blood glucose. In contrast, neither drug individually was found to be associated with such change in the glucose levels.

27.4.2 EMRs

EMRs have emerged as a prominent resource for observational research as they contain not only detailed patient information but also copious longitudinal clinical data. Recently, investigators have begun to explore the use of EMRs for PhV. Data in EMR data bases are typically in two types of formats: (1) structured (e.g., laboratory data) and (2) narrative clinical notes.

27.4.2.1 Structured Data

Several groups have employed computational methods on structured or coded data in EMRs to identify specific ADR signals [62,63]. Jin et al. [64] proposed a new interestingness measure called residual leverage for association rule mining to identify ADR signals from health care administrative databases. Ji et al. [65] introduced potential causal association rules to generate potential causal relationships between a drug and international classification diseases, 9th revision (ICD-9) coded signs or symptoms in EMRs. Noren et al. [66,67] introduced a new measure of temporal association to contrast the observed-to-expected ratio in a time period of interest to that in a predefined control period. Schildcrout et al. [68] analyzed the relationship between insulin infusion rates and blood glucose levels in patients in an intensive care unit (ICU). Zorych et al. [69] investigated various DPA methods for PhV by mapping EMR data to drug-condition 2 × 2 contingency tables.

Yoon et al. [70] demonstrated laboratory abnormality to be a valuable source for PhV by examining the odds ratio of laboratory abnormalities between a drug-exposed and a matched unexposed group using 10 years of EMR data. Evaluation of their algorithm on 470 randomly selected drug-and-abnormal-lab-event pairs produced a positive predictive value of 0.837 and negative predictive value of 0.659. Liu et al. [71] also designed a study to correlate abnormal laboratory results with specific drug administrations by comparing the outcomes of a drug-exposed group and a matched unexposed group using 12 years of EMR data. The authors assessed the relative merits of six DPA methods typically used for analyzing SRS data on their EMR data by systematically evaluating the methods on two independently constructed reference standard data sets of drug–event pairs.

Furthermore, Duan et al. [72] proposed a likelihood ratio model and a Bayesian network model for ADR discovery using the Observational Medical Outcomes Partnership (OMOP) [73] simulated data set. OMOP has designed and developed a procedure to generate fictional persons, drug exposure, and adverse event occurrences with predefined associations between drugs and outcomes. Although the data are simulated, it is part of the OMOP's effort in providing data close to real clinical observational data for larger computational communities to develop and evaluate their methods. The study [73] has shown that the proposed methods performed better by 23.83% over the standard baseline algorithm, the chi-square test.

27.4.2.2 Unstructured Data

Data in narrative clinical notes is not readily accessible for data mining; thus, a natural language processing (NLP) technique is required to extract the needed information. General-purpose NLP systems (e.g., medical language extraction and encoding system [MedLEE] [74], MetaMap [75], and clinical text analysis and knowledge extraction system [cTAKES] [76]), as well as specialized systems (e.g., MedEx [77] for medication information extraction), have been developed for clinical text. Wang et al. [78] first employed NLP techniques to extract drug–ADR candidate pairs from narrative EMRs and then applied the chi-square test with an adjusted volume test to detect ADR signals. Evaluation on seven selected drugs and their known ADRs produced an overall precision and recall of 0.31 and 0.75, respectively.

Similarly, Wang et al. [79] developed other methods based on mutual information (MI) and data processing inequality (DPI) to characterize drug-and-ADR pairs extracted from EMRs. Evaluation on a random sample of two drugs and two diseases indicated an overall precision of 0.81. Furthermore, Wang et al. [80] investigated the use of filtering by sections of reports to improve the performance of NLP extraction for clinically meaningful drug-and-ADR relations. Their evaluation indicated that applying filters improved recall from 0.43 to 0.75 and precision from 0.16 to 0.31.

Sohn et al. [81] proposed a hybrid system to extract physician-asserted ADRs from clinical narratives of psychiatry and psychology patients. They first developed a rule-based system to recognize relationships between individual side effects and causative drugs. The C4.5 decision tree model was then applied to extract sentences containing pairs of side effects and causative drugs using the keyword features and expression patterns of side effects. The hybrid system had an *F*-score of 0.75 in identifying side-effect sentences.

27.4.3 Nonconventional Data Sources

27.4.3.1 Biomedical Literature

Biomedical literature can be used as a complementary resource for prioritizing drug–ADR associations generated from SRSs. Shetty and Dalal [82] retrieved articles (published between 1949 and 2009) that contain mentions of a predefined list of drug-and-ADR pairs (38 drugs and 55 ADRs) from PubMed. The authors then constructed a statistical document classifier to remove irrelevant articles with mentions of treatment relations. Finally, DPA was applied to identify statistically significant pairs from the thousands of pairs in the remaining articles. Evaluation showed that the method identified true associations with 0.41 precision and 0.71 recall.

27.4.3.2 Health Forums

Data posted by users on health-related Web sites may also contain valuable drug safety information. Leaman et al. [83] described a system to mine drug-and-ADR relationships as reported by consumers in user comments to health-related Web sites like DailyStrength (http://www.dailystrength.org/). System evaluation was conducted on a manually annotated set of 3,600 user posts corresponding to 6 drugs. The system was shown to achieve 0.78 in precision and 0.70 in recall.

Chee et al. [84] explored the use of an ensemble classifier on data from online health forums to identify potential watch-list drugs that have an active FDA safety alert. The authors aggregated individuals' opinions and reviews of drugs and used an NLP technique to group drugs that were discussed in similar ways. Interestingly, withdrawn drugs were successfully identified based on messages even before they were removed from the market.

27.4.4 Integrative Approach

Traditionally, postmarketing ADR detection methodologies have focused on data from a single source, but the emerging belief is that combining information across sources can lead to more effective and accurate ADR signals. Matthews et al. [85] combined SRS data and literature findings to build quantitative structure–activity relationship models for several serious ADRs. Vilar et al. [86] proposed an approach to prioritize ADRs generated from SRSs based on chemical structure similarity. Harpaz et al. [87] leveraged EMRs and explicitly combined them with the SRS data to facilitate uniformed, hypothesis-free signal detection, while Vilar et al. [88] attempted to enhance the ADR signals derived from EMRs by molecular structure similarity.

27.5 Available Resources

Despite the progress made toward PhV, there is still little empirical evidence to support the use of one method or data source over another. As a result, research initiatives have been implemented so that methods and data sources can be assessed on a solid scientific footing. The following sections cover various initiatives and public resources useful for method development.

27.5.1 National Initiatives

27.5.1.1 The Sentinel Initiative

In 2007, the US Congress passed the FDA Amendments Act (FDAAA) mandating the FDA to establish an active surveillance system. In response, the FDA implemented the Sentinel Initiative [89] to develop a proactive system that will complement existing systems that FDA has in place to monitor the safety of its regulated products. The Sentinel System enables the FDA to query diverse health care databases such as EMR and administrative and insurance claims data bases so that safety issues of medical products can be evaluated quickly and securely. Since the announcement of the initiative, the FDA has supported a broad public forum to examine a range of subjects and granted pilot projects to explore many technical and policy issues in implementing such surveillance system.

27.5.1.2 OMOP

OMOP [73] is a public–private partnership launched by the Foundation for the National Institutes of Health with Pharmaceutical Research and Manufacturers of America (PhRMA) and the FDA, aiming to identify the most reliable methods for analyzing huge volumes of data drawn from heterogeneous sources. OMOP has designed experiments to test a variety of analytical methodologies in a range of data types to look for well-known drug impacts. In 2010, OMOP organized a competition for developing signal detection methods based on simulated data. A key finding from the 2010 OMOP Cup is that the heterogeneity of data sources and methods strongly affects results. Therefore, robust methods that can be applied to multiple data sources are desired.

27.5.1.3 National Centers for Biomedical Computing

The National Centers for Biomedical Computing (NCBC) are centers funded under cooperative agreement awards with multiple NIH agencies that are intended to be the core of networked national efforts to build computational infrastructure for biomedical computing [90]. There are currently eight such funded centers, and two of these large initiatives are particularly relevant to medication surveillance.

27.5.1.4 Informatics for Integrating Biology and the Bedside

While the focus of Informatics for Integrating Biology and the Bedside (I2B2) is on data harmonization into the common I2B2 schema and information extraction with NLP and other tools, this type of infrastructure has been adopted by the Clinical and Translational Science Award (CTSA) program that currently is in operation at 60 academic medical institutions across the United States. This center and the participating consortium, particularly as the focus of integrating genomic and phenomic data into a common data platform across a large number of institutions, is providing a powerful source of data for conducting postmarketing surveillance [91].

27.5.1.5 Integrating Data for Analysis, Anonymization, and Sharing

Integrating Data for Analysis, Anonymization, and Sharing (iDASH) is the newest NCBC, funded in 2010 by the NIH [92]. The iDASH center aims to develop algorithms and tools for sharing data in a privacy-preserving manner to enable global collaborations anywhere and anytime. It will provide the biomedical researchers a needed data resource for new hypothesis generation and testing in areas like postmarketing surveillance.

27.5.1.6 European Union—Adverse Drug Reaction

European Union—Adverse Drug Reaction (EU-ADR) [93] is a Research AND development project initiated in Europe by the European Commission. Similar to OMOP, EU-ADR, aims to develop computerized systems to detect ADRs in supplementing existing SRSs by exploiting EMRs of over 30 million patients from several European countries. The project employs a variety of text mining, epidemiological, and computational techniques to analyze the clinical data to detect ADR signals. These new developments rely on the expanded secondary use of electronic health care data that typically contain time-stamped interventions, procedures, diagnoses, medications, medical narratives, and billing codes.

27.5.1.7 MLI

The MLI is motivated by two NIH missions: (1) develop new approaches to determine function and therapeutic potential of all genes in the newly sequenced human genome and (2) accelerate the translation of basic research discoveries to new therapeutics for the benefits of public health [14]. The MLI research focuses on screening, cheminformatics, and technology development, and the small-molecule research tools produced would accelerate validation of new drug targets and enable new drug development.

27.5.2 Public Databases

27.5.2.1 DrugBank

The DrugBank database [94–96] contains detailed drug (i.e., chemical, pharmacological, and pharmaceutical) data and comprehensive drug target (i.e., sequence, structure, and pathway) information. Table 27.5 shows the contents in DrugBank (accessed October 2012).

27.5.2.2 PubChem

PubChem is part of the NIH's Molecular Libraries Roadmap Initiative [27,28] to provide information on biological activities of small molecules. It is organized as three linked databases—PubChem Substance, PubChem Compound, and PubChem BioAssay—within the National Center of Biotechnology Information (NCBI) Entrez information retrieval system. The PubChem Substance database contains substance information submitted by depositors, which include any chemical structure information, chemical names, and comments. The PubChem Compound database is composed of a nonredundant set of standardized and validated chemical structures. The PubChem BioAssay database contains bioactivity screens of chemical substances. All this information can be used in various ways to develop ADR detection methods in the preapproval stage.

27.5.2.3 SIDER

SIDER is a side-effect resource [34] developed to aggregate dispersed public information on drug side effects. It contains information on marketed medicines and their recorded ADRs extracted from public documents and package inserts. In the version released in October 2012, SIDER contains 996 drugs, 4,192 side effects, and 99,423 drug–ADR pairs.

27.5.2.4 Pharmacogenomics Knowledge Base

The Pharmacogenomics Knowledge Base (PharmGKB) [97] is an interactive tool for researchers to investigate how genetic variation affects drug responses. PharmGKB

TABLE 27.5

DrugBank Statistics, Accessed in October 2012

Entries	Statistic
Drug entries	6711
FDA-approved small-molecule drugs	1447
FDA-approved biotech (protein/peptide) drugs	131
Nutraceuticals	85
Experimental drugs	5080
Nonredundant protein (i.e., drug target, enzyme, transporter, carrier)	4227

captures complex relationships between genes, variants, drugs, diseases, and pathways. Data in PharmGKB have been curated from multiple sources and can be used to facilitate genome-wide pharmacogenomics (PGx) studies, predict gene–drug relationships and support data-sharing consortia investigating clinical applications of PGx.

27.5.2.5 Connectivity Map

The Connectivity Map [98] is a comprehensive research effort for using genomics in a drug discovery framework by generating a detailed map that links gene patterns associated with disease to corresponding patterns produced by drug candidates. The ultimate goal is to connect human diseases with the genes that underlie them and drugs that treat them. The Broad Institute of the Massachusetts Institute of Technology (MIT) and Harvard brought together specialists in biology, genomics, computing science, pharmacology, chemistry, and medicine to build the Connectivity Map.

27.5.2.6 NIH Chemical Genomics Center Pharmaceutical Collection

The NIH Chemical Genomics Center (NCGC) was established to create a national resource for chemical probe development. The NCGC Pharmaceutical Collection is a comprehensive repository of approved and investigational drugs for high-throughput screening and provides a valuable resource for both validating new disease models and better understanding the molecular basis of disease pathology and intervention [99].

27.5.2.7 Therapeutic Target Database

The Therapeutic Target Database (TTD) was developed to provide comprehensive information about known and explored therapeutic protein and nucleic acid targets, targeted disease, pathway information, and corresponding drugs directed at each of these targets to facilitate target-oriented drug discovery [100]. The database, accessed in November 2012, contains 2025 targets and 17,816 drugs.

27.6 Future Perspectives and Challenges

In this chapter, we have provided a general overview of the rich and diverse applications of AI approaches with respect to different perspectives of PhV. We have seen more and more opportunities emerging as a result of new data generated from various platforms, including high-throughput experiments, EMRs, literature, and self-reported health forums.

It is evident that a new trend of computational approaches for PhV is to link preclinical data from the experimental platform with human safety information observed in the postmarketing phase [101]. From the systems biology perspective, drugs are small molecules that can induce perturbations to biological systems, which involve various molecular interactions such as protein–protein interactions, signaling pathways, and pathways of drug action and metabolism. The body's response to a drug is a complex phenomenological observation that includes both favorable and unfavorable reactions. When a drug is absorbed into the body and interacts with its intended targets, favorable effects are expected. However, a drug often binds to other protein pockets with varying affinities (off-target interactions), leading to observed side effects. Hence, to understand ADRs, it is desirable to incorporate various data sources into one framework.

Moreover, it is critical to identify multi-item ADR associations as they may suggest drug interactions. For instance, if a patient is taking two drugs at the same time and one increases the effect of the other, the patient may experience an overdose. Similarly, if the action of a drug is inhibited by the other, the intended therapeutic effect may be reduced. Drug interactions may also increase the risk of ADRs. Statistical analysis works well with the identification of single drug-and-ADR signals but is not suitable for drug interaction identification. Alternatively, data mining algorithms such as *a priori* algorithms and clustering algorithms are applicable and useful. This provides an excellent opportunity for computer scientists to develop new algorithms for drug interaction detection.

Furthermore, EMRs have become an obvious data choice for PhV. However, many challenges exist in mining EMRs for ADR prediction. Much detailed and useful information is embedded in the narrative notes, making data extraction difficult. There have been studies using NLP techniques to extract drug and ADR concepts from narrative notes for association analysis. Wang et al. [80] have shown that filtering information based on note sections improves the identification of drug-and-ADR relations. Despite the current success, further investigation of other methods, for example, more sophisticated statistical methods and temporal models, is needed.

As yet, few studies have explored the automatic construction of large-cohort or case–control studies from EMRs for ADR prediction. There are many issues to consider in the NLP-based cohort/case–control study construction, for instance, how to extract adverse event concepts from narrative notes. It is common for multiple concepts to describe the same outcome/phenotype. Since most current practices focus on a single outcome at a time, phenotype is usually defined manually by experts. However, for large-scale ADR studies, how do we automatically define the phenotypes? Also, a key distinguishing feature of EMR-based methods is the use of temporal information to define time frames as surveillance windows to identify ADRs; for example, adverse outcomes must occur at certain time periods after drug exposure. However, to accurately determine the temporal relations between events in the narrative text remains challenging. In fact, each of the above-mentioned questions is an active area of research.

After overcoming the above hurdles in study design, one must keep in mind the confounding problem during analysis. For instance, the basic concept behind the cohort design is to partition a population into those who are "exposed" (taking a specific drug) and "unexposed" (taking a comparator drug or not taking a specific drug). A drug is determined to be associated with a specified outcome when the outcome occurs more often in the exposed group than in the unexposed group. Since the group assignment is not random, increased attention must be given when selecting the "unexposed" group. A common technique to minimize the issues caused by confounding and bias is to match patient groups based on a set of basic covariates such as gender, age, and comorbidities. On the other hand, case–control designs divide the study population into those who experienced the outcome ("case") and those who did not experience the outcome ("control"). If the drug exposure occurs more frequently in the cases than in the controls, the drug is said to be associated with the outcome. The same issues with confounding apply to case–control studies. Matching two groups before analysis is usually a good idea.

Lastly, it is important to note that most of the existing computational methodologies for PhV involve assessment of association between a drug and an ADR. However, association does not necessarily imply causation. Intuitively, causation requires not only correlation but also a counterfactual dependence. Inferring cause-and-effect relationships is an intrinsically hard problem in data mining and needs to be further investigated for PhV applications.

A young but rapidly advancing field of health care is personalized medicine—customizing health care decisions and practices to individual patients. The nature of diseases (e.g., their onset and course) is usually as unique as the patients who have them. Every person may respond to drugs or interventions differently because of their genetic makeup, clinical conditions, and environmental information. Classen et al. [102] estimated that about 50% of ADRs are likely to be related to genetic factors. PGx research can impact this problem by linking inherited differences to variable drug responses. Recent success stories of PGx studies include more accurate and safe dosing of warfarin following genotyping at *CYP2C9* [103] and *VKORC1* [104] and recognizing clopidogrel resistance with *CYP2C19* variants [105–107]. Despite significant progress, our knowledge of genetic factors contributing to ADRs is still limited [108]. To accelerate PGx discovery and knowledge validation, more efficient computational approaches are required, presenting the informatics community ample opportunities and challenges.

References

1. Ledley RS. Digital electronic computers in biomedical science. *Science*. 1959 Nov 6;**130**(3384):1225–34.
2. Ledley RS, Lusted LB. Reasoning foundations of medical diagnosis; symbolic logic, probability, and value theory aid our understanding of how physicians reason. *Science*. 1959;**130**(3366):9–21.
3. Shortliffe EH. The adolescence of AI in medicine: will the field come of age in the '90s? *Artif Intell Med*. 1993 Apr;**5**(2):93–106.
4. Patel VL, Shortliffe EH, Stefanelli M, Szolovits P, Berthold MR, Bellazzi R, Abu-Hanna A. The coming of age of artificial intelligence in medicine. *Artif Intell Med*. 2009;**46**(1):5–17.
5. Pirmohamed M, Breckenridge AM, Kitteringham NR, Park BK. Adverse drug reactions. *BMJ*. 1998;**316**(7140):1295–8.
6. Budnitz DS, Pollock DA, Weidenbach KN, Mendelsohn AB, Schroeder TJ, Annest JL. National surveillance of emergency department visits for outpatient adverse drug events. *JAMA*. 2006;**296**(15):1858–66.
7. Lazarou J, Pomeranz BH, Corey PN. Incidence of adverse drug reactions in hospitalized patients: a meta-analysis of prospective studies. *JAMA*. 1998;**279**(15):1200–5.
8. Moore TJ, Cohen MR, Furberg CD. Serious adverse drug events reported to the Food and Drug Administration, 1998–2005. *Arch Intern Med*. 2007;**167**(16):1752–9.
9. Giacomini KM, Krauss RM, Roden DM, Eichelbaum M, Hayden MR, Nakamura Y. When good drugs go bad. *Nature*. 2007;**446**(7139):975–7.
10. Leone R, Sottosanti L, Luisa Iorio M, Santuccio C, Conforti A, Sabatini V, Moretti U, Venegoni M. Drug-related deaths: an analysis of the Italian spontaneous reporting database. *Drug Saf*. 2008;**31**(8):703–13.
11. van der Hooft CS, Sturkenboom MC, van Grootheest K, Kingma HJ, Stricker BH. Adverse drug reaction-related hospitalisations: a nationwide study in The Netherlands. *Drug Saf*. 2006;**29**(2):161–8.
12. Paul SM, Mytelka DS, Dunwiddie CT, Persinger CC, Munos BH, Lindborg SR, Schacht AL. How to improve R&D productivity: the pharmaceutical industry's grand challenge. *Nat Rev Drug Discov*. 2010;**9**(3):203–14.
13. Hopkins AL. Network pharmacology: the next paradigm in drug discovery. *Nat Chem Biol*. 2008;**4**(11):682–90.
14. Austin CP, Brady LS, Insel TR, Collins FS. NIH Molecular Libraries Initiative. *Science*. 2004;**306**(5699):1138–9.

15. Whitebread S, Hamon J, Bojanic D, Urban L. Keynote review: *in vitro* safety pharmacology profiling: an essential tool for successful drug development. *Drug Discov Today*. 2005;**10**(21):1421–33.
16. Liebler DC, Guengerich FP. Elucidating mechanisms of drug-induced toxicity. *Nat Rev Drug Discov*. 2005;**4**(5):410–20.
17. Blagg J. Structure–activity relationships for *in vitro* and *in vivo* toxicity. *Annu Rep Med Chem*. 2006;**41**(2006):353–68.
18. Fliri AF, Loging WT, Thadeio PF, Volkmann RA. Analysis of drug-induced effect patterns to link structure and side effects of medicines. *Nat Chem Biol*. 2005;**1**(7):389–97.
19. Campillos M, Kuhn M, Gavin AC, Jensen LJ, Bork P. Drug target identification using side-effect similarity. *Science*. 2008;**321**(5886):263–6.
20. Scheiber J, Chen B, Milik M, Sukuru SC, Bender A, Mikhailov D, Whitebread S, Hamon J, Azzaoui K, Urban L, Glick M, Davies JW, Jenkins JL. Gaining insight into off-target mediated effects of drug candidates with a comprehensive systems chemical biology analysis. *J Chem Inf Model*. 2009;**49**(2):308–17.
21. Fukuzaki M, Seki M, Kashima H, Sese J. Side effect prediction using cooperative pathways. In *IEEE International Conference on Bioinformatics and Biomedicine*. Washington DC; 2009. p. 142–7.
22. Xie L, Li J, Bourne PE. Drug discovery using chemical systems biology: identification of the protein–ligand binding network to explain the side effects of CETP inhibitors. *PLoS Comput Biol*. 2009;**5**(5):e1000387.
23. Brouwers L, Iskar M, Zeller G, van Noort V, Bork P. Network neighbors of drug targets contribute to drug side-effect similarity. *PLoS One*. 2011;**6**(7):e22187.
24. Pouliot Y, Chiang AP, Butte AJ. Predicting adverse drug reactions using publicly available PubChem BioAssay data. *Clin Pharmacol Ther*. 2011;**90**(1):90–9.
25. Lounkine E, Keiser MJ, Whitebread S, Mikhailov D, Hamon J, Jenkins JL, Lavan P, Weber E, Doak AK, Cote S, Shoichet BK, Urban L. Large-scale prediction and testing of drug activity on side-effect targets. *Nature*. 2012;**486**(7403):361–7.
26. Brown EG, Wood L, Wood S. The medical dictionary for regulatory activities (MedDRA). *Drug Saf*. 1999;**20**:109–17.
27. Chen B, Wild D, Guha R. PubChem as a source of polypharmacology. *J Chem Inf Model*. 2009;**49**(9):2044–55.
28. Bolton E, Wang Y, Thiessen PA, Bryant SH. PubChem: integrated platform of small molecules and biological activities. Chapter 12 in *Annual Reports in Computational Chemistry*. Washington DC: American Chemical Society; 2008.
29. Bender A, Scheiber J, Glick M, Davies JW, Azzaoui K, Hamon J, Urban L, Whitebread S, Jenkins JL. Analysis of pharmacology data and the prediction of adverse drug reactions and off-target effects from chemical structure. *ChemMedChem*. 2007;**2**(6):861–73.
30. Scheiber J, Jenkins JL, Sukuru SC, Bender A, Mikhailov D, Milik M, Azzaoui K, Whitebread S, Hamon J, Urban L, Glick M, Davies JW. Mapping adverse drug reactions in chemical space. *J Med Chem*. 2009;**52**(9):3103–7.
31. Yamanishi Y, Kotera M, Kanehisa M, Goto S. Drug-target interaction prediction from chemical, genomic and pharmacological data in an integrated framework. *Bioinformatics*. 2010;**26**(12):i246–54.
32. Hammann F, Gutmann H, Vogt N, Helma C, Drewe J. Prediction of adverse drug reactions using decision tree modeling. *Clin Pharmacol Ther*. 2010;**88**(1):52–9.
33. Pauwels E, Stoven V, Yamanishi Y. Predicting drug side-effect profiles: a chemical fragment-based approach. *BMC Bioinformatics*. 2011;**12**:169.
34. Kuhn M, Campillos M, Letunic I, Jensen LJ, Bork P. A side effect resource to capture phenotypic effects of drugs. *Mol Syst Biol*. 2010;**6**:343.
35. Huang LC, Wu X, Chen JY. Predicting adverse side effects of drugs. *BMC Genomics*. 2011;**12 Suppl 5**:S11.

36. Ashburner M, Ball CA, Blake JA, Botstein D, Butler H, Cherry JM, Davis AP, Dolinski K, Dwight SS, Eppig JT, Harris MA, Hill DP, Issel-Tarver L, Kasarskis A, Lewis S, Matese JC, Richardson JE, Ringwald M, Rubin GM, Sherlock G. Gene ontology: tool for the unification of biology. The Gene Ontology Consortium. *Nat Genet*. 2000;**25**(1):25–9.
37. Cami A, Arnold A, Manzi S, Reis B. Predicting adverse drug events using pharmacological network models. *Sci Transl Med*. 2011;**3**(114):114ra27.
38. Liu M, Wu Y, Chen Y, Sun J, Zhao Z, Chen XW, Matheny ME, Xu H. Large-scale prediction of adverse drug reactions using chemical, biological, and phenotypic properties of drugs. *J Am Med Inform Assoc*. 2012;**19**:e28–e35.
39. Hazell L, Shakir SA. Under-reporting of adverse drug reactions : a systematic review. *Drug Saf*. 2006;**29**(5):385–96.
40. O'Shea JC, Kramer JM, Califf RM, Peterson ED. Part I: Identifying holes in the safety net. *Am Heart J*. 2004;**147**(6):977–84.
41. Montastruc JL, Sommet A, Bagheri H, Lapeyre-Mestre M. Benefits and strengths of the disproportionality analysis for identification of adverse drug reactions in a pharmacovigilance database. *Br J Clin Pharmacol*. 2011;**72**(6):905–8.
42. Bate A, Evans SJ. Quantitative signal detection using spontaneous ADR reporting. *Pharmacoepidemiol Drug Saf*. 2009;**18**(6):427–36.
43. Hauben M, Madigan D, Gerrits CM, Walsh L, Van Puijenbroek EP. The role of data mining in pharmacovigilance. *Expert Opin Drug Saf*. 2005;**4**(5):929–48.
44. Evans SJ, Waller PC, Davis S. Use of proportional reporting ratios (PRRs) for signal generation from spontaneous adverse drug reaction reports. *Pharmacoepidemiol Drug Saf*. 2001;**10**(6):483–6.
45. Szarfman A, Machado SG, O'Neill RT. Use of screening algorithms and computer systems to efficiently signal higher-than-expected combinations of drugs and events in the US FDA's spontaneous reports database. *Drug Saf*. 2002;**25**(6):381–92.
46. Ahmed I, Haramburu F, Fourrier-Reglat A, Thiessard F, Kreft-Jais C, Miremont-Salame G, Begaud B, Tubert-Bitter P. Bayesian pharmacovigilance signal detection methods revisited in a multiple comparison setting. *Stat Med*. 2009;**28**(13):1774–92.
47. DuMouchel W. Bayesian data mining in large frequency tables, with an application to the FDA spontaneous reporting system. *Am Stat*. 1999;**53**(3):177–202.
48. Almenoff JS, Pattishall EN, Gibbs TG, DuMouchel W, Evans SJ, Yuen N. Novel statistical tools for monitoring the safety of marketed drugs. *Clin Pharmacol Ther*. 2007;**82**(2):157–66.
49. DuMouchel W, Smith ET, Beasley R, Nelson H, Yang X, Fram D, Almenoff JS. Association of asthma therapy and Churg-Strauss syndrome: an analysis of postmarketing surveillance data. *Clin Ther*. 2004;**26**(7):1092–104.
50. Gould AL. Accounting for multiplicity in the evaluation of "signals" obtained by data mining from spontaneous report adverse event databases. *Biom J*. 2007;**49**(1):151–65.
51. Bate A, Lindquist M, Edwards IR, Olsson S, Orre R, Lansner A, De Freitas RM. A Bayesian neural network method for adverse drug reaction signal generation. *Eur J Clin Pharmacol*. 1998;**54**(4):315–21.
52. Lindquist M, Edwards IR, Bate A, Fucik H, Nunes AM, Stahl M. From association to alert—a revised approach to international signal analysis. *Pharmacoepidemiol Drug Saf*. 1999;**8 Suppl 1**:S15–25.
53. Lindquist M, Stahl M, Bate A, Edwards IR, Meyboom RH. A retrospective evaluation of a data mining approach to aid finding new adverse drug reaction signals in the WHO international database. *Drug Saf*. 2000;**23**(6):533–42.
54. An L, Fung KY, Krewski D. Mining pharmacovigilance data using Bayesian logistic regression with James-Stein type shrinkage estimation. *J Biopharm Stat*. 2010;**20**(5):998–1012.
55. Ahmed I, Dalmasso C, Haramburu F, Thiessard F, Broet P, Tubert-Bitter P. False discovery rate estimation for frequentist pharmacovigilance signal detection methods. *Biometrics*. 2010;**66**(1):301–9.

56. Ahmed I, Thiessard F, Miremont-Salame G, Begaud B, Tubert-Bitter P. Pharmacovigilance data mining with methods based on false discovery rates: a comparative simulation study. *Clin Pharmacol Ther*. 2010;**88**(4):492–8.
57. Caster O, Noren GN, Madigan D, Bate A. Large-scale regression-based pattern discovery: the example of screening the WHO global drug safety database. *Stat Anal Data Min*. 2010;**3**(4):197–208.
58. Hopstadius J, Noren GN. Robust discovery of local patterns: subsets and stratification in adverse drug reaction surveillance. In *Proceedings of the 2nd ACM SIGHIT International Health Informatics Symposium*. Miami, FL; 2012. pp. 265–73.
59. Harpaz R, Chase HS, Friedman C. Mining multi-item drug adverse effect associations in spontaneous reporting systems. *BMC Bioinformatics*. 2010;**11 Suppl 9**:S7.
60. Harpaz R, Perez H, Chase HS, Rabadan R, Hripcsak G, Friedman C. Biclustering of adverse drug events in the FDA's spontaneous reporting system. *Clin Pharmacol Ther*. 2011;**89**(2):243–50.
61. Tatonetti NP, Denny JC, Murphy SN, Fernald GH, Krishnan G, Castro V, Yue P, Tsao PS, Kohane I, Roden DM, Altman RB. Detecting drug interactions from adverse-event reports: interaction between paroxetine and pravastatin increases blood glucose levels. *Clin Pharmacol Ther*. 2011;**90**(1):133–42.
62. Brown JS, Kulldorff M, Chan KA, Davis RL, Graham D, Pettus PT, Andrade SE, Raebel MA, Herrinton L, Roblin D, Boudreau D, Smith D, Gurwitz JH, Gunter MJ, Platt R. Early detection of adverse drug events within population-based health networks: application of sequential testing methods. *Pharmacoepidemiol Drug Saf*. 2007;**16**(12):1275–84.
63. Berlowitz DR, Miller DR, Oliveria SA, Cunningham F, Gomez-Caminero A, Rothendler JA. Differential associations of beta-blockers with hemorrhagic events for chronic heart failure patients on warfarin. *Pharmacoepidemiol Drug Saf*. 2006;**15**(11):799–807.
64. Jin HD, Chen J, He HX, Williams GJ, Kelman C, O'Keefe CM. Mining unexpected temporal associations: applications in detecting adverse drug reactions. *IEEE Trans Inf Technol Biomed*. 2008;**12**(4):488–500.
65. Ji YQ, Ying H, Dews P, Mansour A, Tran J, Miller RE, Massanari RM. A potential causal association mining algorithm for screening adverse drug reactions in postmarketing surveillance. *IEEE Trans Inf Technol Biomed*. 2011;**15**(3):428–37.
66. Noren GN, Bate A, Hopstadius J, Star K, Edwards IR. Temporal pattern discovery for trends and transient effects: its applications to patient records. In *Proceedings of the 14th International Conference on Knowledge Discovery and Data Mining SIGKD 2008*. Las Vegas, NV; 2008. pp. 963–71.
67. Noren GN, Hopstadius J, Bate A, Star K, Edwards IR. Temporal pattern discovery in longitudinal electronic patient records. *Data Min Knowl Discov*. 2010;**20**(3):361–87.
68. Schildcrout JS, Haneuse S, Peterson JF, Denny JC, Matheny ME, Waitman LR, Miller RA. Analyses of longitudinal, hospital clinical laboratory data with application to blood glucose concentrations. *Stat Med*. 2011;**30**(27):3208–20.
69. Zorych I, Madigan D, Ryan P, Bate A. Disproportionality methods for pharmacovigilance in longitudinal observational databases. *Stat Methods Med Res*. 2013;**22**(1):39–56.
70. Yoon D, Park MY, Choi NK, Park BJ, Kim JH, Park RW. Detection of adverse drug reaction signals using an electronic health records database: Comparison of the Laboratory Extreme Abnormality Ratio (CLEAR) algorithm. *Clin Pharmacol Ther*. 2012;**91**(3):467–74.
71. Liu M, McPeek Hinz ER, Matheny ME, Denny JC, Schildcrout JS, Miller RA, Xu H. Comparative analysis of pharmacovigilance methods in the detection of adverse drug reactions using electronic medical records. *J Am Med Inform Assoc*. 2013;**20**(3):420–6.
72. Duan L, Kohoshneshin M, Street W, Liu M. Adverse drug effect detection. *IEEE J of Biomed Health Inform*. 2013;**17**(2):305–11.
73. Stang PE, Ryan PB, Racoosin JA, Overhage JM, Hartzema AG, Reich C, Welebob E, Scarnecchia T, Woodcock J. Advancing the science for active surveillance: rationale and design for the Observational Medical Outcomes Partnership. *Ann Intern Med*. 2010;**153**(9):600–6.
74. Friedman C, Alderson PO, Austin JH, Cimino JJ, Johnson SB. A general natural-language text processor for clinical radiology. *J Am Med Inform Assoc*. 1994;**1**(2):161–74.

75. Aronson AR, Lang FM. An overview of MetaMap: historical perspective and recent advances. *J Am Med Inform Assoc*. 2010;**17**(3):229–36.
76. Savova GK, Kipper-Schuler K, Buntrock JD, Chute CG. UIMA-based clinical information extraction system. LREC 2008: towards enhanced interoperability for large HLT systems: UIMA for NLP; 2008, 39.
77. Xu H, Stenner SP, Doan S, Johnson KB, Waitman LR, Denny JC. MedEx: a medication information extraction system for clinical narratives. *J Am Med Inform Assoc*. 2010;**17**(1):19–24.
78. Wang X, Hripcsak G, Markatou M, Friedman C. Active computerized pharmacovigilance using natural language processing, statistics, and electronic health records: a feasibility study. *J Am Med Inform Assoc*. 2009;**16**(3):328–37.
79. Wang X, Hripcsak G, Friedman C. Characterizing environmental and phenotypic associations using information theory and electronic health records. *BMC Bioinformatics*. 2009;**10 Suppl 9**:S13.
80. Wang X, Chase H, Markatou M, Hripcsak G, Friedman C. Selecting information in electronic health records for knowledge acquisition. *J Biomed Inform*. 2010;**43**(4):595–601.
81. Sohn S, Kocher JP, Chute CG, Savova GK. Drug side effect extraction from clinical narratives of psychiatry and psychology patients. *J Am Med Inform Assoc*. 2011;**18 Suppl 1**:i144–9.
82. Shetty KD, Dalal SR. Using information mining of the medical literature to improve drug safety. *J Am Med Inform Assoc*. 2011;**18**(5):668–74.
83. Leaman R, Wojtulewicz L, Sullivan R, Skariah A, Yang J, Gonzalez G. Towards internet-age pharmacovigilance: extracting adverse drug reactions from user posts to health-related social networks. In *Workshop on Biomedical Natural Language Processing*, Uppsala, Sweden: Association for Computational Linguistics; 2010. pp. 117–25.
84. Chee BW, Berlin R, Schatz B. Predicting adverse drug events from personal health messages. In *AMIA Annu Symp Proc*. Washington DC; 2011. pp. 217–26.
85. Matthews EJ, Kruhlak NL, Benz RD, Aragones Sabate D, Marchant CA, Contrera JF. Identification of structure–activity relationships for adverse effects of pharmaceuticals in humans: Part C: use of QSAR and an expert system for the estimation of the mechanism of action of drug-induced hepatobiliary and urinary tract toxicities. *Regul Toxicol Pharmacol*. 2009;**54**(1):43–65.
86. Vilar S, Harpaz R, Chase HS, Costanzi S, Rabadan R, Friedman C. Facilitating adverse drug event detection in pharmacovigilance databases using molecular structure similarity: application to rhabdomyolysis. *J Am Med Inform Assoc*. 2011;**18 Suppl 1**:i73–80.
87. Harpaz R, Vilar S, Dumouchel W, Salmasian H, Haerian K, Shah NH, Chase HS, Friedman C. Combing signals from spontaneous reports and electronic health records for detection of adverse drug reactions. *J Am Med Inform Assoc*. 2013;**20**(3):413–9.
88. Vilar S, Harpaz R, Santana L, Uriarte E, Friedman C. Enhancing adverse drug event detection in electronic health records using molecular structure similarity: application to pancreatitis. *PLoS One*. 2012;**7**(7):e41471.
89. Platt R, Wilson M, Chan KA, Benner JS, Marchibroda J, McClellan M. The new Sentinel Network—improving the evidence of medical-product safety. *New Engl J Med*. 2009;**361**(7):645–7.
90. National Centers for Biomedical Computing. Available from: http://www.ncbcs.org/summary.html
91. Informatics for Integrating Biology & the Bedside. Available from: https://www.i2b2.org/about/index.html
92. Ohno-Machado L, Bafna V, Boxwala AA, Chapman BE, Chapman WW, Chaudhuri K, Day ME, Farcas C, Heintzman ND, Jiang X, Kim H, Kim J, Matheny ME, Resnic FS, Vinterbo SA. iDASH: integrating data for analysis, anonymization, and sharing. *J Am Med Inform Assoc*. 2012;**19**(2):196–201.
93. Coloma PM, Schuemie MJ, Trifiro G, Gini R, Herings R, Hippisley-Cox J, Mazzaglia G, Giaquinto C, Corrao G, Pedersen L, van der Lei J, Sturkenboom M. Combining electronic healthcare databases in Europe to allow for large-scale drug safety monitoring: the EU-ADR Project. *Pharmacoepidemiol Drug Saf*. 2011;**20**(1):1–11.

94. Knox C, Law V, Jewison T, Liu P, Ly S, Frolkis A, Pon A, Banco K, Mak C, Neveu V, Djoumbou Y, Eisner R, Guo AC, Wishart DS. DrugBank 3.0: a comprehensive resource for 'omics' research on drugs. *Nucleic Acids Res*. 2010;**39**(Database issue):D1035–41.
95. Wishart DS, Knox C, Guo AC, Cheng D, Shrivastava S, Tzur D, Gautam B, Hassanali M. DrugBank: a knowledgebase for drugs, drug actions and drug targets. *Nucleic Acids Res*. 2008;**36**(Database issue):D901–6.
96. Wishart DS, Knox C, Guo AC, Shrivastava S, Hassanali M, Stothard P, Chang Z, Woolsey J. DrugBank: a comprehensive resource for in silico drug discovery and exploration. *Nucleic Acids Res*. 2006;**34**(Database issue):D668–72.
97. McDonagh EM, Whirl-Carrillo M, Garten Y, Altman RB, Klein TE. From pharmacogenomic knowledge acquisition to clinical applications: the PharmGKB as a clinical pharmacogenomic biomarker resource. *Biomark Med*. 2011;**5**(6):795–806.
98. Lamb J, Crawford ED, Peck D, Modell JW, Blat IC, Wrobel MJ, Lerner J, Brunet JP, Subramanian A, Ross KN, Reich M, Hieronymus H, Wei G, Armstrong SA, Haggarty SJ, Clemons PA, Wei R, Carr SA, Lander ES, Golub TR. The Connectivity Map: using gene-expression signatures to connect small molecules, genes, and disease. *Science*. 2006;**313**(5795):1929–35.
99. Huang R, Southall N, Wang Y, Yasgar A, Shinn P, Jadhav A, Nguyen DT, Austin CP. The NCGC pharmaceutical collection: a comprehensive resource of clinically approved drugs enabling repurposing and chemical genomics. *Sci Transl Med*. 2011;**3**(80):80ps16.
100. Zhu F, Shi Z, Qin C, Tao L, Liu X, Xu F, Zhang L, Song Y, Zhang J, Han B, Zhang P, Chen Y. Therapeutic target database update 2012: a resource for facilitating target-oriented drug discovery. *Nucleic Acids Res*. 2012;**40**(Database issue):D1128–36.
101. Harpaz R, Dumouchel W, Shah NH, Madigan D, Ryan P, Friedman C. Novel data-mining methodologies for adverse Drug event discovery and analysis. *Clin Pharmacol Ther*. 2012;**91**(6):1010–21.
102. Classen DC, Pestotnik SL, Evans RS, Lloyd JF, Burke JP. Adverse drug events in hospitalized patients. Excess length of stay, extra costs, and attributable mortality. *JAMA*. 1997;**277**(4):301–6.
103. Caraco Y, Blotnick S, Muszkat M. CYP2C9 genotype-guided warfarin prescribing enhances the efficacy and safety of anticoagulation: a prospective randomized controlled study. *Clin Pharmacol Ther*. 2008;**83**(3):460–70.
104. Rost S, Fregin A, Ivaskevicius V, Conzelmann E, Hortnagel K, Pelz HJ, Lappegard K, Seifried E, Scharrer I, Tuddenham EG, Muller CR, Strom TM, Oldenburg J. Mutations in VKORC1 cause warfarin resistance and multiple coagulation factor deficiency type 2. *Nature*. 2004;**427**(6974):537–41.
105. Simon T, Verstuyft C, Mary-Krause M, Quteineh L, Drouet E, Meneveau N, Steg PG, Ferrieres J, Danchin N, Becquemont L. Genetic determinants of response to clopidogrel and cardiovascular events. *New Engl J Med*i. 2009;**360**(4):363–75.
106. Mega JL, Close SL, Wiviott SD, Shen L, Hockett RD, Brandt JT, Walker JR, Antman EM, Macias W, Braunwald E, Sabatine MS. Cytochrome p-450 polymorphisms and response to clopidogrel. *New Engl J Med*. 2009;**360**(4):354–62.
107. Collet JP, Hulot JS, Pena A, Villard E, Esteve JB, Silvain J, Payot L, Brugier D, Cayla G, Beygui F, Bensimon G, Funck-Brentano C, Montalescot G. Cytochrome P450 2C19 polymorphism in young patients treated with clopidogrel after myocardial infarction: a cohort study. *Lancet*. 2009;**373**(9660):309–17.
108. Ingelman-Sundberg M. Pharmacogenomic biomarkers for prediction of severe adverse drug reactions. *New Engl J Med*. 2008;**358**(6):637–9.

28

Artificial Intelligence Resources: Publications and Tools

Arvin Agah

CONTENTS

28.1 Introduction

This chapter presents resources in artificial intelligence (AI), including related publications (journals and conferences) and online tools to utilize AI. It should be noted that the resources listed in this chapter are selected representatives of numerous AI resources that are available. The resources listed are a starting point for those interested in AI.

28.2 Publications

Select current journals in AI are listed in Table 28.1, including the publisher and the URL. These journals and others provide an extensive range of ongoing research projects in AI.

Select current conferences in AI are listed in Table 28.2, along with their sponsors. Proceedings from these conferences are a great source for learning more about research in AI and the potential applications to medicine.

28.3 Tools

A number of tools are available online for those interested in using AI and its applications to a variety of domains, including medicine. These include online software that enables researchers to conduct AI experiments using their own data, open-source systems for data mining, software packages for statistical learning and logic inference, programming frameworks for AI techniques (e.g., neural networks), and companies that provide AI services as products for a fee. A few select tools are listed in Table 28.3, including the URL, the institution that developed the tools, and some remarks.

TABLE 28.1

Select Artificial Intelligence Journals

Journal	Publisher	URL
ACM Transactions on Intelligent Systems and Technology	ACM	tist.acm.org
ACM Transactions on Knowledge Discovery from Data	ACM	www.cs.uic.edu/~tkdd
ACM Transactions on Speech and Language Processing	ACM	tslp.acm.org
Applied Artificial Intelligence: An International Journal	Taylor & Francis	www.tandfonline.com/toc/uaai20/current
Applied Intelligence	Springer	www.springerlink.com/content/100236
Applied Soft Computing	Elsevier	www.journals.elsevier.com/applied-soft-computing
Artificial Intelligence: An International Journal	Elsevier	www.journals.elsevier.com/artificial-intelligence
Artificial Intelligence in Medicine	Elsevier	www.sciencedirect.com/science/journal/09333657
Artificial Intelligence Review	Springer	www.springer.com/computer/ai/journal/10462
Artificial Life	MIT Press	www.mitpressjournals.org/loi/artl
Autosoft Journal: Intelligent Automation and Soft Computing	TSI Press	wacong.org/autosoft/auto
BioMedical Engineering OnLine	BioMed Central	www.biomedical-engineering-online.com
Bioinformatics	Oxford University Press	bioinformatics.oxfordjournals.org
BMC Bioinformatics	BioMed Central Ltd	www.biomedcentral.com/bmcbioinformatics
Computational Intelligence	Wiley	onlinelibrary.wiley.com/journal/10.1111/(ISSN)1467-8640
Computer Vision and Image Understanding	Elsevier	www.journals.elsevier.com/computer-vision-and-image-understanding
Computerized Medical Imaging and Graphics	Elsevier	www.journals.elsevier.com/computerized-medical-imaging-and-graphics
Cybernetics and Systems	Taylor & Francis	www.tandfonline.com/toc/ucbs20/current
Data and Knowledge Engineering	Elsevier	www.journals.elsevier.com/data-and-knowledge-engineering
Data Mining and Knowledge Discovery	Springer	www.journals.elsevier.com/data-and-knowledge-engineering
Decision Support Systems	Elsevier	www.journals.elsevier.com/decision-support-systems
European Journal for Biomedical Informatics	European Federation for Medical Informatics	www.ejbi.org/en/about
Evolutionary Computation	MIT Press	www.mitpressjournals.org/loi/evco
Evolutionary Intelligence	Springer	www.springer.com/engineering/computational+intelligence+and+complexity/journal/12065
Expert Systems: The Journal of Knowledge Engineering	Wiley	onlinelibrary.wiley.com/journal/10.1111/(ISSN)1468-0394
Expert Systems with Applications	Elsevier	www.journals.elsevier.com/expert-systems-with-applications

(continued)

TABLE 28.1 (Continued)

Select Artificial Intelligence Journals

Journal	Publisher	URL
Fuzzy Sets and Systems	Elsevier	www.journals.elsevier.com/fuzzy-sets-and-systems
IEEE Journal of Biomedical and Health Informatics	IEEE	bme.ee.cuhk.edu.hk/JBHI
IEEE Transactions on BioMedical Engineering	IEEE	tbme.embs.org
IEEE Transactions on Evolutionary Computation	IEEE	cis.ieee.org/ieee-transactions-on-evolutionary-computation.html
IEEE Transactions on Fuzzy Systems	IEEE	cis.ieee.org/ieee-transactions-on-fuzzy-systems.html
IEEE Transactions on Image Processing	IEEE	www.signalprocessingsociety.org/publications/periodicals/image-processing
IEEE Transactions on Medical Imaging	IEEE	www.ieee-tmi.org
IEEE Transactions on Neural Networks and Learning Systems	IEEE	cis.ieee.org/ieee-transactions-on-neural-networks-and-learning-systems.html
IEEE Transactions on Pattern Analysis and Machine Intelligence	IEEE	www.computer.org/portal/web/tpami
IEEE/ACM Transactions on Computational Biology and Bioinformatics	IEEE and ACM	www.computer.org/portal/web/tcbb
IET Computer Vision	Institution of Engineering and Technology	digital-library.theiet.org/content/journals/iet-cvi
Image and Vision Computing	Elsevier	www.journals.elsevier.com/image-and-vision-computing
Interdisciplinary Sciences: Computational Life Sciences	Springer	www.springer.com/life+sciences/systems+biology+and+bioinformatics/journal/12539
International Journal of Computer Vision	Springer	www.springer.com/computer/image+processing/journal/11263
International Journal of Imaging Systems and Technology	Wiley	onlinelibrary.wiley.com/journal/10.1002/(ISSN)1098-1098
International Journal Of Intelligent Computing In Medical Sciences and Image Processing	Taylor & Francis	www.tandfonline.com/toc/tmed20/current
International Journal of Intelligent Systems	Wiley	onlinelibrary.wiley.com/journal/10.1002/(ISSN)1098-111X
International Journal of Medical Informatics	Elsevier	www.journals.elsevier.com/international-journal-of-medical-informatics
International Journal on Artificial Intelligence Tools	World Scientific	www.worldscientific.com/worldscinet/ijait
Journal of the American Medical Informatics Association	American Medical Informatics Association	jamia.bmj.com
Journal of Artificial Intelligence Research	AI Access Foundation	www.jair.org

(*continued*)

TABLE 28.1 (Continued)

Select Artificial Intelligence Journals

Journal	Publisher	URL
Journal of Biomedical Informatics	Elsevier	www.journals.elsevier.com/journal-of-biomedical-informatics
Journal of Experimental and Theoretical Artificial Intelligence	Taylor & Francis	www.tandfonline.com/toc/teta20/current
Journal of Intelligent Systems	de Gruyter	www.degruyter.com/view/j/jisys
Journal of Machine Learning Research	Microtome Publishing	jmlr.csail.mit.edu
Knowledge-Based Systems	Elsevier	www.journals.elsevier.com/knowledge-based-systems
Machine Learning	Springer	www.springer.com/computer/ai/journal/10994
Machine Vision and Applications	Springer	www.springer.com/computer/image+processing/journal/138
Medical and Biological Engineering and Computing	Springer	www.springer.com/biomed/human+physiology/journal/11517
Medical Engineering and Physics	Elsevier	www.journals.elsevier.com/medical-engineering-and-physics
Medical Image Analysis	Elsevier	www.journals.elsevier.com/medical-image-analysis
Neural Computing and Applications	Springer	link.springer.com/journal/521
Neural Networks	Elsevier	www.journals.elsevier.com/neural-networks
Pattern Analysis and Applications	Springer	www.springer.com/computer/image+processing/journal/10044
Pattern Recognition	Elsevier	www.journals.elsevier.com/pattern-recognition

(*continued*)

TABLE 28.2

Select Artificial Intelligence Conferences

Conference	Sponsor
AAAI Conference on Artificial Intelligence	Association for the Advancement of Artificial Intelligence (AAAI)
American Medical Informatics Annual Symposium	American Medical Informatics Association (AMIA)
Asian Conference on Computer Vision	Asian Federation of Computer Vision Societies (AFCV)
Australasian Joint Conference on Artificial Intelligence	National Committee for Artificial Intelligence and Expert System, the Australian Computer Society
Canadian Conference on Artificial Intelligence	Canadian Artificial Intelligence Association
Conference on Artificial Intelligence in Medicine	European Society for Artificial Intelligence in Medicine (AIME)
Conference on Medical Image Analysis and Understanding	British Machine Vision Association
Conference on Neural Information Processing Systems	Neural Information Processing Systems (NIPS) Foundation
European Conference on Artificial Intelligence	European Coordination Committee for Artificial Intelligence (ECCAI)
European Conference on Artificial Life	International Society of Artificial Life (ISAL)
European Conference on Computer Vision	European Conference on Computer Vision
Genetic and Evolutionary Computation Conference	Association for Computing Machinery Special Interest Group on Genetic and Evolutionary Computation (SIGEVO)
IEEE Conference on Healthcare Informatics, Imaging, and Systems Biology	IEEE
IEEE Computer Vision and Pattern Recognition	IEEE Computer Society
IEEE International Conference on Computer Vision	IEEE
IEEE International Conference on Computer Vision and Pattern Recognition	IEEE
IEEE International Conference on Robotics and Automation	IEEE
IEEE International Conference on Systems, Man, and Cybernetics	IEEE
IEEE International Conference on Tools with Artificial Intelligence	IEEE
IEEE International Symposium on Computational Intelligence and Informatics	IEEE
IEEE International Symposium on Computer-Based Medical Systems	IEEE
IEEE/RSJ International Conference on Intelligent Robots and Systems	IEEE and Robotics Society of Japan (RSJ)
International Conference of the European Federation for Medical Informatics	European Federation for Medical Informatics
International Conference of the IEEE Engineering in Medicine and Biology Society	IEEE
International Conference on Agents and Artificial Intelligence	Institute for Systems and Technologies of Information Control and Communication
International Conference on Artificial Intelligence and Soft Computing	Polish Neural Network Society
International Conference on Autonomous Agents and Multi-Agent Systems	International Foundation for Autonomous Agents and Multiagent Systems (IFAAMAS)

(*continued*)

TABLE 28.2 (Continued)

Select Artificial Intelligence Conferences

Conference	Sponsor
International Conference on Bioinformatics and Computational Biology	International Society for Computers and Their Applications
International Conference on Case Based Reasoning	International Conference on Case Based Reasoning
International Conference on Health and Medical Informatics	World Academy of Science, Engineering and Technology
International Conference on Image Analysis and Recognition	International Conference on Image Analysis and Recognition (ICIAR)
International Conference on Image Processing Theory, Tools and Applications	IEEE and European Association for Signal Processing (EURASIP)
International Conference on Machine Learning	International Machine Learning Society (IMLS)
International Conference on Computer Vision Theory and Applications	Institute for Systems and Technologies of Information, Control and Communication (INSTICC)
International Joint Conference on Artificial Intelligence	International Joint Conferences on Artificial Intelligence (IJCAI)
International Joint Conference on Biomedical Engineering Systems and Technologies	Institute for Systems and Technologies of Information, Control and Communication (INSTICC)
International Joint Conference on Neural Networks	International Neural Network Society (INNS) and IEEE Computational Intelligence Society (IEEE-CIS)
International Workshop on Artificial Neural Networks in Pattern Recognition	International Neural Network Society (INNS) and International Association for Pattern Recognition (IAPR)
Pacific-Rim Symposium on Image and Video Technology	Pacific-Rim Symposium on Image and Video Technology (PSIVT)
World Congress on Medical and Health Informatics	International Medical Informatics Association (IMIA)

TABLE 28.3

Select Artificial Intelligence Tools

Tool	Developed at	Description	URL
Alchemy	Department of Computer Science and Engineering, University of Washington, Seattle, WA, USA	A software package with algorithms for statistical relational learning and probabilistic logic inference, used for classification, prediction, etc.	alchemy.cs.washington.edu
AITopics	The Association for the Advancement of Artificial Intelligence, Palo Alto, CA, USA	An information portal provided for science and applications of AI	aitopics.org
Artificial Intelligence in Medicine (AIM)	Cedars-Sinai Medical Center, Los Angeles, CA, USA	A research program to develop software for nuclear cardiology to automate the process and analyze three-dimensional heart images, using artificial intelligence techniques	www.cedars-sinai.edu/Patients/Programs-and-Services/Medicine-Department/Artificial-Intelligence-in-Medicine-AIM
Artificial Intelligence in Medicine Inc.	Artificial Intelligence In Medicine Inc., Toronto, Ontario, Canada	A company developing software systems for cancer control, using artificial intelligence for improved utility and efficiency	www.aim.ca
Artificial Medical Intelligence	Artificial Medical Intelligence, Eatontown, NJ, USA	Medical software products for medical documentation and infrastructure	www.artificialmed.com/
Data Applied	Data Applied, Sammamish, WA, USA	Products for decision making using data mining and information visualization	www.data-applied.com
ELKI	Database and Information Systems, Ludwig Maximilians Universität München, Munich, Germany	Use of data management tools for data mining tasks	elki.dbs.ifi.lmu.de
Encog Machine Learning Framework	Heaton Research, Inc., Chesterfield, MO, USA	An open-source machine learning initiative for neural networks and bot framework, including a variety of AI algorithms	www.heatonresearch.com/encog
Expertmaker	Expertmaker, San Francisco, CA, USA	A software company selling AI products (tool kit with different AI technologies) to add reasoning and learning, and searching for discovery and recommendation systems	www.expertmaker.com
Mathematica	Wolfram Research, Champaign, IL, USA	Wolfram SystemModeler	www.wolfram.com
MATLAB	MathWorks, Natick, MA, USA	Bioinformatics, fuzzy logic, global optimization, image processing, and neural network toolboxes	www.mathworks.com

(continued)

TABLE 28.3 (Continued)

Select Artificial Intelligence Tools

Tool	Developed at	Description	URL
Neuroph	University of Belgrade, Belgrade, Serbia	An open-source programming framework to develop common neural network architectures	neuroph.sourceforge.net
OpenCog	OpenCog Foundation, Rockville, MD, USA	An open-source software project on artificial general intelligence (AGI), using mathematics, biology, and software engineering techniques	opencog.org
Orange	Bioinformatics Laboratory, University of Ljubljana, Ljubljana, Slovenia	An open-source data mining and machine learning software suite, for data analysis and visualization	orange.biolab.si
RapidMiner	Rapid-I Inc., Burlington, MA, USA	An open-source system for data mining, either stand-alone or integrated into products	rapid-i.com
Slicer	Brigham and Women's Hospital, Harvard Medical School, Boston, MA, USA	An open-source software for visualization and image analysis	www.slicer.org
Sourceforge	SourceForge.net, Geeknet Media. Dice Holdings, Inc., New York, NY, USA	Dedicated to open-source projects with a section dedicated to AI	sourceforge.net/directory/science-engineering/ai
Tools for Learning Artificial Intelligence	Laboratory for Computational Intelligence, University of British Columbia, Vancouver, Canada	Tools for learning and exploring concepts in artificial intelligence	www.aispace.org
Video Games and Artificial Intelligence	Applied Games, Microsoft Research Cambridge, Cambridge, UK	A tutorial on applying AI techniques in learning, searching, and planning to video games	research.microsoft.com/en-us/projects/ijcaiigames/
Weka	Machine Learning Group, University of Waikato, Hamilton, New Zealand	A widely used software package incorporating several machine learning techniques for data mining, with tools for preprocessing, classification, regression, clustering, association rules, and visualization	www.cs.waikato.ac.nz/ml/weka

Index

Page numbers followed by f and t indicate figures and tables, respectively.

B

C

E

F

G

H

M

N

O

Q

R